Renewable Energy in Marine Environment

Renewable Energy in Marine Environment

Special Issue Editor

Eugen Rusu

MDPI • Basel • Beijing • Wuhan • Barcelona • Belgrade

Special Issue Editor
Eugen Rusu
Department of Applied
Mechanics, Head of Laboratory
of Computations and Modeling
in Applied Mechanics, "Dunarea
de Jos" University of Galati
Romania

Editorial Office
MDPI
St. Alban-Anlage 66
4052 Basel, Switzerland

This is a reprint of articles from the Special Issue published online in the open access journal *Energies* (ISSN 1996-1073) from 2018 to 2019 (available at: https://www.mdpi.com/journal/energies/special_issues/marine).

For citation purposes, cite each article independently as indicated on the article page online and as indicated below:

LastName, A.A.; LastName, B.B.; LastName, C.C. Article Title. *Journal Name* **Year**, *Article Number*, Page Range.

ISBN 978-3-03928-528-0 (Pbk)
ISBN 978-3-03928-529-7 (PDF)

Contents

About the Special Issue Editor

Eugen Rusu received a diploma in Naval Architecture (1982) and a Ph.D. in Mechanical Engineering (1997). Between 1999 and 2004, he worked as a post-doc fellow at the at Hydrographical Institute of the Portuguese Navy, where he was responsible for wave modelling and provided environmental support for some major situations such as: the accident of the M/V Prestige (2002) and the NATO exercises Unified Odyssey (2002) and 'Swordfish' (2003). Eugen Rusu worked as a consulting Scientist at the NATO Centre for Maritime Research and Experimentation, La Spezia, Italy (2005), having modelling coastal waves and surf zone processes as main tasks.

Starting in 2006, in parallel with his activity as professor at the University Dunarea de Jos of Galati, he has been working as a professor collaborator at CENTEC (Centre for Marine Technology and Ocean Engineering), University of Lisbon, Portugal. Since 2012, he has also acted as an expert for the European Commission. Eugen Rusu has published more than 150 works in the fields of renewable energy and marine engineering and he received awards including Doctor Honoris Causa (2015) at the Maritime University of Constanta, Romania Outstanding Contribution in Reviewing, *Renewable Energy* (2015) and *Ocean Engineering* (2016) journals; and Top 1% World Reviewers in the field of engineering (2018). He is also the President of the Council of the Doctoral Schools in Galati University and of the Romanian National Commission of Mechanical Engineering. Since 2018, he has been a corresponding member of the Romanian Academy, the highest level scientific and cultural forum in Romania.

Article

Analyzing the Near-Field Effects and the Power Production of an Array of Heaving Cylindrical WECs and OSWECs Using a Coupled Hydrodynamic-PTO Model

Philip Balitsky *, Nicolas Quartier, Gael Verao Fernandez, Vasiliki Stratigaki and Peter Troch

Department of Civil Engineering, Ghent University, Technologiepark 60, B-9052 Ghent, Belgium; Nicolas.Quartier@UGent.be (N.Q.); Gael.VeraoFernandez@UGent.be (G.V.F.); Vicky.Stratigaki@UGent.be (V.S.); Peter.Troch@UGent.be (P.T.)
* Correspondence: philip.balitsky@ugent.be; Tel.: +32-9-264-54-33

Received: 21 November 2018; Accepted: 11 December 2018; Published: 14 December 2018

Abstract: The Power Take-Off (PTO) system is the key component of a Wave Energy Converter (WEC) that distinguishes it from a simple floating body because the uptake of the energy by the PTO system modifies the wave field surrounding the WEC. Consequently, the choice of a proper PTO model of a WEC is a key factor in the accuracy of a numerical model that serves to validate the economic impact of a wave energy project. Simultaneously, the given numerical model needs to simulate many WEC units operating in close proximity in a WEC farm, as such conglomerations are seen by the wave energy industry as the path to economic viability. A balance must therefore be struck between an accurate PTO model and the numerical cost of running it for various WEC farm configurations to test the viability of any given WEC farm project. Because hydrodynamic interaction between the WECs in a farm modifies the incoming wave field, both the power output of a WEC farm and the surface elevations in the 'near field' area will be affected. For certain types of WECs, namely heaving cylindrical WECs, the PTO system strongly modifies the motion of the WECs. Consequently, the choice of a PTO system affects both the power production and the surface elevations in the 'near field' of a WEC farm. In this paper, we investigate the effect of a PTO system for a small wave farm that we term 'WEC array' of 5 WECs of two types: a heaving cylindrical WEC and an Oscillating Surge Wave Energy Converter (OSWEC). These WECs are positioned in a staggered array configuration designed to extract the maximum power from the incident waves. The PTO system is modelled in WEC-Sim, a purpose-built WEC dynamics simulator. The PTO system is coupled to the open-source wave structure interaction solver NEMOH to calculate the average wave field η in the 'near-field'. Using a WEC-specific novel PTO system model, the effect of a hydraulic PTO system on the WEC array power production and the near-field is compared to that of a linear PTO system. Results are given for a series of regular wave conditions for a single WEC and subsequently extended to a 5-WEC array. We demonstrate the quantitative and qualitative differences in the power and the 'near-field' effects between a 5-heaving cylindrical WEC array and a 5-OSWEC array. Furthermore, we show that modeling a hydraulic PTO system as a linear PTO system in the case of a heaving cylindrical WEC leads to considerable inaccuracies in the calculation of average absorbed power, but not in the near-field surface elevations. Yet, in the case of an OSWEC, a hydraulic PTO system cannot be reduced to a linear PTO coefficient without introducing substantial inaccuracies into both the array power output and the near-field effects. We discuss the implications of our results compared to previous research on WEC arrays which used simplified linear coefficients as a proxy for PTO systems.

Keywords: WEC array; WEC farm; PTO system; PTO system tuning; linearization; hydraulic; WEC-sim; PTO-sim; model coupling; BEM; NEMOH

---

1. Introduction

Ocean Wave Energy is a potential source of clean electricity that can make a significant contribution to the de-carbonization of the world's electricity supply. However, for it to follow the path of offshore wind and become a commercially viable power source, significant cost reductions need to be made. Because of practical limitations on the physical size of an individual Wave Energy Converter (WEC), these devices must be placed in close proximity to benefit from economies of scale such as those witnessed in the offshore wind industry. Such agglomerations of WECs are commonly termed wave farms. To match the power output of offshore wind farms, WEC farms need to consist of hundreds of WECs. How these WECs are grouped and arranged within a wave farm to maximize profitability while minimizing detrimental effects is still an open question.

Due to hydrodynamic interactions between individual WECs and closely spaced groups of WECs, determining the power output of a WEC farm is not a trivial matter. Unlike the case of wind farms, the interactions can be both beneficial and deleterious and correctly modelling them can make or break the financial viability of a WEC farm. As experimental studies are costly and time consuming, the chief design tool for assessing WEC farms is numerical modelling. There are many variables influencing the estimated power output, among them the site wave climate and bathymetry, WEC farm layout and the Power Take-off (PTO system) of each WEC. Modelling them in parallel leads to significant demands on computational power, and often leads to unclear conclusions. An additional complication for the numerical modelers is that many of the aforementioned variables are interdependent; it is, therefore, essential to understand the significance of each of the variables underlying the chosen numerical model.

For a given WEC type and for a given incident wave, a critical parameter that influences the WEC motion and the power output of a WEC farm is the PTO system. Because of the variety of technical solutions and the complexity of modeling the inherently non-linear behavior of most viable PTO systems in WECs, a plurality of previous investigations has assumed a simple mechanical damper as a proxy for the PTO system. Some examples for farms of heaving cylindrical WECs are found in [1–5] and for Oscillating Surging Wave Energy Converters (OSWECs) in [6–9]. Concurrently, due to step improvements in hydrodynamic modelling software, there has been a jump in the number of numerical investigations that have modelled single WECs [10–13] and small farms of WECs [14,15] with fully non-linear hydrodynamics. Yet, as pointed out in Penalba et al. [16] for the case of heaving point absorbers and in [8] for OSWECs, the errors due to a simplified PTO model can override any improvements made by more accurate hydrodynamic models. A particular concern with many existing PTO modelling efforts is that the most common PTO system type developed for commercial WEC prototypes, a hydraulic PTO system, is inherently non-linear [17,18]. A few recent studies, notably [16–22] have implemented realistic hydraulic PTO models with non-linear dynamics. However, these studies were limited in their scope to single WECs and not WEC farms, furthermore, many of the models quite complicated in their implementation.

In this paper, our aim is to implement a realistic hydraulic PTO model for two types of promising WEC technologies, namely heaving cylindrical WECs and OSWECs, in an array composed of 5 WECs. Although the terms WEC farm and WEC array are used interchangeably, we will follow the precedent set in [23] and term a small farm of closely spaced WECs a *WEC array*. The impact of the hydraulic PTO system on the power output and the 'near-field' surface elevations of the 5-WEC array is compared to that of the base case of a linear PTO system. Both PTO systems are simulated using WEC-Sim [6], a dynamical simulator for WECs built in the MATLAB Simulink platform. The PTO model is coupled to the open-source wave-structure interaction solver NEMOH [24] using the perturbed wave field imparted by the motion of the WECs in WEC-Sim. Previously, a similar approach was presented in [25–27] for the case of a wave-structure interaction solver coupled to a wave propagation model using a basic linear PTO model. WEC-Sim has been used in modelling hydraulic PTOs in several recent studies [20,22]. In the present paper, only the near-field zone is simulated with a future goal of coupling to a wave propagation model to model the impact of a WEC farm (consisting of one or multiple WEC arrays) in the 'far-field'. In referencing the near-field we refer to the area inside the WEC array immediately surrounding the WECs, while the far-field can refer to areas outside the immediate area of the WEC array up to several km away. The modifications of the wave field in the presence of multiple bodies are referred to as 'array effects', that are synonymous with 'farm' or 'park effects' used in some literature [2,28–30]. We begin by listing the underlying theory and assumptions in Section 2. Then we provide the details on the two numerical PTO system interpretations used in the study in Section 3 and specify the regular wave test matrix of the simulations in Section 4.2. We then present the results for a single WEC for the power in Section 5.1, the near-field $|\eta|$ in Section 5.2 and compare the performance and effects of a hydraulic PTO system to a linear PTO system in Section 5.2.3. Next, we present the corresponding results for the 5-WEC arrays of heaving cylindrical WECs and OSWECs in Sections 6.4, 6.5 and 6.5.3. Finally, we highlight the key messages of the research in the discussion in Section 7 and make conclusions with a view toward a continuation of the work undertaken in this paper in Section 8.

2. Hydrodynamic Model Description

2.1. Linear Potential Flow

This investigation assumes linear potential flow theory [31], a subset of linear wave theory that allows the fluid velocity, v, to be expressed as the gradient of the time dependent potential Φ, (Equation (1)).

$$v = \nabla \Phi \tag{1}$$

The assumptions underlying potential flow are the following:

- the fluid is inviscid;
- the fluid is incompressible; and
- the flow is irrotational.

The standard assumption of linear theory that the motion amplitudes of the bodies are much smaller than the wavelength also applies. Linear potential flow theory has hitherto been used in most of the investigations into WEC array modelling, for example see [3,29,30,32]. In further assuming that all time-varying quantities oscillate with the same angular frequency ω, we can separate out the time dependence from the time-independent velocity potential ϕ,

$$\phi(x, y, z, t) = \Re \left\{ \phi(x, y, z) e^{-i\omega t} \right\} \tag{2}$$

where ϕ is the complex velocity potential. Due to application of the principle of superposition, linear potential theory allows for the separation of the total velocity potential into the following components (Equation (3)):

$$\phi_t(x, y, z) = \phi_i + \phi_d + \sum_i^6 \phi_r \tag{3}$$

where ϕ_t is the total velocity potential, ϕ_i is the incident wave velocity potential, ϕ_d the diffracted wave velocity potential and $\sum_i^6 \phi_r$ is the sum of the radiated wave velocity potentials for each Degree of Freedom (DoF) of the WEC. In our investigation we only model 1 DoF for each WEC, namely heave for the cylindrical WEC and pitch for the OSWEC. We also introduce the term perturbed wave to denote the wave resulting from sum of the diffracted and radiated velocity potentials.

2.2. Boundary Element Method Solver

In our coupling approach the 'array' effects, induced by the hydrodynamic interaction between the WECs, are resolved by simulating the WEC motions using the open-source potential flow Boundary Element Method (BEM) solver NEMOH [24]. Given Equation (1), NEMOH solves the Laplace equation, Equation (4), for the complex velocity potential, ϕ:

$$\nabla \phi = 0 \tag{4}$$

given a set of boundary conditions on the wetted body surface, the free surface, sea bottom and far-field. The equations of motion are solved using the method of Green's functions, as explained in [24]. An important restriction imposed by the method is the assumption that the water depth h is constant throughout the WEC array domain. The free surface elevation η is calculated by taking the real part of the complex surface elevation $\bar{\eta}$ that is in turn obtained in NEMOH from the free surface boundary condition Equation (5). From the superposition principle of Equation (3), free surface elevations η can be obtained separately for the WEC motions due to the diffracted and the radiated potentials:

$$\eta = -\frac{1}{g} \left(\frac{\partial \phi}{\partial t} \right)_{z=0} \tag{5}$$

where g is the acceleration due to gravity and $z = 0$ is the undisturbed free surface. NEMOH also calculates the coefficients of the added mass $A(\omega)$, hydrodynamic damping $B(\omega)$, and hydrodynamic restoring force or buoyancy force $K(\omega)$ which are used to calculate the WEC motions in Section 3.1.

3. PTO Model Development

3.1. Equations of Motion

To model the WECs with a given PTO system, in this investigation we use the open source mechanical solver WEC-Sim developed by Sandia National Laboratory in collaboration with the National Renewable Energy Laboratory (NREL) in the US [6]. WEC-Sim operates within the MATLAB Simulink environment. For 1 DoF WEC displaced a distance z from equilibrium, WEC-Sim solves for the WEC motion in the time domain using the Cummins Equation (6):

$$M_t \ddot{z}(t) = f_e(t) + f_{rad}(t) + f_{hs}(t) + f_{PTO}(t) + f_v(t) + f_m(t) \tag{6}$$

In the case of a floating WEC oscillating in heave, $M_t = M + A_{33\infty}$ where M is the generalized mass matrix and $A_{33\infty}$ is the asymptotic value of the heave added mass. On the right hand side, $f_e(t)$ is the excitation force, $f_{PTO}(t)$ is the PTO force, $f_{hs}(t)$ is the hydrostatic force, $f_{rad}(t)$ is the force vector of radiation, $f_v(t)$ are the forces that can be modelled as viscous or friction losses in the system, and $f_m(t)$ is the force vectors resulting from the mooring connections. The excitation force is calculated as $f_e(t) = \mathcal{F}^{-1} \{ F_e(\omega) \eta(\omega) \}$, where $\eta(\omega)$ is the Fourier transform of the surface elevation and $F_e(\omega)$ is the frequency domain exciting force transfer function. $f_{hs}(t)$ is the hydrostatic force which is equal to $K_{33} Z(\omega)$ where K_{33} represents the hydrostatic stiffness and $Z(\omega)$ the frequency domain displacement of the heaving cylindrical WEC. The hydrodynamic coefficients representing A, the added mass of the device, B, the hydrodynamic damping and K, the hydrodynamic spring or stiffness, are calculated in the frequency domain in NEMOH for each relevant degree of freedom for the given WECtype. Please note that henceforth all capital letters represent frequency domain complex quantities while small case letter real-valued time-domain quantities. For the regular waves simulated herein, the radiation force $f_{rad}(t)$ can be calculated in the steady state form for a given frequency ω by the following Equation (7):

$$f_{rad}(t) = -A(\omega)\ddot{z} - B(\omega)\dot{z}. \tag{7}$$

In this paper, we do not model $f_v(t)$ and $f_m(t)$ since they are assumed to be negligible, therefore those terms are set equal to zero. The OSWEC described in Section 4.1 is simulated using the same Equation (6), with the substitution of torques for the forces and the pitch angular displacement $\theta(t)$ for the heave displacement $z(t)$ and the coefficients in heave for the coefficients in pitch. Two different types of power take-off systems will be further discussed: a linear and hydraulic PTO system, the former being the most popular way of simplifying a PTO system while the latter being one of the most used PTO systems in commercial WEC designs.

3.1.1. Linear PTO System

The most common way of simulating the effect of the PTO system of a wave energy converter is by modelling its dynamics as linear. This means the PTO system is modelled as a spring-damper-mass system with stiffness coefficient K_{PTO} and damping coefficient B_{PTO}. However, because of the practical difficulty of changing the mass of the PTO system in real-time, it is often assumed the mass is unchangeable, resulting in the spring-damper system as represented in Figure 1 for the heaving cylindrical WEC. For practical reasons, a variable spring system is often difficult to implement, therefore a further simplification is warranted where we set the stiffness coefficient K_{PTO} to zero. In the following calculations, the PTO system will be modelled as linear damper, resulting in the following expression for the PTO force:

$$f_{PTO,l}(t) = -B_{PTO,l}\dot{z}(t) \tag{8}$$

with $B_{PTO,l}$ the linear PTO damping term. The linear PTO influences the dynamics of the heaving cylindrical WEC: it exerts a force, $f_{PTO,l}(t)$, oppositely directed to the WEC's velocity, $\dot{z}(t)$. The instantaneous power $P_{inst,l}$ absorbed by the linear PTO system is calculated as:

$$P_{inst,l}(t) = -f_{PTO,l}(t)\dot{z}(t) = B_{PTO,l}\dot{z}^2(t) \tag{9}$$

When assuming that the waves are sinusoidal the motion of the WEC can be expressed as the real part of a complex value: $\Re\left\{Z(\omega)e^{-i\omega t}\right\}$, where from this point capital letters will represent the complex form of a certain quantity. The average power P_l absorbed by a heaving cylindrical WEC with a linear PTO system in one wave period is given as

$$P_l = \frac{1}{2}B_{PTO,l}|Z(\omega)|^2\omega^2 \tag{10}$$

The expression above is used to find the optimum value for $B_{PTO,l}$ resulting in the maximum average absorbed power P. This leads to

$$B_{PTO,l} = \sqrt{B_{33}^2 + \left(\omega(m + A_{33}) - \frac{K_{33}}{\omega}\right)^2} \tag{11}$$

with m the WEC's mass, A_{33} the added mass in heave, B_{33} the heave component of the hydrodynamic damping and K_{33} the hydrostatic stiffness in heave. The same procedure can be repeated for the OSWEC with a linear PTO system: the PTO-torque $T_{PTO,l}$ is calculated as follows:

$$T_{PTO,l}(t) = -B_{PTO,l}\dot{\theta}(t) \tag{12}$$

with $B_{PTO,l}$ the linear damping coefficient in [Nm/(rad/s)] for the OSWEC and $\dot{\theta}(t)$ the pitch velocity of the OSWEC [rad/s]. The optimal value for $B_{PTO,l}$, resulting in the maximum average absorbed power, is given by

$$B_{PTO,l} = \sqrt{B_{55}^2 + \left(\omega(I + A_{55}) - \frac{K_{55}}{\omega}\right)^2}. \tag{13}$$

Here, I represents the OSWEC's moment of inertia about its hinge, A_{55} represents the added moment of inertia in pitch, B_{55} the pitch component of the hydrodynamic damping and K_{55} the flap buoyancy torque. The average absorbed power by an OSWEC with a linear PTO system is then expressed as:

$$P_l = \frac{1}{2} B_{PTO,l} |\Theta(\omega)|^2 \omega^2 \tag{14}$$

with $|\Theta(\omega)|$ the amplitude of the pitch motion.

3.1.2. Hydraulic PTO System

Although the linear damper is a convenient way of modelling the effects of the PTO system, it is in some cases an oversimplified representation of the realistic PTO system. Realistic full scale WECs are often equipped with a hydraulic PTO system, which can be modelled numerically using WEC-Sim for both a heaving cylindrical WEC and an OSWEC. A schematic representation of a heaving cylindrical WEC equipped with a hydraulic PTO system is given in Figure 1.

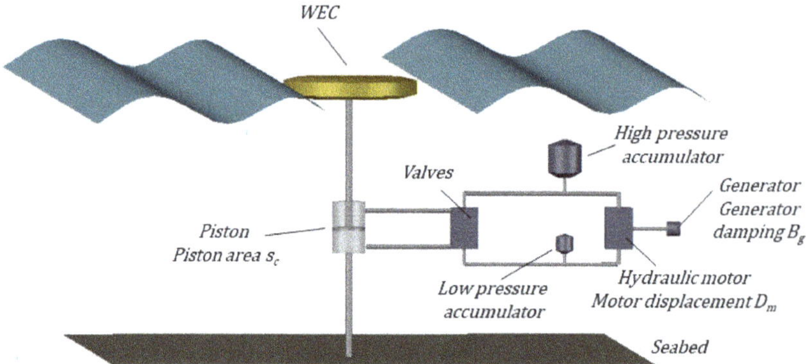

Figure 1. Schematic representation hydraulic PTO for heaving cylindrical WEC.

In the case of a heaving cylindrical WEC, the hydraulic PTO system converts the heaving motion in a pressurized fluid flow. This fluid flow is translated in rotational energy by the variable displacement motor. The motor's axle is connected to a generator's axle, which generates electricity [20]. The provided model calculates the hydraulic PTO force, $f_{PTO,h}$ with:

$$f_{PTO,h}(t) = -sign(\dot{z}(t)) \cdot (p_h(t) - p_\ell(t)) s_c \tag{15}$$

with p_h and p_ℓ respectively the pressure in the high- and low-pressure accumulator, whereas s_c represents the piston area. Accumulators smoothen the peak flows into a quasi-constant flow towards the hydraulic motor [33]. The PTO-force exerted by the hydraulic PTO system always has the opposite sign as the velocity of the heaving cylindrical WEC. The volume flow Q_{piston}, resulting from the up- or downward piston movement is given by:

$$Q_{piston}(t) = s_c \dot{z}(t) \tag{16}$$

Rectifying valves ensure unidirectional flow further in the hydraulic system. This makes fluid flow from the piston into the high-pressure accumulator and then further to the hydraulic motor. Fluid leaving the hydraulic motor flows towards the low-pressure accumulator. The incoming volume flow in the high-pressure accumulator, Q_{in}, is calculated as:

$$Q_{in} = Q_{piston} + Q_{motor}. \tag{17}$$

with Q_{motor} originating from the hydraulic motor - see Equation (20). The total fluid volume inside the accumulator at time t_j equals $V_{in}(t_j)$ and is calculated with:

$$V_{in}(t_j) = V_{in}(t_{j-1}) + Q_{in}(t_j) \cdot dt. \tag{18}$$

It is assumed that initially there is no fluid inside the accumulator, so $V_{in}(0)$ equals 0. The total volume of the accumulator equals V_0, which allows the calculation of the pressure inside the accumulator as follows, according to an isentropic process:

$$p_h(t_j) = \frac{p_{precharge}}{(1 - \frac{V_{in}(t_j)}{V_0})^\gamma} \tag{19}$$

with $p_{precharge}$ the initial pre-charge pressure in the accumulator and γ the adiabatic index, set equal to 1.4. The compressibility of the fluid is neglected. The calculation of the pressure in the low-pressure accumulator, $p_l(t_j)$, is done similarly. The fluid volume flow originating from the motor is determined by:

$$Q_{motor}(t) = \omega_m(t)\alpha D_m \tag{20}$$

In this formula, ω_m represents the angular velocity of the hydraulic motor, whereas α represents the swashplate angle which is the instantaneous motor displacement divided by the maximum motor displacement. D_m represents the nominal motor displacement. The product αD_m represents the volume needed for one revolution of the hydraulic motor, expressed in [m^3/rad]. In MATLAB Simulink, the angular velocity of the hydraulic motor, ω_m, is calculated by integrating the following expression:

$$\dot{\omega}_m(t) = \frac{(p_h(t) - p_\ell(t))\alpha D_m - T_g(t) - T_f(t)}{I_{mg}}, \tag{21}$$

where T_g is the generator torque, T_f the torque due to friction, and I_{mg} the total mass moment of inertia of the motor/generator. The generator torque changes linearly with the motor's angular velocity, ω_m, with a damping coefficient of the generator, B_g:

$$T_g(t) = B_g\omega_m(t). \tag{22}$$

It is assumed that this damping coefficient B_g is constant. The efficiency of the generator depends on its torque T_g and its angular velocity ω_m. A table for the generator efficiency is provided by WEC-Sim for different combinations of T_g and ω_m. The average absorbed power by the hydraulic PTO of a heaving cylindrical WEC over one wave period T is expressed as:

$$P_h = -\frac{1}{T} \int_0^T f_{PTO}(t) \cdot \dot{z}(t) dt \tag{23}$$

The Equation (23) is the absorbed power without taking into account losses in the hydraulic motor and electric generator. The average electrical power will be less than the power at the piston, P_h, since friction in the hydraulic motor and the efficiency of the generator are taken into account in WEC-Sim. In Section 4.4 and further, only the average absorbed power at the piston P_h will be considered. WEC-Sim also provides the ability to implement a hydraulic PTO system for an OSWEC. The principle of a hydraulic PTO system applied to a pitching flap is sketched in Figure 2. In Figure 2, a positive pitching angle θ corresponds to a clockwise movement of the flap, which implies a shortening of the

PTO-bar equipped with the PTO system. This shortening in its turn creates a pressure difference on both sides of the piston. The pitching motion thus induces a linear movement in the piston. Once this linear motion is calculated in Simulink, the force f_{PTO} can be calculated and will be multiplied with the lever arm length ℓ around the hinge to find the torque $\mathcal{T}_{PTO}$:

$$\mathcal{T}_{PTO}(t) = f_{PTO}(t) \cdot \ell(t).\tag{24}$$

How the force f_{PTO} is calculated is explained in Section 3.1.2, in Equation (15), since the hydraulic PTO system for the OSWEC mainly contains the same components as the one for the heaving cylindrical WEC. How the pitching motion of the flap is converted in a linear movement of the piston is briefly explained below. This conversion involves some geometric parameters—see Figure 2 for definitions:

- $\theta(t)$, the varying pitch angle
- g, the offset height of the PTO-bar connection with the seabed
- c, the distance between the flap-hinge and connection with the PTO-bar
- $b(t)$, the length of the PTO-bar, varying in time; for $\theta = 0$, $b = b_{ini}$
- $r(t)$, the vertical distance between the connection points of the PTO-bar, varying in time
- $\beta(t)$, the angle between the PTO-bar and the vertical direction, varying in time
- $\ell(t)$, the length of the lever arm (or the distance of the hinge to the PTO-bar), variable in time.

In Figure 2 the length r varies in time and is evaluated by $r(t) = c \cdot \cos(\theta(t)) - g$, while angle $\beta(t)$ can be calculated as $\beta(t) = \arccos(r(t)/b(t))$. The length of the lever arm ℓ, i.e., the perpendicular distance from the PTO-bar to the hinge can be determined using:

$$\ell(t) = \sin(\theta(t) + \beta(t)) \cdot c\tag{25}$$

The instantaneous absorbed power can be either determined by multiplying $\mathcal{T}_{PTO}$ with the angular velocity $\dot{\theta}$ or by multiplying f_{PTO} with the linear velocity of the piston at each time step, as in Equation (23). As with the heaving cylindrical WEC, only the total absorbed power P_h at the piston will be considered. The average absorbed power by the hydraulic PTO system of an OSWEC over one wave period is expressed as:

$$P_h = -\frac{1}{T} \int_0^T \mathcal{T}_{PTO}(t) \cdot \dot{\theta}(t)dt\tag{26}$$

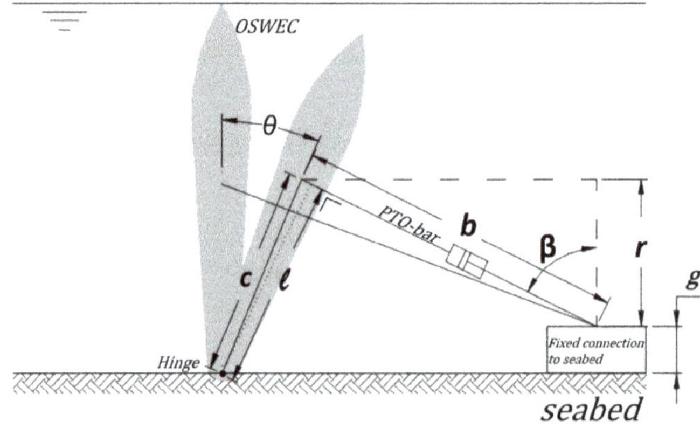

Figure 2. Hydraulic PTO system working principle of a generic OSWEC.

4. Modelled WECs and Input Wave Conditions

In this paper, we present the results for full scale WECs for a series of regular waves of varying heights and periods. The WEC types are outlined in Section 4.1 and the input wave conditions are shown in Table 1 in Section 4.2.

4.1. Modelled WEC Types

The two types of full-scale WECs modelled in this study are a heaving cylindrical buoy and a pitching bottom fixed flap, which is often termed OSWEC [34]. The heaving cylindrical WEC type is a flat cylinder with radius (r) of 10 m and a draft (h_z) of 2.0 m (see Figure 3). The shape was selected based on its overall dimensions being similar to several promising WEC technologies, namely that of Carnegie Wave [35] and SINN Power [36]. Moreover, as noted in a recent study, [37], such a flat disk shape provides a balance between the power absorption, WEC bandwidth, and material cost considerations. Please note that in our case the buoy is not fully submerged as in the case of the Carnegie CETO™ and is instead floating at equilibrium position with a draft of h_z = 2.0 m. The natural or resonance period of the WEC in heave , $T_{r,33} \equiv 5.46$ s. The second is a bottom-fixed surface-piercing OSWEC with a width (w) of 20 m, a height (h) of 12 m, a draft (h_z) of 10 m, and a thickness (δx) of 1.0 m (see Figure 3). The OSWEC is similar to several pre-commercial WEC technologies, specifically the WaveRoller, developed by Finnish company AW-Energy. The natural pitch period of the OSWEC, $T_{r,55} \equiv 17$ s.

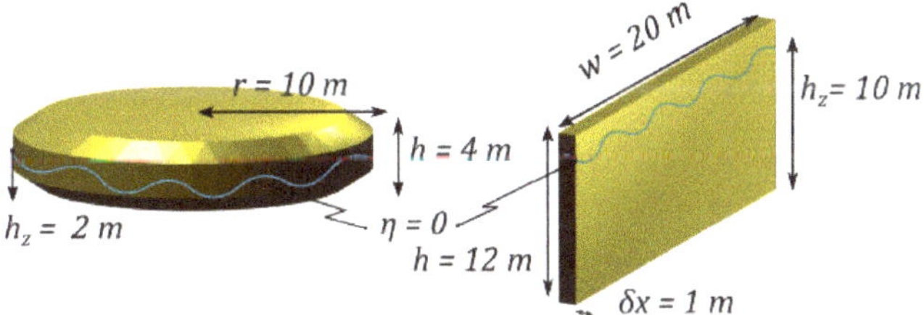

Figure 3. Heaving cylindrical WEC (**left**) and pitching OSWEC (**right**) schematic. The wavy line indicates the undisturbed free surface elevation η.

4.2. Input Wave Conditions

To demonstrate the utility of the presented PTO model coupling, regular waves of two wave heights and four wave periods are simulated as shown in Table 1.

Table 1. Test matrix of regular wave conditions.

Wave Height, H (m)	Wave Period, T (s)			
1.0	6.0	8.0	10.0	12.0
2.0	6.0	8.0	10.0	12.0

Each PTO system configuration presented in Section 4.1 and each WEC type in Section 4.1 is modelled for all wave conditions. In the following Sections 4.3 and 4.4, we determine the optimal PTO system coefficient for each WEC and PTO system type for each wave condition defined in Table 1.

4.3. Optimal PTO System Coefficients: Linear PTO

In Section 3.1.1 it was stated that an optimal value exists for the linear PTO system damping coefficient $B_{PTO,l}$, resulting in the maximum average absorbed power. These damping coefficients are first calculated for the specific case of the heaving cylindrical WEC with Equation (11), for the dimensions described above. The theoretically found values are summarized in Table 2. To calculate the corresponding coefficients for the OSWEC, (13) is applied for the OSWEC with the prescribed dimensions of Figure 3. Results for the optimal linear PTO damping coefficients are given in Table 2.

Table 2. Optimal linear B_{PTO} coefficients for a heaving cylindrical WEC ($10^6 \times$ kg/s) and OSWEC ($10^6 \times$ (kg· m^2)/s).

T (s)	6	8	10	12
Heaving Cylindrical WEC	1.12	2.25	3.46	4.65
OSWEC	128.0	98.40	69.70	51.0

4.4. Optimal PTO System Coefficients: Hydraulic PTO

4.4.1. Optimal Hydraulic PTO System Coefficients for a Heaving Cylindrical WEC

It was proven that an optimal linear damping coefficient exists when a linear PTO system is applied. Since the PTO-force of a hydraulic PTO system, $f_{PTO,h}$ is no longer linearly dependent on the velocity of the heaving cylindrical WEC, no straightforward relationship for an optimal configuration of the hydraulic PTO system can be expressed. To find optimal PTO system parameters, a similar approach as in [38] is followed: a hydraulic PTO system damping term $B_{PTO,h}$ is defined and it is checked for an optimum value. Note however that this damping coefficient $B_{PTO,h}$ cannot be used to calculate the PTO-force $f_{PTO,h}$ by multiplying $B_{PTO,h}$ with the WEC's velocity. It is a coefficient that takes into account the different parameters of the hydraulic PTO system that influence the performance of the WEC, with the same dimensions as the linear damping term $B_{PTO,l}$ [kg/s]:

$$B_{PTO,h} = \left(\frac{s_c}{D_m}\right)^2 B_g \tag{27}$$

$B_{PTO,h}$ can be changed by modifying the piston area, s_c, the motor displacement, D_m or the generator damping, B_g, see Figure 1. In practice it is most convenient to alter the motor displacement D_m [38], e.g., by installing a variable displacement motor as hydraulic motor. It is assumed that the swashplate angle α equals one. Since only D_m will be varied in the following procedure, it is assumed that s_c and B_g are constant: s_c is set as 0.0707 m^2 and B_g as 6 $\frac{\text{Nm}}{\text{rad/s}}$, respectively, based on a prior analysis. Figure 4 proves the existence of an optimal value for $B_{PTO,h}$ for different wave periods in regular waves. As with the linear PTO system, the optimal value for $B_{PTO,h}$ increases with increasing wave period T. Due to the inherent non-linearities of the hydraulic PTO system, a different optimal value for $B_{PTO,h}$ could be expected for a different wave height H at the same wave period T. However, only a small change was observed in the optimal value for $B_{PTO,h}$ when altering the wave height H from 1.0 m to 2.0 m. The same conclusion was made in [38]. Since the average absorbed power P_h stays rather constant close to the optimal value for $B_{PTO,h}$, the effect of a small change in $B_{PTO,h}$ close to its optimum value on P_h is negligible. Therefore, the $B_{PTO,h}$ coefficients summarized in Table 3 will be used for both $H = 1.0$ m and for $H = 2.0$ m.

The optimal hydraulic PTO system damping coefficients for the heaving cylindrical WEC for the studied wave conditions are summarized in Table 3.

Table 3. Optimal hydraulic damping coefficients $B_{PTO,h}$ for a heaving cylindrical WEC ($10^6 \times$ kg/s) and OSWEC ($10^6 \times$ m^2· kg/s).

WEC Type	T (s)	6	8	10	12
Heaving Cylindrical WEC	$H = 1.0$ m	1.5	3.25	4.7	8.3
OSWEC	$H = 1.0$ m	275	175	121	95

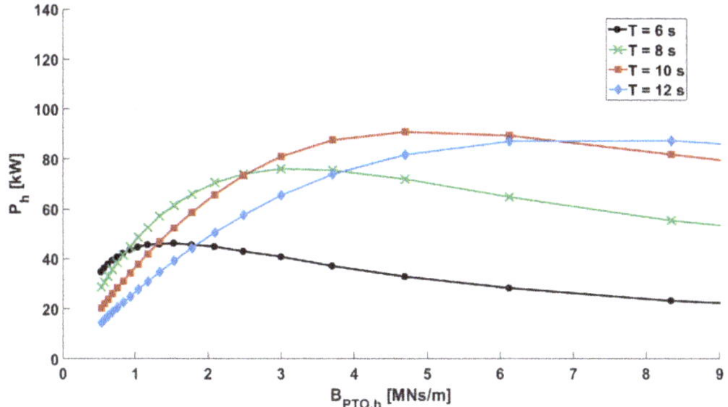

Figure 4. Average absorbed power P_h as function of hydraulic damping coefficient $B_{PTO,h}$ for the heaving cylindrical WEC for four different wave periods and for a wave height $H = 1.0$ m.

4.4.2. Optimal Hydraulic PTO System Coefficients for the OSWEC

Section 3.1.2 also described the application of a hydraulic PTO system to an OSWEC. As with the heaving cylindrical WEC, optimal hydraulic parameters will be found for the OSWEC with dimensions as given in Section 4.1. The piston area was set equal to $s_c = 0.1257$ m^2 while the generator damping B_g is set to 10 $\frac{Nm}{rad/s}$, both values resulting from a prior analysis. Please note that additional geometric parameters must be considered when studying the optimal configuration for an OSWEC with a hydraulic PTO system—see Section 3.1.2 and Figure 2. The hydraulic PTO system applied to the OSWEC exerts a torque, $T_{PTO}(t) - f_{PTO}(t) \cdot \ell(t)$, depending on the PTO force f_{PTO} and the lever arm ℓ, calculated as in Equation (25). The latter depends on the following geometric parameters: g, c and b as defined in Figure 2. This implies that, contrary to the case of the heaving cylindrical WEC, not only the characteristics of the hydraulic PTO system, but also the initial geometric parameters g, c and b_{ini} must be chosen carefully. The reasoning followed in the procedure of optimizing the hydraulic PTO system will briefly be explained below. It is firstly assumed that an optimal PTO-torque exists for each wave period, $T_{PTO,opt}$. When then e.g., c increases, ℓ will increase as well, keeping all other parameters constant. This will result in a lower $f_{PTO,opt}$ to achieve the same $T_{PTO,opt}$. f_{PTO} can be lowered by increasing D_m. Changing the motor displacement will result in a different pressure difference between the accumulators and a different motor speed. The geometric configuration of the hydraulic PTO system for the OSWEC can thus be chosen in such a way that allows the most convenient hydraulic motor parameters. It may be expedient to limit the motor speed or the pressure difference to a certain value, which can be realized by adapting the motor displacement accordingly. A brief numerical analysis has shown that higher values for c and thus higher motor optimal displacements D_m result in lower pressure differences. However, this distance c will probably have to be limited as well due to practical considerations. When looking at sketches of the WaveRoller OSWEC, the hydraulic PTO system seems to be very close to the seabed. After a brief analysis, it was chosen to put c equal to 3.0 m, g to 1.5 m and b_{ini} to 5.0 m. D_m was varied to find an optimal value that results in the maximum P_h. To express an equivalent $B_{PTO,h}$ for the OSWEC (similarly as was done for the heaving cylindrical WEC), following formula is used, resulting in a coefficient with the same dimensions as the linear damping term for the OSWEC:

$$B_{PTO,h} = c \cdot b_{ini}\left(\frac{s_c}{D_m}\right)^2 B_g. \tag{28}$$

Figure 5 shows the average absorbed power P_h for different values of $B_{PTO,h}$ for the four considered wave periods described in Section 5.2 and a wave height $H = 1.0$ m. The optimal value for $B_{PTO,h}$ decreases with increasing wave period, for this range of wave periods. The same conclusion was made for the OSWEC with a linear PTO system: the optimal value for $B_{PTO,l}$ decreases with increasing wave period.

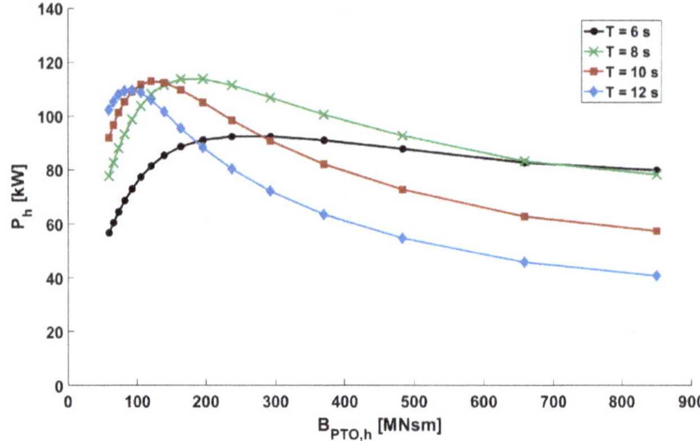

Figure 5. Average absorbed power P_h as function of hydraulic damping coefficient $B_{PTO,h}$ for the OSWEC for four different wave periods and for a wave height $H = 1.0$ m.

5. Comparing the Effects of a Linear to a Hydraulic PTO System for a Single Heaving Cylindrical WEC and a Single OSWEC

5.1. Comparing the Average Power Output for Each WEC vs. Type of PTO System

The average power output for a single WEC of each type is calculated via Equation (10) or Equation (14) for the linear PTO system and via Equation (23) or (26) for the hydraulic PTO system. Please note that for the latter PTO system type the losses in the generator will not be taken into account to provide a fair comparison with the linear results, as noted in Section 3.1.2. The B_{PTO} settings used are described in Sections 4.3 and 4.4. Results for the modelled wave conditions of Table 1 are shown in Table 4. We note that the results for $H = 2.0$ m are almost exactly 4 times the results for $H = 1.0$ m, indicating that the non-linear influence of the hydraulic PTO system in these operational wave conditions is minimal. Therefore, we will focus on the results for a $H = 1.0$ m wave, which we plot in the bar chart in Figure 6.

$$\frac{P_h - P_l}{P_l} \cdot 100. \tag{29}$$

The percent difference is defined by Equation (29). We observe that for the heaving cylindrical WEC, the average power output is always greater with the hydraulic PTO system than with the linear PTO system while, in comparison, for the OSWEC the situation is reversed.

Table 4. Average power output for a single WEC for a linear and hydraulic PTO system. Heaving cylindrical WEC: top two rows. OSWEC: bottom two rows.

WEC Type	Wave Height	Average Power Output Linear P_l (kW)				Average Power Output Hydraulic P_h (kW)			
		Wave Period T (s)				Wave Period T (s)			
	H (m)	6.0	8.0	10.0	12.0	6.0	8.0	10.0	12.0
heaving	1.0	47.98	65.94	72.86	72.04	50.15	77.40	90.05	85.34
cylindrical WEC	2.0	191.91	263.78	291.46	288.14	200.61	311.36	364.59	344.15
OSWEC	1.0	106.47	132.75	131.55	126.83	92.56	114.34	113.08	109.43
	2.0	425.87	531.03	526.49	508.78	367.98	452.59	447.95	434.39

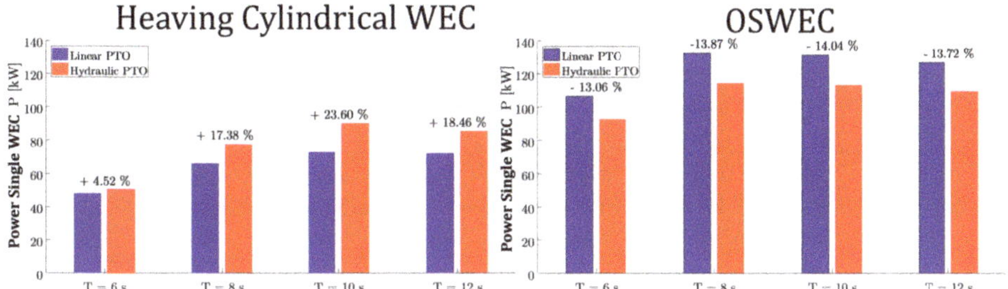

Figure 6. Bar chart showing the power output for one WEC with linear PTO system (P_l) (purple) and hydraulic PTO system (P_h) (red) with the percentage difference between the two. Results for the heaving cylindrical WEC shown on the left and for the OSWEC on the right.

It can be seen that there is a notable increase in the average power output for the hydraulic PTO system (P_h) versus the linear (P_l) for the case of the heaving cylindrical WEC for periods $T \geq 8.0$ s. For these wave conditions, the hydraulic PTO system can damp the motion of the WEC to match the phase of the incident wave condition more effectively. Such is not the case with the OSWEC, where the natural pitching period of the WEC is higher than the investigated wave periods and the hydraulic PTO system is not performing optimally, i.e., it cannot 'speed up' the relative motion. We must note, however, that the linear PTO system for the OSWEC, although it shows on average a 15% improvement in the power performance of the WEC, may be making unrealistic assumptions about the motion of the OSWEC that may result in an artificially increased average power output. Observe that in all cases the average power output for the OSWEC is much higher than for the heaving cylindrical WEC, indicating that the OSWEC is more efficient in absorbing the power of the incoming waves; how this power absorption affects the wave field will be explored in the next Section 5.2.

5.2. Analyzing the Wave Field around One WEC

5.2.1. Calculating the Total and Perturbed Wave Fields

To calculate the wave field around a single WEC for a wave height H, we sum the complex incident unidirectional regular wave field, calculated at each point via Equation (30)

$$\zeta(x, y) = \frac{H}{2} e^{-i(kx)} \tag{30}$$

to the perturbed wave field consisting of the radiated and diffracted wave fields. Both are calculated from their respective potentials via the kinematic free surface boundary condition Equation (5). The radiated wave field is given by Equation (31)

$$\eta_r = -\frac{ZH}{\zeta} \frac{i\omega\phi_r}{2g}. \tag{31}$$

Here ϕ_r is the radiated wave potential and the ratio of the body displacement Z to the wave amplitude ζ is the response amplitude operator (RAO) which is calculated in Equation (32):

$$\frac{Z}{\zeta} = \frac{\bar{F}_e}{-\omega^2(M + A)^2 - i\omega(B_{PTO} + B) + K} \tag{32}$$

The modulus of the complex RAO calculated in Equation (32) is the amplitude of the WEC's position divided by the wave amplitude:

$$|RAO| = |\frac{Z}{\zeta}|$$ (33)

Equation (32) is only valid when modelling a WEC with a linear PTO system. In Equation (32) F_e is the excitation force, M the mass of the device, and A, B and K, the added mass, hydrodynamic damping, and hydrodynamic spring or stiffness coefficients, respectively, determined in NEMOH for each of the relevant degrees of freedom. The B_{PTO} is the linear $B_{PTO,l}$ coefficient in Table 2 for each wave period and WEC type. It is important to mention that the use of the hydraulic PTO coefficient $B_{PTO,h}$, as described in Sections 4.4.1 and 4.4.2, in Equation (32) will lead to incorrect results. The coefficient $B_{PTO,h}$ was composed to combine all significant factors influencing the average absorbed power P_h, to check if an optimum value of the average absorbed power exists and to study the trend of this coefficient over a range of periods. Since the RAO for a WEC with a hydraulic PTO system cannot be calculated analytically, this RAO is determined using numerical time-domain simulations. For a given wave period and wave height, the WEC's displacement is calculated numerically using WEC-Sim. The modulus of the RAO is calculated with Equation (33), whereas the RAO's phase is determined as in Equation (34):

$$\varphi = \omega \cdot \Delta t,$$ (34)

where Δt represents the time shift between the WEC's displacement profile and the surface elevation profile. Since the WEC's position $z(t)$ is not sinusoidal when equipped with a hydraulic PTO system, the following method is used for the calculation of the time shift Δt:

$$\Delta t = \operatorname*{argmax}_{\tau} \int_0^T z(t) \cdot \zeta(t - \tau) dt.$$ (35)

The RAO phase φ will be positive since the WEC's motion is delayed with respect to the incoming wave ($\Delta t > 0$). The complex value of the RAO is now determined as:

$$\frac{Z}{\zeta} = |\frac{Z}{\zeta}|e^{i\varphi}.$$ (36)

The diffracted wave amplitude η_d is given by Equation (37)

$$\eta_d = -\frac{i\omega\phi_d H}{g},$$ (37)

where ϕ_d is the diffracted wave potential. We calculate the wave field around a single WEC for each of the incident wave conditions presented in Table 1. In the two sections following, Sections 5.2.2 and 5.2.3, we show representative results from the 24 cases simulated. Please note that the tally takes into account the fact that for the linear PTO system the result for $H = 1.0$ m and $H = 2.0$ m are the same.

5.2.2. The Influence of the WEC Type on the Wave Field

Before diving into the complicated patterns seen in the 'near-field' η of the array, we model a single WEC in the numerical domain to clarify the impact of WEC type and PTO system type on the wave field. The two WEC types presented in Section 4.1 have a substantially different impact on the incoming waves as witnessed in the plots of the modulus of total wave field $|\eta|$, in Figure 7 for one heaving cylindrical WEC (left) and an OSWEC (right) for a linear PTO system for the same incident wave of $H = 1.0$ m and $T = 6.0$ s. The total $|\eta|$ is the modulus of sum of the complex perturbed $\tilde{\eta}$ and the complex incident wave $\bar{\eta}$. We see right away that the perturbation effect for the OSWEC is much greater than that of the heaving cylindrical WEC, both in magnitude and extent away from the WEC.

This difference is largely a consequence of the diffraction potential of the OSWEC since it presents a barrier to the entire water column compared to the small-draft heaving WEC which presents much less resistance to the incoming waves. As an example, we can observe this difference in Figure 8 for a $H = 1.0$ m, $T = 10.0$ s wave where the diffraction is plotted for a heaving cylindrical WEC on the left and for an OSWEC on the right.

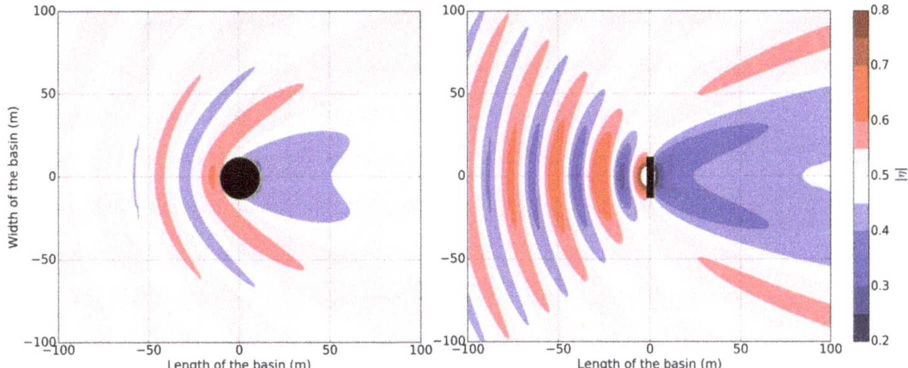

Figure 7. Modulus of the total surface elevation $|\eta|$ for a heaving buoy WEC (**left**) and OSWEC (**right**). Incident wave of $H = 1.0$ m, $T = 6.0$ s propagating from the left.

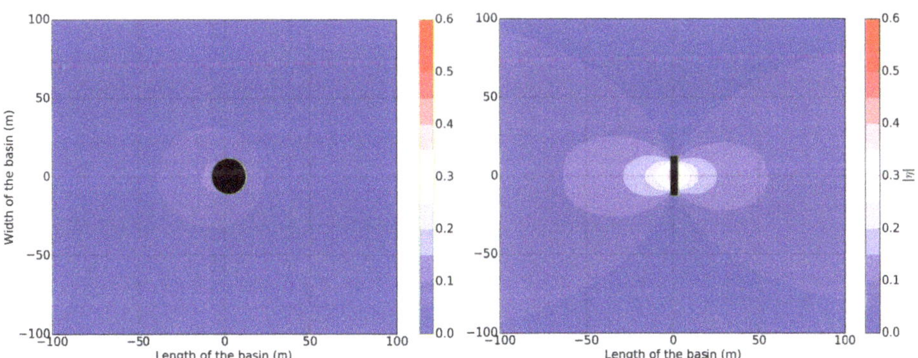

Figure 8. Modulus of the diffracted surface elevation $|\eta|$ for a heaving cylindrical WEC (**left**) and an OSWEC (**left**). Incident wave of $H = 1.0$ m, $T = 10.0$ s propagating from the left.

Moreover, the difference in the radiated wave field is significant as well, especially as we move to higher wave periods, where the OSWEC responds more to the incoming wave whereas the heaving cylindrical WEC is essentially riding on top of the water column. This is significant in our study because it is indeed the radiation which we can influence throughout the PTO model as will be witnessed in the next subsection.

5.2.3. The Influence of the PTO System Type on the Wave Field for a Single WEC

As mentioned in the previous paragraph in Section 5.2.2, the discrepancy between the radiation of the two WECs is less than the difference in diffraction for a given wave. However, it is still significant, and as the radiated wave field is a function of the PTO system as well as the WEC type, we do see a divergence in the perturbed wave field between the different PTO system types. This is noted in a plot of the percent difference between the total $|\eta|$ for the linear and the hydraulic PTO system first for the heaving cylindrical WEC (left) and the OSWEC (right) in Figure 9 for a H = 1.0 m, T = 8.0 s wave.

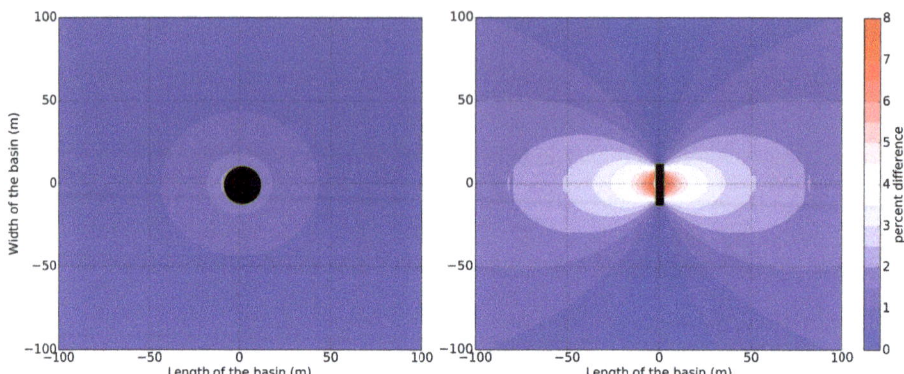

Figure 9. Percent difference (Equation (29)) in the total wave field between the hydraulic and linear PTO system for a heaving cylindrical WEC (**left**) and OSWEC (**right**). Incident wave of H = 1.0 m, T = 8.0 s propagating from the left.

We observe that the variability between the two PTO system types is less than 5 % for the heaving cylindrical WEC while that for the OSWEC is closer to 10 % in the region near the device. This is not demonstrated in the results in the power output (P) however, where in Table 4 in Section 5.1 for the H = 1.0 m, T = 8.0 s wave, the difference between P_l and P_h is 17% and only 14% for the OSWEC. Moreover, $P_h - P_l$ is positive for the heaving cylindrical WEC while the addition of a hydraulic PTO system actually reduces the power output for an OSWEC. This situation is mirrored for the other wave periods where the increase in the perturbed wave field for the OSWEC compared to that of the heaving cylindrical WEC does not induce an increase in the power output of the OSWEC P_h.

6. Analyzing the Power Production and the Near-Field Effects for an Array of 5 WECs With a Hydraulic PTO System

6.1. WEC Array Layout

As we have seen in the results for a single WEC in Section 5, the perturbed wave field around a single WEC strongly depends on both the WEC type and the PTO system modelled. In this section, we extend our results to an array of 5 WECs with a view toward the modelling of a commercial scale WEC farm consisting of multiple WEC arrays. To this end we model two different 5-WEC arrays: one consisting of heaving cylindrical WECs (Figure 10) and the other of pitching OSWECs (Figure 11). The WEC-WEC separation distances d_x and d_y are set to 40 m, which is the 2× the diameter of the heaving buoy WEC and the width of the OSWEC. The array configurations of both WEC types are staggered, an arrangement that was clearly shown to be power-maximizing in several numerical and experimental studies such as in [4,23,26,28,39–41]. In this investigation the water depth is set at 30 m for the heaving buoy and 10.0 m for the OSWEC.

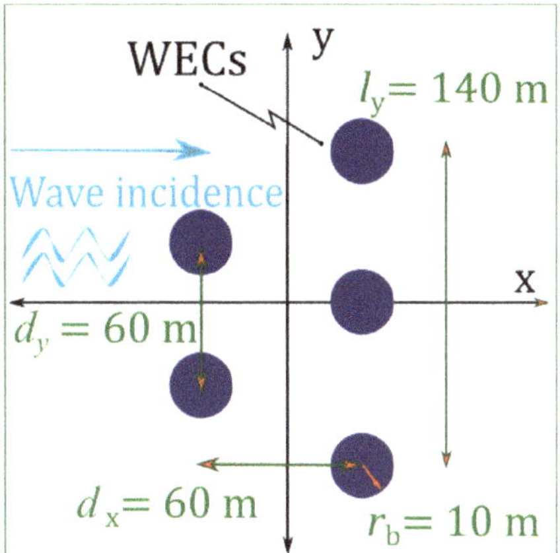

Figure 10. Plan view of the array layout for five heaving cylindrical WECs. The incident wave propagates from the left.

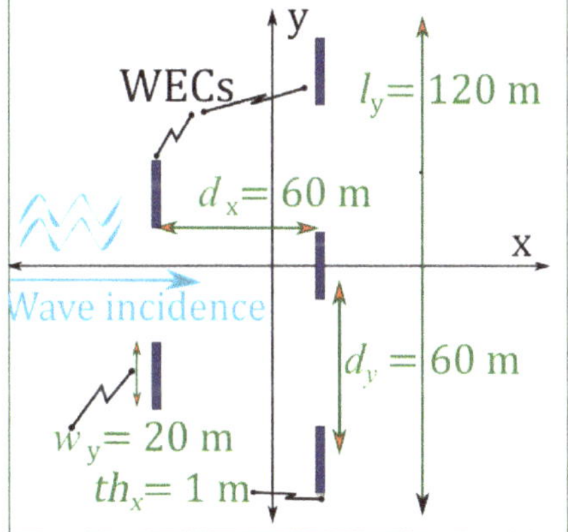

Figure 11. Plan view of the array layout for five pitching OSWECS. The incident wave propagates from the left.

6.2. 1st Order Approximation for the WEC Array Near-Field

To assess the effects of multiple WECs in a WEC array or multiple WEC arrays in a WEC farm on the power output (**P**) of the farm, we need to calculate the total perturbed wave field in the *near-field domain*. As we assume linear theory in our work, we can use the superposition principle to sum up the total wave field by using an iterative approach first developed in [42]. The technique employed is illustrated in Figure 12. The initial step (Step 1) is to propagate the incident wave in the empty numerical basin (no WEC present) to obtain the undisturbed surface elevation. In Step 2 the incident surface elevation is used as input into NEMOH whence the 1st order perturbed wave of WEC Array I, p_{1i}, is evaluated. In Step 3, the average wave amplitude at the location of p_{1i} is used as input into NEMOH to calculate the 1st order perturbed wave of WEC Array II, p_{1ii}. In Step 4, the process in Step 2 is repeated, with p_{1ii} as the new input perturbed wave. Finally, in Step 5, the same process is performed for the 2nd perturbed wave of WEC Array I, p_{2i}. Since the input perturbed wave in each subsequent step after step is reduced by approximately an order of magnitude, for all practical purposes this process can be terminated at Step 4 without any appreciable loss in accuracy, even for the case where interaction is maximized. Therefore, Step 5 is only displayed for a complete description of the proposed iterative method. To calculate the perturbed η, in this paper we limit the summation to the 1st order perturbed waves from each WEC in the array. The power for each WEC is calculated using the average surface elevation that is the sum of the incident wave and the perturbed waves from the nearest two WECs.

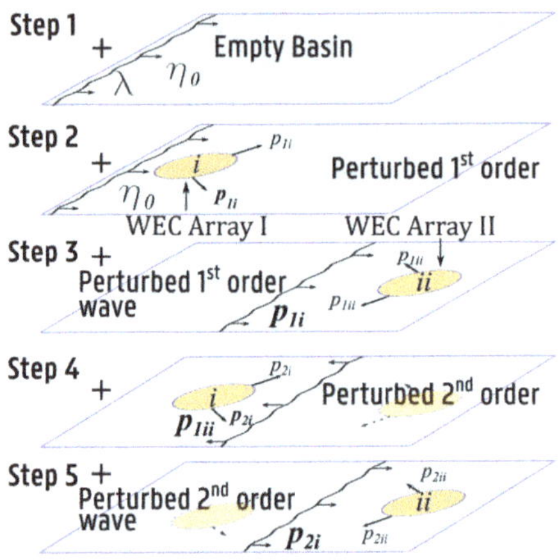

Figure 12. Step by step procedure for determining the perturbed field for regular incident wave propagating from the left [42].

6.3. Power Output Calculation for an Array of 5 WECs

In evaluating the influence of the 5-WEC array interaction effects on the performance of a wave farm, we compute the total power output by the two WEC arrays, after having obtained the modified wave field in the WEC array using the approach outlined in Section 6.2. The power of each array is calculated by the following equations depending on the PTO system and WEC type. For the linear PTO system, we extend Equations (10) and (14) to $\mathcal{M}$ WECs operating in one DoF to Equation (38),

$$\mathbf{P}_l = -\frac{1}{2}\mathbf{B}_{PTO,l}|\mathbf{Z}(\omega)|^2\omega^2, \tag{38a}$$

$$\mathbf{P}_l = -\frac{1}{2}\mathbf{B}_{PTO,l}|\boldsymbol{\theta}(\omega)|^2\omega^2, \tag{38b}$$

where $\mathbf{Z}$ and $\boldsymbol{\theta}$ indicate an $\mathcal{M} \times 1$ column vector of the WEC's position or angular displacement, respectively. $\mathbf{B}_{PTO,l}$ represents an $\mathcal{M} \times \mathcal{M}$ diagonal matrix with the B_{PTO} coefficients for each WEC on the diagonal. For the hydraulic PTO system, the equations equivalent to (23) and (26) are given in Equation (39):

$$\mathbf{P}_h = -\frac{1}{T}\int_0^T \mathbf{F}_{PTO}(t) \cdot \dot{\mathbf{z}}(t)dt, \tag{39a}$$

$$\mathbf{P}_h = -\frac{1}{T}\int_0^T \mathcal{T}_{PTO}(t) \cdot \dot{\boldsymbol{\theta}}(t)dt. \tag{39b}$$

Here as in Equation (38), the boldface quantities represent $\mathcal{M} \times 1$ column vectors of the forces and velocities of the individual heaving cylindrical WECs of the torques and angular displacements of the individual OSWECs. As mentioned in Section 6.2, for each WEC in the array, the motions and the forces used in Equations (38) and (39) are calculated with the input wave equal to the incident wave plus the 1st order WEC array perturbed wave at the location of the given WEC. The magnitude of the

η used for calculating the power **P** in Equations (38) and (39) is taken as the average of the 1st order modified η on a region immediately surrounding the WEC. In addition to calculating the power of each array, we also introduce the 'q value', defined as the ratio of the power of the $\mathcal{M}$-WEC array to the power produced by the sum of $\mathcal{M}$ WECs as if they were operating in isolation:

$$q = \frac{\mathbf{P}_{Array}}{\sum P}. \tag{40}$$

where P is the power output of the linear or hydraulic PTO WEC given by equations (Equations (10), (14), (23) and (26)) for the heaving cylindrical WEC or the OSWEC, respectively. The q value is a commonly used metric in wave energy literature to assess the strength of array effects, we find it used in [30,39,40,43], for example.

6.4. Power Output for an Array of 5 WECs

The 5-WEC array power output for the two PTOs and for the data for the two WEC types is displayed in Table 5 for the modelled wave periods from Table 1 for $H = 1.0$ m and in the bar chart Figure 13. The q value for the various configurations, defined in Equation (40), is displayed in the third and sixth data row. As we have witnessed in Section 5.1, the deviation from linear behavior due to the increase from $H = 1.0$ m to $H = 2.0$ m is very small, therefore we will focus our attention in this and the following sections on the results for $H = 1.0$ m with the knowledge that the results for $H = 2.0$ m show similar patterns and behaviors. As in the single WEC case, we observe a significant increase in the power output of the 5-OSWEC array versus a 5-heaving cylindrical WEC array with the power of the former producing up to 3× more power for a wave period of 8.0 s. Please note that as in the single WEC case analyzed in Section 5.1, the heaving cylindrical WEC array produces more power with increasing wave period while in the case of the OSWEC array, the peak power occurs for $T = 8.0$ s, with a decrease for higher wave periods. Please note that this reduction is more significant in the array case than in the single wave case, a fact that is reflected in the decreasing q values as the period increases. This behavior can be directly linked to the increase in the $|\eta|$ in the 'near-field' zone, as we will observe in Section 6.5.2. For the heaving cylindrical WEC, the q values are also decreasing for wave periods greater than 8.0 s, but with the difference that each q value is consistently below unity. It is clear from the data that in the case of the modelled 5-WEC array configuration, placing the OSWECs in an array is much more advantageous to their performance than for the heaving cylindrical WECs. We must remark however, that in realistic wave conditions with frequency and directional spreading it is near impossible to achieve the phase relationships that lead to high q values and consequently, we expect the relative difference in the array power output between the two types of WECs to diminish.

Table 5. Average power output for an array of 5 WECs for a linear and hydraulic PTO system. heaving cylindrical WEC: top three rows. OSWEC: bottom three rows.

WEC Type	Value	Wave Height	Average Power Output Linear P_l				Average Power Output Hydraulic P_h			
			Wave Period T (s)				Wave Period T (s)			
		H (m)	6	8	10	12	6	8	10	12
heaving	ARRAY	$H = 1.0$	234.54	325.48	315.33	304.44	245.7	387.28	389.27	358.25
cylindrical	SINGLE × 5	$H = 1.0$	239.9	329.72	364.32	360.18	250.77	386.99	450.26	426.69
WEC	q	$H = 1.0$	0.98	0.99	0.87	0.85	0.98	1.00	0.86	0.84
	ARRAY	$H = 1.0$	1001.7	1736.6	1282.6	854.33	867	1617.7	1237.8	868.79
OSWEC	SINGLE × 5	$H = 1.0$	532.34	663.75	657.75	634.15	462.8	571.7	565.4	547.2
	q	$H = 1.0$	1.88	2.62	1.95	1.35	1.87	2.83	2.19	1.59

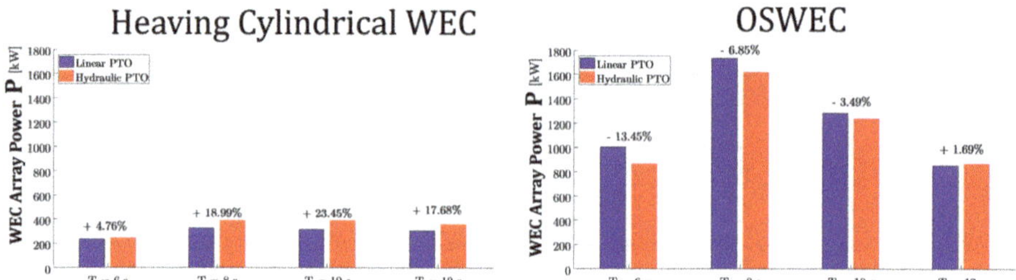

Figure 13. Bar chart showing the power output for a 5-WEC heaving cylindrical WEC array (**left**) and OSWEC array (**right**)with linear PTO system (purple) and hydraulic PTO system (red) with the percentage difference between the two calculated by Equation (29).

6.5. The Near-Field $|\eta|$ for an Array of 5 WECs

In this section, we present the results for the *near-field* wave field for an array of 5 heaving cylindrical WECs, arranged in the configurations displayed in Figures 10 and 11 for the wave periods listed in Table 1 for a wave height H of 1.0 m. The results are presented in Sections 6.5.1 and 6.5.2 as the modulus of the surface elevation $|\eta|$. Using this metric, we show both the total wave field to see the connection between the surface elevation and the array power output, and the perturbed wave field which only displays the array effects, that is deviations from the incident wave field brought about by the interactions with the WEC arrays. Because of the quantitative differences in the wave fields for a heaving cylindrical WEC and an OSWEC, the presentation of the results is split into two Sections 6.5.1 and 6.5.2, where in each subsection we take an in-depth look at the 'near-field' wave amplitude η.

6.5.1. The Perturbed $|\eta|$ for an Array of Heaving Cylindrical WECs

First thing, we look at the wave field of an array of 5 heaving cylindrical WECs for a linear PTO system for $T = 6.0$ s and $T = 8.0$ s. In Figures 14 and 15 the total (left) and perturbed (right) fields are plotted for the named wave periods. Notice that the magnitude of the changes in the total $|\eta|$ due to the presence of the array are much greater for the case of $T = 6.0$ s. This can be seen even more clearly in a comparison of the perturbed $|\eta|$ for the same two wave periods, where the perturbed wave field is nearly $2\times$ greater in magnitude near the WECs. However, it would be incorrect to assume that this difference is linearly proportional to the difference in the power output **P** of the array at these wave periods, as will be elaborated on in Section 7.

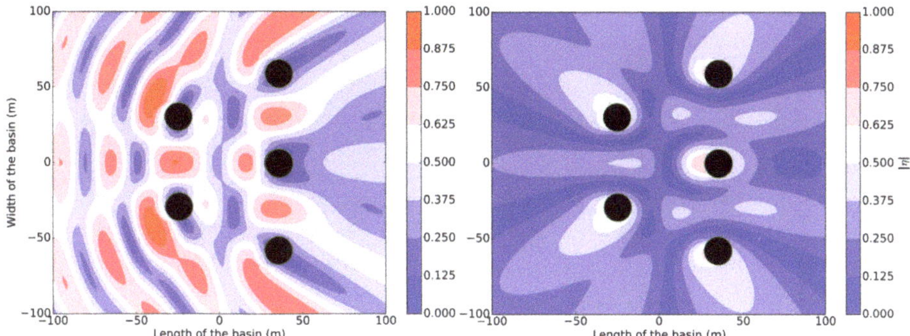

Figure 14. The total (**left**) and perturbed (**right**) $|\eta|$ for a heaving cylindrical WEC for a wave of $H = 1.0$ m, $T = 6.0$ s for a linear PTO system. Incident wave propagating from the left.

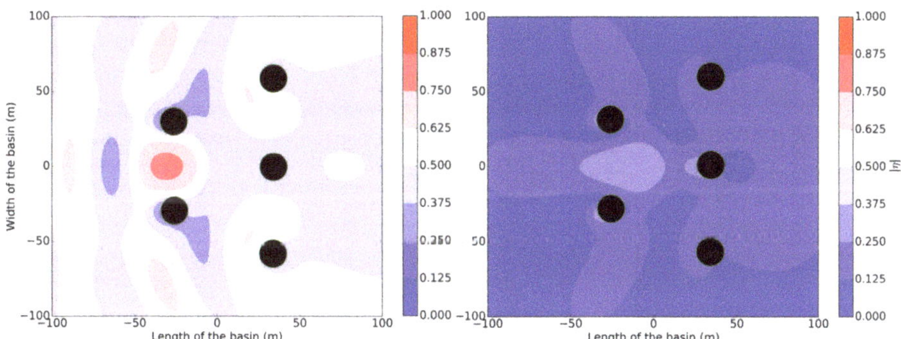

Figure 15. The total (**left**) and perturbed (**right**) $|\eta|$ for a heaving cylindrical WEC for a wave of $H = 1.0$ m, $T = 8.0$ s for a linear PTO system. Incident wave propagating from the left.

Moving on to the two higher wave periods, $T = 10.0$ s and $T = 12.0$ s, the interaction of the incident wave field with the WEC array markedly decreases. We can observe this in a contour plot of the total and the perturbed wave field for $T = 10.0$ s for the heaving cylindrical WEC array with a linear PTO system in Figure 16. We note that although the perturbed wave field is barely perceptible, it does result in a slight enhancement of the total wave field which creates an area of higher total $|\eta|$ in front of the array. For $T = 12.0$ s the shape of the interaction zones is similar to those of $T = 10.0$ s but the magnitude of the array effects is minimal and consequently, these wave fields are not displayed in the interest of brevity.

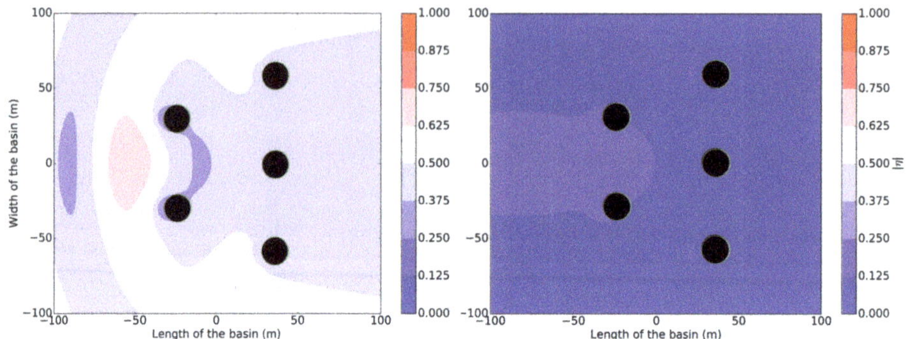

Figure 16. The total (**left**) and perturbed (**right**) $|\eta|$ for a heaving cylindrical WEC for a wave of $H = 1.0$ m, $T = 10.0$ s for a linear PTO system. Incident wave propagating from the left.

6.5.2. Results for an Array of OSWECs

We next move on to explore the results of the simulations for the 5-OSWEC Array. Analogous to Section 6.5.1 we first look at the total near-field $|\eta|$ for $T = 6.0$ s and $T = 8.0$ s, which are the wave periods with the greatest 'array effect' and the highest power output **P**. In Figures 17 and 18 we plot the total $|\eta|$ (left) and the perturbed $|\eta|$ for the two wave periods in question. Observe that the magnitude of both fields is much greater than that of the heaving cylindrical WEC shown in Figures 14 and 15 for both $T = 6.0$ s and $T = 8.0$ s. Moreover, we observe a large difference in the locations of 'hot spots' and 'cold spots', which are areas of strong positive or negative anomalies in $|\eta|$ between $T = 6.0$ s and $T = 8.0$ s. In other words, the areas with destructive interference between the incident and the perturbed wave leads to a decrease in $|\eta|$ or vice versa with constructive interference between the incident and perturbed waves. This is important in understanding the interaction between the wave period and the power output **P** that we will discuss in Section 7.

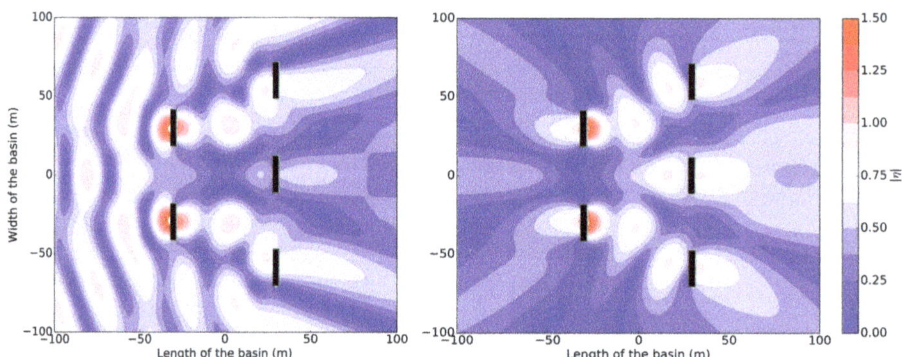

Figure 17. The total (**left**) and perturbed (**right**) $|\eta|$ for an array of heaving cylindrical WECs for a wave of $H = 1.0$ m, $T = 6.0$ s for a linear PTO system. Incident wave propagating from the left.

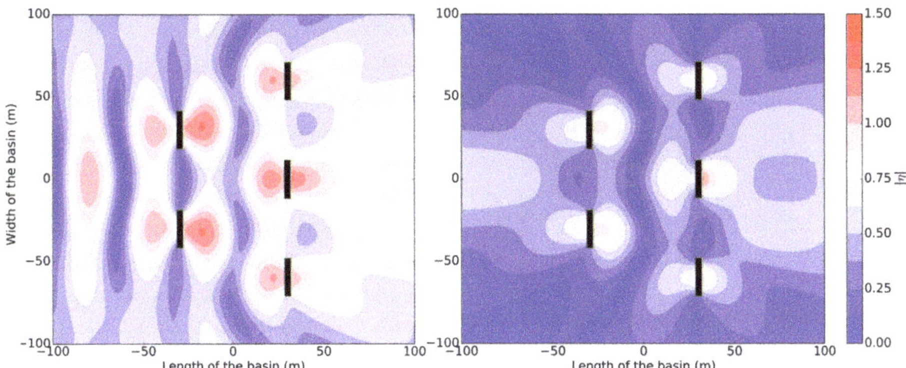

Figure 18. The total (**left**) and perturbed (**right**) $|\eta|$ for an array of heaving cylindrical WECs for $H = 1.0$ m, $T = 8.0$ s for a linear PTO system. Incident wave propagating from the left.

As with the heaving cylindrical WEC, the two largest wave periods $T = 10.0$ s and $T = 12.0$ s display smaller perturbations in the near-field zone. Unlike for the heaving cylindrical WEC, however, they are still significant, as we can witness in Figures 19 and 20 in the plots of the $|\eta|$ for an OSWEC with a linear PTO system. This perturbation effect is mirrored in the positive q values in Table 5 for both $T = 10$ and $T = 12$, unlike in the case of the heaving cylindrical WECs. Again, notice the strong change in the locations of the positive and negative anomalies in the total wave field between Figures 19 and 20. As we will see in the next section Section 6.5.3, these are the two wave periods where the hydraulic PTO system power performance in a OSWEC array is close to or slightly exceeding the linear PTO system WEC array case, unlike the single WEC case in Section 5.1 where the reverse is true.

6.5.3. Comparing the Effect of a Linear PTO System to a Hydraulic PTO System for a Wave Field around a 5-WEC Array

In this section, we compare the effect of the linear and hydraulic PTO system on the near-field of the array. As in Sections 6.5.1 and 6.5.2, we show the outcomes for both the total and perturbed wave fields, but instead of plotting $|\eta|$, we plot the percent difference between the $|\eta|$ of the WEC with hydraulic and the linear PTO system similar to Figure 9 for the a single WEC. We start by looking at the effect of the hydraulic PTO system for the case of the heaving cylindrical WEC. In Figure 21 we plot the difference between the total $|\eta|$ for a heaving cylindrical WEC for the 4 modelled wave periods as defined by Equation (29).

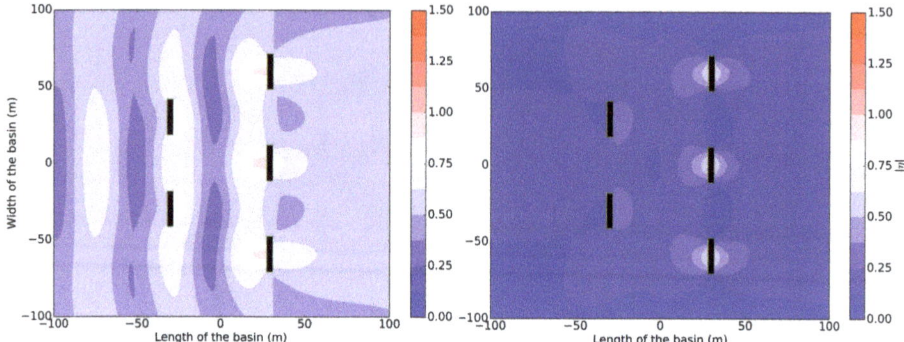

Figure 19. The total (**left**) and perturbed (**right**) $|\eta|$ for an OSWEC for a wave of $H = 1.0$ m, $T = 10.0$ s for a linear PTO system. Incident wave propagating from the left.

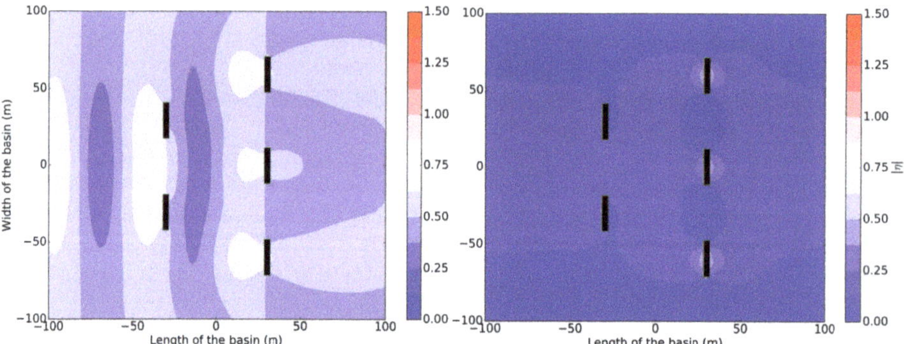

Figure 20. The total (**left**) and perturbed (**right**) $|\eta|$ for an OSWEC for a wave of $H = 1.0$ m, $T = 12.0$ s for a linear PTO system. Incident wave propagating from the left.

The first observation we make is the marked decrease in the η difference as we increase T from 6.0 s to 12.0 s. While for the 6.0 s wave the difference barely exceeds 15 % for areas on the perimeter of the body, for the rest of the wave periods the differences are considerably less, dipping below the 5% threshold of the $T = 12.0$ s case. Please note that whereas for the two shorter wave periods the areas of positive and negative change have a complicated pattern based on the interaction between the radiated waves of each body, for the $T = 10.0$ s and $T = 12.0$ s cases there is a general trend of a higher $|\eta|$ for the hydraulic PTO system for the front rows and lower for the back row, especially for the back middle WEC. Observe that this slight overall decrease in $|\eta|$ does not adversely affect the heaving cylindrical WEC array performance, as we saw in Table 5 in Section 6.4, where the performance of the heaving cylindrical WEC array is significantly better than that of the single WECs.

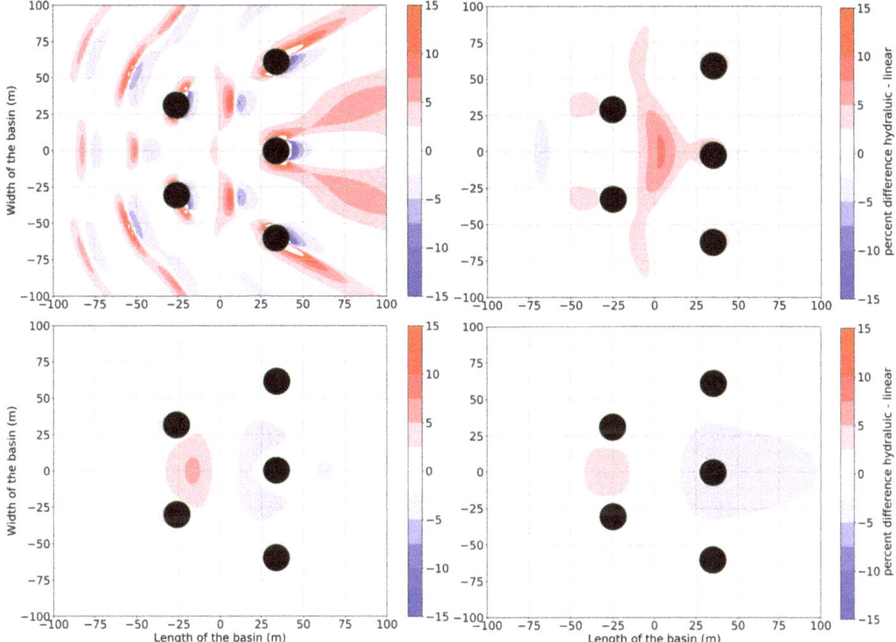

Figure 21. Percentage difference between the $|\eta|$ produced by a heaving cylindrical WEC with a hydraulic PTO vs. a linear PTO system for a wave of $H = 1.0$ m and wave periods of $T = 6.0$ s (**top left**) $T = 8.0$ s (**top right**) $T = 10.0$ s (**bottom left**) $T = 12.0$ s (**bottom right**). Incident wave propagating from the left.

Contrary to the heaving cylindrical WEC array, the difference of the PTO system greatly modifies the wave field of the 5-OSWEC array. In comparing Figure 22 to Figure 21, we see that the percent difference is much greater, in fact more than 100% for the 8.0 s case. We also observe that, unlike for the heaving cylindrical WEC array example, the differences in $|\eta|$ do not markedly decrease with increasing wave period. We see that difference is the greatest for $T = 8$ s but that it is also greater for $T = 12.0$ s than for $T = 10.0$ s. What we see then is that there is a strong effect the hydraulic PTO system on the WEC array wave field, and by comparing the contour plots in Figure 22 to the values for the average absorbed power of the OSWEC array in Table 5, we also notice that the difference in $|\eta|$ is not always proportional to the difference in power. For example, we notice that the magnitude and extent of the positive anomalies for $T = 12.0$ s is greater than that for $T = 10.0$ s but that the hydraulic PTO system 5-OSWEC array produces less power for the higher period. In general, we see that the difference from linear to hydraulic PTO system has a strong effect on the total wave field, but that the quality of the difference is greatly dependent on the wave period. We note that for the $T = 8.0$ s case in particular, there is an overall reduction in the surface elevation in lee of the array for the hydraulic PTO system compared to the linear PTO system, a fact that is reflected in the increase of the q value from 2.62 to 2.83. We can also observe that for the $T = 10.0$ s and especially the $T = 12.0$ s case that there is a net increase in $|\eta|$ inside the array area and a slight decrease outside of it. Again, we see this confirmed in the q values in Table 5 where they increment from 1.95 to 2.19 for the $T = 10.0$ s and from 1.35 to 1.59 for the $T = 12.0$ s wave.

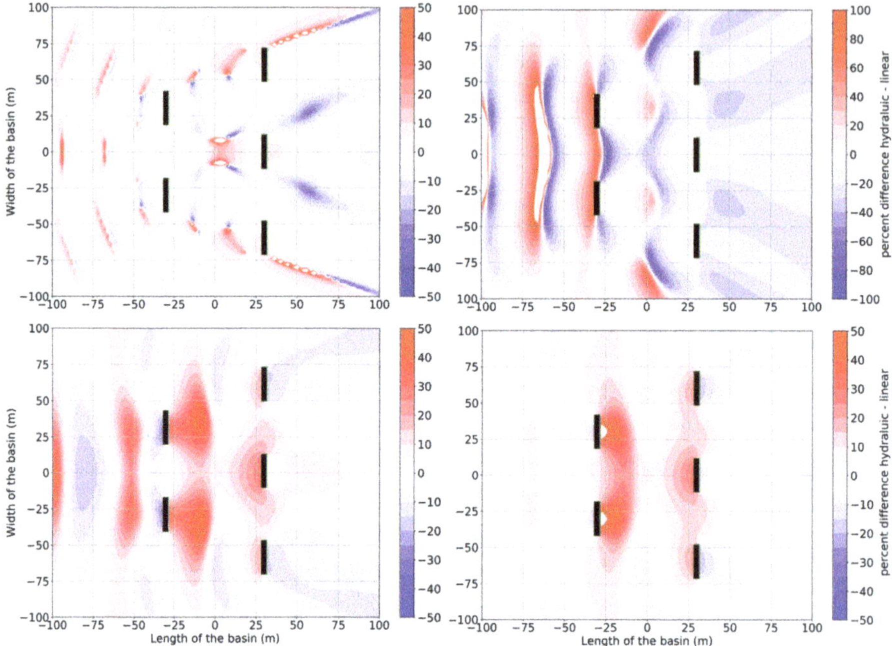

Figure 22. Percentage difference between the $|\eta|$ produced by an OSWEC with a hydraulic PTO vs. a linear PTO system for a wave of $H = 1.0$ m and wave periods of $T = 6.0$ s (**top left**) $T = 8.0$ s (**top right**) $T = 10.0$ s (**bottom left**) $T = 12.0$ s (**bottom right**). Incident wave propagating from the left.

7. Discussion

In the results for the 5-WEC arrays in Section 6 we have seen the interplay between the efficacy of the WEC array from the point of view of average absorbed power and the array wave fields η. The primary determination we can make is that the array effects are much stronger for the 5-OSWEC array case than for the heaving cylindrical WEC array case. Consequently, the effect of the change of the PTO system on the near-field surface elevations is much more significant for the OSWEC than for the heaving cylindrical WEC as highlighted in Figures 21 and 22 in Section 6.5.3. As remarked in Section 6.4 in Figure 13, the effect of the change in PTO on the power output of the array is likewise quite different between the 5-heaving cylindrical WEC array and the 5-OSWEC array, but is not strictly related to the change in the magnitude of the array effects. The addition of a hydraulic PTO significantly increases the power output of the heaving WEC array, especially at the higher wave periods. Meanwhile for the OSWEC array, there is a net decrease in the array power output with a change from a linear to a hydraulic PTO system for all periods except for $T = 12$ s. The interplay between the impact of the PTO systems of the two WEC types placed a closely spaced WEC array on the array power and on the near-field η are conceptualized in the flow chart in Figure 23. The arrow thickness represents the relative magnitude of the effect of each PTO type on the phenomena where the arrows are directed.

As with the magnitude, the location of the greatest changes in the near-field η differs between the 5-heaving cylindrical WEC and the 5-OSWEC array. Observe that the areas of positive and negative % difference in η are very distinct, with the hydraulic PTO system increasing the apparent η behind the heaving cylindrical WEC array while for the OSWEC array the change from a linear to a hydraulic PTO reduces the η behind the WECs. This is not a surprise given that the OSWEC, which operates in shallow water and fills the entire wave column, presents a bigger obstacle to the oncoming waves that results in much greater wave diffraction as observed in Figure 9. It is also the case that the OSWEC produces a stronger radiated wave field. The sum of the two effects results in strong areas of constructive and destructive interference that we observe in the surface elevations in the single OSWEC case in Figure 7 (left) and in the array of 5 OSWECs in Figures 17–20. Note especially the enhancement in the wave fields for $T = 8.0$ s for the OSWEC where the perturbed field is up to 50 % greater around the bodies. This is manifested in the power output **P** of the 5-OSWEC array at this wave period in Table 5: **P** is the highest value among all wave periods for the OSWEC and also with the highest q value, for both PTOs. In contrast, the perturbed wave field for the heaving cylindrical WEC for $T = 8.0$ s is quite small, only differing by a few centimeters from the undisturbed $|\eta|$ as we see in the right panel in Figure 15. Moreover, the impact of the change in the PTO system of the heaving cylindrical WEC is not necessarily reflected in the power output of the heaving cylindrical WEC array. As an example, the 5-heaving cylindrical WEC array outputs the most power at a wave period of $T = 8.0$ s for a linear PTO system, for a hydraulic PTO system the power is higher for wave period of $T = 10.0$ s. Since the near-field array effects and the power output of a 5-heaving cylindrical WEC array are not directly linked, these changes are not reflected in Figure 21 where we see a relative decrease in the near-field $|\eta|$ between the case of a wave period of $T = 8$ s and $T = 10$ s.

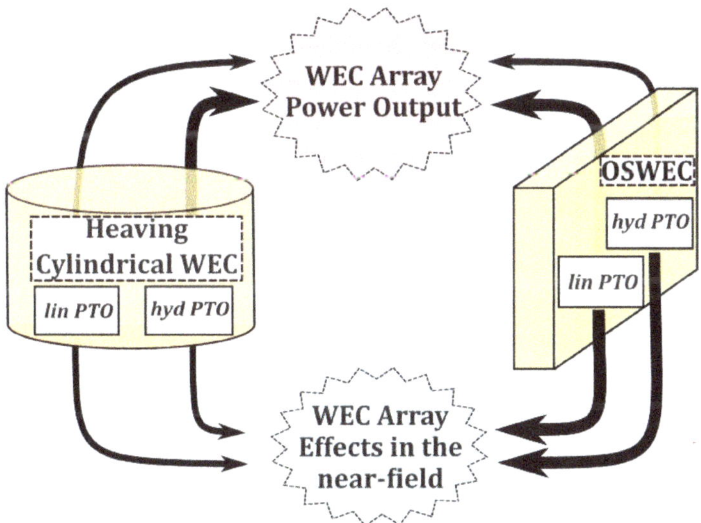

Figure 23. Schematic diagram showing the relationship between the PTO system impacts of the two types of WECs in an array. Thick arrows represent strong influences on the indicated parameters while thin arrows represent weak influences.

When we observe the areas of positive or negative change based on a substitution of a linear for a hydraulic PTO system, in Figure 21, we note a decrease in the change in η for the higher wave periods, indicating that the hydraulic PTO system is indeed extracting more energy from the wave field than the linear PTO system. However, the magnitude of these effects is close to the 5% threshold and can essentially be neglected in a 1st order modelling approach. Conversely, we have noted in Section 6.5.3 that the addition of a hydraulic PTO system to the OSWECs in an array tends to 'pull' in the energy from the surrounding areas to the 'near field'. This is especially true for higher wave periods and is reflected in the relative increase in the power output of an OSWEC array with a hydraulic PTO system compared to the same isolated WEC. In the case of the OSWEC array, the effects are an order of magnitude stronger. We note the present results show the same differences in the strength of the array effects between the heaving cylindrical WEC and the OSWEC arrays as in those presented in [44].

We presume that such contrasting behavior reflects the differences in the underlying hydrodynamics of the WEC-PTO system of the 2 WEC types. For the heaving cylindrical WEC case, the primary driver of an increment in the power output of a hydraulic PTO system is the increase in the PTO system force, while for the OSWEC the hydraulic PTO system has a greater impact on the WEC motion. Indeed, the η of near-field area of the OSWEC array increases with the hydraulic PTO system, especially for the long wave periods $T = 10.0$ s and $T = 12.0$ s. Unlike the heaving cylindrical WEC case, we also note that a change in the PTO system reduces the η in lee of the WEC array, augmenting the areas of destructive interference. This might be important in considering the impact on surrounding WEC arrays and coastal processes.

Still, a change in PTO system for the OSWEC results in an improvement for only the $T = 12.0$ s, with a relative decrease in the power output for the other wave periods compared to the linear PTO system case. We must remark here that for the OSWEC case, for both the single WEC and the array, our linear PTO model can exaggerate the performance of the OSWEC since we are not taking into account the strong non-linearities inherent in the dynamics of this WEC type. This has been pointed out in [8,45] among others. Therefore, if we were to choose a more sophisticated model for the OSWEC, the relative 'underperformance' of the hydraulic PTO system might disappear.

It must be mentioned here that in this paper we are using a linear hydrodynamic model in simulating the WEC arrays for regular waves from a single direction. It has been shown in literature and in our own research that these assumptions would tend to overestimate both the power output and the perturbations in the near-field η. We note that a heaving cylindrical WEC, being axi-symmetric, is much less sensitive to changes in the direction of the incoming wave than the OSWEC. We also remark that for the case of the OSWEC, the linear PTO model might lead to an overestimation of the power and the differences we observe between a 5-OSWEC array power output with a linear and hydraulic PTO model might be in part be due to such assumptions. Therefore, we use the linear model more as a 'benchmark' to compare with previous studies such as [2,5,23,44] rather than a realistic PTO system representation to include in an OSWEC array simulation.

8. Conclusions

In this paper, we have presented a model of arrays of 5 WECs of two WEC types with contrasting hydrodynamics, a heaving cylinder WEC and an OSWEC driven by the surge component of the wave force. In our wave structure interaction-PTO model, we simulated single WECs and arrays with linear and hydraulic PTO systems, calculating both the power output of the WEC array and near-field η of the WEC array using an original iterative method that enables a fast calculation of both quantities. We have elaborated on the distinct hydrodynamic behavior of the heaving cylindrical WEC and the OSWEC.

We noted the differing effects of changing of a WEC PTO system between a single WEC case and an array case as summarized by Figure 23. Pertaining to power output P for the single heaving cylindrical WEC case, we conclude that the hydraulic PTO system brings a significant increase in the power output compared to a linear PTO system with up to 25% improvement for a $H = 1.0$ m, $T = 10.0$ s wave. For an array of 5 heaving cylindrical WECs the result is similar, with the increase due to the hydraulic PTO system mirroring that of the single WEC case. In both cases the impact of the heaving cylindrical WEC array on the near-field is minimal, with the only significant modification of the wave field at a wave period of $T = 6.0$ s. By extension then, a change in the PTO type for a heaving cylindrical WEC array produces no substantial changes to the near-field surface elevations. Therefore, if these effects are the primary target of a given investigation, a hydraulic PTO system can be modelled as a linear PTO system without loss of fidelity.

Conversely, for a single OSWEC, a hydraulic PTO system tempers the performance, with a reduction in the power output P across all wave period around 14%. Intriguingly, the situation for a 5-OSWEC array is different, with the hydraulic PTO system only having a strong negative effect on power output for a $T = 6.0$ s wave. For the other wave periods the change in PTO system does not results in a decrease in the power output, indeed for $T = 12.0$ s it slightly increases. We can see, therefore, that for the case of a 5-OSWEC array the array effects play a strong role in modifying the WEC array power output. There is a two-fold conclusion then for modelling the OSWECs. Firstly, a single OSWEC with a specific PTO system cannot be expected to reflect the behavior of said PTO system in an array. Secondly, the difference between the two types of PTO systems modelled is great enough such that one cannot substitute one PTO system for another without introducing substantial error. As a practical consideration, most existing models of array PTO systems are simulated as linear PTO systems. Although a hydraulic PTO system is more difficult to model in practice, our results have shown that for the case of the OSWEC array with a hydraulic PTO system, it cannot be simplified down without introducing substantial error into both the array power output and the near-field effects. In both aforementioned cases, the WEC array modeler can use the conceptual schematic introduced in Figure 23 as a guideline for choosing which assumptions to make.

It is part of our ongoing research to gradually increase the complexity and sophistication of both the hydrodynamic and the PTO models with the counterbalance of having a fast and intuitive solution for WEC array modelling. It is the next step of our research to use the presented coupled models in a realistic WEC farm simulation using real sea states to test the limitations of the present research mentioned at the end of the discussion in Section 7. Furthermore, out research aim is to expand the calculation of the perturbed wave field to the 'far-field' area away from the WEC farms to study coastal effects and interactions with a changing bathymetry.

Author Contributions: P.B. generated the idea of the paper, set up the numerical simulations, wrote and edited the manuscript. N.Q. set up numerical simulations and co-wrote the manuscript. G.V.F. assisted in the numerical simulations and provided manuscript feedback. V.S. edited and proofread the text and provided the funding resources. P.T. proofread the text and provided the funding resources.

Funding: This research is supported by the Research Foundation Flanders (FWO), Belgium—FWO.OPR.2.01—FWO research project No. 3G029114.

Acknowledgments: The work in this paper was inspired in part by past work of Dr. Tim Verbrugghe within the Dept. of Civil Engineering and Ghent University.

Conflicts of Interest: The authors declare no conflict of interest.

Abbreviations

The following abbreviations are used in this manuscript:

DoF	Degree of Freedom
OSWEC	Oscillating Surge Wave Energy Converter
PA	Point Absorber
PTO	Power Take-off
RAO	Response Amplitude Operator
WEC	Wave Energy Converter

Nomenclature

$A(\omega)$	added mass (kg) or (kg · m^2)
β	angle of incidence of the incoming wave to the x-axis (°)
d_x, d_y	WEC-WEC separation distances in the x and y direction (m)
$B(\omega)$	hydrodynamic damping (kg/s^2)
$B_{PTO,l}$	power-take-off linear damping coefficient (kg/s^2)
$B_{PTO,h}$	power-take-off hydraulic damping equivalent coefficient (kg/s^2)
D_m	variable motor displacement (rev/s)
K_{PTO}	power take-off linear stiffness coefficient ($\frac{\text{N}}{\text{m}}$)
$\mathcal{M}$	number of bodies in the WEC array

$\lvert\eta\rvert$	absolute value of the complex free surface elevation η (m)
$f_{PTO,l}$	PTO-force for linear PTO system
$f_{PTO,h}$	PTO system-force for hydraulic PTO system
p_{ij}	perturbed wave of order j for array i (-)
P_l	mechanical power produced by the WEC with a linear PTO system
P_h	mechanical power produced by the WEC with a hydraulic PTO system
P_{array}	total power output of an isolated WEC array (kW)
q	q-value or gain factor, defined as ratio of power of the $\mathcal{M}$-WEC array to the power produced by the sum of $\mathcal{M}$ isolated WECs
s_c	piston area [m^2]
T_r	resonance or natural period of an oscillating body (s)
$T_{PTO,l}$	PTO-torque for linear PTO system
$T_{PTO,h}$	PTO-torque for hydraulic PTO system
Z	complex amplitude of heave displacement
$z(t)$	heave displacement in time domain (m)
ζ	wave amplitude (m)
Θ	complex amplitude of pitch angular displacement
$\theta(t)$	pitch angular displacement in time domain (rad)
ω	wave frequency (rad/s)
	'array effects' = the hydrodynamic effects of WECs in an array that produce a perturbation in the incident wave field
	'near-field' referring to wave field modification effects in the general location of the WECs inside an array
	'far-field' referring to wave field modification effects outside the immediate area of the WEC array(s)
	'perturbed wave' = radiated + diffracted wave

References

1. Venugopal, V.; Smith, G. Wave Climate Investigation for an Array of Wave Power Devices. In Proceedings of the 7th European Wave and Tidal Energy Conference, Porto, Portugal, 11–14 September 2007; p. 10.
2. Charrayre, F.; Peyrard, C.; Benoit, M.; Babarit, A. A Coupled Methodology for Wave-Body Interactions at the Scale of a Farm of Wave Energy Converters Including Irregular Bathymetry. In Proceedings of the ASME 2014 33rd International Conference on Ocean, Offshore and Arctic Engineering, San Francisco, CA, USA, 8–13 June 2014.
3. Göteman, M.; Engström, J.; Eriksson, M.; Isberg, J. Optimizing wave energy parks with over 1000 interacting point-absorbers using an approximate analytical method. *Int. J. Mar. Energy* **2015**, *10*, 113–126, doi:10.1016/j.ijome.2015.02.001. [CrossRef]
4. Ruiz, P.M.; Ferri, F.; Kofoed, J.P. Experimental Validation of a Wave Energy Converter Array Hydrodynamics Tool. *Sustainability* **2017**, *9*, 115. [CrossRef]
5. Ruiz, P.M.; Nava, V.; Topper, M.B.R.; Minguela, P.R.; Ferri, F.; Kofoed, J.P. Layout Optimisation of Wave Energy Converter Arrays. *Energies* **2017**, *10*, 1262. [CrossRef]
6. Yu, Y.; Lawson, M.; Ruehl, K.; Michelen, C. Development and Demonstration of the WEC-Sim Wave Energy Converter Simulation Tool. In Proceedings of the 2nd Marine Energy Technology Symposium (METS 2014), Seattle, WA, USA, 15–17 April 2014.
7. Zhao, H.T.; Sun, Z.L.; Hao, C.L.; Shen, J.F. Numerical modeling on hydrodynamic performance of a bottom-hinged flap wave energy converter. *China Ocean Eng.* **2013**, *27*, 73–86, doi:10.1007/s13344-013-0007-y. [CrossRef]
8. Schmitt, P.; Asmuth, H.; Elsäßer, B. Optimising power take-off of an oscillating wave surge converter using high fidelity numerical simulations. *Int. J. Mar. Energy* **2016**, *16*, 196–208, doi:10.1016/j.ijome.2016.07.006. [CrossRef]
9. Henry, A.; Folley, M.; Whittaker, T. A conceptual model of the hydrodynamics of an oscillating wave surge converter. *Renew. Energy* **2017**, *118*, 965–972, doi:10.1016/j.renene.2017.10.090. [CrossRef]

10. Paredes, G.M.; Eskilsson, C.; Palm, J.; Bergdahl, L.; Leite, L.M.; Taveira-Pinto, F. Experimental and Numerical Modelling of a Moored, Generic Floating Wave Energy Converter. In Proceedings of the 10th European Wave and Tidal Energy Conference, Aalborg, DK, USA, 2–5 September 2013.

11. Schmitt, P.; Elsaesser, B. On the use of OpenFOAM to model oscillating wave surge converters. *Ocean Eng.* **2015**, *108*, 98–104, doi:10.1016/j.oceaneng.2015.07.055. [CrossRef]

12. Devolder, B.; Rauwoens, P.; Troch, P. Numerical simulation of a single floating point absorber wave energy converter using OpenFOAM. In Proceedings of the 2nd International Conference on Renewable Energies Offshore, Lisbon, Portugal, 24–26 October 2016; pp. 197–205.

13. Verbrugghe, T.; Domínguez, J.M.; Crespo, A.J.; Altomare, C.; Stratigaki, V.; Troch, P.; Kortenhaus, A. Coupling methodology for smoothed particle hydrodynamics modelling of non-linear wave-structure interactions. *Coast. Eng.* **2018**, *138*, 184–198. [CrossRef]

14. Devolder, B.; Stratigaki, V.; Troch, P.; Rauwoens, P. CFD simulations of floating point absorber wave energy converter arrays subjected to regular waves. *Energies* **2018**, *11*, 1–23. [CrossRef]

15. Bharath, A. Numerical Analysis of Arrays of Wave Energy Converters. Ph.D. Thesis, University of Tasmania, Hobart, Australia, 2018.

16. Penalba, M.; Davidson, J.; Windt, C.; Ringwood, J.V. A high-fidelity wave-to-wire simulation platform for wave energy converters: Coupled numerical wave tank and power take-off models. *Appl. Energy* **2018**, *226*, 655–669. [CrossRef]

17. de O. Falcão, A.F. Phase control through load control of oscillating-body wave energy converters with hydraulic PTO system. *Ocean Eng.* **2008**, *35*, 358–366.

18. Folley, M.; Whittaker, T. The control of wave energy converters using active bipolar damping. *J. Eng. Marit. Environ.* **2009**, *223*, 479–487, doi:10.1243/14750902JEME169. [CrossRef]

19. Cargo, C.J.; Plummer, A.R.; Hillis, A.J.; Schlotter, M. Determination of optimal parameters for a hydraulic power take-off unit of a wave energy converter in regular waves. *Proc. Inst. Mech. Eng. Part A J. Power Energy* **2012**, *226*, 98–111, doi:10.1177/0957650911407818. [CrossRef]

20. So, R.; Casey, S.; Kanner, S.; Simmons, A.; Brekken, T.K.A. PTO-Sim: Development of a power take off modeling tool for ocean wave energy conversion. In Proceedings of the 2015 IEEE Power Energy Society General Meeting, Denver, CO, USA, 26–30 July 2015; pp. 1–5, doi:10.1109/PESGM.2015.7285735. [CrossRef]

21. Sell, N.; Plummer, A.; Hillis, A.; Chandel, D. Modelling and calibration of a direct drive hydraulic PTO. In Proceedings of the Twelfth European Wave and Tidal Energy Conference, Cork, Ireland, 27 August–1 September 2017; pp. 886–872, ISSN 2309-1983.

22. Yu, Y.H.; Tom, N.; Jenne, D. Numerical Analysis on Hydraulic Power Take-Off for Wave Energy Converter and Power Smoothing Methods. In Proceedings of the 37th International Conference on Ocean, Offshore and Artic Engineering, Madrid, Spain, 17–22 June 2018; p. V010T09A043, doi:10.1115/OMAE2018-78176. [CrossRef]

23. Balitsky, P.; Verao Fernandez, G.; Stratigaki, V.; Troch, P. Assessment of the Power Output of a Two-Array Clustered WEC Farm Using a BEM Solver Coupling and a Wave-Propagation Model. *Energies* **2018**, *11*, doi:10.3390/en11112907. [CrossRef]

24. Babarit, A.; Delhommeau, G. Theoretical and numerical aspects of the open source BEM solver NEMOH. In Proceedings of the 11th European Wave and Tidal Energy Conference, Nantes, France, 6–11 September 2015.

25. Stratigaki, V. Experimental Study and Numerical Modelling of Intra-Array Interactions and Extra-Array Effects of Wave Energy Converter Arrays. Ph.D. Thesis, Ghent University, Ghent, Belgium, 2014.

26. Balitsky, P.; Verao Fernandez, G.; Stratigaki, V.; Troch, P. Coupling methodology for modelling the near-field and far-field effects of a Wave Energy Converter. In Proceedings of the ASME 36th International Conference on Ocean, Offshore and Arctic Engineering (OMAE 2017), Trondheim, Norway, 25–30 June 2017.

27. Verbrugghe, T.; Stratigaki, V.; Troch, P.; Rabussier, R.; Kortenhaus, A. A Comparison Study of a Generic Coupling Methodology for Modeling Wake Effects of Wave Energy Converter Arrays. *Energies* **2017**, *10*, doi:10.3390/en10111697. [CrossRef]

28. Beels, C.; Troch, P.; Backer, G.D.; Vantorre, M.; Rouck, J.D. Numerical implementation and sensitivity analysis of a wave energy converter in a time-dependent mild-slope equation model. *Coast. Eng.* **2010**, *57*, 471–492, doi:10.1016/j.coastaleng.2009.11.003. [CrossRef]

29. Babarit, A. On the park effect in arrays of oscillating wave energy converters. *Renew. Energy* **2013**, *58*, 68–78. [CrossRef]

30. Penalba, M.; Touzón, I.; Lopez-Mendia, J.; Nava, V. A numerical study on the hydrodynamic impact of device slenderness and array size in wave energy farms in realistic wave climates. *Ocean Eng.* **2017**, *142*, 224–232, doi:10.1016/j.oceaneng.2017.06.047. [CrossRef]

31. Alves, M. Wave Energy Converter modelling techniques based on linear hydrodynamic theory. In *Numerical Modelling of Wave Energy Converters*; Folley, M., Ed.; Elsevier: Amsterdam, The Netherlands, 2016; Chapter 1, pp. 11–65.

32. Garcia Rosa, P.B.; Bacelli, G.; Ringwood, J. Control-informed optimal layout for wave farms. *IEEE Trans. Sustain. Energy* **2015**, *6*, 575–582. [CrossRef]

33. Verbrugghe, T.; Kortenhaus, A.; De Rouck, J. Numerical modelling of control strategies and accumulator effect of a hydraulic power take-off system. In Proceedings of the OCEANS 2015, Genova, Italy, 18–21 May 2015; pp. 1–9.

34. Yu, Y.H.; Li, Y.; Hallett, K.; Hotimsky, C. Design and Analysis for a Floating Oscillating Surge Wave Energy Converter. In Proceedings of the ASME 2014 33rd International Conference on Ocean, Offshore and Arctic Engineering OMAE2014, San Francisco, CA, USA, 8–13 June 2014, doi:10.1115/OMAE2014-24511. [CrossRef]

35. Carengie Clean Energy. Available online: https://www.carnegiece.com/wave/ (accessed on 30 October 2018).

36. SINN Power. *SINN Power Achieves Breakthrough in Energy Supply by Ocean Waves*; SINN Power: Gauting, Germany, 2018.

37. Shadman, M.; Estefen, S.F.; Rodriguez, C.A.; Nogueira, I.C. A geometrical optimization method applied to a heaving point absorber wave energy converter. *Renew. Energy* **2018**, *115*, 533–546, doi:10.1016/j.renene.2017.08.055. [CrossRef]

38. Cargo, C. Design and Control of Hydraulic Power Take-Offs for Wave Energy Converters. Ph.D. Thesis, University of Bath, Bath, UK, 2013.

39. Child, B.; Venugopal, V. Optimal Configurations of wave energy devices. *Ocean Eng.* **2010**, *37*, 1402–1417. [CrossRef]

40. Child, B.; Cruz, J.; Livingstone, M. The Development of a Tool for Optimising of Arrays of Wave Energy Converters. In Proceedings of the 9th European Wave 1 Tidal Energy Conference, Southampton, UK, 5–9 September 2011.

41. Stratigaki, V.; Troch, P.; Stallard, T.; Forehand, D.; Kofoed, J.; Folley, M.A.; Benoit, M.; Babarit, A.; Kirkegaard, J. Wave Basin Experiments with Large Wave Energy Converter Arrays to Study Interactions between the Converters and Effects on Other Users. *Energies* **2014**, *7*, 701–734. [CrossRef]

42. Balitsky, P.; Verao Fernandez, G.; Stratigaki, V.; Troch, P. Assessing the impact on power production of WEC array separation distance in a wave farm using one-way coupling of a BEM solver and a wave propagation model. In Proceedings of the 12th European Wave and Tidal Energy Conference, Cork, Ireland, 27 August–1 September 2017; pp. 1176–1186.

43. Borgarino, B.; Babarit, A.; Ferrant, P. Impact of the separating distance between interacting wave energy converters on the overall energy extraction of an array. In Proceedings of the 9th European Wave and Tidal Energy Conference, Southampton, UK, 5–9 September 2011.

44. Verao Fernandez, G.; Balitsky, P.; Tomey Bozo, N.; Stratigaki, V.; Troch, P. Far-field effects by arrays of oscillating wave surge converters and heaving point absorbers: A comparative study. In Proceedings of the 12th European Wave and Tidal Energy Conference (EWTEC2017) , Cork, Ireland, 27 August–1 September 2017; pp. 1030–1039.

45. Giorgi, G.; Ringwood, J.V. Comparing nonlinear hydrodynamic forces in heaving point absorbers and oscillating wave surge converters. *J. Ocean Eng. Mar. Energy* **2018**, *4*, 25–35. [CrossRef]

Article

Irregular Wave Validation of a Coupling Methodology for Numerical Modelling of Near and Far Field Effects of Wave Energy Converter Arrays

Gael Verao Fernández *, Vasiliki Stratigaki and Peter Troch

Department of Civil Engineering, Ghent University, Technologiepark 60, B-9052 Zwijnaarde, Belgium; vasiliki.stratigaki@ugent.be (V.S.); peter.troch@ugent.be (P.T.)
* Correspondence: gael.veraofernandez@ugent.be; Tel.: +32-9-264-5489; Fax: +32-9-264-5837

Received: 17 December 2018; Accepted: 30 January 2019; Published: 8 February 2019

Abstract: Between the Wave Energy Converters (WECs) of a farm, hydrodynamic interactions occur and have an impact on the surrounding wave field, both close to the WECs ("near field" effects) and at large distances from their location ("far field" effects). To simulate this "far field" impact in a fast and accurate way, a generic coupling methodology between hydrodynamic models has been developed by the Coastal Engineering Research Group of Ghent University in Belgium. This coupling methodology has been widely used for regular waves. However, it has not been developed yet for realistic irregular sea states. The objective of this paper is to present a validation of the novel coupling methodology for the test case of irregular waves, which is demonstrated here for coupling between the mild slope wave propagation model, MILDwave, and the 'Boundary Element Method'-based wave–structure interaction solver, NEMOH. MILDwave is used to model WEC farm "far field" effects, while NEMOH is used to model "near field" effects. The results of the MILDwave-NEMOH coupled model are validated against numerical results from NEMOH, and against the WECwakes experimental data for a single WEC, and for WEC arrays of five and nine WECs. Root Mean Square Error (RMSE) between disturbance coefficient (Kd) values in the entire numerical domain ($RMSE_{K_d,D}$) are used for evaluating the performed validation. The $RMSE_{K_d,D}$ between results from the MILDwave-NEMOH coupled model and NEMOH is lower than 2.0% for the performed test cases, and between the MILDwave-NEMOH coupled model and the WECwakes experimental data $RMSE_{K_d,D}$ remains below 10%. Consequently, the efficiency is demonstrated of the coupling methodology validated here which is used to simulate WEC farm impact on the wave field under the action of irregular waves.

Keywords: numerical modeling; numerical coupling; wave propagation; MILDwave; wave–structure interaction; near field; far field; experimental validation; WECwakes project; wave energy converter arrays

---

1. Introduction

Ocean waves are an enormous marine renewable energy source with the potential to contribute to a reduction in the world's fossil fuel dependency. The exploitation of wave energy is a complex and expensive process that takes place in a rough environment. As a result, a large number of Wave Energy Converters (WECs) technologies are under development [1], with none of them yet reaching a commercial stage. In addition, many WECs have to be deployed and arranged in WEC farms to produce large amounts of electricity and to have economically viable wave energy projects.

The overall wave power absorption of a WEC farm will affect the surrounding wave field creating areas of reduced wave energy (areas of decreased wave height) in the lee of the WEC farm as seen in [2–8]. The hydrodynamic problem of wave power absorption between the WECs within a farm, and between the WECs and the incident wave field is characterized by three different problems namely: wave reflection, diffraction and radiation. The superposition of the reflected, diffracted and radiated wave fields results in a perturbed wave field. The perturbed wave field close to the WECs of the farm caused both by WEC–WEC and wave–WEC interactions is often referred to in literature as the "'near field" effects while the propagation of this perturbed wave field at a larger distance from the WEC farm e.g., in the coastal zone, is referred to as the "far field" effects [9–16].

Substantial numerical research has been carried out to study the "'near field" effects in WEC farms, focusing on optimizing the WEC farm layout and maximizing the power output by employing wave–structure numerical models. Typically, numerical models based on potential flow theory have been used either for calculating semi-analytical coefficients [17–19] or by means of Boundary Elements Method based models (BEMs) [20–22]. The aforementioned numerical models are suited to resolve more accurately the details of WEC (farm) "'near field" effects. However, they are not able to account for the physical processes that influence the "far field" effects such as wave propagation over a varying bathymetry and wave breaking. Furthermore, the numerical simulation time can increase considerably when increasing the number of WECs modelled and the size of the numerical domain. In recent years, the use of non-linear numerical models based on Computational Fluid Dynamics (CFD) [23,24] and Smoothed Particle Hydrodynamics (SPH) [12,25,26] has increased as these models can take into account non-linear effects for wave–structure interactions. Nonetheless, the use of these models is restricted to a small spatial and temporal scale and to an even more limited number of WECs, which makes them also not suitable to study WEC (farm) "far field" effects in a large numerical domain due to high computational cost.

"Far field" effects are traditionally studied in a computationally cost-efficient way using wave propagation models. In [2–4,7,8,27–29], phase-averaging spectral models are used to obtain the wave field in the lee of a WEC farm. The WEC farms in these studies are simplified as obstacles which have been assigned a fixed transmission (and thus wave power absorption) coefficient. In a similar way, Refs. [30,31] used a time-dependent mild slope equation model and simplified each WEC as a wave power absorbing obstacle. To obtain the frequency-dependent wave power absorption coefficient for phase-averaging spectral models and the wave power absorption coefficient (assigned to obstacles/structures) for time-dependent mild slope equation models, wave tank testing or numerical modeling are required. Therefore, the simplified parametrization of the wave power absorbed by WECs is not taking into account the wave–structure interactions of diffraction and radiation of the different WECs modelled [32]. This inaccuracy may lead to an overestimation or underestimation of the WEC farm power absorption and consequently an unrealistic estimation of the "far field" effects in the coastal zone.

From the aforementioned studies, it is clear that modeling the perturbed wave field around a farm of WECs is a complex process. Usually "near field" and "far field" effects are approached separately due to the difficulties in using a single numerical model to obtain a fast and accurate solution for both effects. To rectify these limitations, different coupling methodologies between wave–structure interaction solvers and wave propagation models have been developed in the recent years [9–15]. This allows higher precision in the estimation of "far field" effects, by using a wave–structure interaction solver to obtain an accurate solution of the wave field in a limited area around the WECs of a farm and propagating this resulting wave field further away using a wave propagation model over a coastal zone.

As pointed out in [12], there are different types of coupling methodologies which use one-way and two-way coupling, respectively. In one-way coupled models, there is information transfer in one direction only, where each numerical model is run independently. Examples of such studies, which present linear simulation of "far field" effects of WEC farms by coupling a wave propagation model and a BEM solver, are carried out by [9–11,13,33,34]. Alternatively, in two-way coupled models, both numerical models are run at the same time with a two-way transfer of information between them. Examples of two-way coupled models are provided by [12] who demonstrated coupling of a non-linear wave propagation model with an SPH wave–structure interaction solver, or by [35] who simulated a submerged buoy using a non-hydrostatic wave-flow model implemented in the wave propagation model SWASH [36].

In the present study, a continuation of the one-way coupling methodology presented in [13,14,37] for regular waves between the wave propagation model MILDwave [10,38] and the wave–structure interaction solver NEMOH [39] is performed. This coupling methodology is based on the work of [9,38], who first presented a coupling between a wave propagation model (MILDwave) and a wave–structure interaction solver WAMIT [40]. In [14] specifically, the step-by-step procedure of this coupling methodology is presented and its application range. Moreover, in [14], the theoretical background of both the coupling methodology and of the employed numerical models (MILDwave and NEMOH) is provided. Furthermore, in [14], experimental data from the "WECwakes" database [41] has been used and more specifically wave field measurements for a 9-WEC array interacting with the incoming waves. The latter was used to perform validation of the coupling methodology for regular waves propagating through the 9-WEC array, obtaining good agreement between the experimental and numerical results regarding the impact of the 9-WEC array on the surrounding wave field. In [14], irregular waves were briefly introduced, yet not validated, without presenting a fully developed coupling methodology for irregular wave simulations.

Here, the novelty of this study is the validation of a fully developed coupling methodology for modeling irregular waves using available experimental data [16,41]. In the present manuscript, the coupling methodology is presented in detail for irregular wave generation. Furthermore, the irregular wave cases of a 9-WEC array, a 5-WEC array and a single WEC are selected from the "WECwakes" database for simulations using the coupling methodology and for validation purposes. Moreover, numerical results of the MILDwave-NEMOH coupled model are compared to NEMOH numerical results and experimental data, showing that the coupled model is able to accurately parse the information between the NEMOH and MILDwave numerical domains in the "near field". This information is then propagated into the "far field" in the MILDwave numerical domain as MILDwave correctly models coastal transformations [42]. Based on the results from [14] and on the current results from the present work, it is demonstrated that the developed and validated coupling methodology can be a useful tool for cost-efficient computational time simulations of coastal impacts of farms of floating structures and WECs over a large coastal zone. In contrast, it should also be noted that, due to the limitations of the numerical models employed here, the resulting MILDwave-NEMOH coupled model cannot be used for non-linear sea states and to model morphological coastal impacts.

The structure of the paper is as follows: Section 1 provides a short overview of the state-of-the-art and problem statement. Section 2 presents a description of the generic coupling methodology. Section 3 illustrates the MILDwave-NEMOH coupled model, including a detailed description of the coupling methodology implementation, the wave propagation solver MILDwave and the wave–structure interaction solver NEMOH. A validation test case is described in Section 4 and the results are presented in Section 5. In Section 6, the capability of the "MILDwave-NEMOH" coupled model to simulate "far field" effects of WEC farms is discussed. Finally, the conclusions of this and future work are drawn in Section 7.

2. Generic Coupling Methodology

In this section, the generic coupling methodology first introduced by [9] is briefly presented. The objective of the coupling methodology is to obtain the total wave field around a (group of) structure(s), as a superposition of the incident wave field and the perturbed wave field (which is a combination of the reflected, diffracted and radiated wave fields). The incident wave field propagation and transformation is calculated over a large domain using a wave propagation numerical model. The perturbed wave field is simulated using a wave–structure interaction solver over a restricted domain around the structure(s), namely the coupling region. As it has been pointed out in [9], this coupling methodology can be applied by employing any wave–structure interaction solver that describes the perturbed wave field, any wave propagation model and any type of oscillating or floating structure(s).

The general strategy for the coupling methodology has been also recently reported and updated in [14], but, for clarity, it is presented here briefly. It consists of four steps. Firstly (Step 1), a wave propagation model is used to obtain the incident wave field at the location of the structure(s) when the structure(s) is (are) not present. Secondly (Step 2), the obtained wave field from Step 1 is used as an input for the wave–structure interaction solver at the location of the structure(s). Then, the motion of the structure(s) is solved and an accurate solution of the perturbed wave fields around the structure(s) is obtained. Thirdly (Step 3), the perturbed wave field is used as an input in the wave propagation model and is propagated throughout a large domain. This is done by prescribing an internal wave generation boundary around the structure location. Finally (Step 4), the total wave field due to the presence of the structure(s) is obtained as the superposition of the incident wave field and the perturbed wave field in the wave propagation model.

3. Application of the Coupling Methodology between the Wave Propagation Model, MILDwave, and the Wave–Structure Interaction Solver NEMOH for Irregular Waves

In this section, the generic coupling methodology presented in Section 2 will be demonstrated for coupling between the wave propagation model MILDwave and the wave–structure interaction solver NEMOH. First, a description of the two numerical models employed is presented. Subsequently, a description of the irregular wave generation for the incident, perturbed and total wave fields is provided.

3.1. The Wave Propagation Model, MILDwave and the Wave–Structure Interaction Solver, NEMOH

The wave propagation model chosen for demonstrating the proposed coupling methodology is the mild slope model MILDwave [10,38], developed at the Coastal Engineering Research Group of Ghent University, in Belgium. MILDwave is a phase-resolving model based on the depth-integrated mild slope equations of Radder and Dingemans [43]. MILDwave allows for solving the shoaling and refraction of waves propagating above mild slope varying bathymetries, and it has been widely used in the modeling of WEC farms [10,11,13,30,31,41,44,45]. The basic MILDwave equations are reported in [10].

The wave–structure interaction solver chosen to solve the diffraction/radiation problem is the open-source potential flow BEM solver NEMOH, developed at Ecole Centrale de Nantes [39]. Linear potential flow theory has hitherto been utilized in a majority of the investigations into WEC array modeling—for example, see [11,19,46,47]. NEMOH is based on linear potential flow theory [48], and the basic equations and assumptions employed are reported in [14].

3.2. Generation of the Incident Wave Field for Irregular Waves

Irregular waves can be generated by applying the superposition principle of a number of different linear regular wave components. The incident wave field for a linear regular wave is generated intrinsically in MILDwave. Moreover, MILDwave allows for solving shoaling and refraction of waves propagating over complex bathymetries. The numerical set-up of MILDwave is illustrated in Figure 1. Waves are generated along a linear offshore wave generation boundary by applying the boundary condition of linear regular waves generation:

$$\eta_{I,reg}(x,y,t) = a\cos(\omega t - k(x\cos(\theta) + y\sin(\theta))), \tag{1}$$

where $\eta_{I,reg}$ is the incident regular wave surface elevation, a is the wave amplitude, ω is the angular frequency, k is the wave number and θ is the wave direction. To minimize unwanted wave reflection, absorption layers are placed down-wave and up-wave in the numerical wave basin.

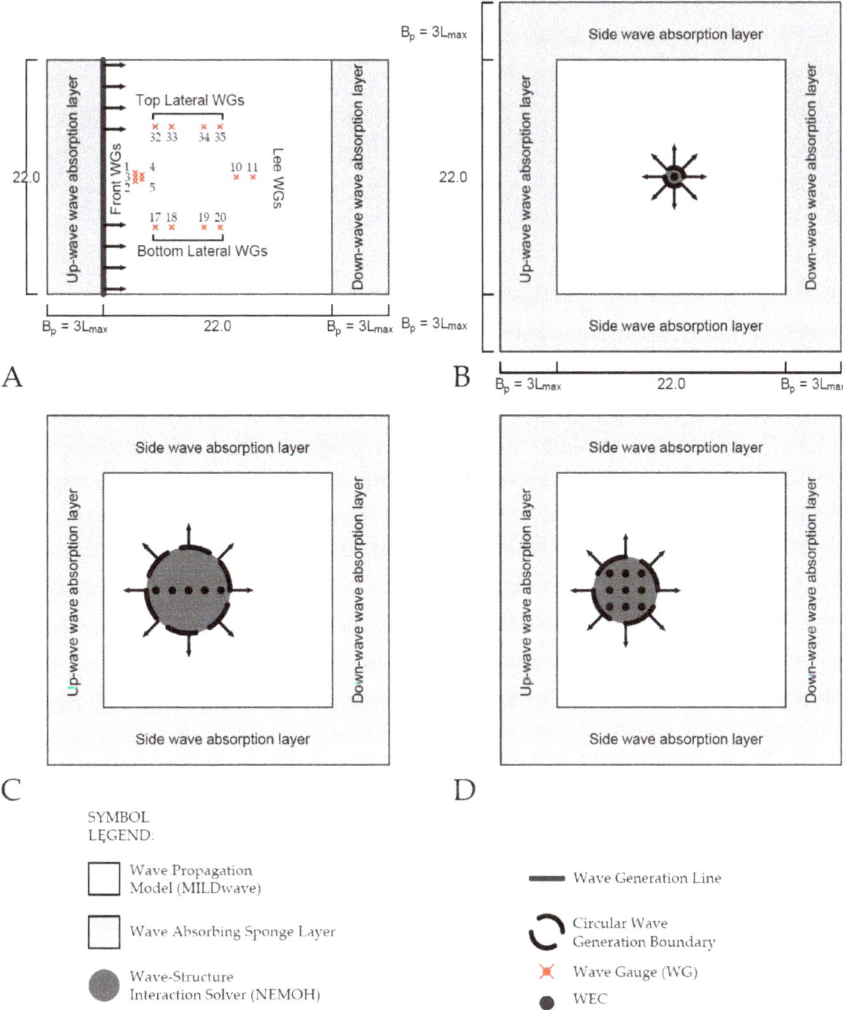

Figure 1. Set-up of the different numerical wave basins used in MILDwave. The wave gauges (WGs) are represented by the x symbol and numbered as they appear in the WECwakes experimental data-set. (**A**) empty numerical wave basin and layout of WGs; (**B**) numerical wave basing with a single WEC; (**C**) numerical wave basin with an array of five WECs (1 column, 1×5); (**D**) numerical wave basin with an array of nine WECs (3 columns and 3 rows, 3×3).

By applying the superposition principle, a first order irregular wave is represented as the finite sum of N regular wave components characterized by their wave amplitude, a_j, and wave period, T_j, derived from the wave spectral density, S_j:

$$\eta_{I,irreg}(x,y,t) = \sum_{j=1}^{N} a_j \cos(\omega_j t - k_j(x \cos(\theta_j) + y \sin(\theta_j)) + \varphi_j), \qquad (2)$$

where

$$a_j = \sqrt{2S_j(f_j) \cdot \Delta f_j},$$ (3)

where $\eta_{I,irreg}$ is the incident irregular wave surface elevation and a_j is the wave amplitude, ω_j is the wave angular frequency, f_j is the wave frequency, k_j is the wave number, θ_j is the wave direction and φ_j is the incident phase, of each wave frequency component. φ_j is selected randomly between $-\pi$ and π to avoid local attenuation of $\eta_{I,irreg}$.

3.3. Generation of the Perturbed Wave Field for Irregular Waves

To calculate the irregular perturbed wave field around a (group of) structure(s) first, it is necessary to obtain the perturbed wave field for each wave frequency as a regular wave. The perturbed wave field in the time domain for a regular wave is obtained in two steps and the generic numerical set-up is illustrated in Figure 1. First, a frequency-dependent simulation is performed using NEMOH to obtain the complex perturbed wave field around the (group of) structure(s). NEMOH resolves the wave frequency-dependent wave radiation problem for each structure(s) and the diffraction (including radiation) over a predetermined numerical grid with the wave phase $\varphi = 0$ at the center of the domain. The resulting radiated and diffracted wave fields for each wave frequency depend on the shape and number of floating structure(s), the number of Degrees of Freedom (DOF) considered, the local constant water depth and the wave period.

The radiated (for each structure) and diffracted (for all structures) complex wave fields in NEMOH are summed up to obtain the perturbed wave field, η_{pert}:

$$\eta_{pert} = \eta_{diff} + \eta_{rad},$$ (4)

where η_{rad} is the radiated wave field and η_{diff} is the diffracted wave field.

Secondly, the perturbed wave field is transformed from the frequency domain to the time domain and imposed onto MILDwave using an internal wave generation boundary (Figure 1). For this study, a circular wave generation boundary is prescribed; however, it can be defined using other shapes as well. Waves are forced away from the circular wave generation boundary by imposing values of free surface elevation $\eta_{circ}(x, y, t)$ as described by Equation (5):

$$\eta_{circ}(x, y, t) = a_c \left| \eta_{pert} \right| \cos(\varphi_{pert,c} - \omega t),$$ (5)

where η_{pert} is the perturbed complex wave field in the circular wave generation boundary, and a_c and $\varphi_{pert,c}$ are the wave amplitude of the incident wave and the wave phase of the perturbed wave at the center of the circular wave generation boundary, respectively. To avoid unwanted wave reflection, wave absorption layers or relaxation zones are implemented up-wave, down-wave and also in the sides of the MILDwave numerical domain (Figure 1).

As in the case for the calculation of the irregular incident wave field, the irregular perturbed wave field is calculated as the finite sum of N regular perturbed wave components characterized at the center of the wave generation boundary by their wave amplitude, $a_{c,j}$, derived from the wave spectrum:

$$\eta_{pert,irreg}(x, y, t) = \sum_{j=1}^{N} a_{c,j} \left| \eta_{pert} \right| \cos(\varphi_{pert,c,j} - \omega_j t),$$ (6)

where

$$a_{c,j} = \sqrt{2S_{c,j}(f_j) \cdot \Delta f_j}$$ (7)

and $\eta_{pert,irreg}$ is the perturbed irregular wave surface elevation where $S_{c,j}$ is the spectral density and $\varphi_{pert,c,j}$ is the perturbed wave phase of each frequency component. $\varphi_{pert,c,j}$ is selected randomly between $-\pi$ and π to avoid local attenuation of the surface elevation.

3.4. Generation of the Total Wave Field for Irregular Waves

The total wave field for irregular waves due to the presence of a (group of) structure(s) is obtained by applying the generic coupling methodology described in Section 2. This is performed by superimposing the irregular incident wave field and the irregular perturbed wave field generated in MILDwave as shown in Sections 3.2 and 3.3, respectively.

Step 1 of the generic coupling methodology is applied N times for irregular waves to calculate the incident wave field for N regular wave components in MILDwave by applying a random phase φ_i for each simulation. From each simulation $a_{c,i}$, and $\varphi_{c,i}$ are obtained at the center of the circular wave generation boundary and are used as input values for NEMOH.

In Step 2, the perturbed wave field is obtained in NEMOH. In NEMOH, $\varphi_{pert,c,j}$ is referenced with respect to the center of the domain (Section 3.3). Therefore, $\varphi_{pert,c,j}$ at the NEMOH numerical domain has to be corrected using the φ_j of the regular incident wave field to assure wave phase matching between the incident and the perturbed waves in MILDwave.

Afterwards, in Step 3, the perturbed wave field is then transformed from the frequency domain to the time domain and propagated into MILDwave for N regular perturbed wave components along the circular wave generation boundary.

Finally, in Step 4, the irregular incident wave field is obtained as the superposition of the N incident regular waves simulations from Step 1. The irregular perturbed wave field is obtained as the superposition of the N perturbed regular wave simulations from Step 3. The total wave field for irregular waves is obtained as the combination of the irregular incident and perturbed wave fields:

$$\eta_{tot,irreg}(x,y,t) = \sum_{j}^{N} \eta_{I,reg,j}(x,y,t) + \sum_{j}^{N} \eta_{pert,reg,j}(x,y,t), \tag{8}$$

where $\eta_{tot,irreg}$ is the total irregular wave surface elevation, and $\eta_{I,reg,j}$ and $\eta_{pert,reg,j}$ are the incident and perturbed wave surface elevations of each wave frequency, respectively.

4. Validation Strategy of the Coupling Methodology between the Wave Propagation Model, MILDwave, and the Wave–Structure Interaction Solver, NEMOH

In this section, a validation test case is presented to validate the MILDwave-NEMOH coupled model against numerical results from NEMOH and experimental data. Showing that the perturbed wave field can be precisely parsed from the NEMOH to the MILDwave domain in the near field of the WEC array. The criteria evaluated for the numerical model validation are also described.

4.1. Validation Test Cases

The validation of the demonstrated generic coupling methodology is carried out by comparing the results from the MILDwave-NEMOH coupled model to those obtained from the numerical model NEMOH and the WEC array experimental data from the WECwakes project [9,16,41].

4.1.1. WECwakes Experimental Data-Set

This section gives a short description of the experimental data-set from the WECwakes project [9,16,41] conducted in the Shallow Water Wave Basin of DHI, Hørsholm (Denmark). In the WECwakes project, arrays up to 25 point absorber type WECs (cylinders of a diameter of 0.315 m) were tested to study "near field" and "far field" effects of heaving point absorber type WECs. A Coulomb friction based damping is used.

The DHI wave basin is 22 m wide and 25 m long and the overall water depth is fixed to 0.7 m. Different WEC array configurations have been tested during the WECwakes project under a wide range of sea states, a large experimental data-set has been generated and is publicly available for numerical validation purposes and for WEC array design guidelines. The wave field around the WECs has been recorded using 41 resistive wave gauges (WGs) distributed in the wave basin.

For the present validation study, three different WEC configurations are selected: a single WEC, an array of five WECs arranged in a 1×5 WEC layout and an array of nine WECs arranged in a 3×3 WEC layout (see Figure 1B–D). A total of 15 wave gauges located in the front, leeward and sides of the WECs array configurations are used to compare the significant wave height, H_s, and the spectral density, S, between the MILDwave-NEMOH coupled model and the experimental data-set. The separating distance between the different WECs is equal to 1.575 m (centre-to-centre distance). The incident irregular wave conditions used to generate waves during the experiments test are defined by a JONSWAP spectrum with $H_s = 0.104$ m and two peak wave periods of $T_p = 1.18$ s and 1.26 s.

4.1.2. "Test Case" Program

The primary objective of the present research is to validate the total wave field around a WEC array obtained using the MILDwave-NEMOH coupled model. For this reason, a "Test Case" (Table 1) program based on the WECwakes experimental data-set has been designed for different irregular wave cases and WEC (array) configurations:

Table 1. "Test Case" program for irregular waves, and different Wave Energy Converter (WEC) (array) configurations.

Test Case Number ♯	Significant Wave Height, H_s (m)	Peak Wave Period, T_p (s)	Water Depth, d (m)	WEC Buoy Motion (-)	WEC (Array) Layout (-)
1	0.104	1.18	0.700	Damped	1×1
2	0.104	1.26	0.700	Damped	1×1
3	0.104	1.26	0.700	No motion (fixed buoy)	1×5
4	0.104	1.26	0.700	Damped	1×5
5	0.104	1.18	0.700	Damped	3×3
6	0.104	1.26	0.700	Damped	3×3

The different "Test Cases" included in Table 1 are performed both using the MILDwave-NEMOH coupled model, and NEMOH. NEMOH simulation results are used: (1) as input for the MILDwave-NEMOH coupled model, and (2) as a benchmark for the validation of the MILDwave-NEMOH coupled model, which is also compared with WECwakes data.

4.1.3. Numerical Set-Up in the Used Models

In MILDwave, simulations are carried out in two types of numerical wave basins (see Figure 1A–D) with an effective domain (area not covered by the wave absorbing sponge layers) of 22 m width and 22 m length, and a constant water depth of 0.700 m. The same dimensions are used in NEMOH. Four equally sized effective numerical domains are used. For the simulations performed to obtain the incident wave field, waves are generated using a linear wave generation line located on the left side of the numerical domain with two equally sized wave absorbing sponge layers placed up-wave (left) and down-wave (right) (see Figure 1A).

For the simulations carried out to obtain the perturbed wave field, waves are generated using an internal circular wave generation boundary (Figure 1B–D). The three different WEC (arrays) configurations of Table 1 are simulated using different coupling radii for the circular wave generation boundary (see Figure 1B–D). Each coupling radius is obtained following the recommendations by [11] as 0.5 times the wave length (L) plus the radius of the WEC or the distance from the centre of the circular area to the most distant WEC for a single WEC and a WEC array, respectively. Four equally sized wave absorbing sponge layers are placed on all sides of the numerical domain.

The dimensions of the total numerical wave basin in MILDwave are not always the same, as the length of the wave absorbing sponge layers (B) is different for each set of wave conditions and depends on L. As irregular waves are obtained as a superposition of N_f regular wave components, B is calculated using L_{max}, which corresponds to T_{max} of the discretized spectra. An increase of B causes a decrease of wave reflection, and as pointed out in [5] for $B = 3 \cdot L_{max}$ wave reflection coefficient drops to 1%.

The total wave field of the MILDwave-NEMOH coupled model is obtained as the superposition of the numerical results from the domains of Figure 1A–D for a single WEC, five WECs and nine WECs, respectively.

In NEMOH, the effect of the WEC's Power Take-Off (PTO) system is taken into account by adding a suitable external damping coefficient, $B_{PTO} = 28.5$ kg/s as defined in [14].

4.2. Criteria Used for the Numerical Model Validation

The accuracy of the obtained numerical results is evaluated in two steps. Firstly, results from the MILDwave-NEMOH coupled model are compared against the NEMOH results. Secondly, results from the MILDwave-NEMOH coupled model are compared against WECwakes experimental data.

The comparison between the MILDwave-NEMOH coupled model and NEMOH is assessed by calculating K_d coefficient values, as defined in Equations (9) and (10), respectively. The K_d coefficient is defined as the ratio between the numerically calculated local total significant wave height, $H_{s,tot}$, and the target incident significant wave height, $H_{s,I}$, imposed along the linear wave generation boundary. In the MILDwave-NEMOH coupled model, the $K_{d,coupled}$ is obtained in the time domain as:

$$K_{d,coupled} = \frac{H_{s,tot}}{H_{s,I}} = \frac{4 \cdot \sqrt{\sum_{t}^{\Delta t}(\eta_{I,irreg,t} + \eta_{pert,irreg,t})^2 \cdot \frac{dt}{\Delta t}}}{H_{s,I}}, \tag{9}$$

where $\eta_{I,irreg,t}$ and $\eta_{pert,irreg,t}$ are the free surface elevations for irregular incident and perturbed waves in each time step dt, from the domains of Figure 1A–D, respectively, and Δt is the time window over which K_d is computed. In NEMOH, the $K_{d,NEMOH}$ is obtained in the frequency domain as:

$$K_{d,NEMOH} = \frac{\left|\eta_{tot,irreg,freq}\right|}{H_{s,I}}, \tag{10}$$

where $\left|\eta_{tot,irreg,freq}\right|$ is the absolute value of free surface elevation for the complex total wave obtained in the frequency domain.

The K_d value is a useful parameter that has been used extensively in literature to study wave field variations [9–11,22,30,31,34,42,45]. $K_d > 1$ and $K_d < 1$ indicate increase and decrease of the local wave height, respectively. When studying WEC arrays, increases in the local wave height indicate the presence of "hot spots" [49], defined as areas of high wave energy concentration. Instead, decrease in the local wave height denotes "wake" effects, which result in an area of reduced wave energy.

To evaluate K_d differences between the MILDwave-NEMOH coupled model and NEMOH, three different outputs have been generated:

1. K_d contour plots of the entire numerical domains;
2. K_d cross-sections along the length of the numerical domains (parallel to the wave propagation direction);
3. Contour plots of the "Relative Difference" between the obtained K_d values (RD_{K_d}) defined as:

$$RD_{K_d,D} = \frac{(K_{d,NEMOH} - K_{d,coupled})}{K_{d,NEMOH}} \cdot 100 \quad \% \quad (-), \tag{11}$$

4. The Root Mean Square Error between K_d values obtained using the MILDwave-NEMOH coupled model and NEMOH for the entire numerical domain ($RMSE_{K_d,D}$):

$$RMSE_{K_d,D} = \sqrt{\frac{\sum_{i=1}^{G}(K_{d,NEMOH} - K_{d,coupled})^2}{G}} \cdot 100 \quad \% \quad (-), \tag{12}$$

where G is the number of grid points of the numerical domain (D).

The validation of results obtained from the MILDwave-NEMOH coupled model against WECwakes experimental data is carried out using data recorded at the 15 numerical and experimental WGs, respectively, as these are illustrated in Figure 1A. For each WG, two different outputs have been generated:

1. Spectral density plots comparing the wave spectra between the MILDwave-NEMOH coupled model and the WECwakes experimental data for the 15 WGs.
2. The Root Mean Square Error between the K_d of the MILDwave-NEMOH coupled model and the $K_{d,WECwakes}$ of the WECwakes experimental data for the 15 WGs, $RMSE_{K_{d,WG}}$:

$$RMSE_{K_{d,WG}} = \sqrt{\frac{\sum_{i=1}^{T}(K_{d,WECwakes} - K_{d,coupled})^2}{C}} \cdot 100 \quad \% \quad (-), \tag{13}$$

where C is the number of Test Cases.

5. Validation Results

In Section 5.1, the results for the irregular wave generation analysis are presented. The comparison between the MILDwave-NEMOH coupled model and NEMOH follows in Section 5.2. First, the results for Test Case 6 are discussed in detail in Section 5.2.1, and then the results for all Test Cases are summarized in Section 5.2.2 in terms of $RMSE_{K_d,D}$ values. The validation between the MILDwave-NEMOH coupled model and the WECwakes experimental data is included in Section 5.3. Similarly, first, the results for Test Case 6 are discussed in detail in Section 5.3.1, while the results for all Test Cases are summarized in Section 5.3.2 in terms of $RMSE_{K_d,D}$ values.

5.1. Sensitivity Analysis for Irregular Wave Generation

Before performing the numerical simulations listed in Table 1, a sensitivity analysis is carried out to ensure a converging result of the irregular wave simulation, while keeping the computational time low. This sensitivity analysis is based on three numerical simulation criteria: (1) the total simulation time Q_{tot}, (2) the number of regular wave components (N_f), and (3) the grid cell size (d_x and d_y) employed in MILDwave. For each criterium, the studied parameter is varied while the other two are kept constant. The numerical domain in Figure 1A is used.

Firstly, different Q_{tot} are considered to ensure a fully developed wave spectrum. Secondly, N_f is modified in order to achieve a wave spectrum close to the theoretical one. Thirdly, $d_x = d_y$ is varied based on wave length, L_p of the incident waves in order to achieve a convergent solution with the theoretical spectral density $S_t(f)$. The numerical spectral density in MILDwave $S_{n,M}(f)$ is obtained at the centre of the domain for the incident wave of "Test case 1" of Table 1. The shortest Q_{tot}, smallest N_f and largest $d_x = d_y$ resulting in an accurate solution of $S_{n,M}(f)$ are selected then to perform the rest of the numerical simulations for the rest of the Test Cases.

The results of the irregular wave generation sensitivity analysis are shown in Figure 2. $S_{n,M}(f)$ for different Q_{tot} is plotted in Figure 2a, while $N_f = 15$ and the $d_x = d_y = 0.08$ m are kept constant. It is clearly observed that, for Q_{tot} of 100 s and 300 s, $S_{n,M}(f)$ does not represent $S_t(f)$. For Q_{tot} of 600 s, there is a good agreement with $S_t(f)$ even for high frequency wave components, without leading to computationally expensive simulations.

$S_{n,M}(f)$ for different N_f is compared to $S_t(f)$ in Figure 2b, while $Q_{tot} = 600$ s and the $d_x = d_y = 0.08$ m are kept constant. Simulations are performed for $N_f = 15$, 20 and 40. There is a good agreement between $S_{n,M}(f)$ and $S_t(f)$ for all three simulations showing a slight amount of spurious energy for high wave frequencies, which is reduced by increasing N_f. Nevertheless, the accuracy gained by increasing N_f from 20 to 40 is not significant as the $S_{n,M}(f)$ peak and the energy contained within the $S_{n,M}(f)$ curve is practically the same. Consequently, it is concluded that increasing N_f is not required and therefore N_f is kept to 20 to reduce the computational time.

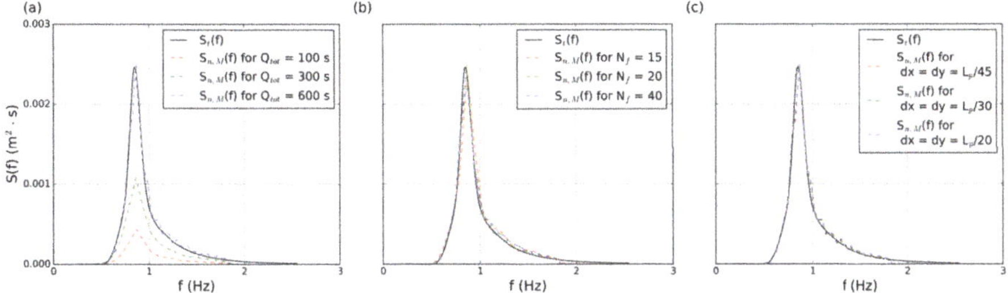

Figure 2. Numerical wave spectrum $S_{n,M}(f)$ generated at the centre of the MILDwave numerical domain for an irregular wave with $T_p = 1.26$ s and $H_s = 0.104$ m and different simulations parameters: (**a**) total simulation time, Q_{tot}; (**b**) number of regular wave components, N_f and (**c**) grid cell size, $d_x(= d_y)$. $S_{n,M}(f)$ is compared in (**a–c**) to the theoretical wave spectrum, $S_t(f)$.

To complete the sensitivity analysis, $S_{n,M}(f)$ for different d_x (where $d_x = d_y$) is compared in Figure 2c, while keeping $Q_{tot} = 600$ s and $N_f = 20$ constant. As recommended for mild slope wave propagation models, a starting value of $d_x = d_y = \frac{L_p}{20} = 0.08$ m is chosen. As seen in Figure 2c, increasing $d_x(= d_y)$ does not result in an improved agreement between $S_{n,M}(f)$ and $S_t(f)$, with T_p and $S_{n,M}(f)$ that appear to be maintained in each case. Thus, a grid size $d_x(= d_y) = 0.08$ m is chosen for all following simulations.

5.2. Comparison between MILDwave-NEMOH Coupled model and NEMOH

5.2.1. Irregular Waves with Wave Period $T_p = 1.26$ s

Using the MILDwave-NEMOH coupled model for Test Cases 2, 4 and 6 from Table 1, the total wave field around one, five and nine WECs, respectively, is simulated using the numerical domain of Figure 1B–D, respectively. K_d results obtained for each considered Test Case are illustrated in Figure 3. The coupling region in the MILDwave-NEMOH coupled model is masked out using a white solid circle and is not considered for the validation. For all three Test Cases, the hydrodynamic behaviour and WEC motions obtained within the coupling region are affecting the incident wave field in the MILDwave-NEMOH coupled model. As a result, a wave reflection pattern is generated in front of the WECs with increased K_d values, while, in the lee of the WECs, "wake effects" appear with reduced values of K_d. The effect of the three different WEC (array) configurations is expressed by an increased impact in terms of wave reflection and wake effects.

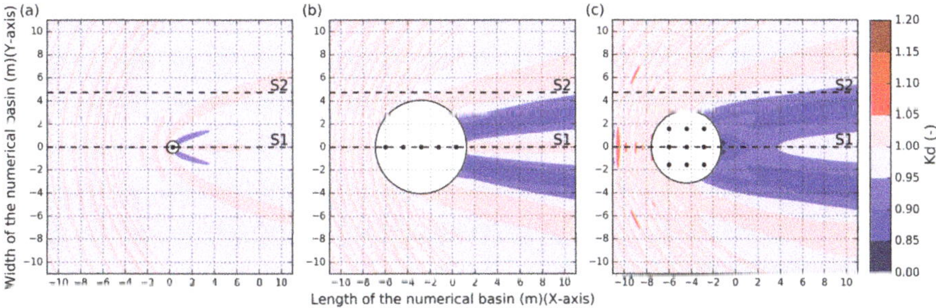

Figure 3. K_d results for an irregular wave with $T_p = 1.26$ s and $H_s = 0.104$ m obtained using the MILDwave-NEMOH coupled model: (**a**) Test Case 2; (**b**) Test Case 4 and (**c**) Test Case 6 of Table 1. Contour levels are set at an interval of 0.05 of K_d value (-). The coupling region is masked out using a white solid circle which includes the WECs (indicated by using black solid circles). Incident waves are generated from the left to the right. S1 and S2 indicate the location of cross-sections.

For the validation, K_d values obtained with the MILDwave-NEMOH coupled model and with NEMOH are compared by means of the RD_{K_d}. Three contour plots for Test Cases 2, 4 and 6 are illustrated in Figure 4a–c, respectively. The MILDwave-NEMOH coupled model provides lower K_d results than NEMOH in the wave reflection zone up-wave of the WECs indicated by positive values of RD_{K_d}, while the extent and magnitude of the wake effects are larger for the MILDwave-NEMOH coupled model as indicated by negative values of RD_{K_d}. The maximum and minimum values of RD_{K_d} are 4% and −4%, respectively, and are obtained for Test Case 6. These differences in the RD_{K_d} between the two models appears close to the coupling region and the wave diffraction zones around the WECs where increased K_d values are observed, and are increased by increasing the number of WECs simulated. Nevertheless, these RD_{K_d} differences are reduced when moving away from the coupling region.

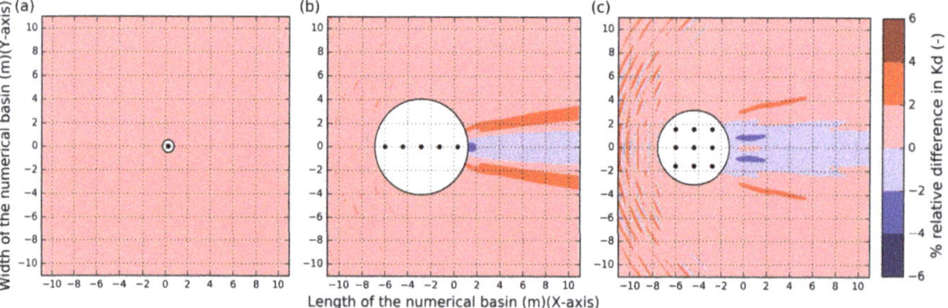

Figure 4. Relative difference (%) in K_d, RD_{K_d}, between the MILDwave-NEMOH coupled model and NEMOH for an irregular wave of T_p = 1.26 s and H_s = 0.104 m: (**a**) Test Case 2; (**b**) Test Case 4; and (**c**) Test Case 6 of Table 1. Contour levels are set at an interval of 2 of relative difference in K_d value (-). The coupling region is masked out using a white solid circle which includes the WECs (indicated by using black solid circles). Incident waves are generated from the left to the right.

To have a closer look at the comparison between the K_d results from the MILDwave-NEMOH coupled model and NEMOH, for Test Cases 2, 4 and 6, two longitudinal cross-sections (indicated in Figure 3) are drawn through: the centre of the domain, at y = 0 m (S1) and through the location of WGs 17,18,19 and 20 (see Figure 1A), at y = 4.75 m (S2). Again, the coupled zone is masked out in cross-section S1 using gray colour. For all considered Test Cases, it can be observed in Figure 5 that there is very good agreement for K_d results between the MILDwave-NEMOH coupled model and NEMOH. For the MILDwave-NEMOH coupled model K_d values are lower in the wave reflection and diffraction regions in front and on the side of the WECs, and higher in the region where wake effects occur in the lee of the WECs, compared to NEMOH.

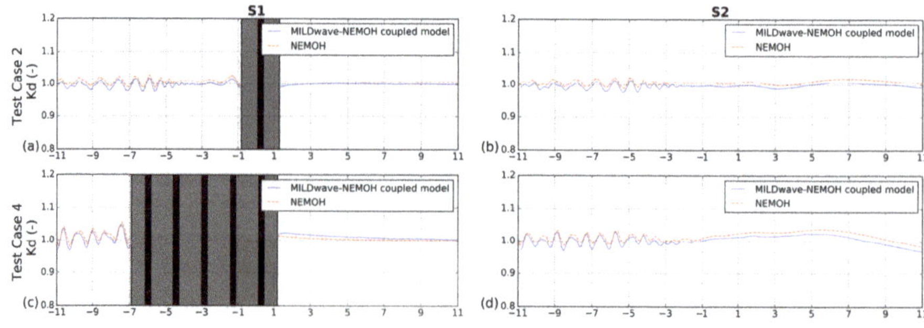

Figure 5. *Cont.*

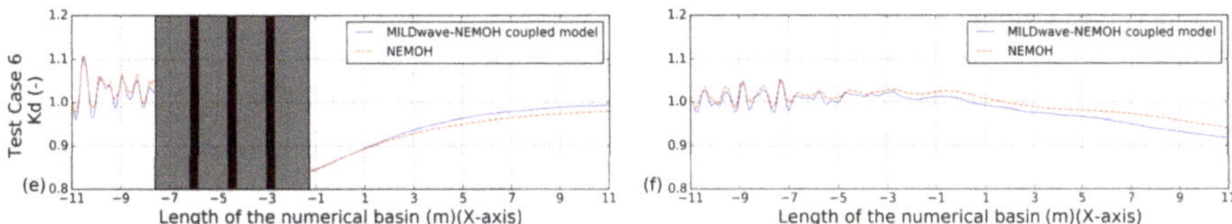

Figure 5. K_d results for the MILDwave-NEMOH coupled model and for NEMOH along two longitudinal cross-sections S1 (left) and S2 (right) as indicated in Figure 3 for: (**a,b**) Test Case 2; (**c,d**) Test Case 4, and (**e,f**) Test Case 6. The coupling region is masked out in gray colour and includes the WECs' cross-sections, which are indicated by black vertical areas.

5.2.2. Comparison Summary

To complete the validation of the MILDwave-NEMOH coupled model, the rest of the Test Cases of Table 1 are presented using the methodology used in Section 5.2.1. Similar conclusions for Test Cases 1, 2 and 5 are drawn: the MILDwave-NEMOH coupled model provides lower K_d results than NEMOH in the wave reflection zone up-wave of the WECs, and increased magnitude of the wake effects down-wave of the WECs indicated by positive and negative values of RD_{Kd}, respectively.

The results for all six Test Cases of Table 1 are then summarized by calculating the $RMSE_{K_d,D}$ over all the grid points of the numerical domain. Figure 6 reports that $RMSE_{K_d,D}$ values remain below 1.60% for the simulated Test Cases.

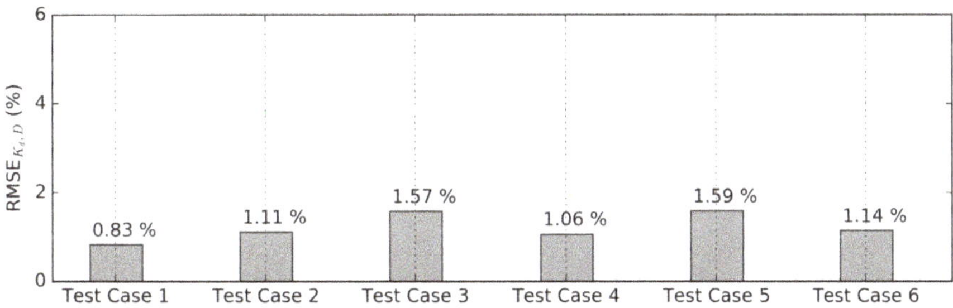

Figure 6. Root-Mean-Square-Error (RMSE) for the K_d, $RMSE_{K_d,D}$, over the entire numerical domain. Comparison between the MILDwave-NEMOH coupled model and NEMOH for all Test Cases of Table 1.

5.3. Comparison between the MILDwave-NEMOH Coupled Model and the WECwakes Experimental Data-Set

5.3.1. Test Case 6

Results for Test Case 6 are shown in Figures 7 and 8 for the 15 WGs shown in Figure 1A. The K_d values from the MILDwave-NEMOH coupled model and from the experimental measurements, $K_{d,coupled}$ and $K_{d,WECwakes}$, respectively, and numerical (using MILDwave-NEMOH coupled model) and experimental results of $S_{n,M-N}(f)$ and $S_{WECwakes}(f)$, respectively, are plotted in Figures 7 and 8. The MILDwave-NEMOH coupled model and the experimental data have a good agreement in the WGs in the lee of the WECs where wake effects take place and in the Bottom Lateral WGs (see Figure 1A) for both the K_d and $S(f)$.

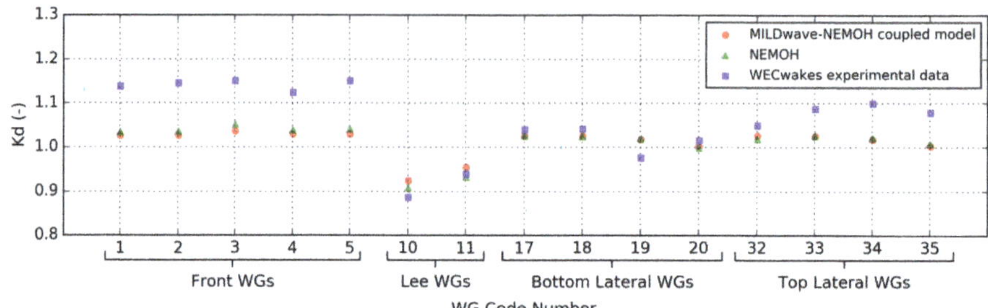

Figure 7. Comparison of the K_d between the MILDwave-NEMOH coupled model and the WECwakes experimental data for all 15 WGs of Figure 1A for Test Case 6 of Table 1.

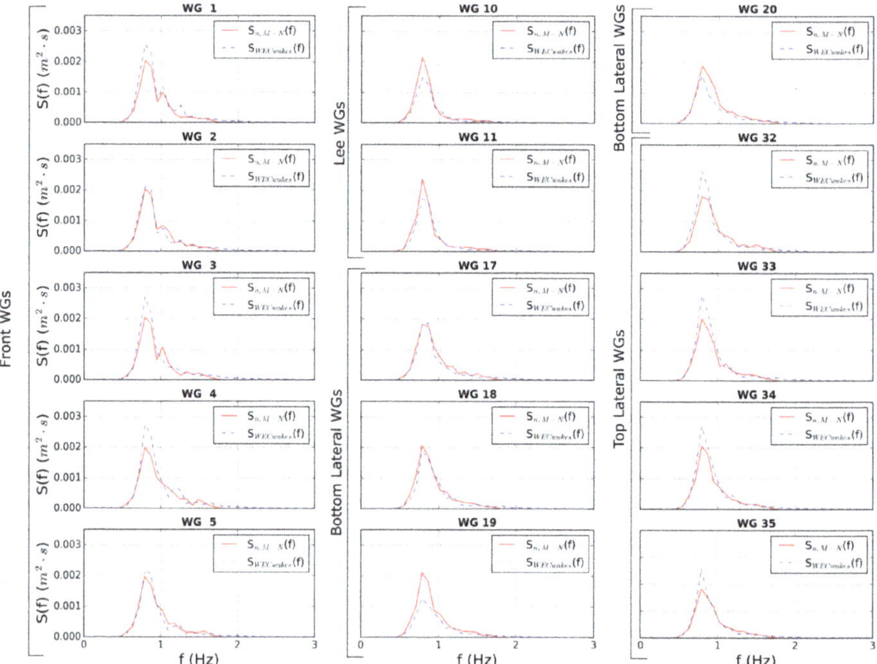

Figure 8. Comparison between the spectral density $S_{n,M-N}(f)$ obtained using the MILDwave-NEMOH coupled model, and the spectral density from the WECwakes experimental data, $S_{WECwakes}(f)$ for all 15 WGs of Figure 1A for Test Case 6 of Table 1.

5.3.2. Comparison Summary

To complete the validation of the MILDwave-NEMOH coupled model against experimental data, the $RMSE_{K_{d,WG}}$ is calculated between the $K_{d,coupled}$ and the $K_{d,WECwavkes}$ for all Test Cases of Table 1. Figure 9 shows the $RMSE_{K_{d,WG}}$ obtained for each WG of Figure 1A. The K_d obtained for the numerical data differs maximal by 10.03% from the experimental data. The $RMSE_{K_{d,WG}}$ ranges between 2.00–10.03%, while the highest agreement is observed at the WGs located in the lee of the WECs and at the Bottom Lateral WGs. The largest $RMSE_{K_{d,WG}}$ are obtained in the front WGs and at the Top Lateral WGs.

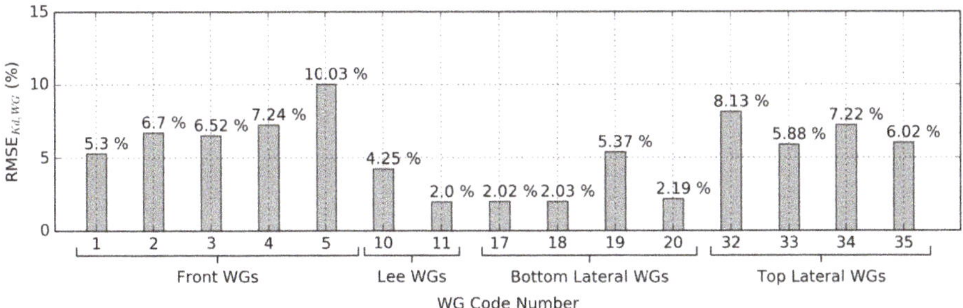

Figure 9. Root-Mean-Square-Error for the K_d for all 15 WGs, $RMSE_{K_{d,WG}}$, of Figure 1A. Comparison between the MILDwave-NEMOH coupled model and the WECwakes experimental data-set.

6. Discussion

An irregular wave generation sensitivity analysis for MILDwave was performed using the different simulation parameters of Section 5.1. The results show that keeping a small N_f for discretizing the irregular wave spectra, using $d_x = d_y = \frac{L_p}{20}$ and Q_{tot} representing 500 waves is sufficient to obtain a good representation of the target irregular long crested sea state. Increasing N_f, decreasing $d_x (= d_y)$ or increasing Q_{tot} will not lead to a significant increase in the accuracy of the obtained results, which also leads to exponential increase of the computational time. This is illustrated for $N_f = 40$ where the computational time is four times higher than the computational time for $N_f = 20$.

Section 5.2 demonstrates that the MILDwave-NEMOH coupled model can accurately propagate the perturbed wave field around different WEC (array) configurations for the linear wave theory based coupling employed here. The results of the MILDwave-NEMOH coupled model are compared against NEMOH results. Small discrepancies between NEMOH and the MILDwave-NEMOH coupled model are found close to the coupling wave generation circle in front of and in the lee of the WEC (array). These discrepancies increase as the number of WECs modelled increases, as shown in Figure 9a, though remaining between ±4%. This shows, as pointed out in [14], that the complexity of the hydrodynamic interactions when modelling the "far field" effects is not influential.

Validation of the MILDwave-NEMOH coupled model against the experimental WECwakes data is performed in Section 5.3 showing a good agreement for the different Test Cases used in this study. An error in predicting the K_d values measured at 15 WGs from the WECwakes tests is quantified in terms of $RMSE_{Kd,WG}$ (%). $RMSE_{Kd,WG}$ values range from 2–10.02% being the WGs in front of the WECs the ones with the least correspondence with the experimental data. On the contrary, for WGs that are further away from the WECs, a better agreement is obtained. The difference within the Front WGs arises due to the non-linear effect of the friction between the WEC shafts and the WEC buoys that cannot be represented with the BEM-based coupling methodology employed, as BEM is based on linear wave theory. This friction is causing the experimental WEC buoy to have smaller motion amplitude than the numerical one obtained in the BEM solver. Thus, the WEC is absorbing less energy from the incoming waves yielding a higher wave reflection in front of the WEC (array). Finally, the asymmetry in the K_d results between the Bottom and the Top Lateral zones is caused by the non-linear behaviour of the WECs in the experimental model and unwanted wave reflection in the wave basin that cannot be modelled in the MILDwave-NEMOH coupled model. In the MILDwave-NEMOH coupled model, all the WECs of the array have an identical behaviour as shown by the symmetric values of K_d given for the top and the bottom lateral zones in Figure 7 and the symmetric total wave field shown in Figure 3. Despite this, the following considerations have to be made: (1) a linear coupled model is compared to experimental data that is inherently non-linear, as confirmed by [11] who reported that the incident wave is a weakly non-linear Stokes second order wave; (2) moreover, the experimental PTO system behaves as a Coulomb damper, yet in the numerical model it is approximated as a linear damper.

For all Test Cases of Table 1, the $RMSE_{K_d,D}$ by comparing the MILDwave-NEMOH coupled model to NEMOH remains below 2%, while, by comparing the $RMSE_{K_d,WG}$ for the 15 WGs of Figure 1A between the MILDwave-NEMOH coupled model and the WECwakes experimental data, this never exceeds 10.02%. Therefore, as there is a good parse of information between the two numerical models, it can be concluded that the coupling methodology can be used to extend the numerical domain for simulating an irregular long crested wave and thus simulate the "far field" effects of WEC farms and arrays in a cost effective way.

However, and as it has already been mentioned in the authors' previous work [14], the coupling of MILDwave and NEMOH has some limitations. Firstly, despite the fact that the computational time for simulating different WEC arrays in this study is reasonable (the longest recorded computational time was that for Test Case 6, which lasted 2 h on 10 cores (Intel(R) Core(TM) i7-8700 CPU@3.2GHz), it can increase considerably when increasing the number of WECs. For an array of J WECs with six DOFs, the computational time for a BEM model increases as σ^{6J}, with increased computational time in larger numerical domains. Secondly, irregular waves are calculated as a superposition of regular waves. It has been proven that it is possible to obtain very good results with a low N_f; however, if a higher resolution of the $S_{n,M-N}(f)$ is needed, depending on the study case requirements, it would lead to an exponential increase of the computational time. Thirdly, NEMOH calculations can only be performed at a constant bathymetry introducing a limitation in that way. Moreover, MILDwave is applied for mild slope bathymetries limiting the MILDwave-NEMOH coupled model to coastal regions with a slope lower than $\frac{1}{3}$. Finally, a realistic modeling of the WEC PTO system is required to maximize the WEC (array) power output and quantify WEC effects on the surrounding wave field [50]. Modeling a resistive PTO system allows us to obtain a cost-efficient simulation regarding computational times, but may result in an overestimation of the incident wave power absorbed by the WEC(s). Realistic PTO systems lead to a reduction of the power output due to losses and differences between the predicted optimum damping and the optimum damping that can be achieved in operational conditions. The control and optimization of the PTO system, however, as shown in [37], does not have a significant influence on the wave field in the "far field".

In terms of limitations of the proposed coupling methodology, these depend each time on the type of models that are coupled [14]. Specifically, for coupling between two linear models such as NEMOH and MILDwave, the resulting coupled model will provide conservative results in study cases when non-linear phenomena are dominating. On the other hand, the above limitations can be overcome when applying the proposed coupling methodology, for non-linear models. However, the use of non-linear models needs to be justified for each specific study case, as they often introduce computational instability and high computational costs.

7. Conclusions

In the present study, the validation of a novel generic coupling methodology for modeling both near and far field effects of floating structures and WECs is presented for the test case of irregular waves. This coupling methodology is demonstrated by employing the models MILDwave and NEMOH, used for generation of irregular long crested waves. The main objective of the coupling methodology is to obtain "far field" effects of WEC arrays at a cost-efficient computational time. To validate the coupling methodology, several Test Cases from the WECwakes experimental data-set have been considered for different WEC (array) configurations and wave conditions, and performed using NEMOH and the MILDwave-NEMOH coupled model.

First, the total wave field evaluated in terms of K_d was compared between the MILDwave-NEMOH coupled model and NEMOH. The MILDwave-NEMOH coupled model showed a good agreement with NEMOH for all the considered test cases, with an $RMSE_{K_d,D}$ below 2%. Next, the model was validated against the experimental WECwakes data obtaining a satisfactory agreement, with a $RMSE_{K_d,WG}$ smaller than 10% for all test cases. Despite some discrepancies between the numerical and experimental results, which are mainly caused due to the inherent non-linear behavior of the experiments, it has been demonstrated that the proposed coupling methodology between the wave propagation model MILDwave and the BEM solver NEMOH can accurately parse the information between the two models and simulate the hydrodynamic behaviour of a WEC array and obtain the modified total wave field in the "near field" for irregular long crested wave conditions. As MILDwave has proven to provide the required level of accuracy for coastal real-world applications, it is possible to extend the numerical domain of the coupled model and simulate "far field" effects over large coastal areas.

Nevertheless, the MILDwave-NEMOH coupled model has some limitations: (1) its applicability is limited to linear and weakly non-linear wave conditions; (2) the computational time can increase considerably if a large number of frequencies and WECs or a complex PTO type is modelled; and (3) the extension of the WEC array is limited to a fixed bathymetry domain.

Regardless of these limitations, based on the results from [14] and on the current results, we can conclude that the MILDwave-NEMOH coupled model introduced has proven to be a reliable tool that can be applied in a fast and efficient way to calculate "far field" effects of WEC arrays. The next step in our modeling work is to extend the methodology to short crested wave conditions.

Author Contributions: G.V.F. set up the numerical experiments; G.V.F. performed the numerical experiments; G.V.F. compared the numerical experiments with experimental data; V.S. and P.T. provided the fundamentals of the coupling methodology; V.S. and P.T. proofread the text and helped in structuring the publication; and V.S. and P.T. provided the experimental WECwakes data.

Funding: This research received no external funding.

Conflicts of Interest: The authors declare no conflict of interest.

Abbreviations

The following abbreviations are used in this manuscript:

WEC	Wave Energy Converter
BEM	Boundary Element Method
CFD	Computer Fluid Dynamics
SPH	Smoothed Particle Hydrodynamics
PTO	Power Take-Off
RAO	Response Amplitude Operator
DHI	Danish Hydraulic Institute
WG	Wave Gauge
RMSE	Root-Mean-Square-Error

References

1. European Marine Energy Centre (EMEC) Ltd. Wave Developers Database. Available online: http://www.emec.org.uk/marine-energy/wave-developers/ (accessed on 13 November 2018).
2. Millar, D.L.; Smith, H.C.M.; Reeve, D.E. Modelling analysis of the sensitivity of shoreline change to a wave farm. *Ocean Eng.* **2007**, *34*, 884–901. [CrossRef]
3. Venugopal, V.; Smith, G. Wave Climate Investigation for an Array of Wave Power Devices. In Proceedings of the 7th European Wave and Tidal Energy Conference, Porto, Portugal, 11–13 September 2007; p. 10.
4. Smith, H.C.M.; Millar, D.L.; Reeve, D.E. Generalisation of wave farm impact assessment on inshore wave climate. In Proceedings of the 7th European Wave and Tidal Energy Conference, Porto, Portugal, 11–14 September 2007.
5. Beels, C. Optimization of the Lay-Out of a Farm of Wave Energy Converters in the North Sea: Analysis of Wave Power Resources, Wake Effects, Production and Cost. Ph.D. Thesis, Ghent University, Ghent, Belgium, 2009.
6. Carballo, R.; Iglesias, G. Wave farm impact based on realistic wave-WEC interaction. *Energy* **2013**, *51*, 216–229. [CrossRef]
7. Iglesias, G.; Carballo, R. Wave farm impact: The role of farm-to-coast distance. *Renew. Energy* **2014**, *69*, 375–385. [CrossRef]

8. Abanades, J.; Greaves, D.; Iglesias, G. Wave farm impact on the beach profile: A case study. *Coast. Eng.* **2014**, *86*, 36–44. [CrossRef]

9. Stratigaki, V. Experimental Study and Numerical Modelling of Intra-Array Interactions and Extra-Array Effects of Wave Energy Converter Arrays. Ph.D. Thesis, Ghent University, Ghent, Belgium, 2014.

10. Troch, P.; Stratigaki, V. Phase-Resolving Wave Propagation Array Models. In *Numerical Modelling of Wave Energy Converters*; Folley, M., Ed.; Elsevier: New York, NY, USA, 2016; Chapter 10, pp. 191–216.

11. Verbrugghe, T.; Stratigaki, V.; Troch, P.; Rabussier, R.; Kortenhaus, A. A comparison study of a generic coupling methodology for modeling wake effects of wave energy converter arrays. *Energies* **2017**, *10*, 1697. [CrossRef]

12. Verbrugghe, T.; Domínguez, J.M.; Crespo, A.J.; Altomare, C.; Stratigaki, V.; Troch, P.; Kortenhaus, A. Coupling methodology for smoothed particle hydrodynamics modeling of non-linear wave–structure interactions. *Coast. Eng.* **2018**, *138*, 184–198. [CrossRef]

13. Balitsky, P.; Fernandez, G.V.; Stratigaki, V.; Troch, P. Coupling methodology for modeling the near-field and far-field effects of a Wave Energy Converter. In Proceedings of the ASME 36th International Conference on Ocean, Offshore and Arctic Engineering (OMAE2017), Trondheim, Norway, 25–30 June 2017.

14. Verao Fernandez, G.; Balitsky, P.; Stratigaki, V.; Troch, P. Coupling Methodology for Studying the Far Field Effects of Wave Energy Converter Arrays over a Varying Bathymetry. *Energies* **2018**, *11*, 2899. [CrossRef]

15. Tomey-Bozo, N.; Babarit, A.; Murphy, J.; Stratigaki, V.; Troch, P.; Lewis, T.; Thomas, G. Wake effect assessment of a flap type wave energy converter farm under realistic environmental conditions by using a numerical coupling methodology. *Coast. Eng.* **2018**. [CrossRef]

16. Stratigaki, V.; Troch, P.; Stallard, T.; Forehand, D.; Folley, M.; Kofoed, J.P.; Benoit, M.; Babarit, A.; Vantorre, M.; Kirkegaard, J. Sea-state modification and heaving float interaction factors from physical modeling of arrays of wave energy converters. *J. Renew. Sustain. Energy* **2015**, *7*, 061705. [CrossRef]

17. Child, B.M.F.; Venugopal, V. Optimal Configurations of wave energy devices. *Ocean Eng.* **2010**, *37*, 1402–1417.

18. Garcia Rosa, P.B.; Bacelli, G.; Ringwood, J.V. Control-informed optimal layout for wave farms. *IEEE Trans. Sustain. Energy* **2015**, *6*, 575–582. [CrossRef]

19. Göteman, M.; McNatt, C.; Giassi, M.; Engström, J.; Isberg, J. Arrays of Point-Absorbing Wave Energy Converters in Short-Crested Irregular Waves. *Energies* **2018**, *11*, 964. [CrossRef]

20. Babarit, A. On the park effect in arrays of oscillating wave energy converters. *Renew. Energy* **2013**, *58*, 68–78.

21. Borgarino, B.; Babarit, A.; Ferrant, P. Impact of wave interaction effects on energy absorbtion in large arrays of Wave Energy Converters. *Ocean Eng.* **2012**, *41*, 79–88. [CrossRef]

22. Sismani, G.; Babarit, A.; Loukogeorgaki, E. Impact of Fixed Bottom Offshore Wind Farms on the Surrounding Wave Field. *Int. J. Offshore Polar Eng.* **2017**, *27*, 357–365. [CrossRef]

23. Devolder, B.; Stratigaki, V.; Troch, P.; Rauwoens, P. CFD simulations of floating point absorber wave energy converter arrays subjected to regular waves. *Energies* **2018**, *11*, 1–23. [CrossRef]

24. Ransley, E.; Greaves, D.; Raby, A.; Simmonds, D.; Hann, M. Survivability of wave energy converters using CFD. *Renew. Energy* **2017**, *109*, 235–247. [CrossRef]

25. Crespo, A.J.C.; Domínguez, J.M.; Rogers, B.D.; Gómez-Gesteira, M.; Longshaw, S.; Canelas, R.; Vacondio, R.; Barreiro, A.; García-Feal, O. DualSPHysics: Open-source parallel {CFD} solver based on Smoothed Particle Hydrodynamics (SPH). *Comput. Phys. Commun.* **2015**, *187*, 204–216. [CrossRef]

26. Crespo, A.; Altomare, C.; Domínguez, J.; González-Cao, J.; Gómez-Gesteira, M. Towards simulating floating offshore oscillating water column converters with Smoothed Particle Hydrodynamics. *Coast. Eng.* **2017**, *126*, 11–26. [CrossRef]

27. Chang, G.; Ruehl, K.; Jones, C.; Roberts, J.; Chartrand, C. Numerical modeling of the effects of wave energy converter characteristics on nearshore wave conditions. *Renew. Energy* **2016**, *89*, 636–648. [CrossRef]

28. Stokes, C.; Conley, D.C. Modelling Offshore Wave farms for Coastal Process Impact Assessment: Waves, Beach Morphology, and Water Users. *Energies* **2018**, *11*, 2517. [CrossRef]

29. Rusu, E. Study of the Wave Energy Propagation Patterns in the Western Black Sea. *Appl. Sci.* **2018**, *8*, 993. [CrossRef]

30. Beels, C.; Troch, P.; De Backer, G.; Vantorre, M.; De Rouck, J. Numerical implementation and sensitivity analysis of a wave energy converter in a time-dependent mild-slope equation model. *Coast. Eng.* **2010**, *57*, 471–492. [CrossRef]

31. Stratigaki, V.; Vanneste, D.; Troch, P.; Gysens, S.; Willems, M. Numerical modeling of wave penetration in ostend harbour. *Coast. Eng. Proc.* **2011**, *1*, 42. [CrossRef]

32. Tuba Özkan-Haller, H.; Haller, M.C.; Cameron McNatt, J.; Porter, A.; Lenee-Bluhm, P. Analyses of Wave Scattering and Absorption Produced by WEC Arrays: Physical/Numerical Experiments and Model Assessment. In *Marine Renewable Energy: Resource Characterization and Physical Effects*; Yang, Z.; Copping, A., Eds.; Springer International Publishing: Cham, Switzerland, 2017; pp. 71–97.

33. Charrayre, F.; Peyrard, C.; Benoit, M.; Babarit, A. A Coupled Methodology for Wave-Body Interactions at the Scale of a Farm of Wave Energy Converters Including Irregular Bathymetry. In Proceedings of the ASME 2014 33rd International Conference on Ocean, Offshore and Arctic Engineering, San Francisco, CA, USA, 8–13 June 2014.

34. Tomey-Bozo, Nicolas and Murphy, Jimmy and Troch, Peter and Lewis, Tony and Thomas, G. Modelling of a flap-type wave energy converter farm in a mild-slope equation model for a wake effect assessment. *IET Renew. Power Gener.*, 2017, *11*, 1142–1152. [CrossRef]

35. Rijnsdorp, D.P.; Hansen, J.E.; Lowe, R.J. Simulating the wave-induced response of a submerged wave-energy converter using a non-hydrostatic wave-flow model. *Coast. Eng.* **2018**, *140*, 189–204. [CrossRef]

36. Zijlema, M.; Stelling, G.; Smit, P. SWASH: An operational public domain code for simulating wave fields and rapidly varied flows in coastal waters. *Coast. Eng.* **2011**, *58*, 992–1012. [CrossRef]

37. Balitsky, P.; Verao Fernandez, G.; Stratigaki, V.; Troch, P. Assessment of the power output of a two-array clustered WEC farm using a BEM solver coupling and a Wave-Propagation Model. *Energies* **2018**, *11*, 2907. [CrossRef]

38. Troch, P. *MILDwave—A Numerical Model for Propagation and Transformation of Linear Water Waves*; Technical Report; Department of Civil Engineering, Ghent University: Ghent, Belgium, 1998.

39. Babarit, A.; Delhommeau, G. Theoretical and numerical aspects of the open source BEM solver {NEMOH}. In Proceedings of 11th European Wave and Tidal Energy Conference, Nantes, France, 6–11 September 2015.

40. WAMIT Inc. *User Manual, Versions 6.4, 6.4 PC, 6.3, 6.3S-PC*; WAMIT: Boston, MA, USA, 2006.

41. Stratigaki, V.; Troch, P.; Stallard, T.; Forehand, D.; Kofoed, J.P.; Folley, M.; Benoit, M.; Babarit, A.; Kirkegaard, J. Wave Basin Experiments with Large Wave Energy Converter Arrays to Study Interactions between the Converters and Effects on Other Users. *Energies* **2014**, *7*, 701–734. [CrossRef]

42. Stratigaki, V.; Vanneste, D.; Troch, P.; Gysens, S.; Willems, M. Numerical modeling of wave penetration in Ostend Harbour. In *Proceedings of Conference on Coastal Engineering*; McKee Smith, J.; Lynett, P., Eds.; Engineering Foundation, Council on Wave Research: New York, NY, USA, 2010, Volume 32, p. 15.

43. Radder, A.C.; Dingemans, M.W. Canonical equations for almost periodic, weakly non-linear gravity waves. *Wave Motion* **1985**, *7*, 473–485. [CrossRef]

44. Tomey-Bozo, N.; Babarit, J.M.A.; Troch, P.; Lewis, T.; Thomas, G. Wake Effect Assesment of a flap-type wave energy converter farm using a coupling methodology. In Proceedings of the ASME 36th International Conference on Ocean, Offshore and Arctic Engineering (OMAE2017), Trondheim, Norway, 25-30 June 2017.

45. Beels, C.; Troch, P.; Kofoed, J.P.; Frigaard, P.; Kringelum, J.V.; Kromann, P.C.; Donovan, M.H.; De Rouck, J.; De Backer, G. A methodology for production and cost assessment of a farm of wave energy converters. *Renew. Energy* **2011**, *36*, 3402–3416. [CrossRef]

46. Penalba, M.; Touzón, I.; Lopez-Mendia, J.; Nava, V. A numerical study on the hydrodynamic impact of device slenderness and array size in wave energy farms in realistic wave climates. *Ocean Eng.* **2017**, *142*, 224–232. [CrossRef]

47. Penalba, M.; Kelly, T.; Ringwood, J. Using NEMOH for Modelling Wave Energy Converters: A Comparative Study with WAMIT. In Proceedings of the 12th European Wave and Tidal Energy Conference (EWTEC 2017), Cork, UK, 27 August–1 September 2017.

48. Alves, M. Wave Energy Converter modeling techniques based on linear hydrodynamic theory. In *Numerical Modelling of Wave Energy Converters*; Folley, M., Ed.; Elsevier: New York, NY, USA, 2016; Chapter 1, pp. 11–65.

49. Iglesias, G.; Carballo, R. Wave energy and nearshore hot spots: The case of the SE Bay of Biscay. *Renew. Energy* **2010**, *35*, 2490–2500. [CrossRef]

50. Child, B.M.F. On the Configuration of Arrays of Floating Wave Energy Converters. Ph.D Thesis, The University of Edinburgh, Edinburgh, UK, 2011.

Article

On the Variation of Turbulence in a High-Velocity Tidal Channel

Charles Greenwood [1],*, Arne Vogler [1] and Vengatesan Venugopal [2]

[1] Marine Energy Research Group, Lews Castle College, University of the Highlands and Islands, Stornoway HS2 0XR, UK; arne.vogler@uhi.ac.uk
[2] Institute for Energy Systems, School of Engineering, The University of Edinburgh, Edinburgh EH9 3DW, UK; v.venugopal@ed.ac.uk
* Correspondence: charles.greenwood@uhi.ac.uk; Tel.: +44-(0)1851-770219

Received: 31 January 2019; Accepted: 15 February 2019; Published: 19 February 2019

Abstract: This study presents the variation in turbulence parameters derived from site measurements at a tidal energy test site. Measurements were made towards the southern end of the European Marine Energy Centre's tidal energy test site at the Fall of Warness (Orkney, Scotland). Four bottom mounted divergent-beam Acoustic Doppler Current Profilers (ADCPs) were deployed at three locations over an area of 2 km by 1.4 km to assess the spatial and temporal variation in turbulence in the southern entrance to the channel. During the measurement campaign, average flood velocities of 2 ms^{-1} were recorded with maximum flow speeds of 3 ms^{-1} in the absence of significant wave activity. The velocity fluctuations and turbulence parameters show the presence of large turbulent structures at each location. The easternmost profiler located in the wake of a nearby headland during ebb tide, recorded flow shielding effects that reduced velocities to almost zero and produced large turbulence intensities. The depth-dependent analysis of turbulence parameters reveals large velocity variations with complex profiles that do not follow the standard smooth shear profile. Furthermore, turbulence parameters based on data collected from ADCPs deployed in a multi-carrier frame at the same location and time period, show significant differences. This shows a large sensitivity to the make and model of ADCPs with regards to turbulence. Turbulence integral length scales were calculated, and show eddies exceeding 30 m in size. Direct comparison of the length scales derived from the streamwise velocity component and along-beam velocities show very similar magnitudes and distributions with tidal phase.

Keywords: turbulence; turbulence intensity; turbulence kinetic energy; ADCP; site measurements; time scale; length scale

---

1. Introduction

As tides pass through narrow channels and around headlands, high-velocity flows are produced. Depending on site-specific characteristics, the turbulence of these flows varies in time and space. When these turbulent flows encounter a tidal stream turbine (TST), the severity of turbulence affects the turbine energy extraction capability by altering blade performance. This results in reduced rotor thrust and torque [1], which causes a detriment in turbine performance of over 10%. Prolonged exposure to high turbulent flows increases fatigue load cycles and blade bending [2], thereby reducing the expected lifespan of a turbine. Mitigation techniques can be applied to keep design tolerances high to account for turbulence related stresses, but this also increases manufacturing costs. The quantification of turbulence remains an important factor in the optimization of energy extraction and turbine durability and thus successful tidal stream developments.

The measurement of turbulence in a fluid, at scales relevant to rotating hydrofoils, by its nature requires high-resolution measurements (>1 Hz). Originally, sensors were developed to determine atmospheric turbulence; these were designed to be unobtrusive, where the measurements do not cause disruption to the flow of the fluid. To achieve this, the principle of Doppler shift was applied. This transmits high-frequency bursts (pings) from three or more transceivers. As the sound is scattered off by suspended moving particles in the fluid, a frequency shift is observed. This shift in frequency uses the Doppler effect formula, where the particle velocity, and therefore fluid velocity, can be calculated. The combination of multiple beams on different axes allows for the velocity calculation in the x, y and z vectors and therefore speed and direction. This was originally done for converging beam sensors [3]. This technology was later applied to the ocean environment [4], where an adapted method for calculating seabed friction velocity was presented for low flow speeds in the presence of waves. Acoustic Doppler Velocimeters (ADVs) offer high-resolution sample rates in the order of 10 Hz assuming a small volume flow homogeneity (5–50 mm) [5]. However, due to the focusing of the beams, a small area can only be observed, and this limits these sensors to small profiles or single point measurements. When the beam angle configuration changes from convergent to divergent, the beams cover a much larger area. If a similar homogenous flow approximation is applied, but now over an area of 1–50 m, then the component flow directions can be calculated for a depth profile. The application of this method has been used for many years in divergent beam ADCPs. Originally operating via "Narrow-Band" acoustic pulses, instantaneous velocity measurements experienced high noise levels, limiting these datasets to time averaged results not suitable for turbulence assessment. The development of "Broadband" acoustic processing has allowed more accurate instantaneous velocity measurements for diverging beam sensors [6,7]. This continues to allow velocity measurements to be made throughout the water column, but with much higher sample rates, without the need for temporal averaging. For some sensors, this means sample rates as high as 10 Hz or above can be used while maintaining low data noise levels.

The number of transceivers on a sensor has increased over the years. While it is possible to get velocity measurements from a single beam, the velocity component is only in the beam direction, providing heavy application restrictions. Three transducers were common for a number of years and this allowed the calculation of directional velocity components. When combined with a pressure sensor, surface waves could also be measured. This used the pressure, u and v velocity (PUV) method to measure the surface elevation using the pressure sensor and the wave directional orbital velocity (u and v) to establish wave direction [8]. The introduction of four beam sensors with opposing angled beams on two perpendicularly aligned axis increased the accuracy of flow measurements. This allowed two measurements to be used to calculate the velocity in each horizontal vector. More recently, five beam sensors have become the new standard. These use the same beam orientation as the four beam sensors; where the introduction of a fifth vertical beam allows a direct measurement of the vertical velocity and surface elevation. While the implementation of a fifth beam and acoustic surface tracking isn't new, it is now more commonplace and integrated as default in a wide range of standard ADCPs.

The calculation of turbulence parameters from beam velocities of a 4-beam ADCP in relatively slow velocities (<1 ms^{-1}) was demonstrated in [8]. This identified the effects of Doppler noise and its implications on turbulence parameters. The Doppler noise is created as a byproduct of the velocity calculation, where a phase difference for the Doppler shift occurs for multiple returns within a spatially determined cell [9]. More recent studies quantifying turbulence in tidal channels compared the velocities between a bed-mounted ADCP and an ADV [10,11]. These studies presented a standardised metric, adopted from the wind energy industry, to quantify tidal turbulence; this is known as the turbulence intensity. The turbulence intensity can be calculated using the velocity fluctuation minus the Doppler noise. If the noise is not accounted for, the instantaneous measurements from ADCPs can overestimate the value of the standard deviation calculated from velocity. This produces higher turbulence values for the same corresponding location. The noise can be described as white noise, where it is distributed evenly across all frequencies. The effects of this noise can be mitigated by averaging; this can either be done in time, where instantaneous velocity fluctuations are averaged over a number of minutes, or in space, where velocity fluctuations can be averaged across the vertical profile. This, however, is not a viable solution when considering turbulence measurements. The quantification of the Doppler noise was improved in [12]; this applied a polynomial least square regression method to extend the inertial range to the high-frequency end of the velocity spectrum, lowering the noise floor. Additional methods for correcting instrument noise were presented in [13]. These use either a Noise Auto-Correlation (NAC) or Proper Orthogonal Decomposition (POD) approach to determine and remove noise. In the case of the NAC method, the noise level is calculated based on the velocity spectrum, restricting the noise reduction to the frequency domain. However, the POD method is capable of reducing noise over the spectrum and within the time domain, providing a more flexible output. In addition, these methods may be more suitable for reducing noise in the presence of waves.

The number of published studies quantifying turbulence levels using in-situ field measurements at tidal energy test sites has increased in recent years [14–20]. Similar studies have been conducted in rivers [21–23], however, due to differences on the onset flow conditions, topography and sedimentation, the results of these studies is of limited relevance. Both the tidal and river studies use divergent beam ADCPs to record the velocity fluctuations. Alternatively, single-beam Doppler profilers, horizontally mounted and aligned with the in-stream velocity direction, have been used for turbulence assessment [18,24]. This configuration provides streamwise velocity fluctuations with no assumption of flow homogeneity, resulting in a more accurate measurement of ambient flow speeds at mid-water depth and turbine hub height. However, additional single beam profilers are required (as installed in [18]) to provide flow information in the transverse and vertical directions. The previous studies provide a basic resource assessment, where flow velocities and directions are first quantified in time and often depth. Then a range of turbulence parameters are presented including the standard deviation of the velocity fluctuations, turbulence intensity, Reynolds stresses, Turbulence Kinetic Energy and velocity spectra. Previous studies also provide benchmarks to other locations that allow the Fall of Warness to be contextualized in terms of other potential tidal test sites.

In this study, data are presented from a sensor network at the southern end of the Fall of Warness, with the locations specified in Figure 1. The diverging acoustic Doppler profilers (D-ADP) were deployed simultaneously to measure wave and currents over a 4 day period. The general resource and turbulence characteristics are calculated and compared for each sensor with depth and time. Specific emphasis is placed on the peak velocities of the flood and ebb tides and the average conditions, where turbulence interactions are likely to be higher. The turbulence intensity, turbulence spectrum, Reynolds stresses and integral time and length scales are presented. A direct comparison between the homogenous approximated length scale and the raw beam length scales are presented for two sensors. The data analysis indicates that along with spatial and temporal variations, sensor make and model specific features also cause variation in observed turbulence parameters. These differences are highlighted and discussed throughout this paper. The presentation of turbulence intensity and length scale provide important boundary conditions required for the simulation of tidal turbines [25]. This is crucial for the long term performance analysis of a single or array of devices.

2. Data Collection and Methodology

2.1. Sensor Deployment

A measurement campaign was devised to simultaneously operate multiple coastal sensors at six locations between the dates of the 13th and 18th of June 2016. To enable the measurement of turbulence parameters, at scales relevant to tidal stream turbines, a continuous time series recording at a sample rate of 0.5 Hz or higher is required. Only four of these sensors were suitable for turbulence analysis and these are discussed in this paper. Work was carried out as part of the Response of Tidal Energy Converters to Combined Tidal Flow, Waves, and Turbulence (FloWTurb) project (EPSRC EP/N021487/1). All sensor locations are shown in Figure 1 and discussed in this paper are SP1, SP3 and SE, where the SE location features multiple sensors deployed in a single frame. Names, depths and sampling information of the relevant sensors are shown in Table 1.

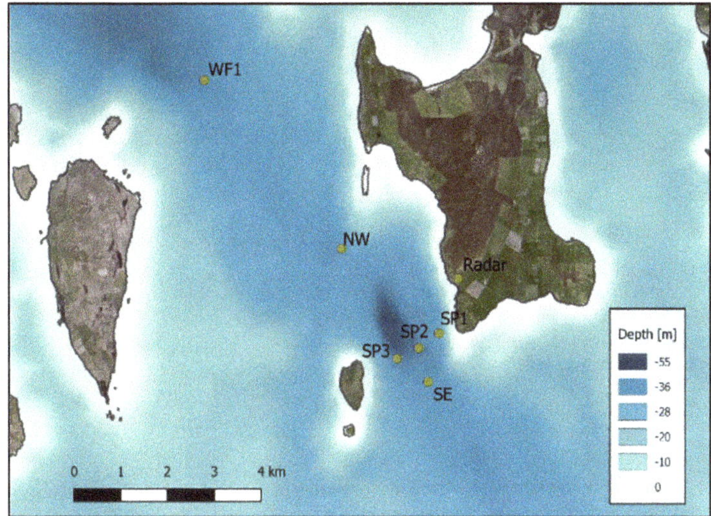

Figure 1. Sensor deployment locations for the Fall of Warness measurement campaign with the bathymetry provided by UK Hydrographic Office.

Table 1. Sensor specifications.

Sensors	Location	Beams	Transmitting Frequency	Beam Angle	Sample Frequency	Vertical Cell Size	Depth
Signature 500	SE	5	500 kHz	25°	4 Hz	0.5 m	37 m
Signature 1000	SP1	5	1000 kHz	25°	4 Hz	0.5 m	37 m
RDI Sentinel 600	SE	4	600 kHz	20°	2 Hz	0.75 m	37 m
RDI Sentinel 600	SP3	4	600 kHz	20°	0.5 Hz	1 m	44 m

The sensors were located at the southern end of the channel within an area of approximately 2 km (along the channel length) by 1.4 km (perpendicular to the channel length). The bathymetry data used in Figure 1 is taken from the UK Hydrographic Office [26]. Each deployment location uses a bottom mounted ADCP housed in a stainless steel frame. A dual-axis gimbal was used to mount each sensor within the frame. The total dry mass of each frame is around 400 kg. In addition, a land-based X-band radar system was used for remote measurement of wave and surface current conditions; further details on the measurement of the flow characteristics using the radar can be found in [27]. The sensors used were the Signature 500 and Signature 1000 manufactured by Nortek (Rud, Norway), and two Workhorse Sentinel 600 ADCPs manufactured by Teledyne RDI (Poway, CA, USA).

The sensor specifications and setup are shown in Table 1. The SP2 sensor was lost on retrieval and therefore no data exists. An additional Waverider DWR-G4 wave buoy (Datawell, Heerhugowaard, The Netherlands) was attached to the SE sensor frame. This provided surface wave information during slack water periods as the ADCP's fifth vertical beam was not enabled. This reduced internal processing time and power consumption for the ADCPs. In order to minimize interference between the four co-located SE sensors (1 MHz Nortek AWAC, 400 kHz Nortek Aquadopp, RDI Sentinel 600 and Nortek Signature 500, the latter two of which are suitable for turbulence characterisation) i.e., sensor crosstalk, each sensor suitable for turbulence characterisation recorded data in a staggered formation, permitting only one sensor operated at any one time. This allowed each sensor to record for 20 min of data, then remain inactive for 40 min while each sensor in turn sampled the water column. When all the data is stitched together from each sensor a continuous measurement of the flow is obtained. The deployment period was scheduled between a neap and a spring tide, where the current velocities were increasing.

2.2. Basic Data Processing

Due to the simultaneous deployment of these sensors, it was not possible to standardise either the sensor type or sampling regime for all the deployment locations. This prevented the use of standard sensor specific processing tools, as the subtle differences in the manufacturer's post-processing procedures may impede the direct comparison of data. To account for this, new software was developed in MATLAB (version R2018a) to read and process raw data from both Teledyne or Nortek sensors. All sensors have a minimum of four diverging beams with beam angles as shown in Table 1. Sensors with an additional fifth beam are capable of surface measurements using acoustic surface tracking. However, the focus of these deployments was on the current data collection, so no vertical beam data were recorded. In order to measure current velocities, the Doppler effect is used based on the frequency shift between emitted and received pulses along each divergent beam. While many previous studies use beam velocities to calculate horizontal velocities and turbulence metrics, the orientation of the beam to the flow is vital, as a misrepresentative value will be recorded and a large difference from the opposing beam will be seen if the beam is not directly in line with the flow direction. This study applies an approach where the velocities from four beams are used to calculate the flow vector in a number of coordinate systems [28,29]. These coordinate systems are shown in Figure 2.

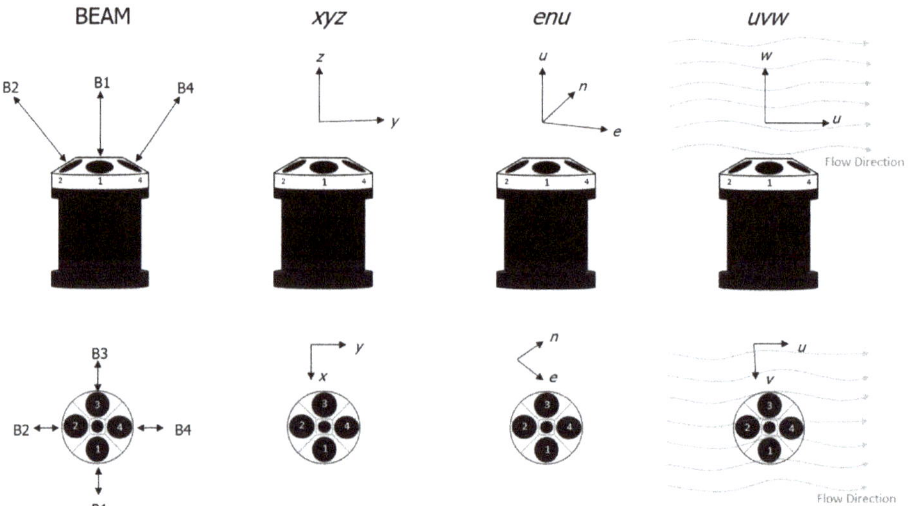

Figure 2. Divergent acoustic Doppler profiler beam coordinates relative to *xyz*, *enu* and *uvw*.

The three coordinate systems used for the calculation of flow and turbulence parameters are *xyz* relative to beam 1, *enu* (East, North and Up) relative to magnetic north, and streamwise relative to the mean flow direction (*uvw*). For this study, a short-term 600 s average was used to calculate mean flow direction. Equation (1) shows the method of calculating the relative directional flow using the sensor orientation (*xyz*) depending on the beam angle (θ):

$$\begin{bmatrix} x \\ y \\ z \end{bmatrix} = \begin{bmatrix} (b1 - b3)/(2\sin(\theta)) \\ (b2 - b4)/(-2\sin(\theta)) \\ (b1 + b2 + b3 + b4)/(4\cos(\theta)) \end{bmatrix} \tag{1}$$

In Equation (1) b1–b4 are the beam velocities from each transducer in a clockwise rotation. This is consistent with Nortek's beam numbering convention, and results from the Teledyne sensors were rearranged to fit this convention. To correct for sensor misalignment on installation and sensor motion during deployment, a coordinate transform was applied to each sampled measurement, utilizing the instrument's onboard tilt sensors and magnetic compass, where the velocities are multiplied by a tilt matrix (R) based on three degrees of freedom, heading (hh), pitch (pp) and roll (rr). This is shown in Equation (2) and the transformation to Earth coordinates is shown in Equation (3):

$$R = \begin{bmatrix} \cos(hh) & \sin(hh) & 0 \\ -\sin(hh) & \cos(hh) & 0 \\ 0 & 0 & 1 \end{bmatrix} \begin{bmatrix} \cos(pp) & -\sin(pp)\sin(rr) & -\cos(rr)\sin(pp) \\ 0 & \cos(rr) & -\sin(rr) \\ \sin(pp) & \sin(rr)\cos(pp) & \cos(pp)\cos(rr) \end{bmatrix} \tag{2}$$

$$enu = R \times xyz \tag{3}$$

This orientates the velocity vectors to Earth coordinates. The *uvw* velocities are calculated based on the same 600 s window, where the average flow direction is subtracted from the *enu* flow directions. This provides a speed and direction relative to the streamwise velocity. The vectors are then calculated using the speed multiplied by the cosine of the new streamwise and transverse directions. For the analysis of flow around a tidal stream turbine, the *uvw* coordinate system is deemed the most relevant, as velocity fluctuations are in line with the turbine hub and perpendicular to the rotor disk. The calculation of the velocity components uses multiple diverging directional beams, this produces a cone-shaped sample area that is narrower at the seabed near the sensor head and expands towards the sea surface based on the transceiver beam angle. In order to calculate the velocity components, it must be assumed that the current flow is homogeneous across the plane of all beams, which is a clear weakness in the use of coordinate transforms for turbulence assessment. The size of this plane increases with distance away from the sensor, which in turn increases the size of the area of the homogeneous approximation. This subsequently increases the size of the minimum resolvable flow structure. While this is a considerable limitation when measuring turbulence, it allows for a simple deployment process where the sensor can be placed on uneven ground without fine positioning to align beam 1 with north. Continuous monitoring of sensor pitch, roll and yaw also allows for compensation of any unexpected frame movements, which are not uncommon in higher velocity environments.

3. Data Analysis and Results

3.1. Flow Characteristics

The average flow characteristics from the high-resolution sensors deployed in the Fall of Warness are shown in Figure 3. The *uvw* velocity vectors are plotted based on a 600 s average, from a start date of 11:00:00 13/07/2017 and are based on data collected over the following four days. The streamwise velocity vector for all sensors shows a periodic fluctuation as a result of the flood and ebb tide, where one complete tidal cycle is composed of two neighbouring peaks in the streamwise velocity. This is indicative of a semidiurnal tidal cycle. The initial peak in the SE—Signature 500 and SP3 Sentinel 600 velocities, depicts the first flood tide, after which the lower magnitude ebb tide follows. This pattern is not the case for the SP1 sensor, where the peaks are shown to occur in the initial part of the flood tides and no obvious peak in the ebb tide is observed. The positive transverse velocity indicates the deviation from the streamwise flow in a clockwise direction and a negative value equates to an anticlockwise deviation. As expected, the *u* velocity far exceeds the *v* velocity for all sensors. The flow directions, relative to magnetic north, and speed at 10 m above the seabed are plotted on the right side of Figure 3. The results indicate similar tidal ellipses for the SE and SP3 sensors. The tidal ellipse shape for these sensors is particularly compressed, indicating an almost bidirectional flow pattern, where the flood tide experiences higher velocities towards a south-south-easterly direction (150°). The tidal ellipse for the SP1 sensor shows much larger directional variation, with a small broad peak for the flood tide and no peak velocity for the ebb tide. This is caused by the close proximity of the sensor to the southern headland of the isle of Eday and related generation of large eddies during ebb tide at the sensor location.

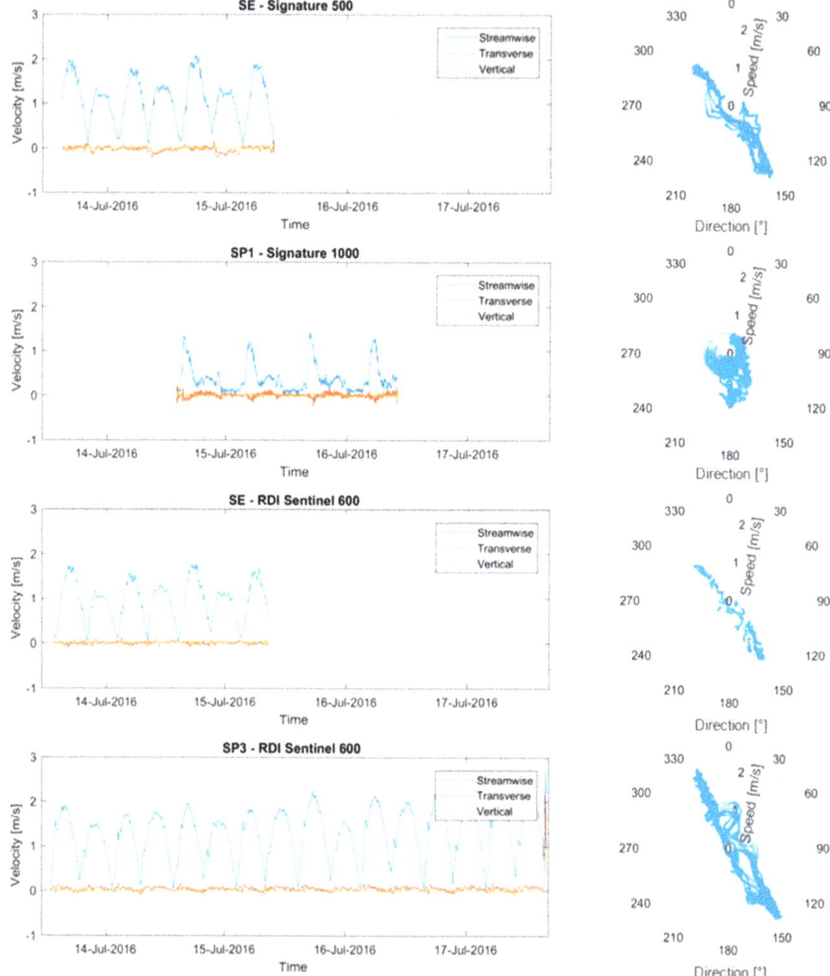

Figure 3. Left: Average flow velocity vectors over 10 min in the streamwise coordinate system for the SE—Signature 500, SP1—Signature 1000, SE—Sentinel V50 and the SP3—Workhorse. Right: Average flow direction (degrees) and velocity (m/s) for the same sensors at a 10 min interval. With the start date time of the 13 July 2016 11:00:00 am. All data provided is taken 10 m above the seabed.

3.2. Depth Profiles

The velocity distribution with depth is presented in Figure 4. This shows the peak flood and ebb tides, based on the maximum velocities over the 600 s windows, for the first 3 respective tidal phases for each sensor. This is conducted for combined speed (uvw), streamwise (u), transverse (v) and vertical (w) velocity components, where the average speed profile indicates the net combined velocity at each depth interval. To visually separate each tidal phase, the flood tide has been plotted as positive (blue) and the ebb tide is shown as negative (red). Distance above the seabed (z) is used to describe the vertical cell position, where the first bin height was set at approximately 2 m for all sensors. In addition to a directional dependence of the velocity magnitudes as shown in Figure 3, the velocity shear profiles in Figure 4 also show reductions in flow speeds toward the sea bed in various degrees of magnitude. The variation in profile is caused as a result of spatial and tidal phase differences, but for the co-located sensor, different flow profiles are seen at the same location between the different sensor types. The SE—Signature 500 sensor shows a typical shear profile from the seabed to 19 m upward, where the current velocity consistently increases. From 19 m to 26 m a considerable reduction in flow speed occurs, and this was also confirmed by looking at the individual beam-by-beam velocities. The co-located SE-RDI Sentinel sensor does not replicate this flow feature and shows a smooth gradually increasing velocity shear profile. Further observational differences identify a substantial variation in the proportion of velocity components, u and v for the SE sensors. This shows the Signature sensor recording a much greater transverse velocity. Both sensor manufacturers have been contacted with regards to the quality of the data sets, the outcome of this suggested that no errors were contained within either data set. Both manufactures were fully satisfied with the quality of their sensor specific data sets. The SP1 sensor shows very different profiles for flood and ebb, where the flood shows a rapid change in the velocity gradient near the seabed and a uniform flow speed of 1.3 ms^{-1} with only a small increase towards the surface. The ebb tide shows very small velocities for the lower part of the water column, and this changes after 13 m upwards, where the velocity increases gradually towards the surface. The SP3 sensor shows a much more conventional shear profile, which experiences a smooth curve from the seabed to the surface, where flow speeds continually increase. The velocity profiles presented provide further details of the flow characteristics around the Fall of Warness. These measurements are key to fully understanding the following analysis of turbulence parameters.

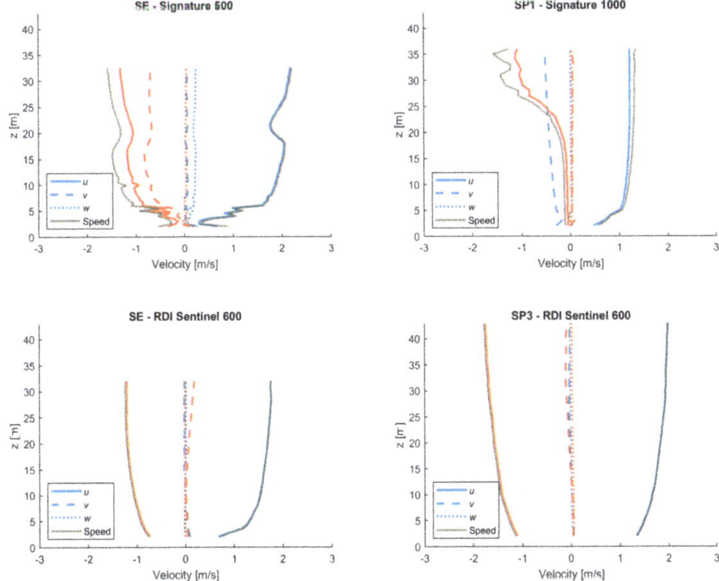

Figure 4. Average peak flood and ebb flow profiles for each sensor for three tidal phases, where blue indicates flood tide and red indicates the ebb tide. The values for the u and speed component of the ebb tide have been inverted to illustrate different flow direction.

3.3. Turbulence Intensity

To quantify the turbulence levels, the commonly used turbulence intensity (I_u) is calculated. The method used in this paper uses Equation (4) as presented in [11]:

$$I_u = \frac{\sqrt{\langle u'^2 \rangle - \sigma_{noise}^2}}{\overline{u}} \tag{4}$$

where u is the streamwise velocity component and $\overline{u}$ is the mean streamwise velocity. $\langle u'^2 \rangle$ is the velocity variance, and σ_{noise}^2 is the Doppler noise from the sensor. The distribution of the Doppler noise is spread evenly across all frequencies. As the high frequency regions of the spectrum have comparably less energy than the lower frequencies, this results in a larger relative noise level for the high frequency turbulence measurements. To account for this increased instrument noise the method described in [12] was applied to quantify the noise variation. This calculates a σ_{noise}^2 based on an extended fit of the turbulence spectrum. This is only possible by applying Kolmogorov's theory of the inertial range; this states that inertial effects are much greater than any viscous effects and suggests a rate of turbulence energy decay of $-5/3$ [30]. This only remains valid in the absence of waves. A Welch's Fast Fourier Transformation (FFT) was applied to quantify the power spectral density of the turbulence time series in the frequency domain. Each continual time series, or ensemble, was separated into window lengths of 200 s, and a Hanning filter was applied to each window. Zero padding was added to increase the number of samples to equal the next power of the total number of samples. If the σ_{noise}^2 value is set to 0, no noise reduction is applied for the turbulence intensity and the calculation becomes the same as the standard calculation [31]. The definition of $\overline{u}$ is sometimes described as the mean flow velocity at the turbine hub height. As this study is not turbine device specific the more flexible definition is applied, this specifies $\overline{u}$ as average streamwise velocity for each depth interval. The calculated turbulence intensity data from the sensor campaign is presented in Figure 5.

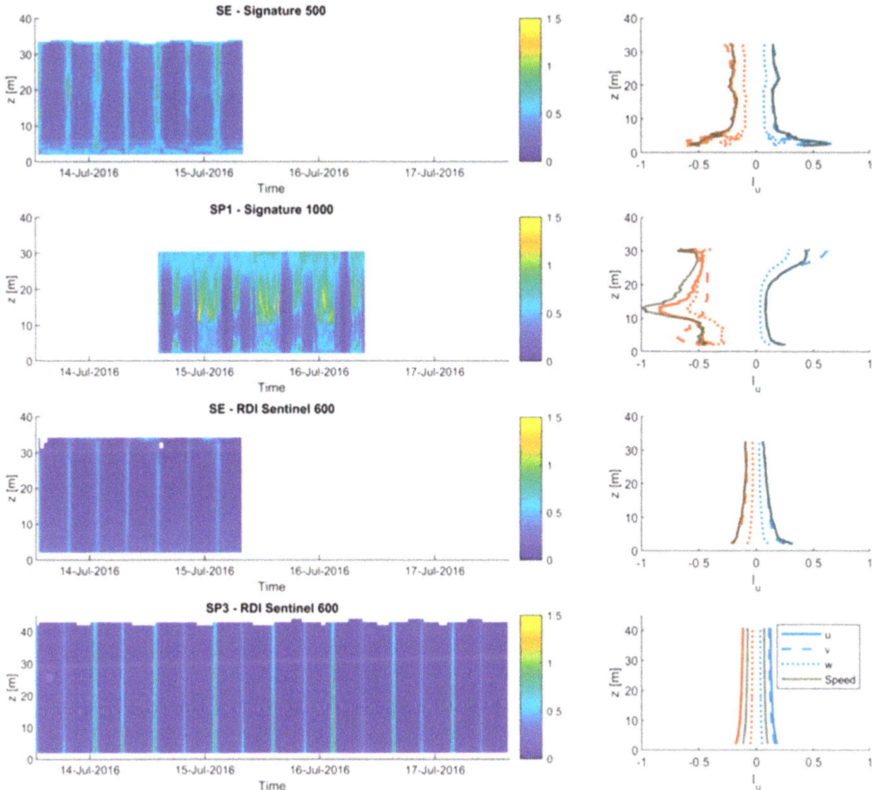

Figure 5. Turbulence intensity. Left: I_u based on 10 min windows for each sensor. Right: Turbulence intensity for the first three peak velocity flood and ebb tides, where the flood is positive (blue) and the ebb tide has been made negative (red).

When the results are reviewed in combination with the velocity time series (Figure 3), the higher turbulence intensities are associated with the slower velocities (near slack water) and the lower turbulence intensities with the flood and ebb phases. This is a result of the standard definition of the turbulence intensity parameter used here. In a different approach, where I_u is not combined with velocity information, but taken only as the velocity variance, a different interpretation of the total turbulence energy can be observed. The turbulence intensity profiles often show larger values for slower flow speeds. This may be misleading in the case of a high turbulence intensity value with a low average velocity, in this case the actual turbulent velocity fluctuations and overall forces involved are actually very low with a negligible impact on tidal turbine performance. This can be seen for the SE and SP3 sensors near the seabed. The SP1 sensor experiences low turbulence levels near the seabed and much larger turbulence intensities near the surface for the flood tide, and upwards from around 18 m for the ebb tide.

3.4. Turbulence Kinetic Energy

The quantification of Turbulence Kinetic Energy (TKE), in comparison to the TI presented in the previous section, offers a more intuitive parameter to quantify turbulence (see Figure 6). The TKE which describes the turbulence within a volume of water, is expressed by:

$$k = \frac{1}{2}\left(\overline{(u')^2} + \overline{(v')^2} + \overline{(w')^2}\right) \tag{5}$$

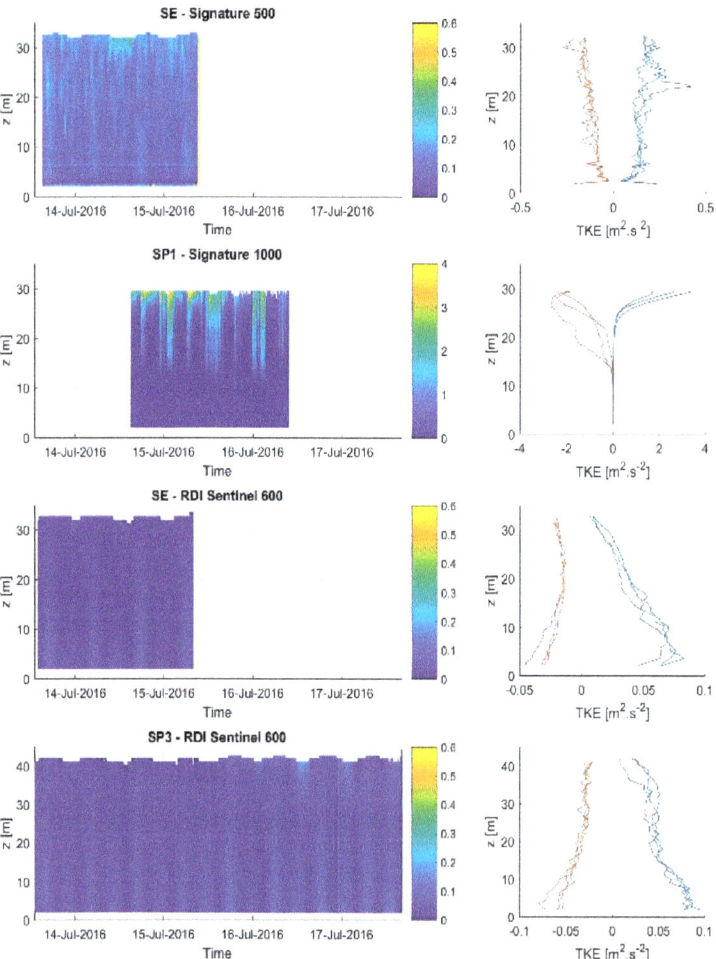

Figure 6. Turbulence kinetic energy. Left: TKE based on 10 min windows for each sensor. Right: TKE for the first three peak velocity flood and ebb tides, where the flood is positive (blue) and the ebb tide negative (red).

The derived TKE values are shown in Figure 6. This indicates larger values for the higher velocity flows, where the flood tides experience larger velocities and therefore increased turbulence magnitudes. The TKE is shown to be depth dependent, with the SE-RDI and SP3-RDI sensors showing larger values towards the seabed. Alternately, the SE—Signature 500 and SP1 sensors show larger values towards the surface. It should be noted that the horizontal scale for the SP1 sensor is much greater compared to the other sensors, which causes values that look close to zero from mid-depths to the seabed. This is due to the large magnitude of turbulent velocities near the surface, which masks the lower water column results.

3.5. Turbulence Spectra

The turbulence velocity spectra were calculated using the combined streamwise and transverse velocity components. These are presented in grey in Figure 7, for a depth of 10 m above the seabed.

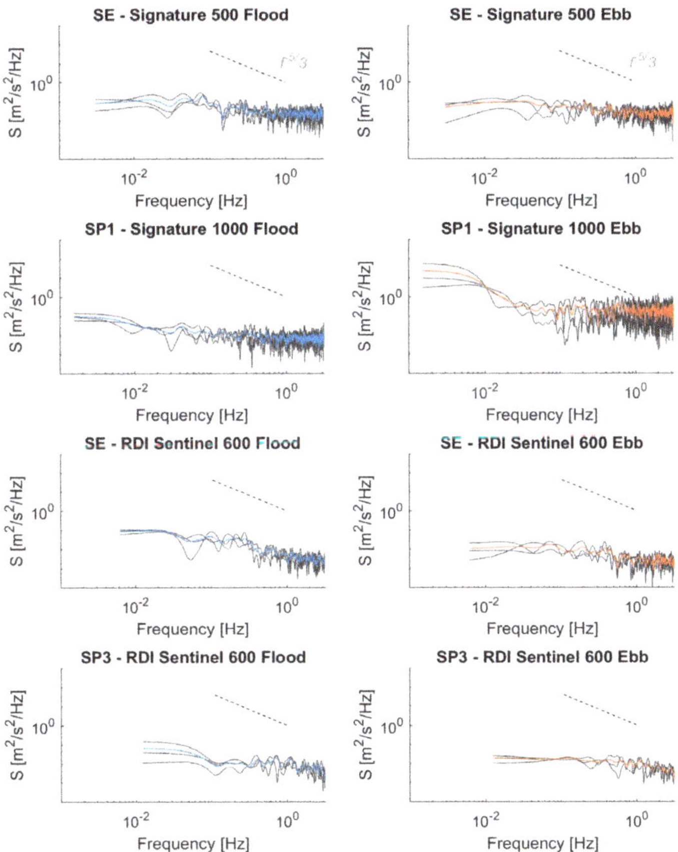

Figure 7. Average turbulence velocity spectra for each sensor during the peak flood (blue) and ebb (red) tide at 10 m from the seabed. The individual spectra for three instances are shown in grey for each sensor during the peak flood and ebb tide. The grey dashed line is included in each subplot to indicate the $-5/3$ gradient of Kolmogorov's turbulence energy cascade.

The peak flood and ebb are presented with the average velocity spectrum for each sensor at the respective tidal phase, where the spectrum is calculated based on a 10 min time period. An additional dotted line is shown with a gradient of $f^{-5/3}$ in each sub-figure as defined by Kolmogorov's model [32], this highlights the gradient of the energy cascade from the inertial sub-range. More recently, the description of the $-5/3$ turbulence energy cascade has been shown to under predict the gradient of the turbulence energy cascade [33]. However, as this study provides measurements for comparisons with other field sites, this study will maintain to use the more conventional $-5/3$

cascade as this provides a better comparison with existing literature. The majority of the spectra show a measured gradient much lower than the $-5/3$ gradient. The results compare well between the SE and SP3 sensors in terms of magnitude and distribution of the energy with frequency. The low-frequency part of the spectrum shows no presence of surface gravity waves, which would be represented by an increased PSD in the region of 0.3 to 0.05 Hz. The lower sample rate of the SP3 and SE RDI Sentinel sensor show less noise variation than both Signature sensors, producing a smoother plot. The SP1 sensor records a much larger turbulence velocity spectrum for the ebb tide, indicating larger turbulence variations, with a heavy weighting toward lower frequencies. The results of the turbulence spectra reflect the conclusion shown in the Turbulence Kinetic Energy results displayed in Figure 7. This shows larger magnitudes of turbulent features for the SP1 sensor, specifically during the ebb tide, where there is an increase in low frequency flow components.

3.6. Reynolds Stresses

The Reynolds shear stresses were calculated for each burst ensemble. The *uw* and the *vw* components were reviewed to provide insight into the stresses travelling from streamwise and transverse directions to the vertical direction. The calculation of the shear stresses is often done on a beam-by-beam basis; however, this does not account for the pitch and roll movement of the sensor or the alignment of the flow to the beam direction. Table 2 shows the mean and the standard deviation in the pitch and roll. This shows small deviations in the mean pitch and roll of less than 5 degrees for all sensors. The standard deviation shows exceptionally small variation in sensor movement for the pitch and roll. A much larger variation in heading is observed for the SP3 sensor, this is due to several frame shifts in the deployment period. These occur over a very short period and the application of the coordinate transformation (shown previously in Section 2.2) has corrected this rotation in the derived enu-framed velocities.

Table 2. Sensor orientation.

Sensor Location	Mean			Standard Deviation		
	Pitch	Roll	Yaw	Pitch	Roll	Yaw
SE (Signature 500)	1.73°	−4.93°	232°	0.07°	0.08°	0.04°
SP1 (Signature 1000)	0.53°	0.21°	164°	0.13°	0.09°	0.37°
SE (RDI Sentinel 600)	−0.92°	−0.50°	233°	0.32°	0.36°	0.60°
SP3 (RDI Sentinel 600)	−0.48°	3.11°	160°	0.45°	0.38°	40.81°

This study uses the conventional approach used in the aeronautical industry, where the velocity components are used to determine the streamwise and transverse shear stresses and are described as:

$$uw = \overline{u'w'} \tag{6}$$

$$vw = \overline{v'w'} \tag{7}$$

where u' is the streamwise velocity fluctuation and v' is the transverse velocity fluctuation. The Reynolds shear stress depth profiles are plotted in Figure 8 for the streamwise and transverse velocities. These are calculated over sample periods of 600 s with an overall average for each flood, ebb, *uw* and *vw* scenario. The two left columns show the mean of the flood tide (blue) and the two right columns show the mean ebb tide results (red), where the individual profiles for the three peak flood and ebb velocities are shown in grey.

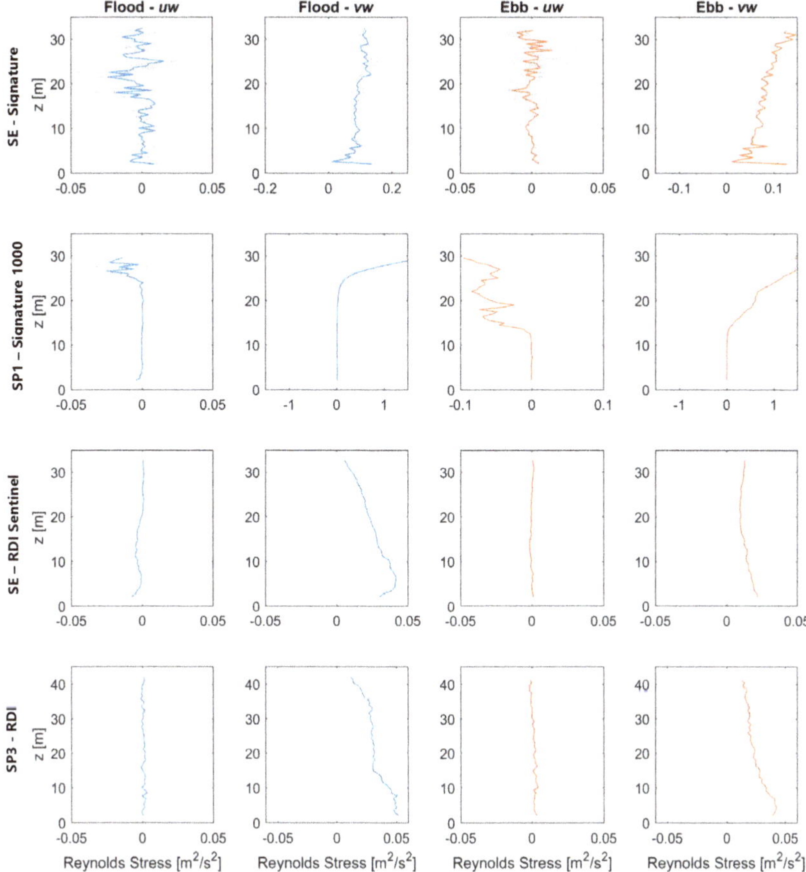

Figure 8. Reynolds stresses depth profiles for streamwise *uw* and transverse *vw* velocity components, where the flood tide is shown in blue and the ebb tide in red.

It should be noted that due to the large variation in stresses between sensors, the scale of the *x*-axis varies. As with TI and TKE, the comparison between sensors and locations shows considerable variations. For the SE—Signature 500 sensor a small amount of stress is transferred between the *uw* direction for both the flood and ebb tides. The transverse results show higher stresses, where more turbulence is transferred in the vertical velocity component. The SP1 sensor experiences similar stress magnitudes as the SE—Signature 500 sensor for the flood and ebb streamwise velocity components. However, the transverse stresses are much greater for the upper part of the water column. The RDI Sentinel sensors at SP3 and SE both show very similar behaviour with small shear stresses for the *uw* component and large vertical energy transfers in the transverse direction. For these profiles, larger stresses are observed towards the seabed. This shows the opposite results for the co-located SE sensors. The larger *vw* stresses indicate there is more vertical movement of turbulence caused by the transverse flow than the streamwise.

3.7. Integral Length and Timescales

The integral scales describe the time and physical length of the turbulent features observed. The time integral (T_i) describes the duration of the largest turbulent features. This is calculated using the integration of the autocorrelation function (R_{uu}) of the velocity fluctuation based on the method presented in [15]:

$$T_i = \int_{\tau=0}^{\tau(R_{uu}(\tau)=0)} R_{uu}(\tau)d\tau \tag{8}$$

$$R_{uu(\tau)} = \frac{R(u(t), u(t+\tau))}{\sigma_u^2} \tag{9}$$

where t is the time and σ_u is the variance of the streamwise velocity fluctuation. The time integral is obtained from the data between $\tau = 0$ to the zero crossing point of the autocorrelation function. If the assumption of Taylor's frozen turbulence hypothesis is invoked the integral length scales are calculated using the following equation where $\overline{u}$ is the mean streamwise velocity for each 600 s time period:

$$L_i = \overline{u}T_i \tag{10}$$

The integral time (left y-axis) and length (right y-axis) scales are presented in Figures 9 and 10 for the streamwise velocity component, with the units of time in seconds and length in meters. These values were calculated based on 10 min velocity fluctuations. Figure 9 shows the integral scales for a depth of 20 m above the seabed. This provides an average length scale of 18 m for the SE—Signature 500 sensor, 22 m for the SE-RDI Sentinel and 45 m for the SP3 sensor. The SP1 sensor experiences more consistent longer and larger flow structures for the ebb tides. Figure 10 plots this data for all depths in the time domain, where the left-hand side plots are the time scale and right-hand side plots are the length scale. This shows the turbulent flow structures as they move through the water column in time. Both SE sensors show a reduced number of turbulence structures during the ebb tide nearer the surface, for these time periods turbulent structures are largest towards the seabed. During the flood tide the SP1 sensor shows large turbulent length scales that are relatively insensitive to the water depth. While, the ebb tide displays a reduction in turbulent structure size for the lower part of the water column, with the larger structures being present in the upper half. The SP3 sensor measured larger turbulent structures that coincide with higher velocity flows i.e., these peaks occur during both the flood and the ebb tide and minimal size structures are recorded during slack water conditions. The vertical distribution of these flow structures shows a parabolic profile with the larger turbulent structures towards the seabed and the surface.

3.8. Length Scales and Homogeneous Assumption

Figures 9 and 10 present the measured integral length scale for all sensors using the streamwise flow vector. This principle is based on the assumption of flow homogeneity over the plane at which the streamwise velocity is calculated. This distance of the homogenous flow assumption is calculated using $d = 2z\sin(\theta)$, where z is the vertical distance from the sensor transceiver to the depth cell of interest and θ is the beam angle of the sensor. For a distance from the transceiver of 13 m and a beam angle of 25 degrees, the length d over which flow homogeneity is assumed equates to 11 m. This leaves a considerable number of the ambient length scales below this value, where they cannot be adequately resolved. This is shown in Figure 11 for a distance of 13 m away from the sensor head, where the length scales below this distance are indicated within the grey shading.

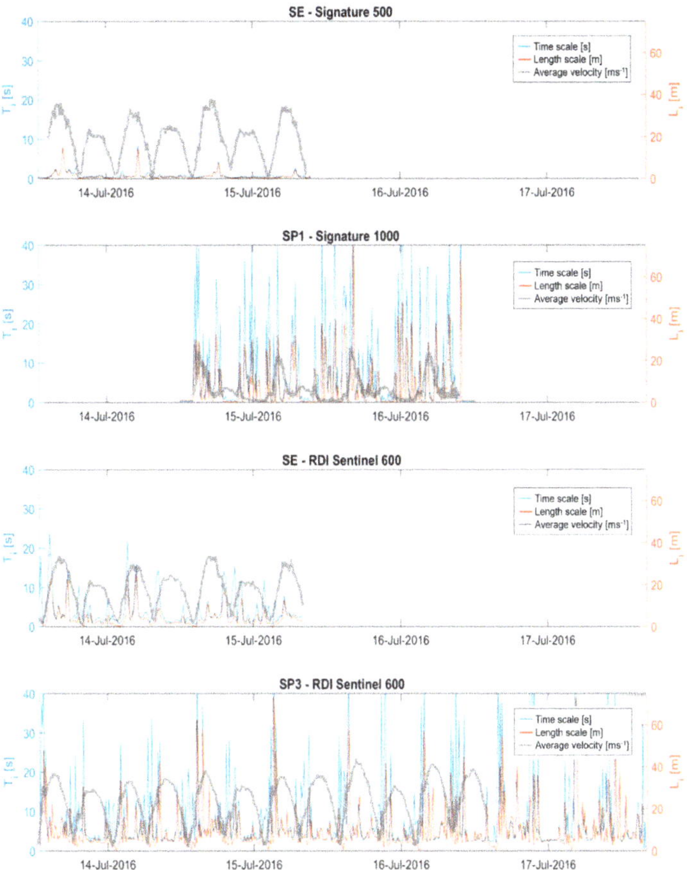

Figure 9. Integral time (blue) and length scales (red) for the streamwise velocity components. Based on measurements 20 m above the seabed for all sensors. The average velocity for each sensor is shown in grey.

The calculation of the individual beam length scales were also plotted as a reference value in Figure 11. This was only possible for the SE sensor location as the sensor head transducers were closely aligned with the flow direction for beams 1 and 3. Beam 1 had a heading of 143 degrees, and the flood equals 150 degrees. Beam 3 had a heading of 323 degrees, and the ebb tide had a flow direction of 320 degrees. It should be noted that the beam-wise velocity is directly taken in its raw state, where the measured speeds are at 25 and 20 degrees off from vertical for the Signature 500 and RDI Sentinel sensors respectively. This is compared to the streamwise flow conditions, which are shown as calculated for the horizontal flow velocity. The average velocity, over the 600 s sample periods, is also shown for each sensor to help indicate length scales relative to the tidal phase. This indicates a much larger length scale for the faster flood tide than for the slower ebb tide. The comparison between the streamwise and beam length scales shows similar values relative to the tidal phase. Large fluctuations in length scales are particularly present in the beam 3 data. The magnitudes of the streamwise and individual beam length scales show a good similarity, suggesting validity of both approaches. A good correlation is observed between the beam 1 and beam 3 length scales for the RDI Sentinel sensor, but less so for the Signature 500 sensor. Gimbal movement may cause a small misalignment in beams 1 and 3, causing the respective beams to sample slightly different regions of the flow, and this may to some extent explain the disparity between the length scales of beams 1 and 3. Table 2 supports this, where the mean pitch and roll is shown to be larger for the Signature 500 device. This means the beams are sampling slightly different regions of the water column, when compared to the RDI instrument. Compared to the Signature 500, the RDI instrument was positioned closer to the vertical axis. This implies a much more valid assumption of Taylor's frozen turbulence hypothesis, as the flow's streamlines pass through both beams at the same distance from the sensor RDI head.

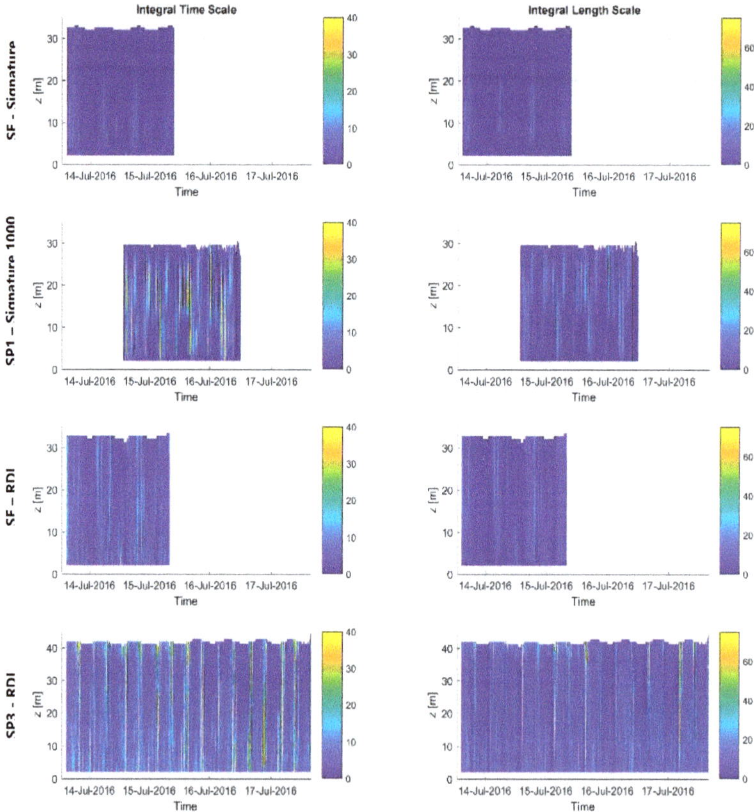

Figure 10. The time (left) and length (right) integral scales presented in time and water depth for all sensors. Colour scale units are presented in seconds for the left column of subplots and meters for the right column of subplots.

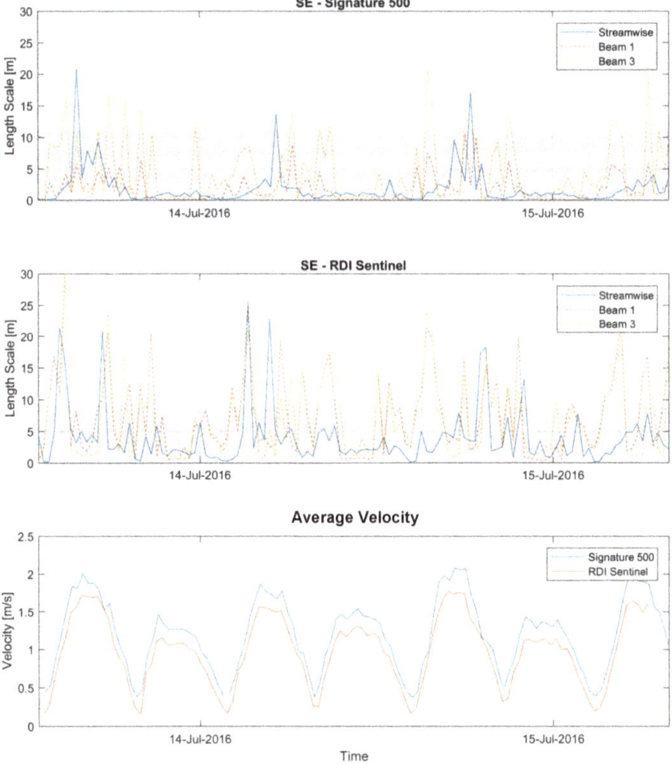

Figure 11. Streamwise, beam 1 and 3 length scales, 13 m above the seabed for both SE co-located sensors. The shaded area indicates length scales shorter than the length over which the homogeneity approximation applies for the given distance from the sensor.

4. Discussion

This study has highlighted the difference between the two co-located SE Signature 500 and Sentinel V50 sensors. These sensors were deployed at less than 1 m distance from each other within the same sensor frame. It was expected that the flow speeds and turbulence parameters for these sensors would record almost identical results. However, this was not the case. While exact sensor setup and sampling regime would provide a more robust comparison, the setups used were within reasonable tolerances conducive to turbulence measurements. These results should yield largely similar outcomes, where only the very high frequency turbulence components would differ. The sensor manufacturers, Nortek and Teledyne, were contacted in regards to data quality and measurement validation. This concluded that both SE located datasets were confirmed as credible data sources, with no significant defects or errors. Both hardware and firmware of both ADCPs has also been confirmed as healthy. Further comparison of the co-located sensor's raw data, showed the presence of the variations in flow profile features, as shown in Figure 4, in the streamwise orientated beam data. This suggests these differences in measurements are a result of the sensors. This measurement campaign does not provide an explanation of the cause of this variation, but brings them to the reader's attention. Further explanation and validation of these measurement variations, e.g., by an attempt to repeat the deployments, is limited due to the nature and cost of field measurements. Additional experimental work would benefit from collecting turbulence measurements in a laboratory flow tank, where more variables can be controlled.

The analysis of the results show the RDI Sentinel sensor having a much smoother velocity profile across the water column, whereas the Signature 500 experiences a much higher velocity variation. The data used in this study purposely extracted the sensor information in its rawest beam velocities format. This seems to be the case for the Signature 500 sensor, whereas the smooth profile of the RDI Sentinel 600 suggests that some form of internal data filtering has occurred. However, this was dismissed when contact was made with the manufacturer. When the average flow conditions are plotted, this initially assumed data cleaning shows very little effect. However, during the turbulence analysis, this apparent smoothing of the velocity variation causes significant effects to the turbulence parameters. A more direct comparison should be carried out, where the sensors sample rate and bin sizes remain constant and the sampling regime remains as similar as possible, with both sensor head aligned in the same orientation, so a direct beam-by-beam comparison is possible. However, the sampling regime will have to remain staggered to avoid sensor crosstalk. Based on the results presented and discussed in the above sections, it is recommended that a standard procedure is applied to the very basic data processing for the output of raw data for all sensor manufactures. This would require a classification of "raw data" that only allows very basic quality controls parameters to be applied. This would allow a reliable use of combined sensor types for turbulence quantification.

The results from the spatially distributed sensors all show turbulence levels that are consistent with previous literature with similar flow speeds for tidal energy test sites [11,15,19,34], where turbulence intensities are of the order of 0.1 for hub height water depths. This is with exception to the SP1 sensor, where a much more turbulent flow is shown due to wake effects from the southern point of Eday. This sensor experiences a very disrupted tidal flow pattern, where flow speeds during the ebb and later part of the flood tide are significantly reduced. This shows the spatial variability of the flow field, where large difference in flow and turbulence characteristics occur over a short distance. Based on these measurements it seems likely that a tidal device could operate without detriment as a result of turbulence, at the SE and SP3 locations. The SP3 sensor experiences significant turbulence and flow disruptions making it a poor location for turbine siting.

Further work should include longer term continuous turbulence measurements, where full spring, neap cycles are measured. Additional considerations may wish to address the impact of seasonality on turbulence characteristics, where winter storms bring large waves to exposed tidal sites, causing alteration to the turbulence depth profile.

5. Conclusions

This study investigates the spatial and temporal differences in turbulence in the southern entry to the Fall of Warness tidal energy test site in Orkney. Four high-resolution bottom-mounted diverging beam ADCPs were deployed to simultaneously collect flow characteristics for several nearby locations. The measurements were made on a waxing lunar phase, with tides midway between neaps and springs. The flow characteristics show an asymmetric velocity distribution, where maximum flood velocities of 2 ms^{-1} at 150° and ebb velocities of 1.2 ms^{-1} at 315° are present within the channel. The SP1 sensor, close to the easterly landmass in Figure 1, experiences flow interference midway through the flood tide that persists into the ebb tide, this provides much lower velocities with a wider directional variation.

The comparison of data from two sensors deployed in the multi-carrier frame at the same location provides different flow results and therefore different turbulence parameters. This provides uncertainty in the accuracy of turbulence measurements using ADCPs. Further comparisons between sensors should be tested in a more controlled environment, e.g., laboratory condition to identify the origins of these variations.

Several turbulence metrics are presented, where the variation in the velocity fluctuations are assessed in depth and time. The turbulence intensities show higher values towards the seabed, relative to the local velocities. The SP1 sensor shows large levels of turbulence near the surface for the flood tide and mid-water column for the ebb tide. This is supported by the presentation of the streamwise (uw) and transverse (vw) Reynold stresses. These stresses indicate larger vertical movement of turbulence in the transverse direction as opposed to the streamwise component; which generally increases towards the seabed. The time and length integral scales show longer and larger turbulent structures during the flood tides. These larger structures are located towards the seabed for the SP1 sensor, whereas for the SP3 sensor these structures are shown nearer the surface. The streamwise and beam length scales were compared at a likely depth of a turbine hub height. This showed that while the streamwise homogenous approximation should prevent the measurement of small (<10 m) length scales, the direct comparison of the individual beam data shows a similar agreement in the results. This shows similar magnitudes length scales across the flood and ebb tide, with some discrepancies occurring, where individual beam data provide larger values and higher frequency of fluctuations. The variation in turbulence between the deployment locations indicates that turbulence values can vary highly over relatively small distances. In order to quantify turbulence for a tidal energy test site for multiple numbers of devices, the deployment of several spatially distributed ADCPs must be required. Caution should be exercised when applying these turbulence conditions to other tidal energy test sites.

Further work focusses on extended deployment durations in obtaining high-resolution datasets. These extended measurement periods will provide more data to allow a more robust quantification of the turbulence parameters for an individual site. The presence of surface gravity waves will also be investigated as well as the application of the fifth vertical beam.

Author Contributions: This study was undertaken as a collaborative work between the authors. The design and implementation of the fieldwork and data acquisition was overseen by A.V. C.G. was leading on the quality control and processing of the data for turbulence and flow characteristics. As Principal Investigator of the FloWTurb project, V.V. has informed specific requirements in relation to fieldwork and processing. The writing of the manuscript was led by C.G., with section specific contributions end editorial support from A.V. and V.V.

Funding: This research was funded by EPSRC grant number EP/N021487/1 under the FloWTurb project.

Acknowledgments: The authors would like to acknowledge the FloWTurb: Response of Tidal Energy Converters to Combined Tidal Flow, Waves, and Turbulence (EPSRC Grant ref. number EP/N021487/1) project for supporting this work. Additional thanks go to Marine Scotland Science who supported the sensor deployment and recovery works with the RV Alba na Mara.

Conflicts of Interest: The authors declare no conflict of interest.

References

1. Blackmore, T.; Myers, L.E.; Bahaj, A.S. Effects of turbulence on tidal turbines: Implications to performance, blade loads, and condition monitoring. *Int. J. Mar. Energy* **2016**, *14*, 1–26. [CrossRef]
2. Milne, I.A.A.; Day, A.H.H.; Sharma, R.N.N.; Flay, R.G.J. The characterisation of the hydrodynamic loads on tidal turbines due to turbulence. *Renew. Sustain. Energy Rev.* **2016**, *56*, 851–864. [CrossRef]
3. Deacon, E.L. The Measurement of Turbulent Transfer in the Lower Atmosphere. *Adv. Geophys.* **1959**, *6*, 211–228.
4. Huntley, D.A. A Modified Inertial Dissipation Method for Estimating Seabed Stresses at Low Reynolds Numbers, with Application to Wave/Current Boundary Layer Measurements. *J. Phys. Oceanogr.* **1988**, *18*, 339–346. [CrossRef]
5. Sellar, B.; Harding, S.; Richmond, M. High-resolution velocimetry in energetic tidal currents using a convergent-beam acoustic Doppler profiler. *Meas. Sci. Technol.* **2015**, *26*, 085801. [CrossRef]
6. Teledyne RDI. *Sentinel V—Next Gen ADCP*; Teledyne RDI: La Gaude, France, 2015.
7. *Nortek Signature 1000/500—A High Performance Scientific Powerhouse*; Vangkroken: Bærum, Norway, 2015.
8. Lu, Y.; Lueck, R.G. Using a Broadband ADCP in a Tidal Channel. Part II: Turbulence. *J. Atmos. Ocean. Technol.* **1999**. [CrossRef]
9. Lohrmann, A. Introduction to AD2CP Turbulence Measurements—YouTube. Available online: https://www.youtube.com/watch?v=_WyTY4N7IvY&t=581s (accessed on 30 March 2018)
10. Thomson, J.; Polagye, B.; Durgesh, V.; Richmond, M.C. Measurements of Turbulence at Two Tidal Energy Sites in Puget Sound, WA. *IEEE J. Ocean. Eng.* **2012**, *37*, 363–374. [CrossRef]
11. Thomson, J.; Polagye, B.; Richmond, M.; Durgesh, V. Quantifying Turbulence for Tidal Power Applications.
12. Richard, J.-B.; Thomson, J.; Polagye, B.; Barda, J. Method for identification of Doppler noise levels in turbulent flow measurements dedicated to tidal energy. *Int. J. Mar. Energy* **2013**, *3–4*, 52–64. [CrossRef]

13. Durgesh, V.; Thomson, J.; Richmond, M.C.; Polagye, B.L. Noise correction of turbulent spectra obtained from acoustic doppler velocimeters. *Flow Meas. Instrum.* **2014**, *37*, 29–41. [CrossRef]

14. Osalusi, E.; Side, J.; Harris, R. Structure of turbulent flow in EMEC's tidal energy test site. *Int. Commun. Heat Mass Transf.* **2009**, *36*, 422–431. [CrossRef]

15. Milne, I.A.; Sharma, R.N.; Flay, R.G.J.; Bickerton, S. Characteristics of the turbulence in the flow at a tidal stream power site. *Philos. Trans. A Math. Phys. Eng. Sci.* **2013**, *371*, 20120196. [CrossRef] [PubMed]

16. Bouferrouk, A.; Hardwick, J.P.; Colucci, A.M.; Johanning, L. Quantifying turbulence from field measurements at a mixed low tidal energy site. *Renew. Energy* **2016**, *87*, 478–492. [CrossRef]

17. Milne, I.A.; Sharma, R.N.; Flay, R.G.J. The structure of turbulence in a rapid tidal flow. *Proc. R. Soc. A Math. Phys. Eng. Sci.* **2017**, *473*, 20170295. [CrossRef] [PubMed]

18. Sellar, B.; Wakelam, G.; Sutherland, D.; Ingram, D.; Venugopal, V. Characterisation of Tidal Flows at the European Marine Energy Centre in the Absence of Ocean Waves. *Energies* **2018**, *11*, 176. [CrossRef]

19. Guerra, M.; Thomson, J.; Guerra, M.; Thomson, J. Turbulence Measurements from Five-Beam Acoustic Doppler Current Profilers. *J. Atmos. Ocean. Technol.* **2017**, *34*, 1267–1284. [CrossRef]

20. McMillan, J.M.; Hay, A.E.; Lueck, R.G.; Wolk, F.; McMillan, J.M.; Hay, A.E.; Lueck, R.G.; Wolk, F. Rates of Dissipation of Turbulent Kinetic Energy in a High Reynolds Number Tidal Channel. *J. Atmos. Ocean. Technol.* **2016**, *33*, 817–837. [CrossRef]

21. Zippel, S.F.; Thomson, J.; Farquharson, G. Turbulence from Breaking Surface Waves at a River Mouth. *J. Phys. Oceanogr.* **2018**, *48*, 435–453. [CrossRef]

22. Konsoer, K.M.; Rhoads, B.L. Spatial–temporal structure of mixing interface turbulence at two large river confluences. *Environ. Fluid Mech.* **2014**, *14*, 1043–1070. [CrossRef]

23. Muste, M.; Yu, K.; Pratt, T.; Abraham, D. Practical aspects of ADCP data use for quantification of mean river flow characteristics; Part II: Fixed-vessel measurements. *Flow Meas. Instrum.* **2004**, *15*, 17–28. [CrossRef]

24. Horwitz, R.M.; Hay, A.E. Turbulence dissipation rates from horizontal velocity profiles at mid-depth in fast tidal flows. *Renew. Energy* **2017**, *114*, 283–296. [CrossRef]

25. *GL Garrad Hassan Tidal Bladed Multibody Dynamics User Manual 2010*; Garrad Hassan and Partners: Bristol, UK, 2010.

26. UK Hydrographic Office INSPIRE portal and MEDIN Bathymetry Data Archive Centre. Available online: https://www.gov.uk/guidance/inspire-portal-and-medin-bathymetry-data-archive-centre (accessed on 2 April 2018).

27. Greenwood, C.; Vögler, A.; Morrison, J.; Murray, A. Spatial Velocity Measurements using X-Band Radar at a High Energy Tidal Test Site. In Proceedings of the European Wave and Tidal Energy Conferences, Cork, Ireland, 30 July 2017.

28. Nortek How Is a Coordinate Transformation Done? Available online: https://www.nortekgroup.com/faq/how-is-a-coordinate-transformation-done (accessed on 29 March 2018).

29. *Teledyne RDI ADCP Coordinate Transformation: Formulas and Calculations*; Teledyne RD Instruments: Poway, CA, USA, 2008.

30. Kolmogorov, A.N. Dissipation of Energy in the Locally Isotropic Turbulence. *Proc. Math. Phys. Sci.* **1991**, *434*, 15–17. [CrossRef]

31. British Standards Institute. *Marine Energy—Wave, Tidal and Other Water Current Converters Part 201: Tidal Energy Resource Assessment and Characterization*; International Electrotechnical Commission: Geneva, Switzerland, 2015.

32. Kolmogorov, A.N. The Local Structure of Turbulence in Incompressible Viscous Fluid for Very Large Reynolds' Numbers. *Dokl. Akad. Nauk SSSR* **1941**, *30*, 301–305. [CrossRef]

33. Kaneda, Y.; Ishihara, T.; Yokokawa, M.; Itakura, K.; Uno, A. Energy dissipation rate and energy spectrum in high resolution direct numerical simulations of turbulence in a periodic box. *Phys. Fluids* **2003**, *15*, L21–L24. [CrossRef]

34. Coles, D.; Greenwood, C.; Vogler, A.; Walsh, T.; Taaffe, D. Assessment of the turbulent flow upstream of the Meygen Phase 1a tidal stream turbines. In Proceedings of the Asian Wave and Tidal Energy Conference, Taipei, Taiwan, 9–13 September 2018.

Article

Implementation of Open Boundaries within a Two-Way Coupled SPH Model to Simulate Nonlinear Wave–Structure Interactions

Tim Verbrugghe [1], Vasiliki Stratigaki [1,*], Corrado Altomare [1], J. M. Domínguez [2], Peter Troch [1] and Andreas Kortenhaus [1]

[1] Department of Civil Engineering, Ghent University, Technologiepark 904, Zwijnaarde B-9052, Belgium; vicky.stratigaki@ugent.be (V.S.); Corrado.altomare@ugent.be (C.A.); peter.troch@ugent.be (P.T.); Andreas.kortenhaus@ugent.be (A.K.)

[2] EPHYSLAB Environmental Physics Laboratory, Universidade de Vigo, AS LAGOAS, Ourense 32004, Spain; jmdalonso@gmail.com

* Correspondence: vicky.stratigaki@ugent.be; Tel.: +32-9-264-54-89; Fax: +32-9-264-58-37

Received: 28 December 2018; Accepted: 15 February 2019; Published: 21 February 2019

Abstract: A two-way coupling between the Smoothed Particle Hydrodynamics (SPH) solver DualSPHysics and the Fully Nonlinear Potential Flow solver OceanWave3D is presented. At the coupling interfaces within the SPH numerical domain, an open boundary formulation is applied. An inlet and outlet zone are filled with buffer particles. At the inlet, horizontal orbital velocities and surface elevations calculated using OceanWave3D are imposed on the buffer particles. At the outlet, horizontal orbital velocities are imposed, but the surface elevation is extrapolated from the fluid domain. Velocity corrections are applied to avoid unwanted reflections in the SPH fluid domain. The SPH surface elevation is coupled back to OceanWave3D, where the originally calculated free surface is overwritten. The coupling methodology is validated using a 2D test case of a floating box. Additionally, a 3D proof of concept is shown where overtopping waves are acting on a heaving cylinder. The two-way coupled model (exchange of information in two directions between the coupled models) has proven to be capable of simulating wave propagation and wave–structure interaction problems with an acceptable accuracy with error values remaining below the smoothing length h_{SPH}.

Keywords: wave–structure interaction; wave propagation model; smoothed particle hydrodynamics; open boundaries; coupling; DualSPHysics; OceanWave3D

---

1. Introduction

Smoothed Particle Hydrodynamics (SPH), a mesh-less method that describes the fluid as a set of discrete elements, named particles, is typically computationally very intensive. However, recent advances using High Performance Computing (HPC) and Graphical Processing Units (GPU) have strongly contributed to significant gains in computational effort [1]. Despite the use of HPC and GPUs, it is still challenging to model realistic engineering problems, which are usually multi-scale problems. This research tries to mitigate this problem statement by studying the possible reduction of the required SPH computational domain. This can be done by coupling SPH to a faster external numerical model which can deal easily with large computational domains. This requires accurate and stable boundary conditions. Both the development of accurate boundary conditions and the coupling of SPH to external models are part of the SPHERIC Grand Challenges [2], which list the key issues to be addressed in order to make SPH a mature method. In literature, there are several research examples where coupling was applied involving SPH methods. A general algorithm for one-way (exchange of information in only one direction) coupling of SPH with an external solution has been proposed in Bouscasse et al. [3]. The interaction between the SPH solver and the external solution is achieved through an interface region containing a so-called "ghost fluid", used to impose any external boundary condition. In Fourtakas et al. [4], a hybrid Eulerian-Lagrangian incompressible SPH formulation is introduced, where two different SPH formulations are coupled, rather than two completely different solvers. The SPH solver DualSPHyics has been coupled in Altomare et al. [5] and Altomare et al. [6], where a one-way coupling was realized with the wave propagation model SWASH [7]. A numerical wave flume has been created to simulate wave impact and run-up on a breakwater. The first part of the used numerical flume is simulated using the faster SWASH model, while the wave impact and run-up are calculated using DualSPHysics. Here, a one-way coupling is sufficient, since there is only interest in the impact of waves on the breakwater. In Kassiotis et al. [8], a similar approach has been adopted, where a 1D Boussinesq-type model is applied for wave propagation in the largest part of the spatial domain, and SPH computations focus on the shoreline or close to off-shore structures, where a complex description of the free-surface is required. In Narayanaswamy et al. [9], the Boussinesq model FUNWAVE [10] was coupled to DualSPHysics, where the key development was the definition of boundary conditions for both models in the overlap zone. A wave generator in SPH moved according to the velocities from the adjacent Boussinesq nodes. Similarly, an incompressible SPH solver has been coupled to a nonlinear potential flow solver QALE-FEM [11] in Fourtakas et al. [12]. In Chicheportiche et al. [13], a one-way coupling between a potential Eulerian model and an SPH solver is realised, applying a non-overlapping method using the unsteady Bernoulli equation at the interface. These studies applied coupling to speed up the simulation time by minimizing the computationally intensive SPH domain. Other studies apply coupling to combine both the benefits of mesh-based and mesh-less Computational Fluid Dynamics (CFD) methods. In Didier et al. [14], the wave propagation model FLUINCO [15] is coupled to an SPH code, and validated with experimental data of wave impact on a porous breakwater. A hybrid multiphase OpenFOAM-SPH model is presented in Kumar et al. [16], where the SPH method is used on free surfaces or near deformable boundaries, whereas OpenFOAM is used for the larger fluid domain. A similar coupling is used, where breaking waves are modelled with SPH and the deeper wave kinematics are modelled with a Finite Volume (FV) method. This has been demonstrated in Marrone et al. [17] for a Weakly-Compressible SPH (WCSPH) solver and in Napoli et al. [18] for an Incompressible SPH (ISPH) solver. This research focuses on applying open boundaries in a two-way coupling methodology (exchange of information in two directions between the coupled models) between the Fully Nonlinear Potential Flow (FNPF) wave propagation solver OceanWave3D [19] and the WCSPH solver DualSPHysics [20]. A first version of this coupling has been introduced in Verbrugghe et al. [21]. However, instead of open boundaries, moving boundaries were applied to transfer the orbital velocities from the wave propagation model to the SPH solver.

Typically, the numerical domain for wave propagation modelling in DualSPHysics is at least 3–4 wavelengths long [5]. Combined with a required small particle size to accurately reproduce the surface elevation, this leads to computationally intensive simulations. This research is aimed at reducing the necessary fluid domain, and providing accurate boundary conditions capable of active wave generation and wave absorption by applying a coupling with a wave propagation model. In this manner, realistic open sea conditions can be simulated where waves enter on the left-hand-side of the fluid domain and exit freely on the right-hand-side of the fluid domain. The WCSPH solver DualSPHysics and the FNPF solver OceanWave3D are used here to demonstrate the coupling methodology, using the recently developed open boundaries [22–24]. The open boundary formulation applies so-called "buffer zones" containing layers of buffer particles, positioned adjacent to the fluid domain. Buffer particles are used to enforce certain conditions in the presence of fluid inlets and outlets. Particularly, the physical information of buffer particles can be imposed by the user a priori or can be extrapolated from the fluid domain with a procedure, which is first-order consistent.

Although these open boundaries are similar to what was presented by Ni et al. [25], there are some key differences making the formulation used here by Tafuni et al. [24] more flexible. Firstly, flow reversion problems can not be simulated with the method by Ni et al. [25]. Secondly, there is no possibility to extrapolate flow quantities using ghost nodes. Thirdly, the method to impose free surface elevation is different. Fourthly, the applied velocity profiles and corrections are depth-averaged. Lastly, only 2nd order wave generation is possible, where the method introduced here is compatible with up to 5th order generation.

Applying open boundaries for wave generation and wave absorption is meant to cover those cases where classical wave generation techniques can fail or are very computationally expensive, e.g., open sea states, simulating floating structures and energy devices, wave breaking conditions, etc. Additionally, the buffer zones in the open boundaries accept physical information from any source: e.g., linear wave theory, nonlinear wave theories, external numerical models such as CFD models, or even measurement data.

The presented two-way coupling methodology expands the current SPH state-of-the art with the following features:

- Accurate wave generation and wave absorption through coupling an SPH solver to an FNPF solver using the open boundary formulation by Tafuni et al. [24],
- Having an online exchange of information in two directions between the SPH solver and the FNPF solver.

This paper is structured as follows: in Section 2, the theoretical background of the SPH method is given, while Section 3 is focusing on the specific boundary conditions available in DualSPHysics which are applied in this research. Section 4 introduces the coupling methodology between the wave propagation model OceanWave3D and the SPH solver DualSPHysics. This methodology is demonstrated and validated in 2D in Section 5 and a proof-of-concept in 3D is demonstrated in Section 6. Finally, some concluding remarks are given in Section 7.

2. Smoothed Particle Hydrodynamics

The solver used for the detailed modelling of the wave–structure interactions is DualSPHysics [20]. DualSPHysics applies the SPH formulation, a mesh-less method that describes the fluid as a set of discrete elements, named particles. The following section explains the general SPH theory behind Dualsphysics, as reported in Crespo et al. [26].

2.1. SPH Fundamentals

The physical properties of a particle a, determined by the Navier–Stokes equations, can be calculated by interpolation of the values of the nearest neighbouring particles. The contribution of the neighbouring particles is weighted, based on their distance to particle a, using a kernel function W, and a smoothing length h_{SPH}. When a particle is at a distance larger than $2h_{SPH}$ away from particle a, the contribution can be neglected.

Fundamentally, any function $F(\mathbf{r})$, defined in the distance between two particles $\mathbf{r}'$, is estimated by integral approximation:

$$F(\mathbf{r}) = \int F(\mathbf{r}')W(\mathbf{r} - \mathbf{r}', h_{SPH})d\mathbf{r}. \tag{1}$$

In order to numerically solve Equation (1), discretization is necessary. In its discrete form, the integral approximation transforms into an interpolation at a given location (or particle a) and a summation over all the particles within the region defined by the kernel support:

$$F(\mathbf{r}_a) \approx \sum_b F(\mathbf{r}_b)W(\mathbf{r}_a - \mathbf{r}_b, h_{SPH})\Delta V_b. \tag{2}$$

Here, ΔV_b is the volume of the neighbouring particle b. If $\Delta V_b = m_b/\rho_b$, with m_b and ρ_b being the mass and density of particle b, then Equation (2) becomes:

$$F(\mathbf{r}_a) \approx \sum_b F(\mathbf{r}_b)\frac{m_b}{\rho_b}W(\mathbf{r}_a - \mathbf{r}_b, h_{\text{SPH}}). \tag{3}$$

The choice of the kernel function has a large influence on the performance of the SPH method. The kernel is expressed as a function of the non-dimensional distance between particles $q = r/h_{\text{SPH}}$. Here, r is the distance between a certain particle a and a particle b. The smoothing length h_{SPH} defines the area around particle a in which the contribution of neighbouring particles is considered. In this work, to ensure stability with a high number of particles, a Quintic kernel is applied [27] with an influence domain of $2h_{\text{SPH}}$, defined as:

$$W(q, h_{\text{SPH}}) = \alpha_D \left(1 - \frac{q}{2}\right)^4 (2q + 1) \qquad 0 \le q \le 2. \tag{4}$$

In DualSPHyics, α_D is set to $7/4\pi h_{\text{SPH}}^2$ (2D).

2.2. Governing Equations

The governing equations to model fluid dynamics are the Navier–Stokes equations. In their SPH formulation, the momentum conservation is expressed as:

$$\frac{d\mathbf{v}_a}{dt} = -\sum_b m_b \left(\frac{P_b}{\rho_b^2} + \frac{P_a}{\rho_a^2} + \Pi_{ab}\right) \nabla_a W_{ab} + \mathbf{g}. \tag{5}$$

In Equation (5), P_a and P_b is the pressure of the particle a and b respectively, while ρ_a and ρ_b are the density of the particle a and b, respectively. The viscosity term Π_{ab} is based on the artificial viscosity scheme, as proposed by Monaghan [28]. It is a common method used in SPH to introduce viscosity, mainly due to its simplicity. It is defined as:

$$\Pi_{ab} = \begin{cases} \frac{-\alpha \overline{c_{ab}} \mu_{ab}}{\overline{\rho_{ab}}} & \mathbf{v}_{ab} \cdot \mathbf{r}_{ab} < 0, \\ 0 & \mathbf{v}_{ab} \cdot \mathbf{r}_{ab} > 0. \end{cases} \tag{6}$$

With the mean density $\overline{\rho_{ab}} = 0.5(\rho_a + \rho_b)$, the distance between particle a and b as $r_{ab} = r_a - r_b$ and the velocity difference as $\mathbf{v}_{ab} = \mathbf{v}_a - \mathbf{v}_b$. The mean speed of sound is denoted $\overline{c_{ab}}$ and α is a coefficient that needs to be set by the user to ensure a proper dissipation of density fluctuations. In this work, the value of α is set to 0.01, based on Altomare et al. [29], where wave propagation and wave loadings on coastal structures were studied.

DualSPHysics applies a WCSPH formulation. This means that the mass m of every particle is kept constant, while only their density ρ fluctuates. These fluctuations are calculated by solving the continuity equation, expressing the conservation of mass. In WCSPH formulation, this is defined by:

$$\frac{d\rho_a}{dt} = \sum_b m_b \mathbf{v}_{ab} \cdot \nabla_a W_{ab}. \tag{7}$$

Using a weighted summation of the mass terms would result in a density decrease in the interface between fluids, near the free surface and close to the boundaries. For this reason, a time differential is used, as suggested in Monaghan [28].

One of the main reasons for large computation times in WCSPH is the necessity for a very small time step Δt due to the inclusion of the speed of sound c. However, the compressibility can be adapted by artificially setting the speed of sound c to a lower value, resulting in a reasonable Δt. This enables the use of an equation of state to determine the pressure P of the fluid. This method is considerably faster than solving the Poisson's equation, appearing in an incompressible SPH (ISPH) approach. Some ISPH benchmark cases can be found in Shao and Lo [30,31]. According to Monaghan [32] and Batchelor [33], the relationship between density ρ and pressure P follows Tait's equation of state; a small density oscillation will lead to large pressure variations:

$$P = B\left[\left(\frac{\rho}{\rho_0}\right)^{\gamma} - 1\right]. \tag{8}$$

Here, B is related to the compressibility of the fluid, while ρ_0 is the reference density of the fluid, which is set to 1000 kg/m^3 in this work. The parameter γ is the polytrophic constant, ranging between 1 and 7. The maximum limit for ρ is set as $B = c^2\rho_0/\gamma$. Consequently, the choice of B is of high importance, since it determines the value of c. As mentioned before, c can be artificially lowered to ensure a reasonable time step [32]. However, in DualSPHysics, c is kept at least 10 times higher than the maximum expected flow velocity **v**.

The time integration of Equations (5) and (7) can be performed using a Verlet scheme or a two-stage Symplectic method. The latter is time reversible in the absence of friction or viscous effects [34]. In this work, both schemes are applied, where the explicit second-order Symplectic scheme has an accuracy in time of $\mathcal{O}(\Delta t^2)$ and involves a predictor and corrector stage. An explicit time integration scheme is applied, depending on the Courant–Friedrichs–Lewy (CFL) number, the force terms and the viscous diffusion term. This results in a variable time step Δt, calculated according to Monaghan and Kos [35].

2.3. Delta-SPH Formulation

The state equation mentioned in Section 2.2 describes a very stiff fluid density field. Unfortunately, this can lead to high-frequency low-amplitude oscillations in the density scalar field [36]. This effect is enlarged by the natural disordering of the Lagrangian particles. In order to mitigate these pressure fluctuations, a delta-SPH formulation can be applied. This is performed by adding a diffusive term to the continuity Equation (7), which was originally introduced by Molteni and Colagrossi [36]:

$$\frac{d\rho_a}{dt} = \sum_b m_b\mathbf{v}_{ab} \cdot \nabla_a W_{ab} + 2\delta_\Phi hc_0 \sum_b (\rho_b - \rho_a)\frac{r_{ab} \cdot \nabla_a W_{ab}}{r_{ab}^2}\frac{m_b}{\rho_b}. \tag{9}$$

Here, δ_Φ is a coefficient that needs to be appropriately selected by the user. Physically, the delta-SPH formulation can be defined as adding the Laplacian of the density field $\nabla^2\rho$ to the continuity equation. The influence of this added term in the continuity equation has been carefully studied by Antuono et al. [37]. There, the convergence was analysed by decomposing the Laplacian operator, together with a linear stability analysis to investigate the influence of δ_Φ. Within the fluid domain bulk, Equation (9) represents an exact diffusive term. However, close to open boundaries such as the free surface, the behaviour changes. There, the kernel is truncated (there are no particles sampled outside of an open boundary), which results in a net first-order contribution [37]. Consequently, a net force is applied to the particles. For non-hydrostatic situations, this force is not considered relevant, since the magnitude is negligible with respect to any other involved forces. Antuono et al. [37] did propose corrections to this effect, but they require a large computational cost since the correction involves the solution of a re-normalization problem for the fluid density gradient. Within this work, the recommended delta-SPH (δ_Φ) coefficient of 0.1 [20] is applied within DualSPHysics.

3. Boundary Conditions in DualSPHysics

3.1. Open Boundary Conditions

The implementation of open boundaries in DualSPHysics is discussed in detail in Tafuni et al. [24]. Inflow and outflow buffer zones are defined near the inlets and outlets of the computational domain of DualSPHysics.

The implemented open boundary condition is illustrated in Figure 1, where a fluid is flowing near a buffer zone, implemented as an open boundary. The buffer zone is located in between the domain edge and the threshold boundary, which separates the fluid domain from the buffer zone. The buffer zone is filled with layers of SPH buffer particles on which boundary conditions can be imposed. The buffer size should be at least equal to or larger than the kernel radius. This is necessary to have full kernel support for the fluid particles close to the threshold boundary. In the present research, the buffer zone width is chosen as $8 \cdot d_p$ in the direction normal to the open boundary, where d_p is the particle size adopted in DualSPHysics. There are two possibilities to provide physical information to the buffer particles: physical quantities are either imposed by the user or extrapolated from the fluid domain using so-called "ghost nodes". As illustrated in Figure 1, the positions of the ghost nodes are calculated through mirroring of the buffer particle locations into the fluid domain, this along a direction which is perpendicular to the open boundary. When the fluid quantities are calculated at the ghost nodes, a standard SPH particle interpolation would not be consistent. Due to the proximity to an open boundary, the kernel would be truncated. Therefore, the method proposed by Liu and Liu [38] is applied to obtain first-order kernel and particle consistency. More specifically, a multi-dimensional, first-order Taylor series approximation of the field function $f(\mathbf{x})$ is multiplied by the kernel function evaluated at particle k, $W_k(\mathbf{x})$. The series approximation and its first order derivatives, $W_{k,\beta}(\mathbf{x})$, are given by:

$$\int f(\mathbf{x})W_k(\mathbf{x})d\mathbf{x} = f_k \int W_k(\mathbf{x})d\mathbf{x} + f_{k,\beta} \int (\mathbf{x} - \mathbf{x}_k)W_k(\mathbf{x})d\mathbf{x}, \tag{10}$$

$$\int f(\mathbf{x})W_{k,\beta}(\mathbf{x})d\mathbf{x} = f_k \int W_{k,\beta}(\mathbf{x})d\mathbf{x} + f_{k,\beta} \int (\mathbf{x} - \mathbf{x}_{k,\beta})W_{k,\beta}(\mathbf{x})d\mathbf{x}. \tag{11}$$

Here, β is an index ranging from 1 to δ, the amount of dimensions. A system of $\delta + 1$ equations in $\delta + 1$ unknowns, f_k and $f_{k,\beta}$ is formed. The solution to the system is given in particle notation:

$$f_k = \frac{\begin{vmatrix} \sum_i f_i W_{ki} \Delta V_i & \sum_i (\mathbf{x}_i - \mathbf{x}_k) W_{ki} \Delta V_i \\ \sum_i f_i W_{ki,\beta} \Delta V_i & \sum_i (\mathbf{x}_i - \mathbf{x}_k) W_{ki,\beta} \Delta V_i \end{vmatrix}}{\begin{vmatrix} \sum_i f(\mathbf{x}) W_{ki} \Delta V_i & \sum_i (\mathbf{x}_i - \mathbf{x}_k) W_{ki} \Delta V_i \\ \sum_i f(\mathbf{x}) W_{ki,\beta} \Delta V_i & \sum_i (\mathbf{x}_i - \mathbf{x}_k) W_{ki,\beta} \Delta V_i \end{vmatrix}}, \tag{12}$$

$$f_{k,\beta} = \frac{\begin{vmatrix} \sum_i W_{ki} \Delta V_i & \sum_i f_i W_{ki} \Delta V_i \\ \sum_i W_{ki,\beta} \Delta V_i & \sum_i f_i W_{ki,\beta} \Delta V_i \end{vmatrix}}{\begin{vmatrix} \sum_i W_{ki} \Delta V_i & \sum_i (\mathbf{x}_i - \mathbf{x}_k) W_{ki} \Delta V_i \\ \sum_i W_{ki,\beta} \Delta V_i & \sum_i (\mathbf{x}_i - \mathbf{x}_k) W_{ki,\beta} \Delta V_i \end{vmatrix}}. \tag{13}$$

Next, the value of f_o at the open boundary can then be found using the corrected values of f_k and $f_{k,\beta}$ at the ghost nodes:

$$f_o = f_k + (\mathbf{r}_o - \mathbf{r}_k) \cdot \widetilde{\nabla} f_k. \tag{14}$$

Here, $\widetilde{\nabla} f_k$ is the corrected gradient at the ghost nodes.

The open boundary algorithm introduces several new features that make SPH more applicable to real engineering problems. The first one is the possibility of using buffer zones to impose time-varying velocity and pressure profiles, as well as pressure and velocity gradients along a specified direction. Next, a variable free-surface elevation can be imposed, which is an essential prerequisite in free-surface flow problems where waves can enter and exit the computational domain. Finally, the buffer zones are characterised by a dual behaviour, allowing both inward and outward flows, making flow reversion possible. Consequently, when flow velocities are extrapolated from the fluid domain, mixed velocity fields are possible where part of the buffer zone contains fluid particles entering the domain, and another part contains fluid particles leaving the domain. This can be specifically important in studying flow problems where modelling of strong rotations or oscillations is necessary. This flexibility is an important distinctive feature. The open boundary algorithm is available on both the parallel CPU and GPU versions of DualSPHysics. This results in considerable speed-ups when the code is running on powerful GPUs or large CPU clusters. This is particularly necessary when simulating real engineering problems. There, a large number of particles is required to simulate high-resolution flow problems with complex geometries, while maintaining a reasonable computation time.

3.2. Fixed Boundary Condition with Pressure Correction

Within DualSPHysics, fixed boundaries are normally realised using a set of particles called dynamic boundary particles [39]. These dynamic boundary conditions (DBC) have the advantage of being applicable to arbitrary 2D and 3D shapes, and provide good validation in many engineering problems. However, they are prone to unphysical fluid density and pressure values. Additionally, they exert high repulsive forces on the fluid particles, resulting in a separation distance. Within this work, a correction is applied to the dynamic boundary particles which resolves these issues. The DBC are here approached as a special case of an open boundary, more specifically one where the velocity of the buffer particles is zero, and the pressure is extrapolated from the fluid domain. This is similar to the approach by Marrone et al. [40]. The applied correction leads to less pressure oscillations in the fluid domain, and solves the occurrence of an unphysical gap between the boundary particles and the fluid. A downside of this method is the larger computation time, since both the continuity and momentum equations need to be solved for DBC and a unit vector, normal to the DBC, needs to be calculated for each boundary particle.

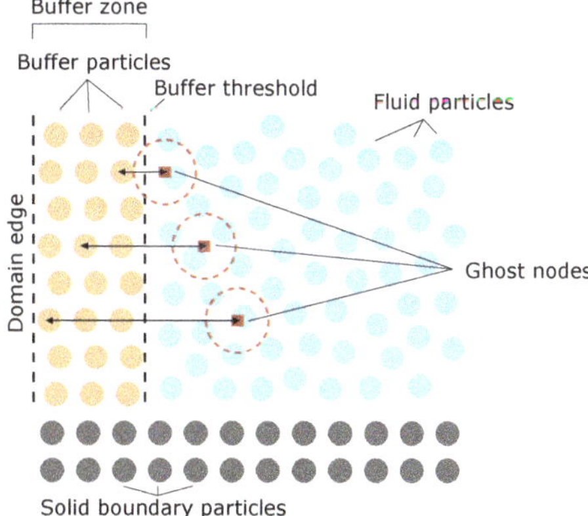

Figure 1. Sketch of the implemented open boundary model, adapted from Tafuni et al. [24].

3.3. Comparison to the State-of-the-Art

A very similar implementation of the open boundaries presented in this work has been introduced by Ni et al. [25], as was already mentioned in the Introduction. This section explains the main differences to the open boundaries as implemented by Tafuni et al. [24] in more detail.

The main similarity between both methods is the use of buffer particles on which physical quantities such as velocity and surface elevation can be imposed. However, there are a few importance differences to be listed. Firstly, according to Ni et al. [25], the open boundaries are either an inlet or an outlet, while the implementation by Tafuni et al. [24] is more flexible and allows flow in both directions within one buffer zone. This makes it applicable to flow-reversion problems. Secondly, there are no ghost nodes to extrapolate physical quantities from the fluid domain to the buffer particles. Thirdly, the method to impose variable free surface in the buffer zone is different. Inter-particle distance is stretched out for fluctuations smaller than dp. When the necessary change is equal to or larger than dp, an extra row of particles is added. Here, the inter-particle distance in the vertical direction is kept constant, and a new row of particles is added when necessary. Fourthly, the applied active wave absorption at the inlet is similar to what is used in this work, but they impose a uniform incident velocity profile instead of correcting a variable profile with a uniform correction. Lastly, only first-order wave theory is used for the regular waves, as well as a simple solitary wave. The method presented here allows Stokes 5th order wave theory to be used as well as external input from OceanWave3D to produce accurate nonlinear waves.

4. Coupling Methodology Using DualSPHysics and OceanWave3D

As mentioned before, SPH simulations are very computationally intensive. The flow and pressure fields required from a wave energy converter (WEC) SPH model are often limited to a zone closely spaced around the floating WEC. However, there is a spatial need to allow for proper wave generation and wave absorption, around 3–4 wavelengths long. This leads to a significant increase in water particles, and thus higher computation times. Moreover, wave generation techniques available in DualSPHysics are limited to first and second order wave generation by using piston-type or flap-type wave paddles. This generation type requires a certain propagation length before the full kinematics and surface elevation are developed. Within this research, the objective is to simulate higher-order (up to fifth-order) regular and irregular long-crested waves in a domain which is as small as possible. In an attempt to answer both the problem of computational performance and the problem of wave generation, a coupling methodology as illustrated in Figure 2 is developed.

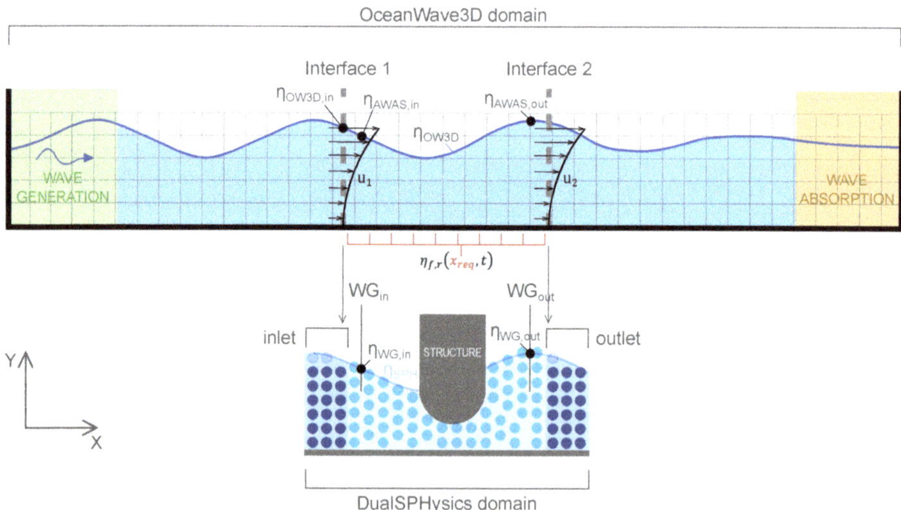

Figure 2. General sketch of coupling methodology between the OceanWave3D domain (**top**) and the DualSPHysics domain (**bottom**).

Here, a DualSPHysics fluid domain with a length of one wave length L_{wav} is chosen, with an inlet on the left-hand-side of the domain and an outlet on the right-hand-side of the domain (see Figure 2). Each buffer zone consists of eight layers of buffer particles. A sensitivity analysis illustrated in Figure 3 has shown that wave propagation results are accurate for buffer zones with at least eight layers. The dimensionless ratio $K_D = \frac{H_{measured}}{H_{imposed}}$, with H the wave height, is shown as a function of the normalized position x/L_{wav} for a Stokes third-order wave. The number of buffer particle layers n_l is varied from 1 to 16 and is doubled at each iteration.

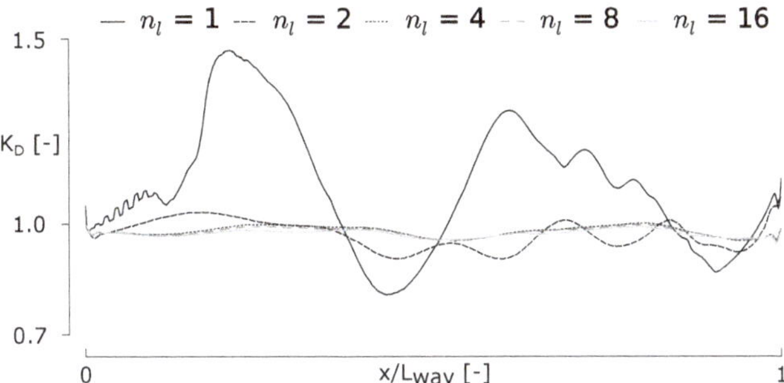

Figure 3. Sensitivity analysis on the number of buffer particle layers n_l necessary for accurate wave propagation, represented by the ratio K_D along the fluid domain.

The OceanWave3D domain is considerably larger and is often chosen as a multitude of wave lengths (e.g., 10). The wave generation zone measures around $2 \cdot L_{wav}$ while the wave absorption zone is taken as at least $3 \cdot L_{wav}$. The grid size d_x is set as a multitude of the SPH inter-particle distance $d_x = n \cdot d_p$. Here, n is set to be equal to 5. The physical quantities imposed on the SPH particle then originate from OceanWave3D. At the inlet, horizontal orbital velocities u_1 and surface elevation $\eta_{OW3D,in}$ originating from interface 1 are imposed on the buffer particles, while the pressure is set to be hydrostatic. At the outlet, only the horizontal orbital velocities u_2 originating from interface 2 are imposed, the surface elevation and pressure are extrapolated from the fluid domain (see Table 1). No vertical orbital velocities are applied, since there is no accuracy benefit by imposing vertical velocities, but there is a negative impact on stability, since the particle spacing increases vertically, leading to a reduced kernel support. To mitigate this problem, we managed to impose both horizontal and vertical velocities to the buffer particles, but only update the particle positions according to the horizontal velocities. However, no difference was noticed in the accuracy of the measured velocity fields and surface elevations. For this reason, only the horizontal orbital velocities were imposed.

Table 1. Imposed and extrapolated quantities for inlet and outlet buffer particles (Imp = imposed, Ext = extrapolated, Hyd = hydrostatic).

Quantity	Horizontal Velocity u	Vertical Velocity w	Surface Elevation η	Pressure p
inlet	Imp	No	Imp	Hyd
outlet	Imp	No	Ext	Ext

By imposing horizontal velocities on both the inlet and outlet, the hydrodynamic problem becomes over-constrained, which can result in unwanted wave reflections in the fluid domain. Additionally, when a floating or fixed structure is positioned in the fluid domain, waves will reflect on the structure and transform around it. The open boundaries should be able to compensate for the reflected waves and the outlet needs to absorb the transformed wave effectively. In this research, this is done by applying velocity corrections at the inlet and the outlet, based on the measured free surface close to the buffer zones, specifically at a distance of $8 \cdot d_p$. This distance has been selected based on a sensitivity analysis, illustrated in Figure 4. The same Stokes third-order wave was simulated, each time varying the wave measurement distance d_{WG} from $1 \cdot d_p$ to $16 \cdot d_p$. For a value of d_{WG} equal to $8 \cdot d_p$, the wave measurement location is close enough to the buffer zone to have a minimal phase difference, but far enough to avoid inaccuracies due to transitional effects between the buffer zone and the fluid domain. In Figure 2, these measuring locations are denoted as WG_{in} (Wave Gauge) and WG_{out}. The applied velocity correction is a shallow water wave correction based on the measured reflection [41], but is implemented differently depending on the inlet or the outlet.

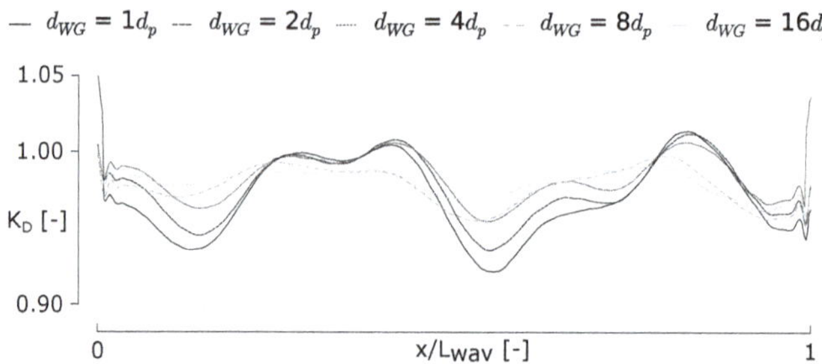

Figure 4. Sensitivity analysis on distance d_{WG} from the wave measurement locations WG_{in} and WG_{out} to the inlet/outlet buffer zone, necessary for accurate wave propagation, represented by the ratio K_D along the fluid domain.

4.1. Inlet Velocity Correction

At the inlet, the objective is to generate always the required incident wave. The surface elevation is measured directly outside of the inlet, and the velocity is corrected to ensure that the generated surface elevation matches the theoretical one. In case a higher surface elevation is measured than what was imposed, the corrected velocity should be lower than the originally imposed profile, in order to compensate the excess of velocity, since that profile leads to wave reflections in the fluid domain. Within the code, this correction is implemented as follows:

$$u_{in}(z,t) = u_1(z,t) - [\eta_{WG,in} - \eta_{AWAS,in}] \cdot \sqrt{\frac{g}{d}}. \tag{15}$$

Here, u_{in} is the horizontal velocity at the inlet, u_1 is the imposed horizontal velocity coming from OceanWave3D, $\eta_{WG,in}$ is the measured free surface elevation near the inlet, $\eta_{AWAS,in}$ is the OceanWave3D free surface measured at the location of WG_{in}, g is the earth's acceleration and d is the water depth. This correction is similar to the active wave absorption applied in Altomare et al. [42], although there it was used to correct the displacement of a piston-type wave maker formed by moving boundary particles instead of buffer particles.

4.2. Outlet Velocity Correction

At the outlet, the objective is to absorb any wave propagating towards the outlet. Technically, the applied open boundaries do not absorb the waves, but rather try to match the velocity field present in the fluid domain as close as possible, creating an 'open door' for the propagating wave. The surface elevation $\eta_{WG,out}$ is measured directly outside of the outlet, and compared to the calculated OceanWave3D free surface $\eta_{AWAS,out}$ at the location of WG_{out}. The velocity u_{out} is then corrected to ensure that the imposed velocities u_2 match the measured ones. In case a higher surface elevation $\eta_{WG,out}$ is measured than what was imposed $\eta_{AWAS,out}$, the corrected velocity u_{out} should be higher than the originally imposed velocity u_2. Otherwise, discontinuities in the velocity field would occur, which would induce unwanted reflected waves into the domain:

$$u_{out}(z,t) = u_2(z,t) - [\eta_{AWAS,out} - \eta_{WG,out}] \cdot \sqrt{\frac{g}{d}}. \tag{16}$$

4.3. Coupling Algorithm

The implementation of the coupling is illustrated in Figure 5. The detailed wave–structure interactions are calculated with the latest version of DualSPHysics (v4.2.030). As mentioned in Section 4, velocity corrections need to be applied for accurate representation of the free-surface elevation and the wave kinematics. For this reason, a two-way coupling is realised between a Python [43] program and DualSPHysics. Rather than a direct coupling between OceanWave3D and DualSPHysics, Python is used as an intermediate communication step. This allows for the coupling to be more generic, and not exclusively applicable to OceanWave3D. Additionally, the use of Python allows the user to monitor the simulation and easily perform accuracy checks on the transferred data. The Transmission Control Protocol (TCP) is used for the data transfer. At the start of the simulation, a dedicated port (50007) is opened to allow communication. At the start of each time step, Python sends both the inlet and outlet velocities and the inlet surface elevation to DualSPHysics. At the end of the time step, DualSPHyics sends back the measured surface elevation near the inlet and outlet. In Python, the velocity corrections are calculated and applied to the originally imposed horizontal velocities. With this coupling methodology, any data can be used to impose horizontal velocities and surface elevation to the inlet and outlet buffer zones. For example, wave theory solutions are an excellent method to provide the model with accurate orbital velocities and surface elevations. Alternative to using wave theory, and as demonstrated in this paper, the wave propagation model OceanWave3D can be coupled to DualSPHysics using the MPI protocol for data transfer. Here, the use of a TCP port was not possible due to compatibility issues with the OceanWave3D Fortran code. Although the coupling methodology is here specifically introduced for ocean wave simulations, it can be generalised to a more generic application. Specifically, the implementation using Python as an intermediate communication process, and using TCP sockets and the MPI protocol to send data from one software package to another, can be applied to other research domains as well. For example, an open channel flow modelling tool can be coupled to SPH to simulate river flow dynamics, or a 1D pipe flow hydraulic model can be coupled to SPH to simulate pressurized flow problems. However, a detailed study of the applicability of the methodology falls outside the scope of this research.

Figure 5. Coupling implementation scheme showing several options to generate and absorb waves within a two-way coupled model: wave theory solutions, the wave propagation model OceanWave3D or any external data.

Apart from the implementation, a schematic representation of the coupling algorithm is illustrated in Figure 6. The most complex coupling is discussed here, being the two-way coupling with OceanWave3D. The coupling communication occurs at the beginning and at the end of an OceanWave3D time step, which is referred to as the "communication time step". Due to the larger grid size and simplified equations in OceanWave3D, the communication time step is considerably larger than the DualSPHysics time step. This means that DualSPHysics performs multiple smaller time steps during one communication time step. At the beginning of a communication time step, the horizontal orbital velocities at the inlet and outlet locations, u_1 and u_2, are calculated in OceanWave3D, as well as the surface elevation of the complete OceanWave3D domain η_{OW3D}, and sent to the Python programming using the MPI protocol. Simultaneously, Python receives the measured surface elevation from DualSPHysics near the inlet and outlet $\eta_{WG,in/out}$. The velocity correction is calculated using $\eta_{AWAS,in}$ and $\eta_{AWAS,out}$, resulting in the corrected orbital velocities $u_{in/out}$ which are sent back to DualSPHysics, together with the surface elevation at the inlet $\eta_{OW3D,in}$. The quantities are imposed within the inlet and outlet buffer zones and the DualSPHysics simulation calculates multiple time steps until the communication time step is reached. The surface elevation from DualSPHysics η_{SPH} is sent to Python, where the signal is filtered and sent back to OceanWave3D as $\eta_{f,r}$ to overwrite the original result. To ensure a smooth transition between the DualSPHysics free surface and the OceanWave3D free surface, a relaxation functions f_{rel} is applied (see Figure 7). The applied function f_{rel} is the same as is applied within OceanWave3D's generation and absorption relaxation zones. Within the relaxation zone, the filtered solution $\eta_{f,r}$ is then obtained as given in Equation (18), with B the length of the zone, x the location within the relaxation zone and x_0 the reference location of the relaxation zone:

$$\eta_{f,r} = (1 - f_{rel}) \cdot \eta_{OW3D} + f_{rel} \cdot \eta_{SPH}, \qquad (17)$$

$$f_{rel} = \left(\frac{x - x_0}{B} \right)^{3.5}. \qquad (18)$$

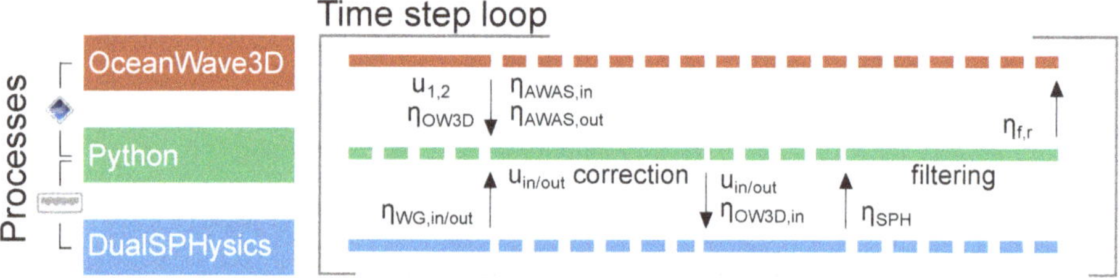

Figure 6. Schematic representation of the coupling algorithm during one communication time step.

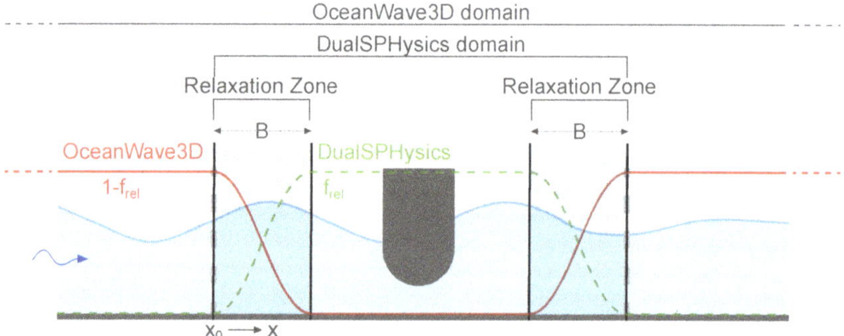

Figure 7. Sketch of 'relaxation zones' providing a smooth transition between the OceanWave3D domain and the DualSPHysics domain.

5. 2D Coupled Model: Validation

First, the two-way coupling between DualSPHysics and OceanWave3D is validated in 2D. The coupling is applied to compare the response of a box, floating in three degrees of freedom (heave, surge and pitch) to experimental data, as described in Ren et al. [44]. The numerical test set-up is illustrated in Figure 8. The full OceanWave3D domain has a length of 20.0 m, while the DualSPHysics domain is 6.12 m long (three wave lengths) and starts at $x = 4.0$ m. The floating box is positioned at $x = 5.5$ m in a water depth of $d = 0.4$ m and has the dimensions 0.3 m × 0.2 m (length × height) with a draft $q_{box} = 0.1$ m. A regular wave with wave height $H = 0.1$ m and wave period $T = 1.2$ s is generated, characterised as a Stokes third-order wave. A particle size of $d_p = 0.01$ m is used. The SPH domain is chosen to be larger than one wavelength L_{wav}, since the box is freely floating and will drift in the x-direction over time.

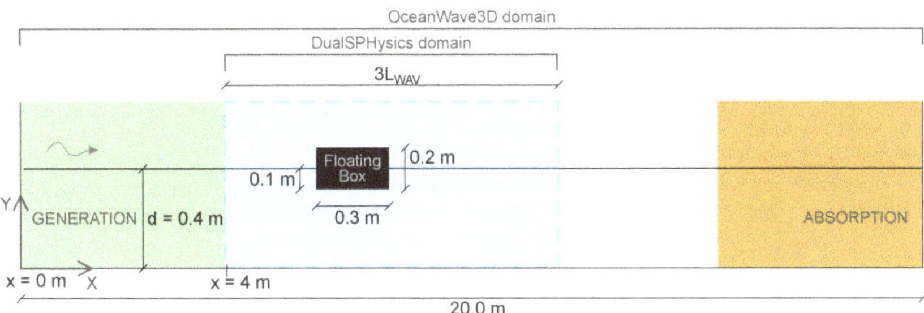

Figure 8. Numerical test set-up for simulation of the response of a floating box to a custom wave signal. The DualSPHysics domain and OceanWave3D domain are indicated. L_{wav} is the wavelength.

5.1. DualSPHysics Solver Options

The efficiency and accuracy of DualSPHysics was investigated in Altomare et al. [45] for wave propagation and absorption showing good agreement between numerical results and experimental data. Accordingly, similar DualSPHysics solver options used for that work will be used here to perform the numerical simulations. The solver options used in this section are summarized in Table 2. The time integration scheme is chosen to be the Verlet scheme, employing a variable time step based on the CFL number. The chosen kernel is the Wendland kernel with a smoothing length of $h_{SPH} = 2.0 \cdot d_p$. The artificial viscosity is set to be $\alpha = 0.01$. Tait's equation of state is applied to calculate the fluid pressures. The boundary conditions are the open boundaries introduced by Tafuni et al. [24] and some numerical diffusivity is added by applying a δ-SPH treatment with a coefficient of $\delta - SPH = 0.1$. These parameters have been thoroughly studied in Verbrugghe [46] and have proven to lead to accurate wave propagation within DualSPHysics. There, a convergence analysis has proven a convergence rate in between first-order and second-order, and a thorough validation study of the wave dynamics was performed.

Table 2. SPH formulation and parameters.

Time Integration Scheme	Verlet
Time Step	Variable (including CFL and viscosity)
Kernel	Wendland
Smoothing Length	$2.0 \cdot d_p$
Viscosity Treatment	Artificial ($\alpha = 0.01$)
Equation of State	Tait equation
Boundary Conditions	Open Boundary Conditions
δ-SPH	Yes (δ-SPH = 0.1)

5.2. Results and Discussion

Both a one-way coupling and two-way coupling are compared to the experimental data, and the corresponding errors are illustrated in Figure 9. In the one-way coupling, OceanWave3D only provides the horizontal orbital velocities to the inlet and outlet zone, and the surface elevations for the inlet and the calculation of the velocity corrections. For the two-way coupling, the surface elevations from the DualSPHysics domain are transferred back to the OceanWave3D domain, where the original solution is overwritten. The part close around the floating box is not coupled back, since the 'measured' SPH free surface elevations are not physical there. The top three graphs of Figure 9 show a direct comparison between the experimental and numerical results for the heave motion, the pitch motion and the surge motion, respectively. Qualitatively, a good correspondence between the numerical and experimental results is shown. Specifically, the heave and pitch motions are accurately reproduced. In order to quantify the accuracy of the results, the difference between the experimental and numerical results is calculated and illustrated in the bottom three graphs of Figure 9. The error on the heave motions remains well below the smoothing length $h = 0.02$ m. Additionally, it can be noticed that the result from the two-way coupling has a lower error than the result from the one-way coupling. The pitch error stays below 3.5 degrees, which is around 10% of the total pitch motion range, and there is no clear distinction in the error between the one-way coupling and the two-way coupling. The error on the surge motion is less satisfactory. Here, both the one-way coupling and two-way coupling are less capable of modelling the cyclic surge path, the net drift in the x-direction is even less accurate for the two-way coupling as for the one-way coupling. However, since the error on the surge motion logically becomes larger in time, the results are still acceptable.

Additionally, the instantaneous surface elevation profile at $t = 15$ s of DualSPHysics and OceanWave3D is compared in Figure 10. Here, the part close to the floating box is masked out since there is no coupling performed there. Three surface elevation profiles are plotted. For the two-way coupling methodology, both the OceanWave3D and DualSPHysics profile are plotted. Normally, these lines are expected to be exactly the same, since the original OceanWave3D surface elevation is overwritten. However, relaxation functions are applied to ensure a smooth transition between the OceanWave3D solution and the DualSPHysics solution. In Figure 10, this can be noticed close to the masked out zone both solutions match perfectly. Behind the masked-out zone, both solutions remain the same until they slightly differ again at the boundary of the SPH zone ($x = 10.12$ m). As a reference, the OceanWave3D solution for a one-way coupling is shown as well. Here, there is no influence from the DualSPHysics solution, and OceanWave3D propagates waves as if there is no floating box present. Nevertheless, this does not impact the accuracy of the results significantly, as proven in Figure 9. In general, it can be concluded that the one-way coupling is slightly more accurate than the two-way coupling. The advice is to only apply the two-way coupling methodology, when there is a significant wave transformation effect around the structure, which needs to be propagated further within the larger OceanWave3D domain. For example, modelling a WEC device that extracts a significant amount of energy from the waves will result in a significant reduction of wave height behind the device. When it is required to model the further propagation of that reduced wave within the larger OceanWave3D domain, a two-way coupling should be applied. Otherwise, when there is no interest in the wave field, far away from the DualSPHysics domain, the one-way coupling is advised.

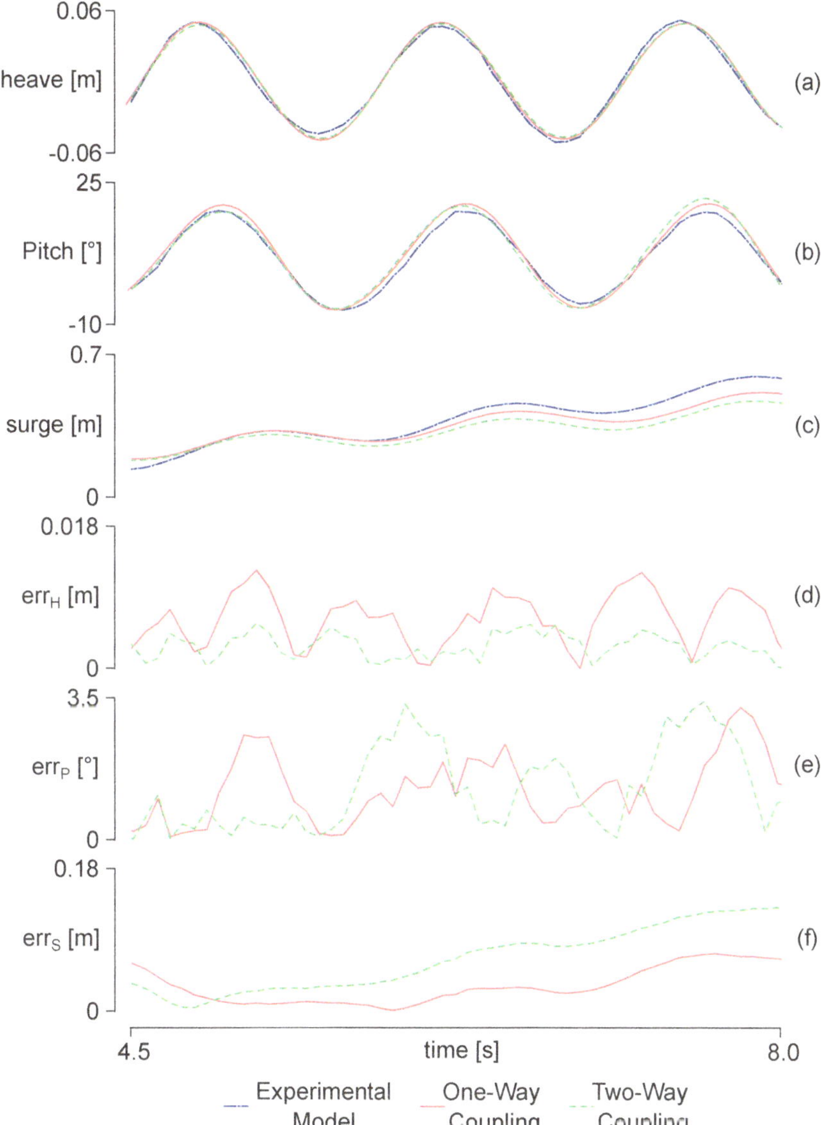

Figure 9. Comparison of time series and error between numerical and experimental results of the three degrees of freedom of the box with heave, pitch and surge motions (top three graphs). The difference between the numerical results and the experimental results is expressed in the error values for heave err_H, pitch err_P and surge err_S (bottom three graphs).

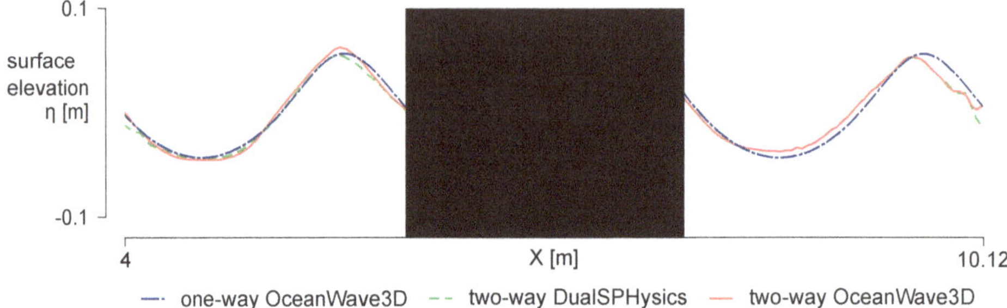

Figure 10. Comparison of the instantaneous surface elevation profile between one-way and two-way coupling results of OceanWave3D and DualSPHysics. The grey masked out zone is the region around the floating box, omitted from the coupling.

6. 3D Coupled Model: Proof-of-Concept

In this section, the introduced coupling methodology is extended to a 3D domain. A one-way coupling with a fifth-order Stokes wave theory is applied. The buffer zones are stretched in the y-direction, perpendicular to the 2D plane of the 2D validation case, and the imposed quantities are constant in that y-direction. This means 3D simulations are restricted to long-crested waves. Consequently, the coupling methodology can be used to model wave flume type experiments where significant 3D effects are present. To demonstrate this, a heaving disk is simulated, impacted by steep nonlinear waves. This results in nonlinear wave surge forces and wave overtopping.

6.1. Experimental Set-up

Experiments have been performed in the large wave flume of Ghent University (see Figure 11). The flume is 30.0 m long and 1.0 m wide. A cylindrical WEC with a diameter of $D = 0.5$ m is positioned at a distance of 10.6 m from the wave paddle. The WEC's motion is restricted to heave by a vertical rod over which it is sliding. Friction losses are minimized by using two sets of PTFE bearings: one on the top and one on the bottom of the cylinder. The WEC consists of 10 glued plastic disks and is waterproofed with a black, silyl-modified (MS) polymer coating. Wave absorption material is installed at the left end of the wave flume to ensure minimal wave reflections. Active wave absorption is applied using two wave gauges about 3.0 m away from the wave paddle. Seven wave gauges (WG) are installed to measure the free surface elevation. The WEC has a draft of $q_{WEC} = 0.113$ m. The motion of the WEC is captured by video tracking using a GoPro Hero 5 [47], filming in Full HD at 120 fps. The typical fisheye distortion is corrected using video processing software. The horizontal surge force $F_{x,WEC}$ is measured by 2 force transducers, installed in a rigid rod behind the WEC to which it is connected. The incident wave has a wave height of $H = 0.12$ m, a wave period of $T = 1.2$ s in a water depth of $d = 0.7$ m. This results in a Stokes third-order wave with a wave length of $L_{wav} = 2.17$ m.

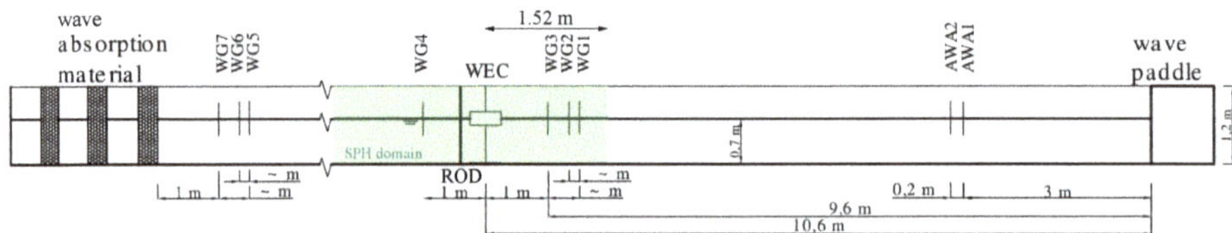

Figure 11. Experimental set-up for tests with a heaving disk type WEC in the large wave flume of Ghent University. The numerical DualSPHysics domain is indicated as a green zone.

6.2. Numerical Set-up

The numerical set-up is illustrated in Figure 12 and summarized in Table 3. It is chosen to set the length of the fluid domain to twice the wave length $L_{SPH} = 2 \cdot L_{wav} = 4.34$ m. This is around 1/7 of the total flume length. The particle size is chosen to be $d_p = 0.01$ m. The Stokes fifth-order wave theory is applied. Since there is always a gap of 1.5 times the smoothing length h_{SPH} between fluid particles and boundary particles, the top of the floating disk is lowered with a value of $1.5 \cdot h_{SPH} = 1.5 \cdot 2.0 \cdot d_p$, in order to get the correct wave overtopping height. A 3D inlet zone and outlet zone are configured, with each eight layers of buffer particles. All other boundaries are fixed boundary conditions with the option to extrapolate the pressure from the fluid domain, in order to avoid local pressure peaks (see Section 3.2). The WEC is positioned at a distance of 1.52 m from the inlet boundary of the numerical domain.

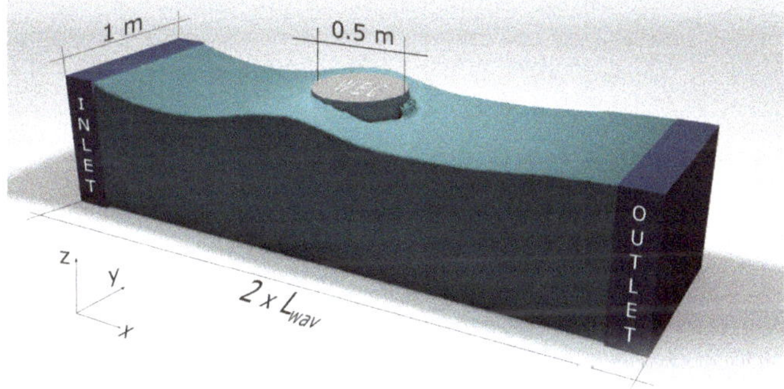

Figure 12. Numerical set-up for 3D modelling of a heaving cylindrical WEC in a two-way coupled model.

Table 3. Simulation parameters for 3D modelling of a heaving cylindrical WEC.

Wave Height H [m]	0.12
Wave Period T [s]	1.2
Water Depth d [m]	0.7
Wave Length L_{wav} [m]	2.17
Particle Size d_p [m]	0.01
Domain Length L_{SPH} [m]	4.34
Domain Width W [m]	1.0
WEC Diameter D [m]	0.5
WEC Draft q_{WEC} [m]	0.113
Wave Theory	Stokes 5th
Time Step Algorithm	Verlet
Artificial Viscosity	0.01
δ-SPH	0.1

6.3. Results and Discussion

First, the numerical surface elevations from the one-way coupled model are compared with experimental results. The signals from WG3 and WG4 (see Figure 11) are compared to the numerical results. WG3 is positioned 1.0 m in front of the WEC, WG4 1.0 m behind the WEC. The comparison with the data from WG3 is shown in plot (a) of Figure 13, while plot (b) shows the comparison with the WG4 data. The surface elevation in front of the WEC is accurately reproduced. There is some initialisation time, but after a while the wave signal is steady and matches the experimental data very well. The error remains well below the smoothing length $h_{SPH} = 0.02$ m. Behind the WEC, the numerical results are less accurate. The one-way coupled model calculates lower wave heights behind the WEC than what was registered in the experiments. The measured free surface also has a steeper nonlinear profile than what was calculated numerically. This could be due to the shortened SPH domain with respect to the full experimental wave flume length. However, the error on the surface elevation is still acceptable since the maximum error at the third wave crest measures 0.0196 m and is still smaller than the smoothing length h_{SPH}.

Second, the horizontal surge force is calculated and compared to the experimental data in plot (c) of Figure 3. Here, it is clear that there is a very good match between the numerical and experimental results. Although both signals have some noise, the overall trend of the data matches very well.

Third, the comparison of the heave motion of the WEC to the experimental data is shown in plot (d) of Figure 3. Again, after initialisation of the surface elevation in the wave flume, a steady regime is obtained in which the calculated WEC motion and measured motion show an excellent agreement, with a maximum error of $0.4h_{SPH}$.

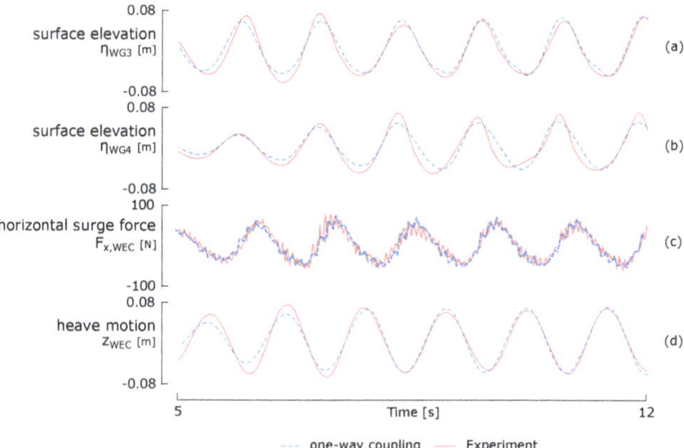

Figure 13. Comparison of a one-way coupling 3D proof-of-concept with experimental data of a heaving cylinder in overtopping nonlinear waves. (**a**) shows the surface elevation 1 m in front of the WEC; (**b**) the surface elevation 1 m behind the WEC; (**c**) the horizontal surge force acting on the WEC and (**d**) the heave motion of the WEC.

6.4. Computational Speed-up

The coupling methodology with open boundaries leads to significant performance gains with respect to typical SPH simulations. In this section, an estimation of the computational speed-up is made, comparing the presented one-way coupled model with a stand-alone numerical model, describing the complete experimental set-up. The latter is thus an SPH simulation where the full wave flume is modelled, including the wave paddle motion and the floating disk, installed at $x = 10.6$ m. The comparison is summarized in Table 4. In the one-way coupled model, the domain length was set at twice the wave length, resulting in a total length of 4.34 m. This is about 1/7 of the wave flume length, which explains the number of fluid particles in the full model, which is about 718% more than in the one-way coupled model. The difference in total number of particles is lower, at only 508%. This is due to the thickness of the bottom boundary in the coupled model, which is thicker to allow for accurate pressure extrapolation. The difference in GPU memory is similar at 505%, with an absolute value of 3941 MB for the full model. Although our GPU card has 8106 MB of memory available, this simulation is incredibly demanding. Computational performance is not only dependent on GPU memory, since the number of CUDA cores and their clock rate are much more indicative of simulation time. The estimated runtime, which is calculated at the start of the simulation, is significantly longer for the full model than for the coupled model. However, the difference of 411% is slightly lower than the difference in particles. The real runtime of the coupled model was 91 h. This remarkable difference in estimated runtime and real runtime can be explained by inaccurate estimations at the beginning of the simulation, in combination with performing other tasks on the computer, slowing down the simulation. It is chosen to not run the full model completely, since the computer becomes unusable during the simulation and it could take up to 375 h or almost 16 days to finish if the same performance as the coupled model is assumed. It is safe to assume that a speed-up of around 400% can be achieved, by applying the coupling methodology.

Table 4. Computational speed-up for 3D proof-of-concept.

	One-Way Coupling	Full Model	Difference
# Particles	5,010,954	25,473,152	508%
# Fluid Particles	2,949,433	21,165,100	718%
GPU Memory [MB]	780	3941	505%
Estimated Runtime [hr]	35	144	411%
Real Runtime [hr]	91	375	411%

6.5. Visual Comparison of Overtopping

During the experiments, some of the steep nonlinear waves were overtopping the cylindrical WEC. However, no overtopping rates or thickness of the overtopping layer were measured, so a correct validation is not possible. For this reason, only a visual comparison of an overtopping event between the experiment and the numerical model is performed. In Figure 14, a time progression of an overtopping wave is shown. Here, the wave height was set at $H = 0.15$ m and the wave period was $T = 1.0$ s. This resulted in steep, highly nonlinear waves with significant overtopping. In addition, for this test case, video images were available to perform a visual comparison. In the left plot, the wave is just about to hit the cylindrical WEC. In the middle plot, 0.4 s later, the overtopping wave is on top of the WEC, and in the final plot, the overtopped volume is flowing from the top back into the flume. Visually, the correspondence between the numerical model and the experimental images is very good, apart from the interactions of the overtopped water with the vertical rod, since the rod was not included in the numerical model.

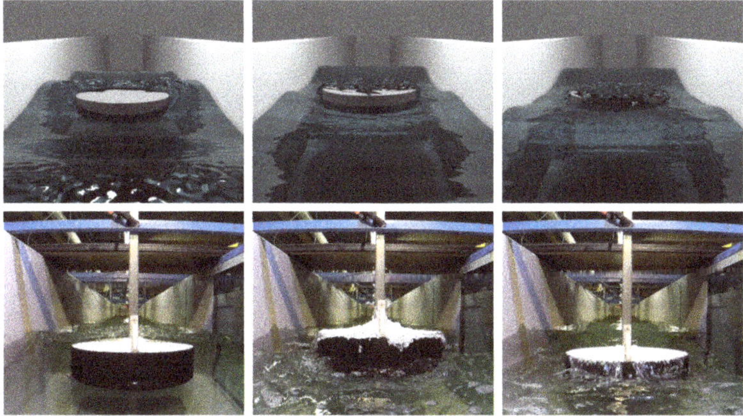

Figure 14. Visual comparison of overtopping waves between the 3D proof-of-concept with experimental data of a heaving cylinder in overtopping nonlinear waves. The plot shows a time progression of the wave, from left to right with a time difference of 0.4 s between each frame.

7. Conclusions

Development of accurate boundary conditions and the coupling of SPH solvers to external models are identified as key topics within the SPHERIC Grand Challenges to let SPH method attain a high level of maturity. The present work is a clear contribution to both issues, proposing a novel two-way coupling methodology that applies the open boundary formulation to accurately generate, propagate and absorb waves within a two-way coupled DualSPHysics-OceanWave3D model. A Python program was used as an intermediate process to calculate and apply the velocity corrections to the inlet and outlet buffer zones, to avoid reflections inside the DualSPHysics fluid domain. This coupling technique allows overcoming one of the main shortcomings of SPH-based solvers, related to computational cost. Despite the increase of HPC and GPU performance during the last years, a large number of particles (i.e., high model resolution) is still necessary, in order to simulate real engineering problems with sufficient accuracy. A coupling between a detailed but computationally expensive mesh-free model and a wave propagation model represents a trade-off between model accuracy and efficiency. This can be done by coupling SPH to a faster external numerical model which can deal easily with large computational domains. A proper two-way coupling scheme, however, requires accurate and stable boundary conditions. In fact, a reduced domain size is useless when the imposed quantities at the boundary are not accurate enough. Through coupling, other software packages (fast-calculating) can be applied to provide physical quantities to the boundary conditions, more accurately than what the built-in methods or analytical solutions can provide. The two-way coupling methodology with OceanWave3D was validated in 2D by comparing the surface elevations and the motions of a floating box to experimental data. Finally, a 3D proof-of-concept was introduced where steep nonlinear waves interact with and overtop a heaving cylindrical WEC. The results all show a good agreement with the experimental data. In the performed tests, the coupled model has at least two to four times less particles to simulate, which directly results into faster computation times. Alternatively, for the same computation time as a stand-alone SPH simulation, there is the possibility of simulating more particles for a higher accuracy. One of the main benefits of using open boundary conditions rather than moving boundaries or relaxation zone techniques is that there are no issues with Stokes drift, and the velocity and pressure field are significantly smoother than those calculated with the coupling methodology applying moving boundaries. However, there are still a number of limitations to this revised coupling methodology. Firstly, only quasi-3D simulations are possible. The buffer zones do not allow non-uniform velocities or surface elevations along the y-directions. This means the 3D simulations are limited to long-crested. waves. However, this can already be very meaningful to simulate typical wave flume experiments, or real engineering problems where there is not much variability in the y-direction. The numerical stability in 3D simulations needs to improve. The open boundaries still have some compatibility issues with the periodic boundary conditions within DualSPHysics and with the boundary pressure extrapolation algorithm. Although good results are obtained, there is a certain loss of particles during the simulations. This will, however, be solved within future releases of the source code.

Author Contributions: Conceptualization, T.V. and C.A.; Methodology, T.V. and C.A.; Software, T.V. and J.M.D.; Validation,T.V.; Formal Analysis, T.V.; Investigation, T.V.; Resources, T.V. and C.A.; Data Curation, T.V.; Writing—Original Draft Preparation, T.V.; Writing—Review & Editing, T.V., V.S. and C.A.; Visualization, T.V.; Supervision, V.S., P.T., A.K. and C.A.; Project Administration, T.V.; Funding Acquisition, T.V., V.S., P.T. and A.K.

Funding: This research was funded by Agency for Innovation by Science and Technology in Flanders (IWT) grant number 141402.

Conflicts of Interest: The authors declare no conflict of interest

References

1. Gotoh, H.; Khayyer, A. On the state-of-the-art of particle methods for coastal and ocean engineering. *Coast. Eng. J.* **2018**, *60*, 79–103. [CrossRef]
2. SPH Numerical Development Working Group. Avaliable online: http://spheric-sph.org/grand-challenges (accessed on 20 February 2019).
3. Bouscasse, B.; Marrone, S.; Colagrossi, A.; Di Mascio, A. Multi-purpose interfaces for coupling SPH with other solvers. In Proceedings of the 8th International SPHERIC Workshop, Trondheim, Norway, 4–6 June 2013.
4. Fourtakas, G.; Stansby, P.; Rogers, B.; Lind, S. An Eulerian-Lagrangian incompressible SPH formulation (ELI-SPH) connected with a sharp interface. *Comput. Methods Appl. Mech. Eng. Comput. Methods Appl. Mech. Eng.* **2018**, *329*, 532–552. [CrossRef]
5. Altomare, C.; Domínguez, J.M.; Crespo, A.J.C.; Suzuki, T.; Caceres, I.; Gómez-Gesteira, M. Hybridisation of the wave propagation model SWASH and the meshfree particle method SPH for real coastal applications. *Coast. Eng. J.* **2016**, *57*, 1–34.
6. Altomare, C.; Tagliafierro, B.; Domínguez, J.; Suzuki, T.; Viccione, G. Improved relaxation zone method in SPH-based model for coastal engineering applications. *Appl. Ocean Res.* **2018**, *81*, 15–33. [CrossRef]
7. Zijlema, M.; Stelling, G.; Smit, P. SWASH: An operational public domain code for simulating wave fields and rapidly varied flows in coastal waters. *Coast. Eng.* **2011**, *58*, 992–1012. [CrossRef]
8. Kassiotis, C.; Ferrand, M.; Violeau, D.; Rogers, B.; Stansby, P.; Benoit, M.; Rogers, B.D.; Stansby, P.K. Coupling SPH with a 1D Boussinesq-type wave model. In Proceedings of the 6th International SPHERIC Workshop, Hamburg, Germany, 8–10 June 2011; pp. 241–247.
9. Narayanaswamy, M.; Crespo, A.J.C.; Gómez-Gesteira, M.; Dalrymple, R.A. SPHysics-FUNWAVE hybrid model for coastal wave propagation. *J. Hydraul. Res.* **2010**, *48*, 85–93. [CrossRef]
10. Kirby, J.T.; Wei, G.; Chen, Q.; Kennedy, A.B.; Dalrymple, R.A. *FUNWAVE 1.0: Fully Nonlinear Boussinesq Wave Model-Documentation And User's Manual*; Research Report NO. CACR-98-06; University of Delaware: Delaware, NJ, USA, 1998.
11. Ma, Q.; Yan, S. QALE-FEM for numerical modelling of nonlinear interaction between 3D moored floating bodies and steep waves. *Int. J. Numer. Methods Eng.* **2009**, *78*, 713–756. [CrossRef]
12. Fourtakas, G.; Stansby, P.K.; Rogers, B.D.; Lind, S.J.; Yan, S.; Ma, Q.W. On the coupling of Incompressible SPH with a Finite Element potential flow solver for nonlinear free surface flows. In Proceedings of the 27th International Ocean and Polar Engineering Conference, International Society of Offshore and Polar Engineers, San Francisco, CA, USA, 25–30 June, 2017.
13. Chicheportiche, J.; Hergault, V.; Yates, M.; Raoult, C.; Leroy, A.; Joly, A.; Violeau, D. Coupling SPH with a potential Eulerian model for wave propagation problems. In Proceedings of the 11th SPHERIC International Workshop, Munich, Germany, 14–16 June 2016; pp. 246–252.
14. Didier, E.; Neves, D.; Teixeira, P.R.F.; Neves, M.G.; Viegas, M.; Soares, H. Coupling of FLUINCO mesh-based and SPH mesh-free numerical codes for the modeling of wave overtopping over a porous breakwater. In Proceedings of the International Short Course on Applied Coastal Research, Lisbon, Portugal, 4–7 June 2013.
15. Teixeira, P.R.D.F.; Awruch, A.M. Numerical simulation of fluid–structure interaction using the finite element method. *Comput. Fluids* **2005**, *34*, 249–273. [CrossRef]
16. Kumar, P.; Yang, Q.; Jones, V.; Mccue-Weil, L. Coupled SPH-FVM simulation within the OpenFOAM framework. *Procedia IUTAM* **2015**, *18*, 76–84. [CrossRef]
17. Marrone, S.; Di Mascio, A.; Le Touzé, D. Coupling of Smoothed Particle Hydrodynamics with Finite Volume method for free-surface flows. *J. Comput. Phys.* **2016**, *310*, 161–180. [CrossRef]
18. Napoli, E.; De Marchis, M.; Gianguzzi, C.; Milici, B.; Monteleone, A. A coupled Finite Volume–Smoothed Particle Hydrodynamics method for incompressible flows. *Comput. Methods Appl. Mech. Eng.* **2016**, *310*, 674–693. [CrossRef]
19. Engsig-Karup, A.P.; Bingham, H.B.; Lindberg, O. An efficient flexible-order model for 3D nonlinear water waves. *J. Comput. Phys.* **2009**, *228*, 2100–2118. [CrossRef]

20. Crespo, A.; Domínguez, J.; Rogers, B.; Gómez-Gesteira, M.; Longshaw, S.; Canelas, R.; Vacondio, R.; Barreiro, A.; García-Feal, O. DualSPHysics: Open-source parallel CFD solver based on Smoothed Particle Hydrodynamics (SPH). *Comput. Phys. Commun.* **2015**, *187*, 204–216. [CrossRef]
21. Verbrugghe, T.; Domínguez, J.M.; Crespo, A.J.; Altomare, C.; Stratigaki, V.; Troch, P.; Kortenhaus, A. Coupling methodology for smoothed particle hydrodynamics modelling of nonlinear wave–structure interactions. *Coast. Eng.* **2018**, *138*, 184–198. [CrossRef]
22. Tafuni, A.; Domínguez, J.; Vacondio, R.; Sahin, I.; Crespo, A. Open boundary conditions for large-scale SPH simulations. In Proceedings of the 11th SPHERIC International Workshop, Munich, Germany, 14–16 June 2016.
23. Tafuni, A.; Domínguez, J.M.; Vacondio, R.; Crespo, A. Accurate and efficient SPH open boundary conditions for real 3D engineering problems. In Proceedings of the 12th International SPHERIC Workshop, Ourense, Spain, 13–15 June 2017; pp. 1–12.
24. Tafuni, A.; Domínguez, J.; Vacondio, R.; Crespo, A. A versatile algorithm for the treatment of open boundary conditions in Smoothed particle hydrodynamics GPU models. *Comput. Methods Appl. Mech. Eng.* **2018**, *342*, 604–624. [CrossRef]
25. Ni, X.; Feng, W.; Huang, S.; Zhang, Y.; Feng, X. A SPH numerical wave flume with non-reflective open boundary conditions. *Ocean Eng.* **2018**, *163*, 483–501. [CrossRef]
26. Crespo, A.; Domínguez, J.; Gómez-Gesteira, M.; Barreiro, A.; Rogers, B. *User Guide for DualSPHysics Code*; The University of Manchester and Johns Hopkins University, University of Vigo: Vigo, Spain, 2018.
27. Wendland, H. Piecewise polynomial, positive definite and compactly supported radial functions of minimal degree. *Adv. Comput. Math.* **1995**, *4*, 389–396. [CrossRef]
28. Monaghan, J.J. Smoothed particle hydrodynamics. *Annu. Rev. Astron. Astrophys.* **1992**, *30*, 543–574. [CrossRef]
29. Altomare, C.; Crespo, A.J.; Domínguez, J.M.; Gómez-Gesteira, M.; Suzuki, T.; Verwaest, T. Applicability of Smoothed Particle Hydrodynamics for estimation of sea wave impact on coastal structures. *Coast. Eng.* **2015**, *96*, 1–12. [CrossRef]
30. Shao, S.; Lo, E.Y. Incompressible SPH method for simulating Newtonian and non-Newtonian flows with a free surface. *Adv. Water Resour.* **2003**, *26*, 787–800. [CrossRef]
31. Lo, E.Y.; Shao, S. Simulation of near-shore solitary wave mechanics by an incompressible SPH method. *Appl. Ocean Res.* **2002**, *24*, 275–286.
32. Monaghan, J.J. Simulating free surface flows with SPH. *J. Comput. Phys.* **1994**, *110*, 399–406. [CrossRef]
33. Batchelor, G.K. *An Introduction to Fluid Dynamics*; Cambridge University Press: Cambridge, UK, 2000.
34. Leimkuhler, B.J.; Reich, S.; Skeel, R.D. Integration methods for molecular dynamics. *IMA Vol. Math. Appl.* **1996**, *82*, 161–186.
35. Monaghan, J.; Kos, A. Solitary waves on a Cretan beach. *J. Waterw. Port Coast. Ocean Eng.* **1999**, *125*, 145–155. [CrossRef]
36. Molteni, D.; Colagrossi, A. A simple procedure to improve the pressure evaluation in hydrodynamic context using the SPH. *Comput. Phys. Commun.* **2009**, *180*, 861–872. [CrossRef]
37. Antuono, M.; Colagrossi, A.; Marrone, S. Numerical diffusive terms in weakly-compressible SPH schemes. *Comput. Phys. Commun.* **2012**, *183*, 2570–2580. [CrossRef]
38. Liu, M.; Liu, G. Restoring particle consistency in smoothed particle hydrodynamics. *Appl. Numer. Math.* **2006**, *56*, 19–36. [CrossRef]
39. Crespo, A.; Gómez-Gesteira, M.; Dalrymple, R.A. *Boundary Conditions Generated by Dynamic Particles in SPH Methods*; CMC-Tech Science Press: Henderson, NV, USA, 2007; Volume 5, p. 173.
40. Marrone, S.; Antuono, M.; Colagrossi, A.; Colicchio, G.; Touzé, D.L.; Graziani, G. Delta-SPH model for simulating violent impact flows. *Comput. Methods Appl. Mech. Eng.* **2011**, *200*, 1526–1542. [CrossRef]
41. Dean, R.G.; Dalrymple, R.A. Wavemaker Theory. In *Water Wave Mechanics for Engineers and Scientists*; World Scientific: Singapore, 1991; pp. 170–186.
42. Altomare, C.; Domínguez, J.; Crespo, A.; González-Cao, J.; Suzuki, T.; Gómez-Gesteira, M.; Troch, P. Long-crested wave generation and absorption for SPH-based DualSPHysics model. *Coast. Eng.* **2017**, *127*, 37–54. [CrossRef]
43. Rossum, G. *Python Reference Manual*; Technical Report; CWI: Amsterdam, The Netherlands, 1995.

44. Ren, B.; He, M.; Dong, P.; Wen, H. Nonlinear simulations of wave-induced motions of a freely floating body using WCSPH method. *Appl. Ocean Res.* **2015**, *50*, 1–12. [CrossRef]

45. Altomare, C.; Suzuki, T.; Domínguez, J.; Barreiro, A.; Crespo, A.; Gómez-Gesteira, M. Numerical wave dynamics using Lagrangian approach: Wave generation and passive & active wave absorption. In Proceedings of the 10th SPHERIC International Workshop, Parma, Italy, 16–18 June 2015.

46. Verbrugghe, T. Coupling Methodologies for Numerical Modelling of Floating Wave Energy Converters. Ph.D. Thesis, Ghent University, Ghent, Belgium, 2018.

47. Learn About Your New HERO5 Black. Avaliable online: https://gopro.com/yourhero5/black (accessed on 20 February 2019).

Article

Accurate and Fast Generation of Irregular Short Crested Waves by Using Periodic Boundaries in a Mild-Slope Wave Model

Panagiotis Vasarmidis *, Vasiliki Stratigaki and Peter Troch

Department of Civil Engineering, Ghent University, Technologiepark 60, 9052 Ghent, Belgium;
Vicky.Stratigaki@UGent.be (V.S.); Peter.Troch@UGent.be (P.T.)
* Correspondence: Panagiotis.Vasarmidis@UGent.be; Tel.: +32-9-264-54-89

Received: 9 February 2019; Accepted: 22 February 2019; Published: 26 February 2019

Abstract: In this work, periodic lateral boundaries are developed in a time dependent mild-slope equation model, MILDwave, for the accurate generation of regular waves and irregular long and short crested waves in any direction. A single wave generation line inside the computational domain is combined with periodic lateral boundaries. This generation layout yields a homogeneous and thus accurate wave field in the whole domain in contrast to an L-shaped and an arc-shaped wave generation layout where wave diffraction patterns appear inside the computational domain as a result of the intersection of the two wave generation lines and the interaction with the lateral sponge layers. In addition, the performance of the periodic boundaries was evaluated for two different wave synthesis methods for short crested waves generation, a method proposed by Miles and a method proposed by Sand and Mynett. The results show that the MILDwave model with the addition of periodic boundaries and the Sand and Mynett method is capable of reproducing a homogeneous wave field as well as the target frequency spectrum and the target directional spectrum with a low computational cost. The overall performance of the developed model is validated with experimental results for the case of wave transformation over an elliptic shoal (Vincent and Briggs shoal experiment). The numerical results show very good agreement with the experimental data. The proposed generation layout using periodic lateral boundaries makes the mild-slope wave model, MILDwave, an essential tool to study coastal areas and wave energy converter (WEC) farms under realistic 3D wave conditions, due to its significantly small computational cost and its high numerical stability and robustness.

Keywords: Mild-slope wave propagation model; MILDwave; periodic lateral boundaries; short crested waves

1. Introduction

Numerical wave propagation models are commonly used as engineering tools for the study of wave transformation in coastal areas. Berkhoff [1] derived the first elliptic mild-slope wave equation and based on this, the parabolic model [2] and the hyperbolic model [3] have been developed to predict the transformation processes of regular waves, such as wave refraction, diffraction, shoaling, and reflection. In the parabolic model, wave reflection and diffraction in the direction of wave propagation are neglected and hence it suffers from low accuracy in cases where these phenomena are significant. On the other hand, the hyperbolic model, which takes into consideration wave reflection and diffraction in the direction of wave propagation, provides higher accuracy, but more computational time compared to the parabolic model. Moreover, to study the transformation of random waves, time dependent mild-slope equations have been developed. Radder and Dingemans [4] suggested a set of canonical equations, which are based on the time dependent mild slope equations and are derived using the

Hamiltonian theory of surface waves. Booij [5] proved that these equations are valid for sea bottom slopes up to 1/3. However, Suh et al. [6] extended the latter equations by including higher order bottom effect terms proportional to the square of the bottom slope and to the bottom curvature to study wave propagation on rapidly varying topography.

For numerical prediction of the wave field in a coastal area, waves are generated along the offshore boundary of the computational domain and propagate towards the coastline. However, to be able to apply a sponge layer to absorb waves reflected towards the wave generation boundary, waves should be generated inside the numerical domain and not along the boundary. This internal wave generation technique in combination with numerical wave absorbing sponge layers was firstly proposed by [7] for Peregrine's [8] classical Boussinesq equations. Later, Lee and Suh [9] achieved wave generation for the mild slope equations of [3] and [4] by applying the source term addition method. The main observation was that the energy of the incident waves can be properly obtained from the viewpoint of energy transport. To generate multidirectional waves, they applied an L-shaped wave generator, which is generally composed of two wave generation lines; one parallel to the x-axis and one parallel to the y-axis of the numerical domain as well as wave absorbing sponge layers behind the two wave generation lines. However, wave diffraction patterns appear inside the computational domain as a result of the intersection of the two wave generation lines and due to the interaction with the lateral sponge layers. To deal with this problem, Lee and Yoon [10] proposed an arc-shaped wave generation line; two parallel lines connected to a semicircle to avoid wave diffraction caused at the intersection of the previous wave generation lines. Further, Kim and Lee [11] used an arc-shaped source band that gives smaller errors than the Lee and Yoon [10] method, especially for a coarse grid size. Recently, Lin and Yu [12] proposed a promising method for non-reflective boundaries in a mild-slope wave model to avoid the use of sponge layers. However, in non-reflective boundaries, the level of re-reflection strongly depends on the initial approximations since the characteristics of the reflected waves (i.e., wave angle, wave celerity) inside the numerical domain cannot be estimated a priori.

In the present paper, a wave generation layout using periodic lateral boundaries is developed where a single internal wave generation line parallel to the y-axis is combined with periodic lateral boundaries at the top and bottom of the domain. With this technique, the information leaving one end of the numerical domain enters the opposite end and thus no lateral sponges are required. In this way, the wave diffraction patterns that appear inside the computational domain as a result of the intersection of the two wave generation lines and due to the interaction with the lateral sponge layers (see Figure 1) are avoided. In Figure 1, the water surface elevation, η, is presented for a regular wave field generated by an L-shaped wave generator (red dashed lines) at an angle, $\theta = 45°$, from the x-axis. The absence of uniformity of the surface elevation along the crests and troughs, which can be observed in the same figure, is a sign of the existence of diffraction patterns inside the numerical domain.

Periodic boundaries are implemented in a time dependent mild-slope equation model, MILDwave, developed at Ghent University [13] in order to accurately generate regular and irregular waves in any direction. In addition, the performance of the periodic boundaries is evaluated for two different wave synthesis methods to generate short crested waves, a method proposed by Miles [14] and a method proposed by Sand and Mynett [15]. The mild-slope equations of Radder and Dingemans [4] without the extension of Suh et al. [6] are the basic equations employed in the phase-resolving model, MILDwave, used to simulate wave transformation processes, such as refraction, shoaling, reflection, transmission, and diffraction, intrinsically [16]. MILDwave has previously been used to predict wave diffraction and wave penetration inside harbours [17] and to study wave power conversion applications [18–20]. One of the challenges in the field of renewable energies is to determine the optimal geometrical layout for wave energy converter (WEC) farms, targeting the maximum possible energy production and the correct assessment of the impact of WEC farms on the wave field. To do so, accurate and detailed numerical modelling of WEC farms under realistic 3D wave conditions is considered crucial. This kind of application requires a homogeneous wave field in the whole numerical domain. To the present authors' knowledge, periodic boundaries, which are commonly used in non-hydrostatic models [21] and

Boussinesq models [22], have not been used before in a mild-slope wave model to study short crested waves. So, the novelty of the present work concerns the capability of accurate and fast generation of such homogeneous wave fields with periodic boundaries in MILDwave. In engineering applications, where non-linearities are not significant, mild-slope wave models are preferred instead of Boussinesq models or non-hydrostatic models due to their significantly smaller computational cost and their high numerical stability and robustness, and thus further development of these models should be encouraged. The implementation of periodic boundaries is important as it introduces noteworthy improvements in mild-slope models, which can then make full use of their benefits for the study of WEC farms under oblique long crested regular and irregular waves or short crested waves (real sea waves).

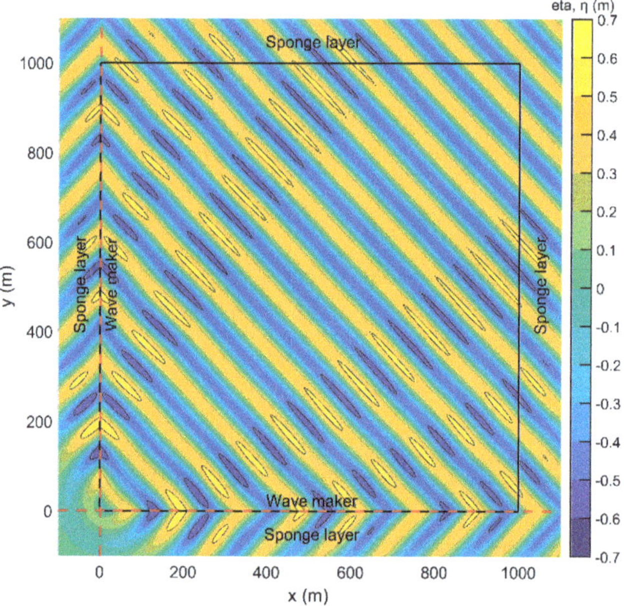

Figure 1. Water surface elevation, η, for regular incident waves with a wave height of H = 1 m, wave period of T = 12 s, and wave propagation angle of θ = 45° affected by diffraction patterns due to the intersection of the two wave generation lines and the interaction with the lateral sponge layers.

In the next section, a short description of the mild-slope wave propagation model, MILDwave, and the methodology of the implemented periodic boundaries are presented. Section 3 provides a detailed overview of the model results for regular and irregular long and short crested waves. In Section 4, the validation results are presented where the accuracy of the developed model is compared with experimental data. The last section provides conclusions and a summary discussion of the present study.

2. Numerical Methodology

2.1. The Mild-Slope Wave Propagation Model, MILDwave

MILDwave is a mild-slope wave propagation model based on the depth integrated time dependent mild-slope equations of Radder and Dingemans [4]. These are given in Equations (1) and (2) and describe the transformation of linear irregular waves with a narrow frequency band over a mildly varying bathymetry (bottom slopes up to 1/3):

$$\frac{\partial \eta}{\partial t} = B\varphi - \nabla \cdot (A\nabla\varphi) \tag{1}$$

$$\frac{\partial \varphi}{\partial t} = -g\eta \tag{2}$$

where $\eta(x, y, t)$ and $\phi(x, y, t)$ are, respectively, the surface elevation and the velocity potential at the still water level, ∇ is the horizontal gradient operator, t is the time, and g is the gravitational acceleration. The coefficients, A and B, are calculated using Equations (3) and (4), respectively:

$$A = \frac{CC_g}{g} \tag{3}$$

$$B = \frac{\omega^2 - k^2 CC_g}{g} \tag{4}$$

with C the phase velocity and C_g is the group velocity for a wave with a wave number, k, angular frequency, ω, wavelength, L_w, and frequency, f. For irregular waves, C, C_g, k, and ω are replaced by the wave characteristics for the carrier frequency, $\bar{f}$.

A finite difference scheme, as described in [23], is used to discretize and solve Equations (1) and (2). The domain is divided in grid cells with dimensions, Δx and Δy, and central differences are used both for spatial and time derivatives. Both η and ϕ are calculated at the center of each grid cell at different time levels, $(n + 1/2)\Delta t$ and $(n + 1)\Delta t$ as:

$$\eta_{i,j}^{n+1/2} \simeq \eta_{i,j}^{n-1/2} + B_{i,j}\phi_{i,j}^n \Delta t - \frac{A_{i+1,j} - A_{i-1,j}}{2\Delta x} \frac{\phi_{i+1,j}^n - \phi_{i-1,j}^n}{2\Delta x} \Delta t \\ - A_{i,j}\frac{\phi_{i-1,j}^n - 2\phi_{i,j}^n + \phi_{i+1,j}^n}{\Delta x^2} \Delta t \\ - \frac{A_{i,j+1} - A_{i,j-1}}{2\Delta y} \frac{\phi_{i,j+1}^n - \phi_{i,j-1}^n}{2\Delta y} \Delta t \\ - A_{i,j}\frac{\phi_{i,j-1}^n - 2\phi_{i,j}^n + \phi_{i,j+1}^n}{\Delta y^2} \Delta t \tag{5}$$

$$\phi_{i,j}^{n+1} \simeq \phi_{i,j}^n - g\eta_{i,j}^{n+1/2}\Delta t \tag{6}$$

where A and B are given by Equations (3) and (4) and the superscripts and subscripts stand for the relevant time step and the relevant cell of the grid, respectively.

In the current configuration of MILDwave, the boundary conditions are formed in such a way that one layer of ghost cells is considered at each boundary. The values at the cells closest to the boundary are copied to the ghost cells and thus the layer of the ghost cells acts as a fully reflective boundary (solid wall). However, the effect of the fully reflective domain boundaries is negligible because absorbing sponge layers are applied at the outside boundaries to dissipate the incoming wave energy.

The grid cell size, $\Delta x = \Delta y$, is chosen so that $L_w/20 \leq \Delta x = \Delta y \leq L_w/10$ (for irregular waves, L_w = shortest wave length (maximum wave frequency)) while the time step meets the Courant-Friedrichs-Lewy criterion to guarantee a stable and consistent result.

In MILDwave, waves are generated along a wave generation line near the offshore boundary by applying the source term addition method proposed by [9] where the source term propagates with the energy velocity. According to this method, additional surface elevation, η^*, is added with the desired energy to the calculated surface elevation, η, at the wave generation line for each time step and is given by:

$$\eta^* = 2\eta^I \frac{C_e \Delta t}{\Delta x} \cos\theta \tag{7}$$

where Δx is the grid size in the x-axis, Δt is the time step, θ is the wave propagation angle with respect to the x-axis, η^I the water surface elevation of incident waves, and C_e is the energy velocity given by Equation (8):

$$C_e = \overline{C_g} \frac{\overline{\omega}}{\omega} \sqrt{1 + \frac{\overline{C}}{\overline{C_g}} \left(\left(\frac{\omega}{\overline{\omega}} \right)^2 - 1 \right)} \tag{8}$$

where the bar indicates that the variable is associated with the carrier angular frequency, $\overline{\omega}$.

It has been proven that the model of [4] can be used to simulate the transformation of long and short crested waves. To generate long crested irregular waves, a parameterized JONSWAP (Joint North Sea Wave Observation Project) spectrum, $S(f)$ (Equation (9)), or a TMA (Texel, Marsen, Arsloe) spectrum (Equation (10)), which can be applied in shallow water, has been used as an input spectrum:

$$S(f) = \frac{0.0624}{0.230 + 0.0336\gamma - \frac{0.185}{1.9+\gamma}} H_s^2 f_p^4 f^{-5} \gamma^{\exp\left(-\frac{(f-f_p)^2}{2\sigma^2 f_p^2}\right)} \exp\left(-\frac{5}{4}\left(\frac{f_p}{f}\right)^4\right) \tag{9}$$

$$S(f) = \frac{\alpha g^2}{(2\pi)^4 f^5} \exp\left(-\frac{5}{4}\left(\frac{f_p}{f}\right)^4 + (\ln\gamma)\exp\left(-\frac{(f-f_p)^2}{2\sigma^2 f_p^2}\right)\right)\phi(f,d) \tag{10}$$

where H_s is the significant wave height, f_p is the peak wave frequency, γ is the peak enhancement factor, α is the Phillips constant, and σ is the spectral width parameter, which depends on the value of the wave frequency:

$$\sigma = \begin{cases} 0.07 & f \leq f_p \\ 0.09 & f \geq f_p \end{cases} \tag{11}$$

The frequency dependent factor, $\phi(f,d)$, which takes into account the effect of a finite water depth, d, is given by:

$$\phi(f,d) = \begin{cases} 0.5\omega_h^2, & \omega_h < 1 \\ 1 - 0.5(2 - \omega_h)^2, & 1 \leq \omega_h \leq 2 \\ 1, & \omega_h > 2 \end{cases}, \quad \omega_h = 2\pi f \sqrt{\frac{d}{g}} \tag{12}$$

To generate short crested waves, two different wave synthesis methods are employed in the present study, a single summation method proposed by [14] and a second single summation method proposed by [15]. In single summation models, each wave component must have a unique frequency and each frequency component can only travel in one direction, while several wave components are travelling in the same direction. According to the Miles method, the surface elevation is defined as follows:

$$\eta(x,y,t) = \sum_{n=1}^{N} \sum_{m=1}^{M} A_{nm} \cos(\omega_{nm}t - k_{nm}(x\cos\theta_m + y\sin\theta_m) + \varepsilon_{nm}) \tag{13}$$

with the wave amplitude, $A_{nm} = \sqrt{2S(f_{nm})D(f_{nm},\theta_m)M\Delta f\Delta\theta}$, the wave angular frequency, $\omega_{nm} = 2\pi f_{nm} = 2\pi(M(n-1) + m)\Delta f + 2\pi f_{min}$, the frequency interval, $\Delta f = \frac{f_{max} - f_{min}}{NM-1}$, the wave propagation angle, $\theta_m = (m-1)\Delta\theta + \theta_0 - \theta_{max}$, the wave propagation angle interval, $\Delta\theta = \frac{2\theta_{max}}{M-1}$, the random phase, ε_{nm}, and the maximum discrete wave direction, θ_{max}. In this way, the directional spreading function is discretized into M equally spaced wave directions ranging from θ_{min} to θ_{max}.

Sand and Mynett [15] proposed a method in which the directional spectrum is decomposed as follows:

$$\eta(x,y,t) = \sum_{n=1}^{N} \sqrt{2S(f_n)\Delta f} \cos(\omega_n t - k_n(x\cos\theta_n + y\sin\theta_n) + \varepsilon_n) \tag{14}$$

In this method, the wave propagation angles are selected at random according to the cumulative distribution function of the directional spreading function, $D(f,\theta)$, and are assigned to each frequency component. The Miles method [14] provides an accurate representation of the targeted spreading function shape, but introduces different localized distortions in the frequency spectrum to obtain a close fit to the spreading function. On the other hand, the Sand and Mynett method [15] yields an exact match to the frequency spectrum.

Several semi-empirical proposed formulations of the directional spreading function, $D(f, \theta)$, have been reported, most of which consider the spreading function to be independent of the wave frequency. Here, an alternative of the well-known spreading function of [24] is employed [25]:

$$D(f, \theta) = \frac{1}{\sqrt{\pi}} \frac{\Gamma(s_1+1)}{\Gamma(s_1+\frac{1}{2})} \cos^{2s_1}(\theta - \theta_0), \quad -\frac{\pi}{2} < \theta - \theta_0 < \frac{\pi}{2} \tag{15}$$

where s_1 is the directional spreading parameter, Γ is the Gamma function, and θ_0 is the wave propagation angle.

To derive the relation between the directional spreading parameter, s_1, and the spreading standard deviation, σ_θ, which is called the directional width, the methodology of [26] is followed:

$$\sigma_\theta = \sqrt{2\left(1 - \sqrt{\left(\int_0^{2\pi} D(f, \theta)\sin\theta d\theta\right)^2 + \left(\int_0^{2\pi} D(f, \theta)\cos d\theta\right)^2}\right)} \tag{16}$$

Substitution of Equation (15) into Equation (16) yields:

$$\sigma_\theta = \sqrt{2 - \frac{2\Gamma^2(s_1 + 1)}{\Gamma\left(s_1 + \frac{1}{2}\right)\Gamma\left(s_1 + \frac{3}{2}\right)}} \tag{17}$$

The relation between s_1 and σ_θ is indicated in Figure 2. Hence, fixed values of s_1 and σ_θ can be determined for wind and swell waves:

$$s_1 = \begin{cases} 1.17 & \rightarrow \sigma_\theta = 30° \quad \text{(wind waves)} \\ 15.8 & \rightarrow \sigma_\theta = 10° \quad \text{(swell waves)} \end{cases} \tag{18}$$

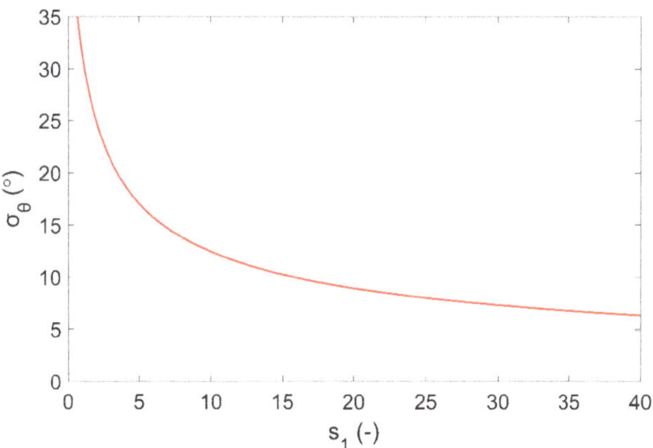

Figure 2. Relation between the directional spreading parameter, s_1, and the spreading standard deviation, σ_θ.

2.2. Periodic Boundaries

To create a homogeneous wave field of oblique long crested regular and irregular waves or short crested waves, periodic lateral boundaries have been implemented. In the present study, a single wave generation line parallel to the y-axis is combined with periodic lateral boundaries at the top and bottom of the domain. In this way, the information leaving one end of the numerical domain enters the opposite end and thus the required model length in this direction is reduced. This technique can lead to a more homogeneous wave field than an L-shaped and an arc-shaped wave generation since no wave diffraction problems are caused by the presence of lateral sponge layers and the intersection of the two wave generation lines (Figure 1).

Figure 3 is a schematic representation of the way that the periodic boundaries are working. Firstly, the left part of the figure presents the numerical domain under investigation, with the squares representing a random number of cells where Nx and Ny is the number of cells in the x and y direction, respectively. A layer of ghost cells is present next to each vertical boundary, acting as a fully reflective wall as described in the previous section. On the other hand, the dashed lines, which are parallel to the x-axis, represent the periodic boundaries.

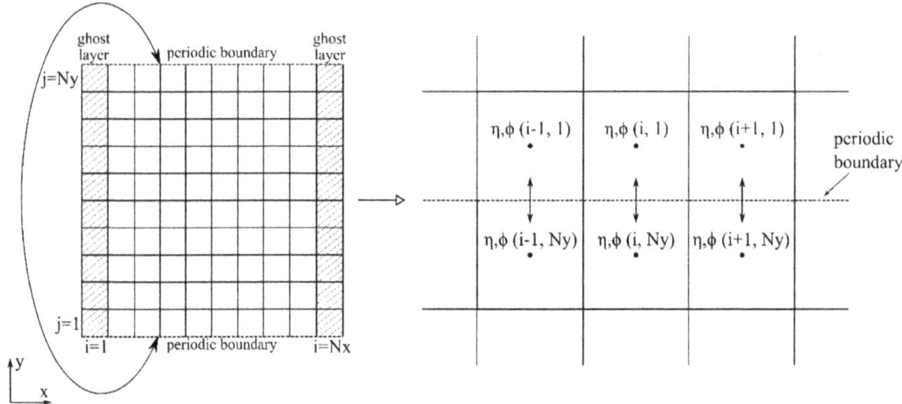

Figure 3. Definition sketch of the implemented periodic lateral boundaries.

At the position of the periodic boundaries, Equation (5) is solved considering that the two boundaries are adjacent, as it can be observed from the right part of Figure 3, yielding Equations (19) and (20) instead of Equation (5) for the bottom (j = 1) and top (j = Ny) layer of cells, respectively:

$$
\begin{aligned}
\eta_{i,1}^{n+1/2} \simeq{} & \eta_{i,1}^{n-1/2} + B_{i,1}\phi_{i,1}^{n}\Delta t \\
& - \frac{A_{i+1,1}-A_{i-1,1}}{2\Delta x}\frac{\phi_{i+1,1}^{n}-\phi_{i-1,1}^{n}}{2\Delta x}\Delta t \\
& - A_{i,1}\frac{\phi_{i-1,1}^{n}-2\phi_{i,1}^{n}+\phi_{i+1,1}^{n}}{\Delta x^{2}}\Delta t \\
& - \frac{A_{i,2}-A_{i,Ny}}{2\Delta y}\frac{\phi_{i,2}^{n}-\phi_{i,Ny}^{n}}{2\Delta y}\Delta t \\
& - A_{i,j}\frac{\phi_{i,Ny}^{n}-2\phi_{i,1}^{n}+\phi_{i,2}^{n}}{\Delta y^{2}}\Delta t
\end{aligned}
\tag{19}
$$

$$
\begin{aligned}
\eta_{i,Ny}^{n+1/2} \simeq{} & \eta_{i,Ny}^{n-1/2} + B_{i,Ny}\phi_{i,Ny}^{n}\Delta t \\
& - \frac{A_{i+1,Ny}-A_{i-1,Ny}}{2\Delta x}\frac{\phi_{i+1,Ny}^{n}-\phi_{i-1,Ny}^{n}}{2\Delta x}\Delta t \\
& - A_{i,j}\frac{\phi_{i-1,Ny}^{n}-2\phi_{i,Ny}^{n}+\phi_{i+1,Ny}^{n}}{\Delta x^{2}}\Delta t \\
& - \frac{A_{i,1}-A_{i,Ny-1}}{2\Delta y}\frac{\phi_{i,1}^{n}-\phi_{i,Ny-1}^{n}}{2\Delta y}\Delta t \\
& - A_{i,j}\frac{\phi_{i,Ny-1}^{n}-2\phi_{i,Ny}^{n}+\phi_{i,1}^{n}}{\Delta y^{2}}\Delta t
\end{aligned}
\tag{20}
$$

This implementation is valid under the assumption that the bathymetry close to the lateral boundaries is uniform for the waves to pass from one boundary to the other without any distortion and thus the continuity to be satisfied. In addition, the distance between the two boundaries should be an integer multiple of the wavelength to avoid phase differences. To ensure that this requirement is valid, the wave propagation angle, θ, of each wave component is slightly adjusted such that Equation (21) is satisfied [27]:

$$k \sin \theta = \alpha \, \frac{2\pi}{L_{per}} \tag{21}$$

where k is the wave number of each wave component, α is an integer, and L_{per} ($L_{per} = \Delta y * Ny$) is the periodicity length.

This means that the simulated wave propagation angle slightly deviates from the input wave angle. To minimize this deviation, the periodicity length, L_{per}, of the computational domain should be carefully chosen to be larger than one periodic wave length, L_y ($L_y = L_w / \sin \theta$) [28]. Hence, in the case of irregular long crested waves, each wave component propagates in a slightly different direction than the mean value and a directional spread of the wave energy occurs. However, it has been verified that the spreading standard deviation, σ_θ, is very small ($\sigma_\theta < 2°$) and the waves can be considered as uni-directional.

3. Results by the Model, MILDwave

In this section, the applicability and accuracy of periodic boundaries to propagate regular and irregular long and short crested waves over a flat bottom is examined. For the evaluation of the numerical simulation results, a common method is followed. The results for a series of test cases are presented in terms of a disturbance coefficient, k_d, for the entire numerical domain. For regular waves, k_d is given by the ratio, H/H_{GB}, where H is the local wave height and H_{GB} is the wave height at the wave generation boundary. In the case of irregular waves, the significant wave heights are used to calculate the k_d value (H_s/H_{sGB}). For all test cases, the numerical basin is 20 L_w long and 20 L_w wide (where L_w is the wave length and is equal to L_w = 100 m) while a uniform water depth, d = 7.5 m, is used. Thus, in the following numerical tests, the target wave field is a homogeneous one. A homogeneous wave field is characterized by the lack of significant fluctuations from the target, k_d, value, which is equal to k_{d_t} = 1.

3.1. Generation of Regular Waves

Firstly, oblique regular waves are examined to compare three different wave generation layouts. The first wave generation layout (Figure 4a) is an L-shaped wave generator, which is composed by two orthogonal lines as proposed by [9]. The two wave generation lines intersect at a point outside the sponge layers and they extend to the computational domain boundary. The hatched areas indicate the areas of the numerical domain covered by wave absorbing sponge layers. These sponge layers are necessary to absorb waves reflected by the numerical domain boundaries and by structures. The second wave generation layout (Figure 4b) is based on the method suggested by [10], who proposed an arc-shaped wave generation layout where two parallel lines are connected to a semicircle. Figure 4c shows the third wave generation layout that is the main research object of the present paper, in which a single wave generation line parallel to the y-axis is combined with periodic lateral boundaries at the top and bottom of the domain. This means that the computational grid is repeated in the y-direction and hence waves are passing freely from the one periodic boundary to the other.

Due to the nature of the governing equations of the numerical model, MILDwave (linear mild-slope equations), and the fact that a flat empty numerical basin is examined, the performance of the model for different wave heights and periods is similar. Thus, regular waves with a wave period of T = 12 s, wave height of H = 1 m, and wave propagation angles from $\theta = 0°$ to $\theta = 45°$ with respect to the x-direction are simulated. The numerical basin consists of an inner computational domain without the sponge layers with dimensions of 10 L_w × 10 L_w for the first two layouts and 10 L_w × 20 L_w for the third one, as well as sponge layers with a width of 5 L_w at the edges of the wave basin. The sponge layer formula as described in [18] is used to minimize reflections, where the free surface elevation is multiplied on each new time step with an absorption function that has the value of 1 at the start of the sponge layer and smoothly decreases until it reaches the value of 0 at the end. As a result, the reflection is zero. The grid cell size is chosen so that $\Delta x = \Delta y = L_w / 20 = 5$ m. To obtain a steady state wave field, waves are generated for a duration of 1000 s with a time step of $\Delta t = 0.25$ s.

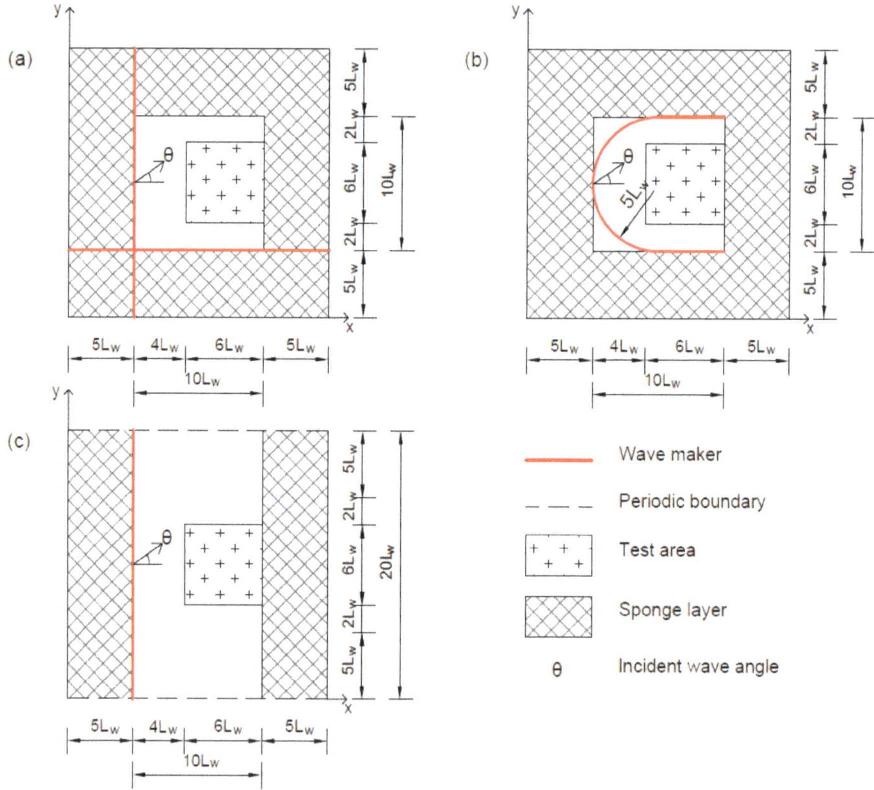

Figure 4. Definition sketch of three wave generation layouts: (**a**) L-shaped, (**b**) arc-shaped, (**c**) line-shaped with periodic lateral boundaries. Dimensions are expressed in the number of wave lengths, $L_w = 100$ m, which corresponds to a wave period, $T = 12$ s, and water depth, $d = 7.5$ m.

Figure 5 shows the plan views of the resulting disturbance coefficient, k_d, and the water surface elevation, η, in the inner computational domain at a time of $t = 80$ T, where T is the wave period, for each wave generation layout for a wave propagation angle of $\theta = 30°$. The black solid contour lines indicate the region where the surface elevation is out of the target range ($\eta > 0.5$ m, $\eta < -0.5$ m). It is observed that the third wave generation layout with the periodic lateral boundaries yields a more homogeneous wave field across the whole domain, which is not disturbed by unwanted wave diffraction patterns in contrast to the other two wave generation layouts. These wave diffraction disturbances are caused by the lateral sponge layers and the intersection of the two wave generation lines, and thus the k_d values fluctuate around the target value ($k_{d_t} = 1$).

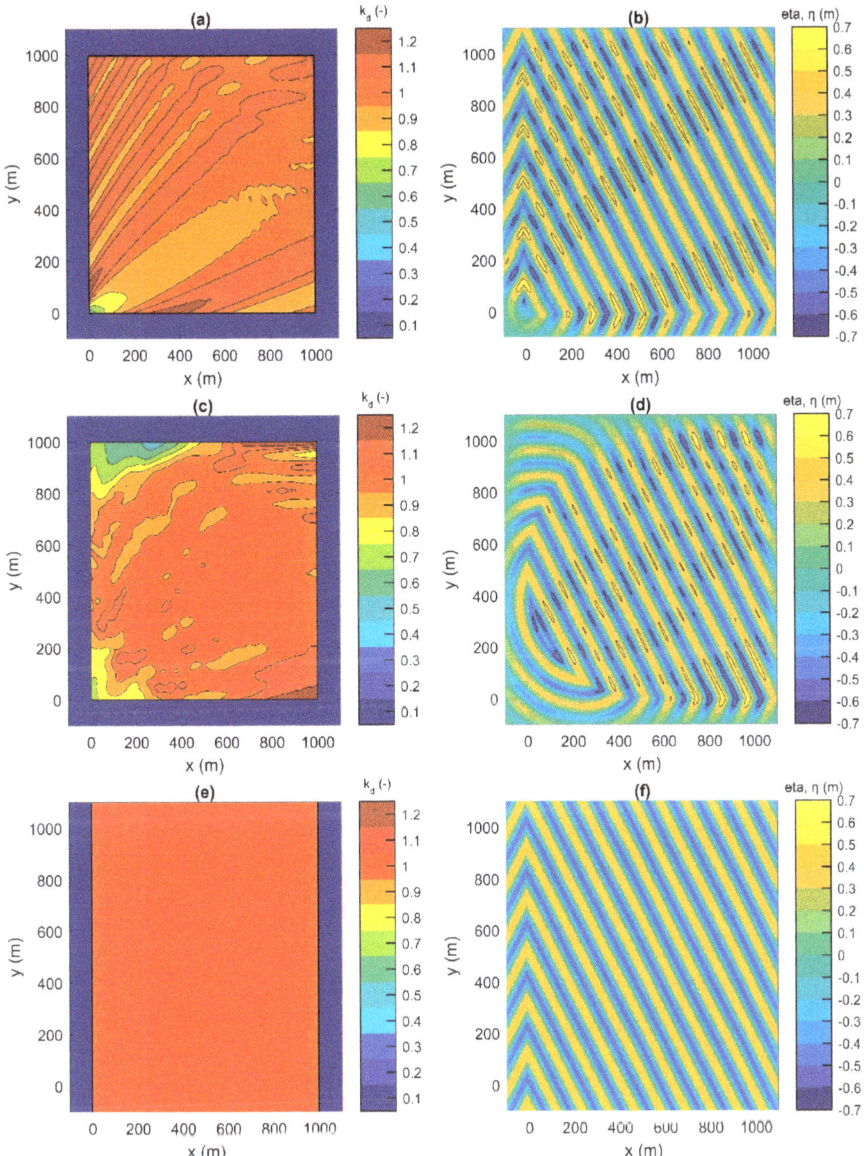

Figure 5. Disturbance coefficient, k_d, and water surface elevation, η, in the inner computational domain at t = 80 T for regular incident waves with H = 1 m, T = 12 s, and θ = 30° generated by an L-shaped wave generation layout (**a,b**), an arc-shaped wave generation layout (**c,d**), and by the use of periodic boundaries (**e,f**).

To obtain a quantified difference between the three different wave generation layouts, the absolute error is calculated in a specific test area far from the sponge layers as in [10]. The error is determined as follows:

$$E = \frac{|H_c - H_t|}{H_t} \tag{22}$$

where H_c is the local resulting wave height at a point of the numerical domain and H_t is the target wave height. The test area in which the absolute error is calculated is 6 L_w long and 6 L_w wide and is indicated in Figure 4 (area of crosses). In Figure 6, the mean (E_{mean}) and the maximum (E_{max}) value of the calculated absolute errors for each wave generation layout for a range of different wave propagation angles are plotted. It is observed, as seen from Figure 5 for $\theta = 30°$, that the periodic boundaries' wave generation layout provides the smallest value of E_{mean} and E_{max} for all θ values. For the L-shaped wave generation layout and the periodic boundaries' wave generation layout, E_{mean} is the lowest for $\theta = 45°$ with values of 0.042 and 0.007, respectively, while for the arc-shaped wave generation layout, E_{mean} is the lowest for $\theta = 30°$ with a value of 0.021. The trend of E_{max} (Figure 6b) is similar to that of the values of E_{mean} from Figure 6a. The E_{max} values of the oblique regular wave field generated using periodic boundaries are significantly smaller than the resulting E_{max} values for the L-shaped and the arc-shaped layouts for all wave directions. In addition, the comparison between the L-shaped wave generation layout and the arc-shaped wave generation layout reveals that in the case of the latter, the wave diffraction due to the intersection of the two generation lines is avoided [10]. However, for wave directions of θ larger than 30°, E_{max} is similar for both layouts with values around 0.105 while for the periodic boundaries' wave generation layout, it is only 0.012. Hence, this proves that a single wave generation line combined with periodic lateral boundaries provides results of a higher accuracy compared to the other two wave generation layouts and is capable of providing a homogeneous wave field of short crested waves with minimal error.

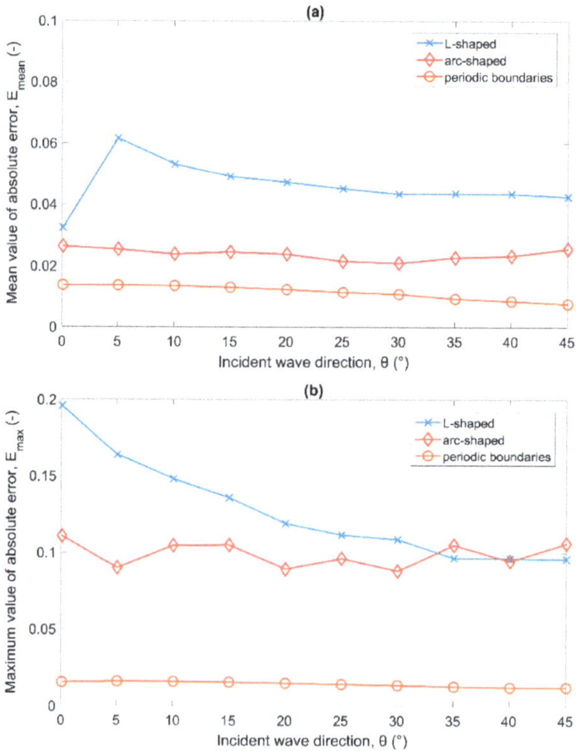

Figure 6. (**a**) Mean (E_{mean}) and (**b**) maximum (E_{max}) value of the absolute errors calculated using Equation (22) for each wave generation layout for a range of different incident wave angles, θ.

3.2. Generation of Irregular Long Crested Waves

We consider a test case of uni-directional irregular waves fitting a JONSWAP spectrum, with a peak wave period, T_p = 12 s, a significant wave height, H_s = 1 m, and a peak enhancement factor, γ = 3.3. The frequency range is confined between 0.75 f_p and 2 f_p, which covers 94% of the total energy, to ensure numerical stability. Periodic lateral boundaries are applied at the top and bottom of the domain similar to Section 3.1, while the length of the sponge layers at both ends of the numerical basin is determined by the longest wave length in the incident wave spectrum. The grid cell size and the time step, Δt, are defined by considering the highest frequency (2 f_p) and the lowest frequency (0.75 f_p), respectively.

A single wave generation line is placed at the left boundary after the sponge layer. To obtain a steady state wave field, time series of the surface elevation of irregular waves are generated for a duration of 7200 s with a time step of Δt = 0.2 s. Figure 7 shows the plan views of the k_d values and the water surface elevations in an inner domain (part of the domain without the left and right sponge layers) of 1 km in both directions, at a time instant of t = 7000 s, for wave propagation angles of θ = $0°$ and θ = $45°$. It is clear that for both wave propagation angles, the k_d is uniform all over the inner domain with values close to k_{d_t} = 1. In addition, it is observed by comparing the surface elevation of the two different wave directions that for the case of θ = $45°$, the surface is not as uniform along the crest as it is for the case of θ = $0°$. This is happening due to the fact that in order to ensure that waves are periodic, the wave propagation angle, θ, of each wave component is slightly adjusted and a directional spread of the wave energy occurs as described in detail in Section 2.2. Hence, the wave propagation angle varies from $43.2°$ to $47.7°$ for the present wave conditions. However, the deviation from the mean wave direction is small and the waves can be considered as uni-directional.

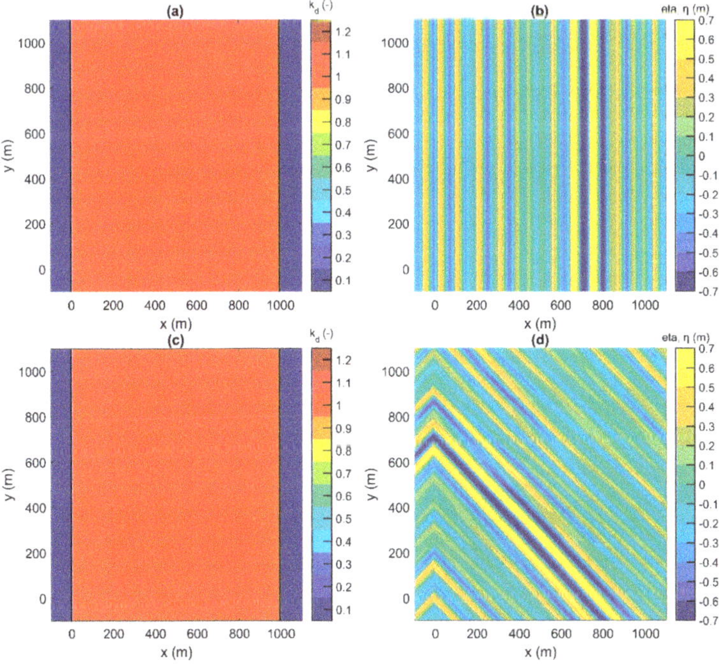

Figure 7. Disturbance coefficient, k_d, and water surface elevation, η, in the inner computational domain at t = 7000 s for irregular long crested waves with H_s = 1 m, T_p = 12 s, and (**a**,**b**) θ = $0°$ and (**c**,**d**) θ = $45°$.

In Figure 8, a comparison is made between the target frequency spectrum (S_t) and the simulated frequency spectra (S_0 and S_{45}) for the cases of $\theta = 0°$ and $\theta = 45°$. The surface elevations, η, at the electronic wave gauges, which are positioned at the centre of the computational domain, are recorded from t = 40 T_p to t = 600 T_p with a sampling interval of 0.2 s. The recorded data are processed in segments of 2048 points per segment. A taper window and an overlap of 20% are applied for smoother and statistically more significant spectral estimates. The resulting frequency spectra agree well with S_t apart from those that correspond to wave frequencies higher than 2 f_p, since there is no energy for these frequencies in the MILDwave model. Furthermore, it is observed that the periodic lateral boundaries do not affect the waves since for both wave propagation angles, the frequency spectra are similar.

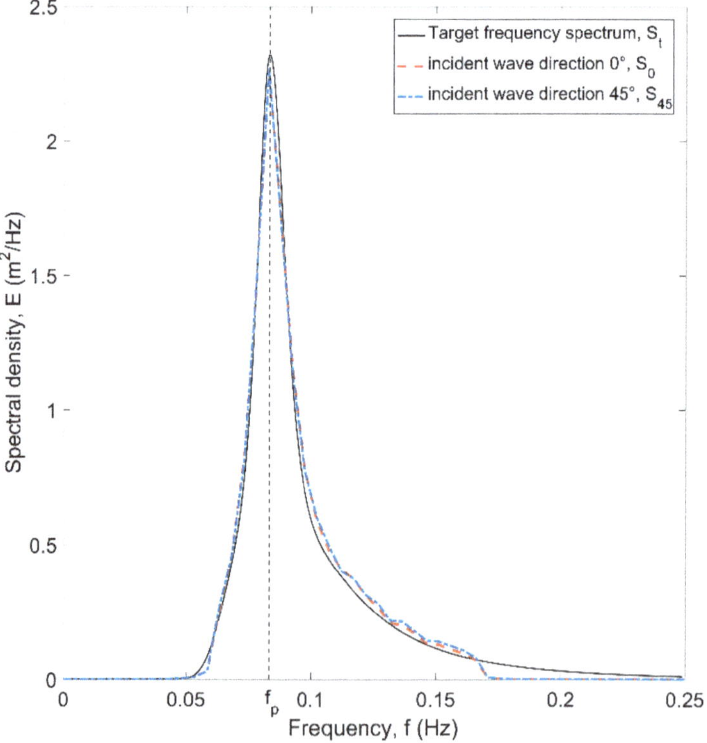

Figure 8. Comparison between the frequency spectra, S_0 and S_{45}, resulting from MILDwave simulations for different incident wave angles ($\theta = 0°$ and $\theta = 45°$, respectively), and the target frequency spectrum, S_t, for irregular waves with $H_s = 1$ m, $T_p = 12$ s, and $\gamma = 3.3$.

3.3. Generation of Irregular Short Crested Waves

Finally, we consider a test case of irregular short crested waves. The parameters of the input frequency spectrum are identical as for the long-crested irregular waves ($H_s = 1$ m, $T_p = 12$ s, $\gamma = 3.3$) described in the previous section. Waves are generated for a duration of 7200 s with $\Delta t = 0.2$ s while to generate short crested waves, two different wave synthesis methods have been employed: A method proposed by [14] and a method proposed by [15] as described in detail in Section 2.1.

The directional wave spectrum was measured by a group of five wave gauges placed in the centre of the computational domain, using a "CERC 5" configuration as introduced by [29]. The recorded normalised spreading function distributions, D_n, are compared with the target distribution, $D_{n,t}$, in Figure 9 for both wave synthesis methods and for the spreading standard deviation, $\sigma_\theta = 10°$ (swell waves) and $\sigma_\theta = 30°$ (wind waves). The normalised spreading function is calculated by integrating the calculated 3D spectrum over all wave frequencies and normalizing it. The difference between the two measured normalised spreading function distributions of the two methods is small and the agreement with the target distribution is very good. The Sand and Mynett method agrees slightly better with the target distribution, $D_{n,t}$, than the Miles method especially for the case of $\sigma_\theta = 30°$.

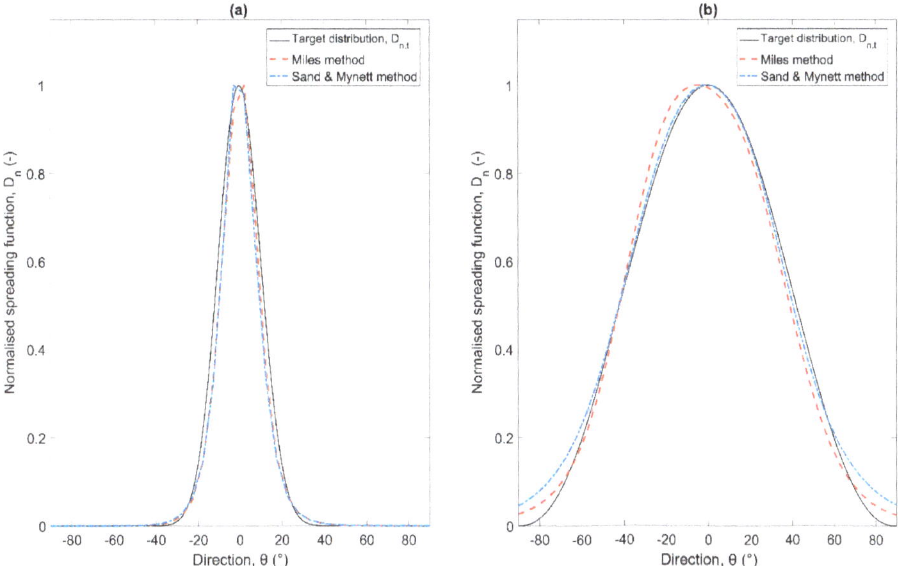

Figure 9. Comparison between the normalised spreading function distributions, D_n, calculated in MILDwave with the Miles and the Sand and Mynett wave synthesis methods and the target distribution, $D_{n,t}$, for irregular short crested waves with a spreading standard deviation, σ_θ, of (a) 10° and (b) 30°.

Figures 10 and 11 show the plan views of the k_d values throughout the inner computational domain and the surface elevations at a time instant of $t = 7000$ s for each wave synthesis method for a mean wave propagation angle of $\theta = 0°$ and spreading standard deviation of $\sigma_\theta = 10°$ and $\sigma_\theta = 30°$. The results are presented in an inner domain with dimensions of $10\,L_p \times 10\,L_p$, where L_p is the wave length that corresponds to the peak wave frequency, f_p. It is clear that the Sand and Mynett method yields a more homogenous wave field than the Miles method. Specifically, the maximum (E_{max}) value of the calculated absolute errors for the Miles method is 0.051 for $\sigma_\theta = 10°$ and 0.085 for $\sigma_\theta = 30°$, respectively, while for the Sand and Mynett method, E_{max} is 0.013 for $\sigma_\theta = 10°$ and 0.029 for $\sigma_\theta = 30°$, respectively. It is worth mentioning that in the Sand and Mynett method, the number of wave components, N_{tot}, is equal to the number of wave frequencies, N. This is happening due to the fact that the wave propagation angles are randomly selected according to the cumulative distribution function of the directional spreading function, $D(f, \theta)$, and are assigned to each wave frequency component. In contrast to that, in the Miles method, the number of the wave components, N_{tot}, is equal to the product of the number of wave frequencies, N, multiplied with the number of wave angles, M. Thus, in the Sand and Mynett method, (i) the computational time, (ii) the number of corrections of the wave propagation angles in order to ensure the periodicity of the waves, and (iii) the non-homogeneity of the generated wave field are much smaller than those for the Miles method, making the Sand and

Mynett method preferable when periodic lateral boundaries are applied in a mild-slope wave model. Specifically, the developed model simulation time for a test case of short crested waves of a duration of 7200 s in a numerical basin, which is 2 km long and 2 km wide, is approximately 3 min on a PC with an Intel Core CPU (Central Processing Unit) 2.90 GHz.

The results show that the MILDwave model with the addition of the periodic boundaries is capable of reproducing a homogeneous short-crested wave field in the whole computational domain as well as the target frequency spectrum and the target directional spectrum. Thus, in future studies, a WEC farm and its effect on the near and far field can be examined under real sea waves by coupling the developed model with a wave-structure interaction solver as described in [19,20]. More precisely, during this coupling methodology, firstly, the short-crested wave field is calculated by the developed model at the position of the WEC farm. Subsequently, this wave field (time series of surface elevation) is used as an input for a wave-structure interaction solver to simulate the diffracted and radiated wave field generated by the presence of the WEC farm. Finally, the resulting wave field is coupled back to MILDwave and propagates throughout the whole numerical domain. As a result, with the coupling methodology in MILDwave, the impact of WEC farms, especially the far field, can be studied using accurate real sea waves.

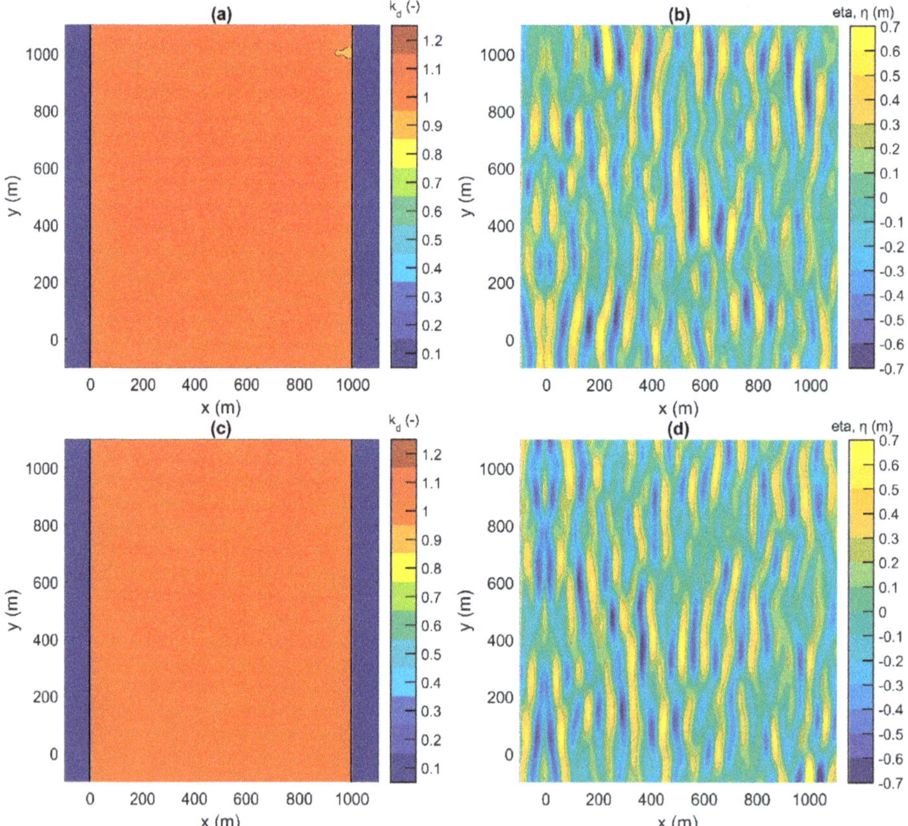

Figure 10. Disturbance coefficient, k_d, and water surface elevation, η, in the inner computational domain at t = 7000 s for irregular short crested waves with H_s = 1 m, T_p = 12 s, θ = 0°, and σ_θ = 10° synthesized by using the Miles method (**a,b**) and the Sand and Mynett method (**c,d**).

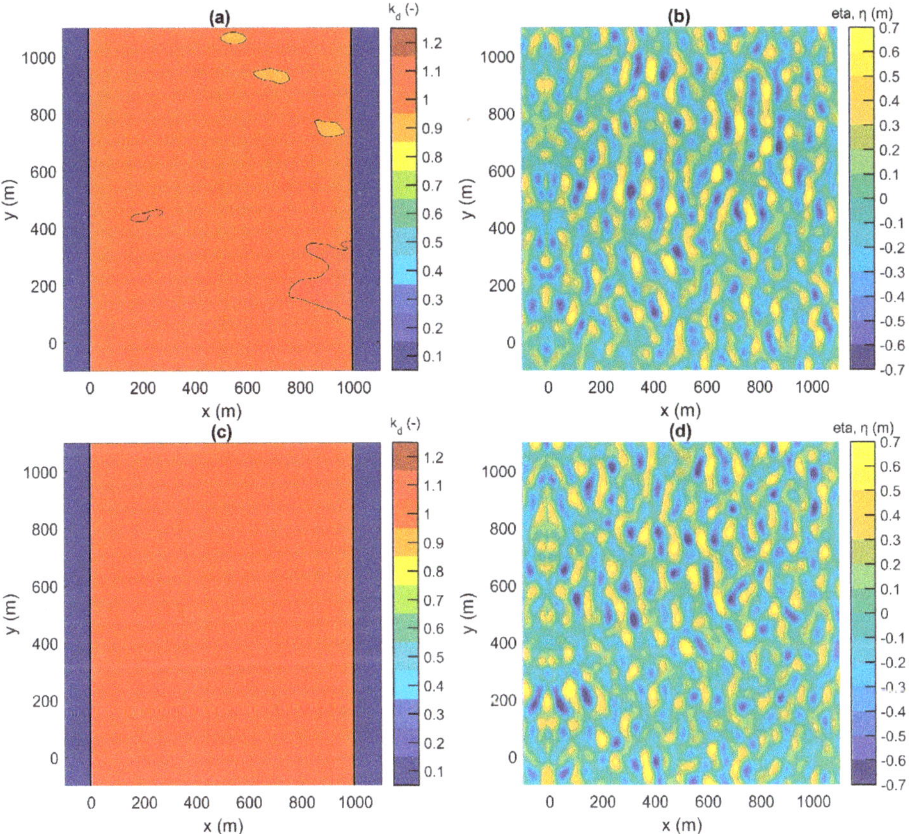

Figure 11. Disturbance coefficient, k_d, and water surface elevation, η, in the inner computational domain at t = 7000 s for irregular short crested waves with H_s = 1 m, T_p = 12 s, θ = 0°, and σ_θ = 30°, synthesized by using the Miles method (**a,b**) and the Sand and Mynett method (**c,d**).

4. Numerical Validation Using the Vincent and Briggs Shoal Experiment

To demonstrate the strategic importance of periodic boundaries in the generation of a short crested wave field over varying water depths, simulations are conducted for waves propagating over a shoal. A study of regular and irregular wave propagation over an elliptic shoal, which has been performed by [30], is used here for validation of the present numerical model.

The bathymetry of the experimental setup of Vincent and Briggs is implemented in MILDwave, as is illustrated in Figure 12, is defined as:

$$\left(\frac{x}{3.05}\right)^2 + \left(\frac{y}{3.96}\right)^2 = 1 \tag{23}$$

$$d_e = -0.4572 + 0.7620\left\{1 - \left(\frac{x}{3.81}\right)^2 + \left(\frac{y}{4.95}\right)^2\right\}^{0.5} \tag{24}$$

where x and y are the coordinates centered at the center of the shoal and d_e is the bottom level at any point inside the shoal area. The shoal is similar to that used in the experiments of [31], with a minimum water depth of d_{min} = 15.24 cm at the center of the shoal while the area around the shoal has a constant water depth of d = 45.72 cm. In addition, the simulated shoal is located at the center of the numerical domain to take advantage of the lateral periodic boundaries and create a fully developed wave field at the region of the shoal. The grid cell size is $\Delta x = \Delta y = 0.05$ m. To obtain a steady state wave field, waves are generated with a duration of 400 s with a time step of $\Delta t = 0.01$ s.

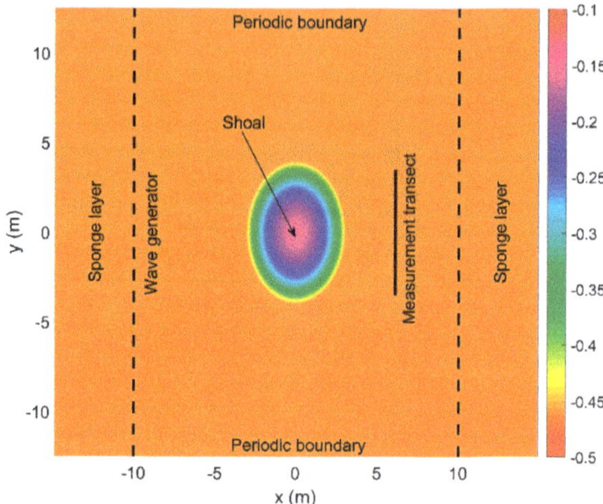

Figure 12. Bottom levels of the experimental setup used by Vincent and Briggs, as it is implemented in the numerical domain in MILDwave.

The experiments of Vincent and Briggs are conducted for three kinds of incident breaking and non-breaking waves, i.e., regular, irregular long crested, and irregular short crested waves. In the present study, only irregular long and short crested waves, where large scale breaking is not involved, are examined. The experimental and numerical input wave parameters for non-breaking series are listed in Table 1.

Table 1. Numerical input wave conditions for non-breaking waves based on the Vincent and Briggs experimental wave conditions.

Test Case ID	Peak Period, T_p (s)	Significant Wave Height H_s (cm)	Phillips Constant, α (-)	Peak Enhancement Factor, γ (-)	Spreading Standard Deviation, σ_θ (°)
U3	1.3	2.54	0.00155	2	0
N3	1.3	2.54	0.00155	2	10
B3	1.3	2.54	0.00155	2	30
U4	1.3	2.54	0.00047	20	0
N4	1.3	2.54	0.00047	20	10
B4	1.3	2.54	0.00047	20	30

Following [30], we use the TMA spectrum [32] as the target wave frequency spectrum in which the spectral energy density, $S(f)$, depends on the parameters, α (Phillip's constant), f_p (peak wave frequency), γ (peak enhancement factor), and σ (spectral width parameter), as described in Section 2.1. The parameter, γ, is assigned values of 2 for the broad frequency spectrum and 20 for the narrow frequency spectrum. Similarly, these frequency spectra are combined with narrow ($\sigma_\theta = 10°$) and broad ($\sigma_\theta = 30°$) directional spreading, while the α value is selected to correspond to the target wave height.

The directional spreading function, $D(f, \theta)$, given in Equation (15) is used instead of the wrapped normal spreading function used by [30]. However, the directional spreading parameter, s_1, is calculated according to Equation (17) in order to achieve the same distribution as the experimental one. Finally, the mean wave propagation angle is $\theta = 0°$ and the number of wave components, N, varies from 50 to 400, with the highest value for the case of broad-banded directional spreading. The measured frequency spectra are compared with the target spectrum (S_t) in Figure 13 for the two values of the peak enhancement factors used here (see Table 1). The agreement is very good especially for the narrow banded spectrum where the measured one replicates the desired target spectrum.

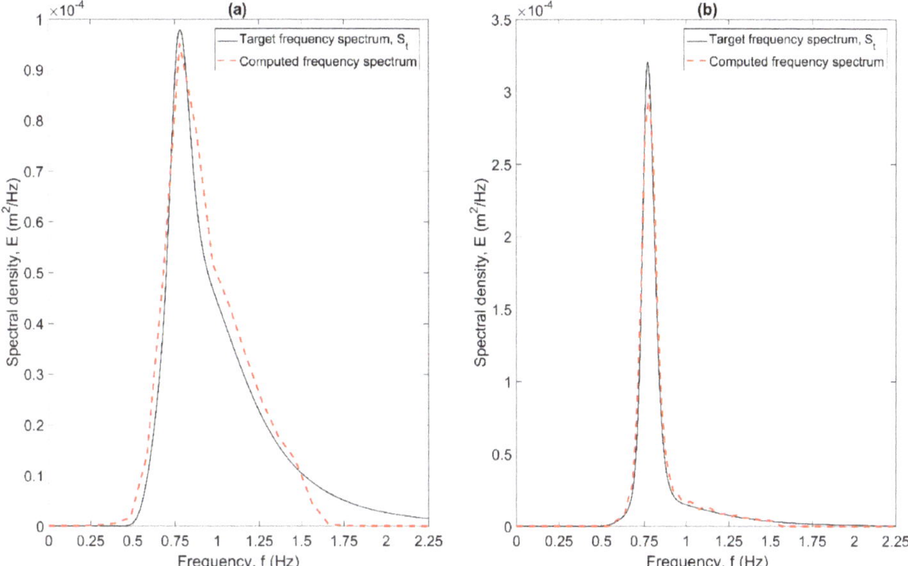

Figure 13. Comparison between the frequency spectra simulated in MILDwave and the target frequency spectrum for irregular waves with H_s = 2.54 cm, T_p = 1.3 s, σ_θ = 0°, and (**a**) γ = 2 and (**b**) γ = 20.

Figure 14 shows the comparison of normalised wave heights, H/H_0, (where H is the local significant wave height and H_0 is the significant wave height at the wave generation boundary) between numerical model results and experimental data along the measurement transect shown in Figure 12 for the test cases of non-breaking waves listed in Table 1. Additionally, to evaluate the model, the root mean square error (RMSE) and the skill factor for the normalised wave heights are calculated as:

$$\text{RMSE} = \sqrt{\frac{\sum_{i=1}^{N}(P_i - O_i)^2}{N}} \quad \text{Skill} = 1 - \sqrt{\frac{\sum_{i=1}^{N}(P_i - O_i)^2}{\sum_{i=1}^{N} O_i^2}} \tag{25}$$

where O and P indicate the observed and predicted values, respectively.

Very good agreement between the numerical model and the experimental model is observed (Figure 14 and Table 2). MILDwave with the addition of the periodic boundaries correctly predicts the wave focusing behind the shoal for the case of irregular long crested waves where the wave height is strongly affected by the change of the water depth. The model gives a maximum normalised wave height of around 2.05 at y = 0 for both the broad frequency spectrum (Test Case U3) and the narrow frequency spectrum (Test Case U4). On the other hand, a broad directional spreading distribution (Test Case B3, Test Case B4) yields much less spatial wave height variation and the effect of wave refraction is significantly diminished.

Table 2. Root mean square error (RMSE) and skill factor of the normalised wave heights.

Test Case ID	U3	N3	B3	U4	N4	B4
RMSE	0.141	0.048	0.068	0.068	0.102	0.070
Skill	0.869	0.955	0.936	0.940	0.905	0.924

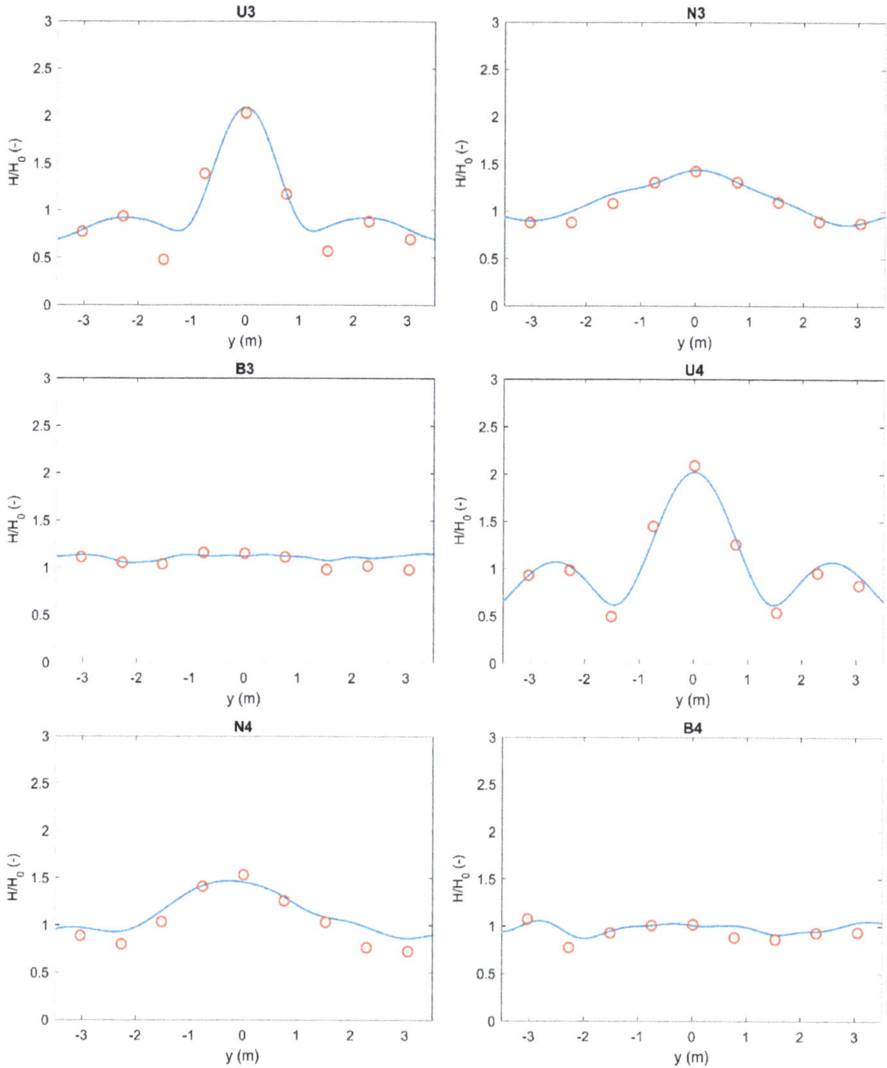

Figure 14. Comparison of normalised wave heights, H/H_0, between numerical model results (blue solid lines) and experimental data (red circles) along the measurement transect of Figure 12. for the test cases of non-breaking waves of Table 1.

5. Conclusions

In the present paper, periodic lateral boundaries were developed in a time dependent mild-slope equation model, MILDwave, for the accurate and fast generation of regular and irregular waves in any direction.

Initially, three different wave generation layouts were examined and compared. The first wave generation layout is an L-shaped wave generator, which is composed of two orthogonal lines. The second wave generation layout is an arc-shaped wave generator where two parallel lines are connected to a semicircle. The third wave generation layout consists of a single wave generation line parallel to the y-axis and periodic lateral boundaries. Numerical experiments were conducted for regular waves with wave propagation angles from $\theta = 0°$ to $\theta = 45°$. The results were presented by means of contour plots, in terms of a disturbance coefficient (k_d) for the calculation domain, allowing easy comparisons between the different layouts. The wave generation layout with the periodic boundaries showed the best results since this layout leads to a homogeneous wave field, which is not disturbed by unwanted wave diffraction patterns in contrast to the other two wave generation layouts.

Then, this wave generation layout was used to simulate irregular long and short crested waves by using two different wave synthesis methods: A method proposed by Miles [14] and a method proposed by Sand and Mynett [15]. The results indicate that the MILDwave model with the addition of the periodic boundaries is capable of reproducing a homogeneous wave field (especially when the Sand and Mynett method is used) as well as the target frequency spectrum and the target directional spectrum. Finally, the developed model was also used to study wave transformation over an elliptic shoal (Vincent and Briggs shoal experiment), where very good agreement was observed between the numerical model and the experimental results. The aforementioned observations indicate that periodic boundaries make the mild-slope wave model, MILDwave, an essential tool to generate multi-directional waves and study their transformation due to its significantly small computational cost and its high numerical stability and robustness.

The next step is to combine the developed model with the coupling methodology described in [19,20] to study a WEC farm under real sea waves, since this paper proves that now MILDwave is able to accurately generate such a kind of wave field.

Author Contributions: P.V. developed the fundaments of the periodic boundaries implementation. P.V. set up the numerical experiments. P.V. compared the numerical experiments with experimental data. V.S. and P.T. supervised, proofread the text and helped in structuring the publication.

Funding: This research was funded by Research Foundation—Flanders (FWO), Belgium, grant number of Ph.D. fellowship 11D9618N.

References

1. Berkhoff, J.C.W. Computation of combined refraction-diffraction. In Proceedings of the Thirteenth Conference on Coastal Engineering, Vancouver, BC, Canada, 10–14 July 1972; pp. 471–490. [CrossRef]

2. Radder, A.C. On the parabolic equation method for water-wave propagation. *J. Fluid Mech.* **1979**, *95*, 159–176. [CrossRef]

3. Copeland, G.J.M. A practical alternative to the "mild-slope" wave equation. *Coast. Eng.* **1985**, *9*, 125–149. [CrossRef]

4. Radder, A.C.; Dingemans, M.W. Canonical equations for almost periodic, weakly nonlinear gravity waves. *Wave Motion* **1985**, *7*, 473–485. [CrossRef]

5. Booij, N. A note on the accuracy of the mild-slope equation. *Coast. Eng.* **1983**, *7*, 191–203. [CrossRef]

6. Suh, K.D.; Lee, C.; Park, W.S. Time-dependent equations for wave propagation on rapidly varying topography. *Coast. Eng.* **1997**, *32*, 91–117. [CrossRef]

7. Larsen, J.; Dancy, H. Open boundaries in short wave simulations—A new approach. *Coast. Eng.* **1983**, *7*, 285–297. [CrossRef]

8. Peregrine, D.H. Long waves on a beach. *J. Fluid Mech.* **1967**, *27*, 815–827. [CrossRef]

9. Lee, C.; Suh, K.D. Internal generation of waves for time-dependent mild-slope equations. *Coast. Eng.* **1998**, *34*, 35–57. [CrossRef]

10. Lee, C.; Yoon, S.B. Internal generation of waves on an arc in a rectangular grid system. *Coast. Eng.* **2007**, *54*, 357–368. [CrossRef]

11. Kim, G.; Lee, C. Internal generation of waves on an arced band. *Ocean Eng.* **2013**, *67*, 77–88. [CrossRef]

12. Lin, X.; Yu, X. A finite difference method for effective treatment of mild-slope wave equation subject to non-reflecting boundary conditions. *Appl. Ocean Res.* **2015**, *53*, 179–189. [CrossRef]

13. Troch, P. *MILDwave—A Numerical Model for Propagation and Transformation of Linear Water Waves*; Internal Report; Department of Civil Engineering, Ghent University: Zwijnaarde, Belgium, 1998.

14. Miles, M.D. A Note on Directional Random Wave Synthesis by the Single Summation Method. In Proceedings of the 23rd IAHR Congress, Ottawa, Canada, 21–25 August 1989; pp. 243–250.
15. Sand, S.E.; Mynett, A.E. Directional Wave Generation and Analysis. In Proceedings of the IAHR Seminar on Wave Analysis and Generation in Laboratory Basins, Lausanne, Switzerland, 1–4 September 1987; pp. 363–376.
16. Troch, P.; Stratigaki, V. Chapter 10—Phase-Resolving Wave Propagation Array Models. In *Numerical Modelling of Wave Energy Converters*; Folley, M., Ed.; Academic Press: Cambridge, MA, USA, 2016; pp. 191–216.
17. Stratigaki, V.; Vanneste, D.; Troch, P.; Gysens, S.; Willems, M. NUMERICAL MODELING OF WAVE PENETRATION IN OSTEND HARBOUR. *Coast. Eng. Proc.* **2011**, *1*, 42. [CrossRef]
18. Beels, C.; Troch, P.; De Backer, G.; Vantorre, M.; De Rouck, J. Numerical implementation and sensitivity analysis of a wave energy converter in a time-dependent mild-slope equation model. *Coast. Eng.* **2010**, *57*, 471–492. [CrossRef]
19. Balitsky, P.; Verao Fernandez, G.; Stratigaki, V.; Troch, P. Assessment of the Power Output of a Two-Array Clustered WEC Farm Using a BEM Solver Coupling and a Wave-Propagation Model. *Energies* **2018**, *11*, 2907. [CrossRef]
20. Verao Fernandez, G.; Balitsky, P.; Stratigaki, V.; Troch, P. Coupling Methodology for Studying the Far Field Effects of Wave Energy Converter Arrays over a Varying Bathymetry. *Energies* **2018**, *11*, 2899. [CrossRef]
21. Zijlema, M.; Stelling, G.; Smit, P. SWASH: An operational public domain code for simulating wave fields and rapidly varied flows in coastal waters. *Coast. Eng.* **2011**, *58*, 992–1012. [CrossRef]
22. Shi, F.; Kirby, J.T.; Harris, J.C.; Geiman, J.D.; Grilli, S.T. A high-order adaptive time-stepping TVD solver for Boussinesq modeling of breaking waves and coastal inundation. *Ocean Model.* **2012**, *43–44*, 36–51. [CrossRef]
23. Brorsen, M.; Helm-Petersen, J. On the Reflection of Short-Crested Waves in Numerical Models. *Int. Conf. Coast. Eng.* **1998**, 394–407. [CrossRef]
24. Mitsuyasu, H.; Tasai, F.; Suhara, T.; Mizuno, S.; Ohkusu, M.; Honda, T.; Rikiishi, K. Observations of the Directional Spectrum of Ocean WavesUsing a Cloverleaf Buoy. *J. Phys. Oceanogr.* **1975**, *5*, 750–760. [CrossRef]
25. Frigaard, P.; Helm-Petersen, J.; Klopman, G.; Stansberg, C.T.; Benoit, M.; Briggs, M.J.; Miles, M.; Santas, J.; Schäffer, H.A.; Hawkes, P.J. IAHR List of Sea Parameteres: an update for multidirectional waves. In Proceedings of the IAHR Seminar Multidirectional Waves and their Interaction with Structures, 27th IAHR Congress, San Francisco, CA, USA, 10–15 August 1997.
26. Kuik, A.J.; van Vledder, G.P.; Holthuijsen, L.H. A Method for the Routine Analysis of Pitch-and-Roll Buoy Wave Data. *J. Phys. Oceanogr.* **1988**, *18*, 1020–1034. [CrossRef]
27. Johnson, D.; Pattiaratchi, C. Boussinesq modelling of transient rip currents. *Coast. Eng.* **2006**, *53*, 419–439. [CrossRef]
28. Dimakopoulos, A.S.; Dimas, A.A. Large-wave simulation of three-dimensional, cross-shore and oblique, spilling breaking on constant slope beach. *Coast. Eng.* **2011**, *58*, 790–801. [CrossRef]
29. Borgman, L.E.; Panicker, N.N. *Design Study for a Suggested Wave Gauge Array off Point Mugu*; Technical Report 1–14; Hydraulic Engineering Laboratory, University of California at Berkeley: Berkeley, CA, USA, 1970.
30. Vincent, C.L.; Briggs, M.J. Refraction—Diffraction of Irregular Waves over a Mound. *J. Waterw. Port Coastal Ocean Eng.* **1989**, *115*, 269–284. [CrossRef]
31. Berkhoff, J.C.W.; Booy, N.; Radder, A.C. Verification of Numerical Wave Propagation Models for Simple Harmonic Linear Water Waves. *Coast. Eng.* **1982**, *6*, 255–279. [CrossRef]
32. Bouws, E.; Günther, H.; Rosenthal, W.; Vincent, C.L. Similarity of the wind wave spectrum in finite depth water: 1. Spectral form. *J. Geophys. Res.* **1985**, *90*, 975. [CrossRef]

Article

Large-Scale Experiments to Improve Monopile Scour Protection Design Adapted to Climate Change—The PROTEUS Project

Carlos Emilio Arboleda Chavez [1,*], Vasiliki Stratigaki [1], Minghao Wu [1], Peter Troch [1], Alexander Schendel [2], Mario Welzel [2], Raúl Villanueva [2], Torsten Schlurmann [2], Leen De Vos [3], Dogan Kisacik [4], Francisco Taveira Pinto [5], Tiago Fazeres-Ferradosa [5], Paulo Rosa Santos [5], Leen Baelus [6], Viktoria Szengel [6], Annelies Bolle [6], Richard Whitehouse [7] and David Todd [7]

[1]   Department of Civil Engineering, Ghent University, 9052 Ghent, Belgium; vicky.stratigaki@ugent.be (V.S.); minghao.wu@ugent.be (M.W.); peter.troch@ugent.be (P.T.)
[2]   Ludwig-Franzius-Institute for Hydraulic Estuarine and Coastal Engineering, 30167 Hannover, Germany; schendel@lufi.uni-hannover.de (A.S.); welzel@lufi.uni-hannover.de (M.W.); villanueva@lufi.uni-hannover.de (R.V.); schlurmann@lufi.uni-hannover.de (T.S.)
[3]   Geotechnics Division, Department of Mobility and Public Works, Flemish Government, 9052 Ghent, Belgium; leen.devos@mow.vlaanderen.be
[4]   Institute of Marine Sciences and Technology, Dokuz Eylül University, Izmir 35340, Turkey; dogan.kisacik@deu.edu.tr
[5]   CIIMAR—Interdisciplinary Centre of Marine and Environmental Research, Faculty of Engineering, Hydraulics, Water Resources and Environment Division, University of Porto, 4200-465 Porto, Portugal; fpinto@fe.up.pt (F.T.P.); tferradosa@fe.up.pt (T.F.-F.); pjrsantos@fe.up.pt (P.R.S.)
[6]   International Marine & Dredging Consultant N.V., 2018 Antwerp, Belgium; leen.baelus@imdc.be (L.B.); viktoria.szengel@imdc.be (V.S.); annelies.bolle@imdc.be (A.B.)
[7]   HR Wallingford Ltd, Howbery Park, Wallingford, Oxfordshire OX10 8BA, UK; r.whitehouse@hrwallingford.com (R.W.); d.todd@hrwallingford.com (D.T.)
*   Correspondence: carlosemilio.arboledachavez@ugent.be

Received: 18 April 2019; Accepted: 1 May 2019; Published: 6 May 2019

Abstract: This study aims to improve the design of scour protection around offshore wind turbine monopiles, as well as future-proofing them against the impacts of climate change. A series of large-scale experiments have been performed in the context of the European HYDRALAB-PLUS PROTEUS (Protection of offshore wind turbine monopiles against scouring) project in the Fast Flow Facility in HR Wallingford. These experiments make use of state of the art optical and acoustic measurement techniques to assess the damage of scour protections under the combined action of waves and currents. These novel PROTEUS tests focus on the study of the grading of the scour protection material as a stabilizing parameter, which has never been done under the combined action of waves and currents at a large scale. Scale effects are reduced and, thus, design risks are minimized. Moreover, the generated data will support the development of future scour protection designs and the validation of numerical models used by researchers worldwide. The testing program objectives are: (i) to compare the performance of single-layer wide-graded material used against scouring with current design practices; (ii) to verify the stability of the scour protection designs under extreme flow conditions; (iii) to provide a benchmark dataset for scour protection stability at large scale; and (iv) to investigate the scale effects on scour protection stability.

Keywords: offshore wind turbines; large scale experiments; scour protection damage; wide-graded materials; climate change conditions; optical measurements; acoustic measurements; waves-current interaction

---

1. Introduction

This study aims to improve the design of scour protection around offshore wind turbine monopiles, as well as future-proofing them against the impacts of climate change. Offshore wind energy conversion units (more commonly called offshore wind turbines) contribute significantly to the electric production matrix and it is expected that their overall production will only increase in the near future. In fact, several concessions have been granted, recently, by countries around the North Sea basin. This intensive industrial development demands that strategic research covering every aspects influencing the useful lifetime of offshore-based wind energy converters is undertaken. Currently, monopiles are the most used support structure for offshore wind energy converters, which are usually arranged in large numbers forming offshore wind farms. Offshore wind farms contribute significantly to the reduction of greenhouse emissions by providing clean and renewable energy, contributing to efforts to deal with climate change challenges. Monopile structures are large cylinders made of steel that are driven into the seabed. The characteristics of the soil and the pile penetration depth provide the stability needed for the monopile to withstand the harsh marine hydrodynamic loads (currents and waves). The interaction between the monopile and the hydrodynamic loads produces an amplification of the stresses applied to the seabed surrounding the monopile, leading to the development of a scour hole which in turn reduces the bearing capacity of the monopile foundation. An overview of the studies dealing with scouring processes can be found in [1,2].

Riprap material has successfully been used as a protective measure against scour and erosion in river and coastal engineering. In river engineering, the study of scour protection made of riprap material has been studied around bridge piers (see [3–5]). In the marine environment, breakwaters [6] and cable protection are some of the specific examples where stone granular material is used as a protective method against hydrodynamic action. In fact, Galay et al. [7] stated that "a stone riprap layer has universal acceptance and proven performance under highly variable flow conditions". Scour protection around offshore monopile foundations, needs to consider different flow conditions (currents and waves), larger water depths and a different density of the fluid (fresh water/sea water).

Five failure mechanisms of scour protection designs in the marine environments are enumerated by Sumer and Fredsøe [2]: disintegration of the riprap layer (scour of the protective layer), winnowing (removal of bed material from underneath the protection layer), edge scour, destabilization by bed-form progression and, sinking of the protection material due to various factors (momentary liquefaction, liquefaction due to build-up of pore pressure, scour below the individual stones, …). Mainly, three of these failure mechanisms have caught the attention of the coastal engineering community, namely, disintegration, winnowing and edge scour.

The disintegration failure mode of an armor layer over a geotextile under different hydrodynamic conditions was studied by De Vos et al. [8]. Static and dynamic stability of the armor layer were tested in a model scale of 1:50 (all the scale factors consider a prototype monopile diameter of 5 m) under different waves, currents and a combined action of both flows. Loosveldt and Vannieuwenhuyse [9] extended the test dataset of De Vos et al. [8] by including larger grain sizes, by varying the water depth and by performing a parametric analysis of the pile diameter (scales of 1:100, 1:50 and 1:40) on the scour protection damage. Nielsen [10] focused on the winnowing of scour protection under different waves and currents. The testing scales used for the unidirectional flow (current) experiments were 1:35, 1:9 and 1:5. Nielsen [10] provided an explanation to the sinking of the scour protections in the "Horns Rev 1" wind farm and gave improved guidelines for the design of filter layers through the mobility parameter. Whitehouse et al. [11] evoked an optimization of scour protection design taking into account rock size, density, number of layers and width of the cover. Finally, [12] performed experiments using physical models with a scale ranging from 1:100 to 1:50 for the study of edge scour under waves and currents.

Schendel et al. [13,14] presented large scale experiments of scour protection design under waves and currents. The scale used for wave tests which included a monopile was 1:5, whereas the scale for current tests without a monopile was 1:1. In the latter case, the material tested as scour protection was the actual prototype material. This work introduced a single armor layer composed of a wide-graded material. 'Wide-graded' refers to a large gradation (ratio D_{85}/D_{15}, where D_{85} and D_{15} account for the diameter larger than 85% and 15% of the mass of the material) of the granular material composing the scour protection ($D_{85}/D_{15} > 1.5$, for wide graded material and $D_{85}/D_{15} > 2.5$ for very wide graded material, see [15] Table 3.4). The usage of this novel technique could reveal itself to be easier to install, as well as cost effective compared to a traditional two-layer scour protection design (filter and armor). Nevertheless, it is concluded that more experiments should be carried out to fully understand the stabilizing process of using wide-graded materials as scour protection. In this direction, [16] studied different compositions of scour protection material in small scale experiments (scales ranging from 1:100 to 1:45) under a unidirectional current.

Deterministic design criteria exist for the classic narrow graded two-layer scour protection ([8,10]) but none has been established for wide-graded materials. Fazeres-Ferradosa et al. [17] proposed a reliability analysis of the scour protection failure and a probabilistic design, without considering the gradation of the scour protection material.

It is clear that there is a lack of (public) data for large-scale experiments of classically designed scour protection solutions under the combined action of waves and current. Furthermore, to the authors' knowledge, the study of the grading of the scour protection material as a stabilizing parameter has never been done under the combined action of waves and currents at large scale. By operating at a large scale, model effects are reduced. This allows design uncertainty during the early stages of wind farm projects to be reduced. To cover this data and knowledge gaps, large scale experiments have been carried out in the fast flow facility (abbreviated as FFF) of HR Wallingford in the United Kingdom. The PROTEUS (Protection of offshore wind turbine monopiles against scouring) testing campaign is a collaborative effort between the Department of Civil Engineering at Ghent University (Belgium), HR Wallingford (UK), the Ludwig Franzius Institute for Hydraulic, Estuarine and Coastal Engineering at the University of Hannover (Germany), the Faculty of Engineering at the University of Porto (Portugal), the Geotechnics division of the Belgian Department of Mobility and Public Works (Belgium), and the International Marine and Dredging Consultants (IMDC nv) (Belgium). PROTEUS is performed in the context of the European HYDRALAB-PLUS program and funded by the European Union's Horizon 2020 Research and Innovation Program. The aim of this manuscript is to present the PROTEUS project and, specifically, to present the experimental setup, the methodology followed throughout the study and quality of the unique dataset acquired during the testing campaign, which addresses the data and knowledge gaps in scour protection studies. The novel PROTEUS experiments, presented further in this paper, test the static and dynamic stability of different scour protection designs including monopiles at two different large scales 1:16 and 1:8, under the combined action of waves and currents. Most importantly, the obtained experimental data will be available publically for the international research community, under HYDRALAB rules. The target outcomes of the experimental campaign include: (i) study of wide grade material performance with respect to narrow graded materials; (ii) study of scale effects in scour protection around monopiles; (iii) analysis of bed shear stresses in wave-current flows; (iv) formalization of methodologies for the assessment of the damage of scour protection. These topics will be the basis of our future work within the PROTEUS project.

2. Stability of Scour Protection Around Monopiles: Governing Physics

2.1. Governing Physics at A Glance

In the marine environment, sea water flows take the form of waves or currents. Such flows apply shear stresses to the seabed (in intermediate and shallow waters in the case of waves). In the presence of a monopile, flows are accelerated in the circumferential direction of the monopile, due to the contraction of flow lines. Furthermore, complex highly turbulent flow structures appear at the base of the monopile and at its wake amplifying the shear stresses [10]. The amplification of the shear stresses in turn increases the erodible potential of the flow. In the case of an unprotected monopile base, scour develops. In the case of a monopile protected by a scour protection, the scour protection material can be removed, eventually leading to the failure of the scour protection. To quantify the amount of material removed by the flow with respect to the pile section and the nominal mean stone diameter, the damage number has been introduced by [8], defined as detailed in Section 5.2.

2.2. Scaling in Experimental Studies

Scaling principles for hydraulic experiments can be found in [18,19]. As stated in [8], the scaling of waves and currents should preserve the Froude, *Fr*, and the Reynolds, *Re*, numbers in experimental studies. The Froude and Reynolds number are defined as follows:

$$Fr = \frac{U}{\sqrt{gL}} \tag{1}$$

$$Re = \frac{UL}{v} \tag{2}$$

where, U is the velocity, g is the acceleration due to gravity, L is a length and v is the kinematic viscosity of the water. Scale effects arise from the impossibility of achieving Froude and Reynolds similarity between model and prototype when scaling geometric lengths. Such scale effect can be reduced by increasing the scale of the model. In hydraulic experiments, Froude similarity is normally considered (see [19]).

To deal with the above described scaling issues, the large-scale tests within the PROTEUS project have been introduced, which are described in the next sections. In the present study, scour protection material, monopile diameter, water depth and wave height are scaled geometrically. Wave period and current velocity are scaled using Froude similarity.

3. Experimental Setup

3.1. Experimental Facility

The FFF experimental facility is a race-track shaped flume (illustrated in Figure 1). It comprises a main working channel, 4.0 m wide and 57.0 m long, and a secondary channel, 2.6 m wide and 50.0 m long. The hinge flat type multi-element wave generator with active wave absorption (located at the left in Figure 1) can deliver significant wave heights up to 0.5 m and a maximum wave height up to 1.0 m, depending on the water depth. The water depth can be set in the range of 0.85–2.00 m. At the opposite side of the wave generator (at the right in Figure 1), a beach made of sponge material passively absorbs the generated wave trains. The axial pumps (located in the secondary channel) can deliver a discharge of up to 3.5 m³/s and their reversible nature can provide a current propagation following or opposing the waves.

A local reference system was established with the origin being at the front of the wave maker, in the middle of the channel on the flume floor. The positive x-axis points into the wave propagation direction (from left to right in Figure 1), the positive y-axis points upwards in the top view in Figure 1 and the positive z-axis follows the gravity vector (see Figure 2). In the sketch of the main channel (Figure 2), the position of the resistive wave gauges (abbreviated as WGs), the acoustic doppler velocity meters (ADVs) and the scale model of a monopile are indicated. Further information on the instrumentation can be found in Section 4.

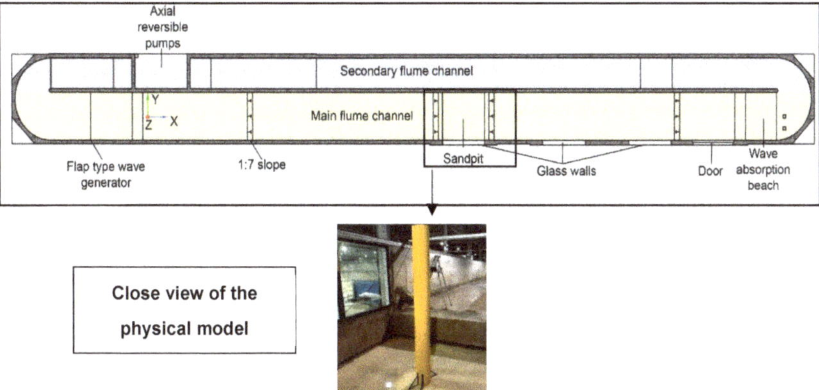

Figure 1. Sketch of the fast flow facility (FFF) flume channels including the position of the scale models.

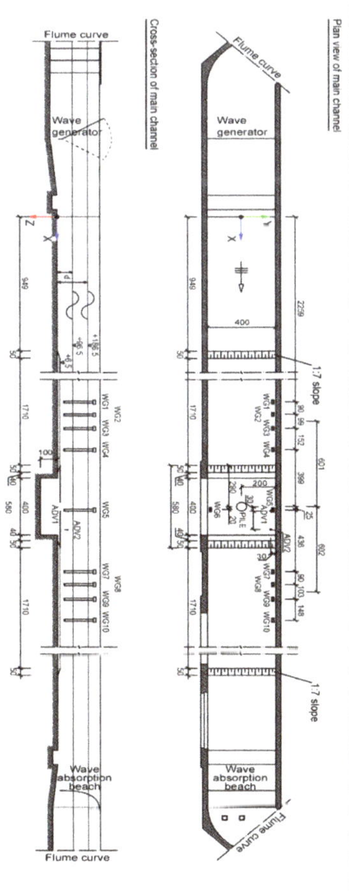

Figure 2. Overview of the FFF main channel, and of the experimental setup including the intrumentation positions, the reference coordinate system and the scale model location.

3.2. Monopile Scale Models

There are two variants of monopile scale models with two different diameters, D_p = 0.3 m and D_p = 0.6 m, which were constructed from thin-walled metal wrapped around wooden cylinders of limited height (Figure 3b). Each monopile model is placed in the wave flume with its center at x = 30 m and y = 0 m, following the local reference system presented in Figures 1 and 2. Each model is attached to a mounting base fixed at the bottom of the sand pit (Figure 3a). Iron wedges provide additional support to the model by fixing it to the facility's concrete floor (Figure 3c).

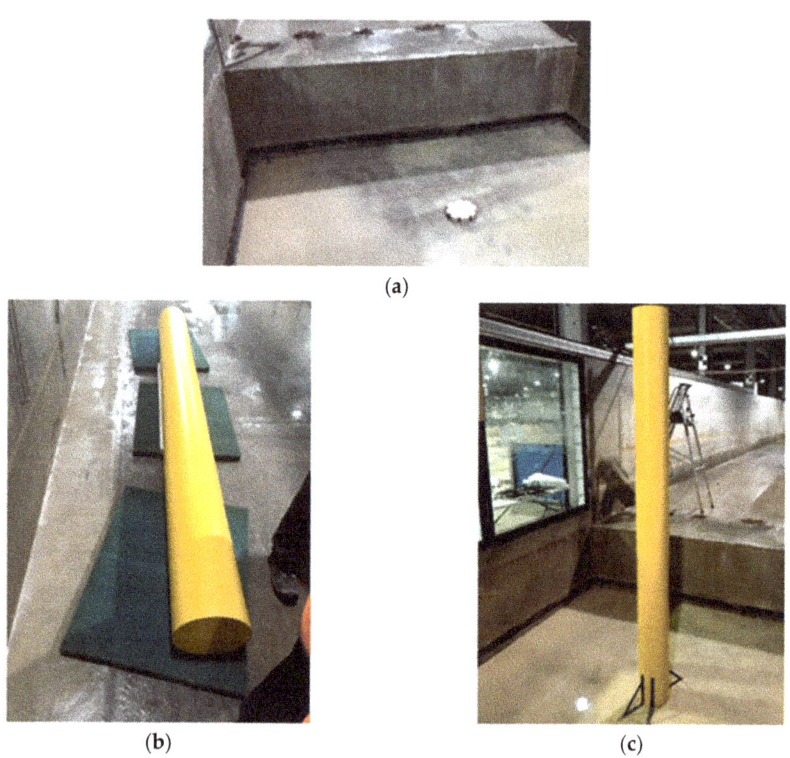

Figure 3. (**a**) Monopile mounting base; (**b**) monopile scale model (scale 1:16) waiting to be installed; (**c**) monopile scale model placed on its support structure and external iron edges installed at the base.

3.3. Scour Protection Models and Sand Pit Preparation

The location of the sand pit (details presented in Figure 3a,c) with respect to the wave generator is shown in Figure 1. The sand pit consists of a 4.0 m long, 4.0 m wide and 1.0 m high box. This sand pit size provides the necessary area for testing large-scale scour protection models over a sand bed, which is composed of uniform sand, mean size diameter 0.21 mm, for all tests. The filling of the sand pit was done after the installation of each of the monopile scale models (Figure 3c). First the sand carrier drops its load on the absorption beach side of the sand pit. Then, every load is evenly distributed along and across the sand pit using shovels. Every two loads (around 20 cm thickness) the sand is compacted using a vibration compactor. This is done to prevent lowering of the sand bed level during the test phase. Once the sand pit is filled, the sand bed is flattened, the geotextile is placed (for the relevant tests), and finally, the installation of the scour protection layer takes place. To ensure the uniformity of the distribution of mass and fraction, the material is mixed and installed using templates as shown in Figure 4a–e. Each composition of scour protection material is prepared from weighting every fractions of the prepared sieved rock material. A portion of the scour protection material around the monopile is painted red, following methodology principles introduced by [8]. The painting is done to allow a good visual assessment of the damage in the scour protection model, as the painted material is placed strategically in the region where the highest hydrodynamic loads are expected. The painted area has diameter of two times the monopile diameter (referred to as the 'inner ring') and is presented in Figure 4.

Figure 4. Installation of a scour protection scale model scale (scale 1:16.7). (**a**) Inner ring templates placed around the monopile; (**b**) outer ring placement using templates; (**c**) finalized scour protection model; (**d**) scour protection model placement over a geotextile; (**e**) placement of a scale model (scale 1:8.3) using larger templates.

The painting procedure is performed in a manner that avoids: (a) (substantial) modification of the specific density of the scour protection material; (b) aggregates of paint and rock material which might appear during the drying process, especially for the smaller fractions; (c) interference of paint pigments with the optical characteristics of the laser scanner during the topography scanning. To obtain this, the looseness of the material is ensured by mixing the stones whilst drying, and the compatibility of the paint and the laser scanner is checked.

3.4. Experiment Execution

After the installation of the scour protection model, pictures of its initial state (before filling the flume with water and before wave or current generation) are taken. The flume is filled and the initial topography is scanned before the initiation of a test. In Figure 5, a flow chart of the tasks that need to be performed for each test is presented.

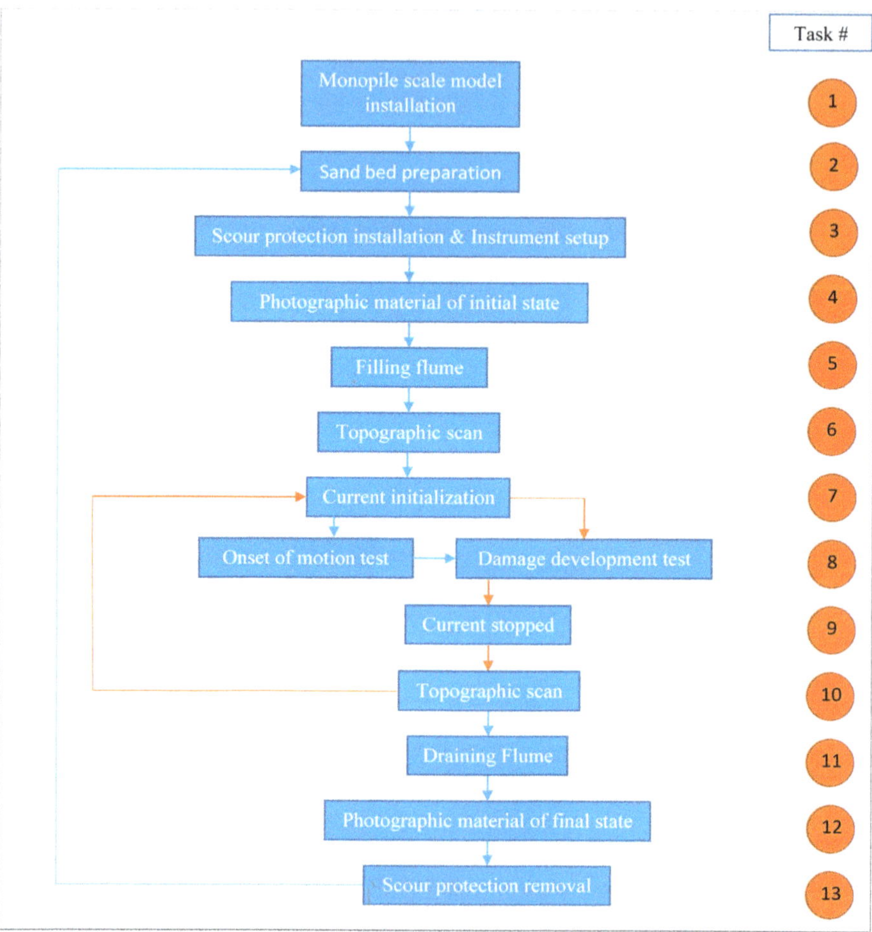

Figure 5. Flow chart of experiment execution. The loop indicated using orange arrows is performed three times for Test No. 14 and twice for the other damage development tests.

4. Instrumentation Setup

During the testing campaign data is recorded in both the main and the secondary flume channels. In the main flume channel, the free surface elevations and 3D flow velocities are recorded at ten and two (point velocity measurements) locations respectively (see Figure 2), to characterize the flow in the vicinity of the monopile. In the secondary channel, the profile of horizontal flow velocities and the free surface elevation are measured at one location to characterize the current characteristics in the facility. The data acquisition system, the free surface elevation measurements and the two point 3D flow velocity measurements start recording with the start of the axial pumps. The profile horizontal flow velocity measurement was initiated manually in a synchronized manner with the initiation of the axial pumps.

Before the onset of motion tests and after every damage development test, the topography of the scour protection model is measured. The topography measurements provide the initial, intermediary and final state of the scour protection model. Photographic material is produced before filling and after draining the flume. The summary of the measured parameters and instrumentation used is shown in Table 1.

Table 1. Measured parameters and instrumentation.

Measured Parameter		Instrumentation
Free surface elevation		Resistive wave gauges (WGs)
Flow velocities	3D point measurements	Acoustic doppler velocity meters (ADVs)
	Profile measurements of the horizontal velocity	Aquadopp profiler
Scour protection model topography		ULS-200 laser scanner
Photographic material		Cameras

4.1. Free Surface Elevation Measurements

Resistive wave gauges (abbreviated as WGs) are used to measure the free surface elevation at a sampling frequency of 100 Hz. These 1.2 m long gauges (shown in Figure 6) are partially immersed in water, and the output voltage is proportional to the immersed portion of the wave gauge. The locations of the 10 WGs along the main flume channel in the (x, y) plane of the local reference system are presented in Table 2. Four WGs are placed in front of the monopile (WG1–WG4), four downstream the monopile (WG7–WG10) and two on each side of the monopile (WG5 and WG6). The vertical positioning of the WGs depends on the water depth, and they are placed such that the middle of the WG length lies at the still water level which is measured in the secondary channel. WGs record the (incoming, reflected, transmitted and diffracted) wave field in the vicinity of the monopile.

Figure 6. Example of resistive wave gauge used during the Protection of offshore wind turbine monopiles against scouring (PROTEUS) project.

Table 2. Positions of the wave gauges following the local coordinate system indicated in Figures 1 and 2.

Wave Gauge No.	Position in x-Direction (m)	Position in y-Direction (m)
WG1	22.59	1.97
WG2	23.49	1.97
WG3	24.48	1.97
WG4	26.00	1.97
WG5	30.24	1.97
WG6	30.19	−1.95
WG7	34.60	1.97
WG8	35.50	1.97
WG9	36.53	1.97
WG10	38.91	1.97

4.2. Velocity Measurements

Velocity measurements are performed using three devices; two Acoustic Doppler Velocity meters (abbreviated as ADVs), Nortek Vectrino II type (Figure 7a) with a sampling frequency of 100 Hz, and an Acoustic Doppler Velocity Profiler (abbreviated as Aquadopp), Nortek Aquadopp HR 2MHz profiler type (Figure 7b), with a sampling frequency of 1.0 Hz. The ADVs measure the three dimensional components of the flow velocity over 3.0 cm (it is considered in this case as a point measurement), while the Aquadopp measures the magnitude of the velocity and its direction parallel to the x-axis. During the calibration stage, the Aquadopp is positioned at x = 30.0 m and y = 0.0 m in the main flume channel, at the planned position of the monopile. During the testing stage, the Aquadopp was moved to the secondary flume channel in order to prevent any flow disturbance near the model. The Aquadopp is placed facing downwards at a height of 0.86 m from the flume floor and provides a measurement of the current undisturbed by the monopile or/and the waves. The spatial resolution for the Aquadopp measurements is 1.0 cm starting from a distance of 11.0 cm from the device's head, which is the blanking distance of the instrument. The Aquadopp discretizes the water column in bins with a size corresponding to the spatial resolution of 1.0 cm. For instance, for Test 04 the last bin, number 73, is located at a distance of 83.0 cm from the Aquadopp head.

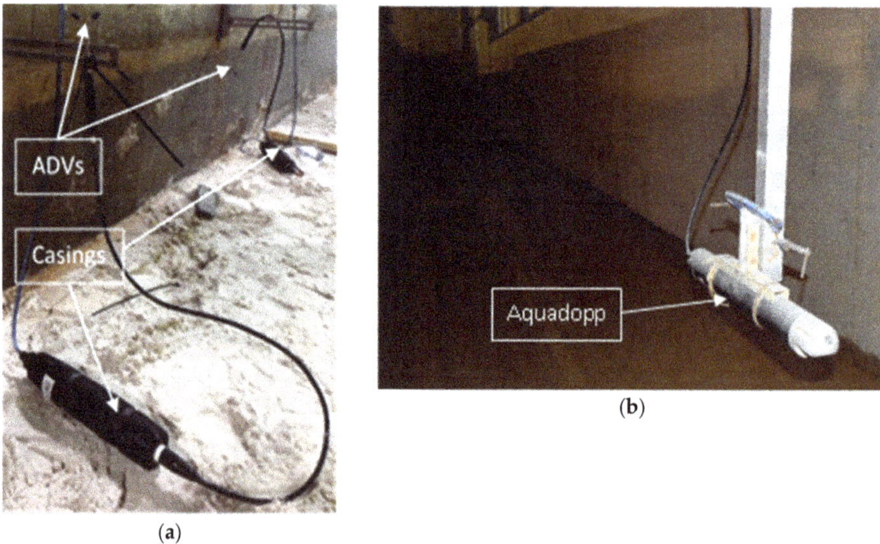

Figure 7. Instrumentation used for velocity measurements: (**a**) two ADVs placed above the sand pit; (**b**) down-facing Aquadopp profiler placed in the secondary flume channel. The Aquadopp vertical position was z = −0.83 m for water depths 0.9 m and 1.2 m and z = −0.91 m for water depths 1.5 m and 1.8 m.

The ADVs are placed above the sand pit at the side of the main flume, just downstream the monopile, outside the influence of its wake (Figure 2, Figure 7a). The ADVs aim to capture the resulting 3D velocity components of currents and the waves' orbital velocity components in the vicinity of the monopile. The housings of the ADVs (black casings in Figure 7a) are hollow and buoyant, and therefore they were buried in the sand and weights were placed over them to ensure that they remain underneath the seabed. Due to their placement, the positive direction in the x-axis is the inverse of the local reference system. The Aquadopp's and the ADVs' locations are summarized in Table 3. The vertical position ($0.4d$, where d is the water depth) of the ADVs was chosen in order to be outside the boundary layer, at this vertical position, the flow velocity is considered characteristic of the flow velocity in the water column near the monopile.

Table 3. Positions of the Aquadopp profiler and the two ADVs following the local reference system, as indicated in Figures 1 and 2.

Instrument	Position in x-Direction (m)	Position in y-Direction (m)	Position in z-Direction (m)
Aquadopp	29.5	0 (middle of secondary flume)	−0.86
ADV1	30.30	1.70	−0.4d
ADV2	31.60	1.70	−0.4d

4.3. Scour Protection Topography Measurements

The topography of the scour protection model is measured using an ULS-200 underwater laser scanner which operates at 7 Hz (7 mm/s) mounted in a traverse system above the scour protection model (Figure 8). The vertical resolution of the ULS-200 is 1 mm. A first topography scan is performed after the placement of the scour protection model, with the flume filled with water, to provide the initial state of the scour protection model. Damage development tests are composed of two or three wave trains. After each wave train the topography of the scour protection model is measured using the laser scanner. The damage is calculated by superimposing the laser scans.

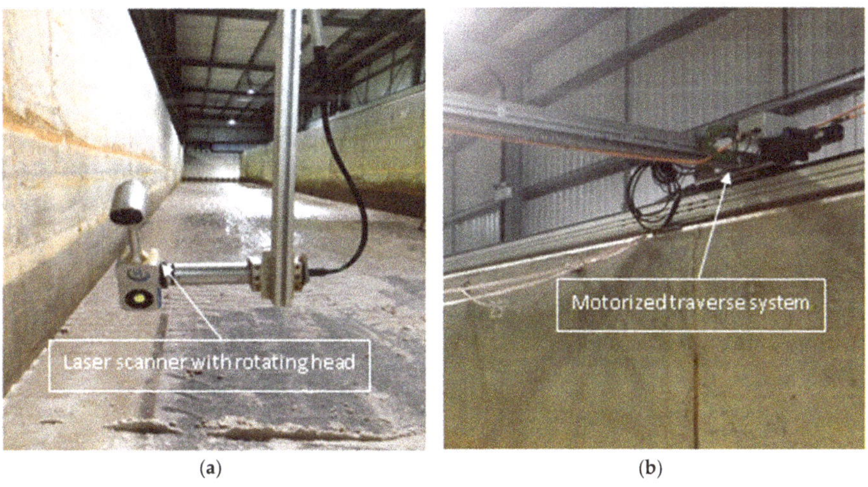

(a)   (b)

Figure 8. (**a**) The ULS-200 laser scanner with rotating head. (**b**) The motorized transverse system.

4.4. Optical Measurements Material

After the installation of the scour protection and before its removal, photographs are taken from different optical angles using cameras. The visual analysis of the photographic material aims to complete the laser scanner data in terms of patterns of stone motion and visual assessment of the scour protection damage. Photos were taken at two different position heights at 7 different locations (C1–C7 in Figure 9). The first camera position (C1–C3) has an average height of 1.55 m above the sand pit, while the second camera position (C4–C7) has a height taken at 2.13 m. The pictures are then merged providing a complete view of the scour protection model.

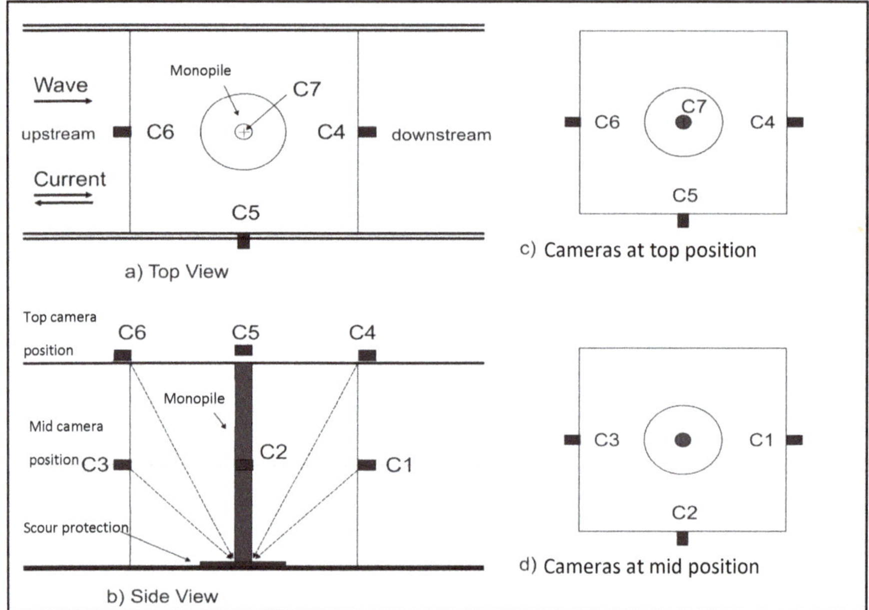

Figure 9. Camera position for photographic material recording.

5. Experimental Test Program

Two types of tests were carried out during this testing campaign, namely, onset of motion and damage development tests for each of the scour protection models. The testing program objectives were: (i) to compare the performance of single-layer wide-graded material used against scouring with current design practices; (ii) to verify the stability of the scour protection designs under extreme flow conditions; (iii) to provide a benchmark dataset for scour protection stability at large scale; and (iv) to investigate the scale effects on scour protection stability. The experimental conditions are summarized in Tables 4–6. In Tables 4 and 5, the experiments' basic hydrodynamic conditions and the hydrodynamic variants are included. The variants of a test are performed successively. The variants of the onset of motion tests test different wave heights and wave periods. The variants of the damage development tests test different number of waves.

5.1. Onset of Motion Tests

During onset of motion tests, short regular wave trains (12 waves) were generated after reaching a stable current velocity. The scour protection is observed throughout the propagation of the wave train through the glass walls of the FFF, see Figure 1, in order to spot the motion of the scour protection material. Motion of scour protection material (stones) refers to the displacement of a stone which size, d_s, is larger or equal to the mean stone diameter ($d_s > D_{50}$) for a distance at least equal to two times the mean stone diameter [8]. Once it has been established that motion of the stones occurred, new wave conditions are tested. The current generation is not interrupted in-between applying different wave conditions. The test conditions for the onset of motion tests are shown in Table 4.

Table 4. Onset of motion measured test conditions. The highlighted conditions are the ones where the motion of scour protection material is spotted. S/N stands for the serial number.

Test No.	Water Depth	Monopile Diameter	Current Velocity	Test Variant	Wave Height	Wave Period
S/N	d (m)	D_p (m)	U_c (m/s)	S/N	H (m)	T (s)
				A	0.22	2.94
				B	0.28	2.94
03	1.2	0.3	−0.25	C	0.27	2.94
				D	0.33	2.47
				E	0.39	2.47
				A	0.20	2.91
				B	0.22	2.93
				C	0.28	2.98
05	1.5	0.3	0.27	D	0.32	2.94
				E	0.35	2.94
				F	0.32	2.51
				G	0.37	2.48
				A	0.25	2.94
07	1.2	0.3	−0.23	B	0.29	2.94
				C	0.33	2.46
				D	0.31	2.46
				A	0.20	2.46
09	0.9	0.3	−0.23	B	0.22	2.06
				C	0.26	2.08
				A	0.50	3.50
				B	0.37	3.48
				C	0.42	3.48
11	1.8	0.6	−0.39	D	0.54	3.48
				E	0.41	2.84
				F	0.46	2.85
				G	0.50	2.83
				H	0.56	2.85

The onset of motion tests with clear motion of the scour protection material are highlighted.

The visibility in the flume was not good when the current was established due to suspended sediment. Once the wave generation started, the sediment transport was enhanced and the turbidity of the water increased substantially. Therefore, the results of the onset of motion test need to be considered with care because of their qualitative nature.

5.2. Damage Development Tests

Damage development tests assess a dynamically stable design of scour protections. Such design allows some motion of the scour protection material. From this perspective, failure is considered if armoring material is removed over a minimum area of four armor units ($4 \times D_{50}^2$, D_{50} is the mean stone diameter of the scour protection model). Such a design of the scour protection allows very little motion of the scour protection material. The criteria for assessing the damage undertaken by the scour protection is the global damage number, S_{3D}. Following the methodology presented by [8], the scour protection model is subdivided into subsections with an area equal to the area of the monopile as shown in Figure 10.

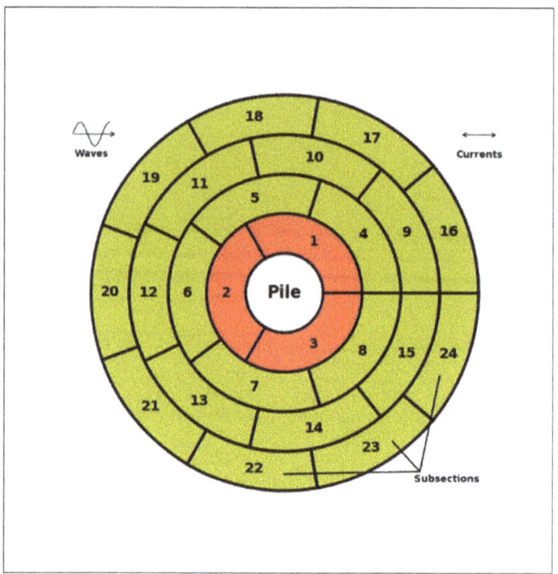

Figure 10. Sketch of the scour protection model around the monopile divided in subsections as in [8], with the inner ring in red. The waves and current propagation directions are also indicated. The decomposition in subsections is made depending on the direction of propagation of the current. The present setup is for a current propagation opposing waves, the setup should be mirrored if the current follows the waves.

The damage number of each of the subsections is calculated from the eroded volume, V_e, the nominal mean diameter, D_{n50}, and the monopile diameter, D_p, using the formula:

$$S_{3D,sub} = \frac{V_e}{D_{n50}\pi\frac{D_p^2}{4}} \tag{3}$$

The global damage number is obtained by considering the maximum damage number of the subsections:

$$S_{3D} = \max\left(S_{3D,sub}\right) \tag{4}$$

According to [8], a conservative value for the maximum acceptable global damage number is 1.1. This value is still debated as shown in [20]. Damage development tests are performed in a similar way as the onset of motion test; when the current has reached the desired velocity, a long wave train is generated (1000 irregular waves). The current is stopped when the wave train is completed and a topography laser scan takes place. Then, the current is restarted and established, and a longer wave train of 2000 irregular waves is generated and, finally, the last laser scan is performed. Test 14 had an additional 2000-wave wave train, followed by a laser scan (Table 5). V_e will be determined by the comparison of the topography laser scans. The (measured) test conditions for the damage development tests are shown in Table 5. The acquisition, calculation and correlation of the quantities presented in Table 5 are discussed in Section 6.

Table 5. Damage development measured conditions.

Test No.	Variant	Significant Wave Height	Peak Wave Period	Main Channel Computed Average Flow Velocity	Mean Flow Velocity Secondary Channel	Mean Flow Velocity ADV1	Mean Flow Velocity ADV2	Number of Waves
S/N	S/N	H_s (m)	T_p (s)	U_{comp} (m/s)	U_{SC} (m/s)	U_{ADV1} (m/s)	U_{ADV2} (m/s)	N (-)
04	A	0.25	2.45	−0.49	−0.70	−0.46	−0.46	1000
	B	0.24	2.48	−0.50	−0.70	−0.46	−0.46	2000
06	A	0.28	2.20	−0.38	0.62	0.39	0.38	1000
	B	0.28	2.20	−0.37	−0.59	0.39	0.38	2000
08	A	0.19	2.44	−0.50	−0.70	−0.46	−0.45	1000
	B	0.19	2.44	−0.50	−0.70	−0.46	−0.45	2000
10	A	0.18	2.05	−0.33	−0.46	−0.30	−0.28	1000
	B	0.16	2.05	−0.33	−0.46	−0.30	−0.29	2000
12	A	0.37	2.81	−0.50	−0.75	−0.51	-	1000
	B	0.38	2.83	−0.51	−0.75	−0.52	-	2000
13	A	0.33	2.34	−0.57	−0.83	−0.63	-	1000
	B	0.34	2.35	−0.57	−0.83	−0.63	-	2000
14	A	0.39	2.83	−0.51	−0.75	−0.49	-	1000
	B	0.41	2.83	−0.51	−0.75	−0.49	-	2000
	C	0.41	2.90	−0.51	−0.76	−0.49	-	2000
15	A	0.41	2.88	−0.49	−0.74	-	-	1000
	B	0.39	2.86	-	-	−0.49	-	2000

The intrinsic properties of the scour protection material, i.e., the mean stone diameter, D_{50}, and the gradation of the material composition, D_{85}/D_{15}, are stated in Table 6.

Table 6. Properties of scour protection composition and indication of usage.

Scour Protection Mixture No.	Test No.	Mean Diameter	Gradation of the Material
S/N	S/N	D_{50} (mm)	D_{85}/D_{15} (-)
1	03/04	12.5	2.48
2	05/06	6.75	2.48
3	07/08/09/10	6.75	2.48
4	11/12/13	13.5	2.48
5	14	13.5	6
6	15	13.5	12
7 (Geotextile)	03/04/07/08	-	-

Mixture 1 is the scale model of a grading 2–80 kg. A wide-graded material with a mean diameter of 110 mm in prototype scale is studied at intermediate model scale by Mixture 2 and 3 and at large scale model by Mixtures 4, 5 and 6. The variable between Mixtures 4, 5 and 6 is the gradation of the material. Figure 11 presents the grain size distribution of the mixtures, as obtained from the fabrication of the mixtures.

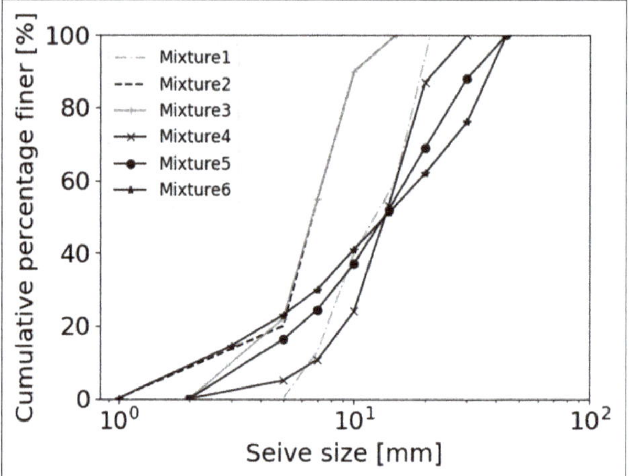

Figure 11. Percentage finer against the sieve size for the 6 tested mixtures.

The use of geotextile as a filter was studied in tests 03/04/07/08 at intermediate model scale.

6. Results and Discussion

Results from Test 04 are presented in the present manuscript. During Test 04, the hydrodynamic conditions represent an extreme condition of a current with a velocity $U_c = 2$ m/s at prototype scale. The scour protection model is subjected to considerable hydrodynamic loads. The visual assessment of the damage is of "Level 2" (dynamically stable conditions) following the criteria presented in [11] used for visual assessment of the damage levels:

- Level 1: Statically stable conditions (no or little movement of the stones);
- Level 2: Dynamically stable conditions (stone motion without failure of the scour protection);
- Level 3: failure of the scour protection.

This test results compared to the damage prediction formula presents the overestimation of damage in scour protection material by current design practices.

Four testing phases are present during a damage development test. In order to depict these testing phases, the free surface measurements of WG1 and the velocity measurements at bin 20 of the Aquadopp are presented in Figure 12, throughout Test 04_B (i.e., variant B of Test 04). The initialization of the tests refers to the phase of the current build up (indicated as "phase A" in Figure 12). In the current stabilization phase, the discharge is maintained during 5–10 minutes, allowing the current's full establishment in the facility (indicated as "phase B" in Figure 12). The wave generation phase is performed with a fully established, constant current (indicated as "phase C" in Figure 12). Finally, during the finalization of the test, the current and wave generation are stopped (indicated as "phase D" in Figure 12).

During "phase B", it is assumed that the scour protection suffered did not suffer damage. The fluctuations of the velocity measurements are due to the turbulent flow.

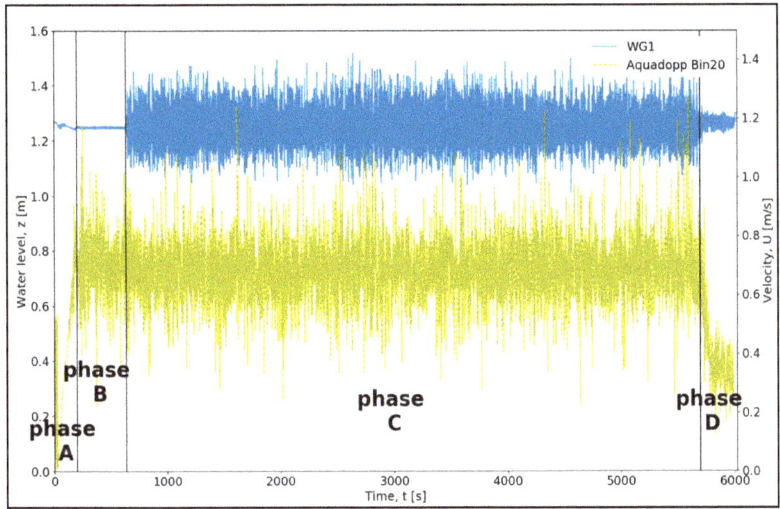

Figure 12. Test 04_B measurement of WG1 (left vertical axis) and Aquadopp at bin 20 (right vertical axis) with test phases: ("phase A") Initialization of the test (time: 0–200 s); ("phase B") Current stabilization (time: 200–633 s); ("phase C") Wave Generation (time: 633–5697 s); ("phase D") Test finalization and water level stabilization (time: 5691–6010 s).

6.1. Velocity Measurements Correlation

The four phases A–D, presented in Figure 12, are present in all the damage development tests. The initialization phase A and the current stabilization phase are only performed for the variant A (TestXX_A) of the onset of motion tests.

The current measurements are performed through means of the Aquadopp and the ADVs, which allows the verification of the accuracy of the measurements. The starting point of the flow analysis is the velocity profile measurements in the secondary channel, presented in Figure 13 during the current stabilization phase B and during the wave generation phase C. From Figure 13, it is observed that the mean velocity profiles of the flow do not differ significantly in the wave generation phase C with respect to the mean velocity profile of the flow in the current stabilization phase B and, therefore, the generated current can be considered constant in a test.

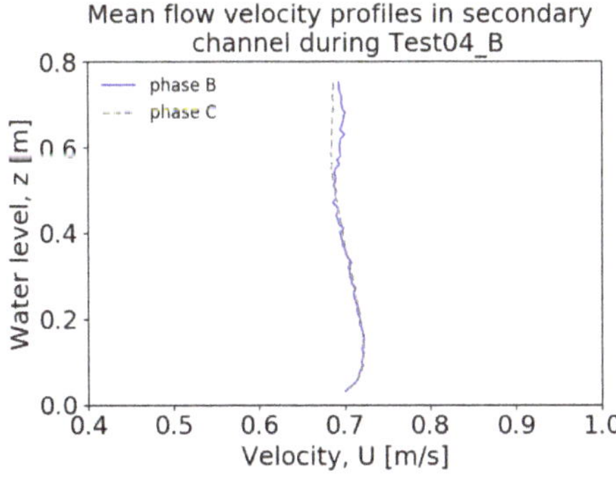

Figure 13. Mean profile velocity from Aquadopp measurements during current stabilization (phase B) and wave generation phase (phase C) of Test 04_B over the water column.

The velocity profile measurements presented in Figure 13 were performed in the secondary channel (Section 4). The main channel's flow velocity can be acquired either by considering conservation of discharge from the secondary channel, or by the point measurements of the flow's velocity done by the ADVs during the current stabilization phase B of Test 04_B. The discharge, Q, is expressed bellow as a function of the section area of the flow, S, and the mean flow velocity, V.

$$Q = S \cdot V \tag{5}$$

In order to consider conservation of discharge, the section area of the flow must be computed, using measurements of the secondary channel water level (SCWL) and the WG measurements (e.g., from WG1), as presented in Figure 14 during the current stabilization phase B.

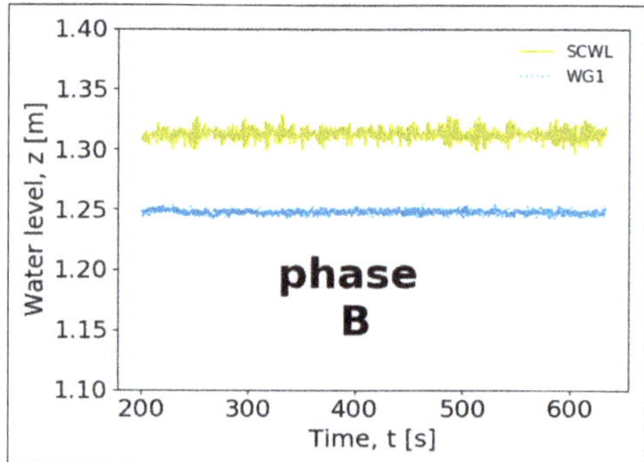

Figure 14. Secondary channel water level (SCWL) and WG1 measurements during the current stabilization phase B of Test 04_B.

In Table 7, the average water level measured in the main channel (MCWL) by the 10 WGs is presented. The average water level in the main channel is found by averaging the mean water levels measured by resistive wave gauges. It needs to be considered that, in the region where the WGs are placed, the facility floor was raised by 6.5 cm (see Figure 1).

Table 7. Average main channel water level (MCWL), secondary channel water level (SCWL) and WG water level measurements for test 04_B.

Avg. SCWL (m)	WG1 (m)	WG2 (m)	WG3 (m)	WG4 (m)	WG5 (m)	WG6 (m)	WG7 (m)	WG8 (m)	WG9 (m)	WG10 (m)	Avg. MCWL (m)
1.312	1.247	1.244	1.244	1.246	1.247	1.246	1.248	1.248	1.246	1.25	1.247

Using the ADVs, the average flow velocity in the main flume can be calculated by averaging the measurements of the velocity x-components over the current stabilization phase B. The mean velocity of the flow measured by the ADVs in the main channel and by the Aquadopp, and the velocity of the flow, in the main channel, computed from the conservation of discharge and the Aquadopp measurements in the secondary channel are presented in Table 8.

Table 8. Average velocity of the flow in the main and secondary channel acquired by different methods and the computed velocity of the flow for Test 04_B in phase B.

Mean Flow Velocity ADV1	Mean Flow Velocity ADV2	Mean Flow Velocity Secondary Channel	Main Channel Computed Average Flow Velocity
U_{ADV1} (m/s)	U_{ADV2} (m/s)	U_{SC} (m/s)	U_{comp} (m/s)
0.46	0.46	0.70	0.50

It is observed that the flow velocity in phase B presents a good correlation with the computed flow velocity. The ADV measured average flow velocity presents a deviation of 8% with respect to the computed flow velocity in the main channel. This deviation between the ADV measurements and the computed flow velocity can be accounted for by the hydrodynamic action of the wave absorbing beach or the monopile itself. This effect will be further studied using numerical modelling or a more in-depth analysis of the measurements of the other tests performed during PROTEUS.

6.2. Free Surface Elevation Spectra

In Figure 15 the spectral densities of the free surface measurements from WG1, WG10 and the target JONSWAP spectrum are presented. The comparison between the measured and the target spectra show good agreement.

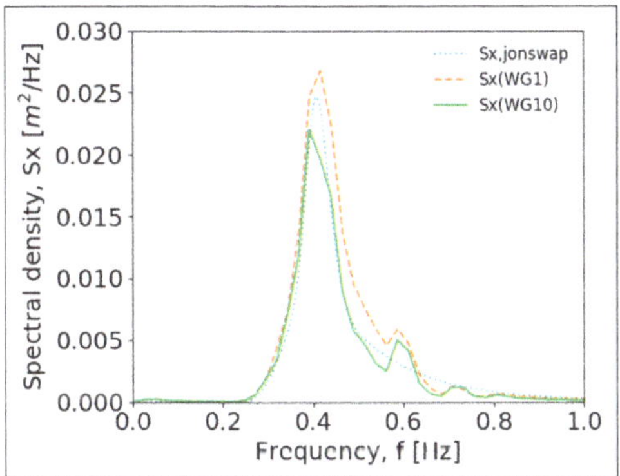

Figure 15. Spectral density of the target JONSWAP spectrum, Sx, jonswap, and free surface measurements of WG1, $Sx(WG1)$, and WG10, $Sx(WG10)$, for test 04_B over the frequency, f.

The characteristic values of the free surface measurements carried out for irregular waves are calculated and compared to the target wave characteristic values. The significant wave height ($H_{m0} = H_s$) is calculated by:

$$H_{m0} = 4\sqrt{m_0} \tag{6}$$

$$T_{m-1,0} = \frac{m_{-1}}{m_0} \tag{7}$$

$$T_{m0,1} = \frac{m_0}{m_1} \tag{8}$$

$$T_{m0,2} = \sqrt{\frac{m_0}{m_2}} \tag{9}$$

where, m_0, m_{-1} and m_2 are spectral moments and $T_{m-1,0}$, $T_{m0,1}$ and $T_{m0,2}$ are characteristic spectral wave period, respectively, wave energy period, first moment wave period, mean zero up-crossing period. The n-th spectral moment is calculated from:

$$m_n = \int_0^\infty S_x(f) \cdot f^n df \tag{10}$$

where $Sx(f)$ is the spectral density function of the frequency, f the frequency and the infinitesimal quantity, df. The peak wave period, T_p, is determined from the frequency bin at which the maximum spectral density, $Sx(f)$max, occurs. The wave characteristics obtained from the spectral analysis are presented in Table 9. From the wave characteristics, it is observed that the difference in H_{m0} between the target and the mean H_{m0} measured value is of 6%. In terms of T_p this difference is 0.8%. It can be concluded that there is a good agreement between the generated spectrum and the target JONSWAP (Joint North Sea Wave Project) spectrum.

Table 9. Wave characteristics from spectral analysis of the WGs measurements and mean values for Test 04_B.

Parameter	Significant Wave Height	Peak Period	Wave Energy Period	First Moment Wave Period	Mean Zero Up-Crossing Period
Symbol	H_{m0}	T_p	$T_{m-1,0}$ (T_e)	$T_{m0,1}$	$T_{m0,2}$ (T_z)
Unit	(m)	(s)	(s)	(s)	(s)
Target value	0.225	2.46	-	-	-
WG1	0.255	2.40	2.49	2.21	2.04
WG2	0.258	2.40	2.50	2.19	1.83
WG3	0.257	2.56	2.50	2.18	1.77
WG4	0.249	2.40	2.50	2.21	1.89
WG5	0.226	2.40	2.53	2.24	2.08
WG6	0.273	2.40	2.47	2.21	1.91
WG7	0.215	2.56	2.54	2.25	2.03
WG8	0.219	2.56	2.52	2.22	1.79
WG9	0.222	2.56	2.51	2.22	1.79
WG10	0.222	2.56	2.57	2.249	1.98
Mean value	0.240	2.48	2.51	2.22	1.91

6.3. Wave Orbital Velocity Spectra

The same spectral treatment is performed to the point velocity measurements provided by ADV1. The normalized spectra acquired from the point velocity measurements in the x- and z-direction, WG1 and the normalized JONSWAP spectrum are presented in Figure 16. It is observed that fluctuations of the point flow velocity measurements follow the same fluctuations in the wave measurements and the target wave JONSWAP spectrum. Even if this result was expected, it is presented here to show that further analysis on the velocity point measurements could provide valuable information on the wave orbital velocity when waves propagate on a unidirectional flow.

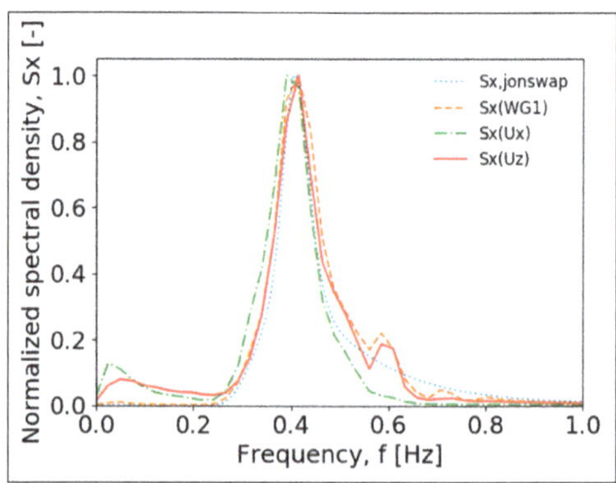

Figure 16. Normalized spectral density of target JONSWAP, Sx, jonswap, measurements of water level, Sx(WG1), and ADV velocity measurements (ADV2) x-components, $Sx(Ux)$, and z-components, $Sx(Uz)$, over the frequency.

6.4. Scour Protection Damage Development Results

The results presented so far aim to show the accuracy in generating hydrodynamic conditions. Here, the study of the dynamic and static stability of the tested scour protection models is introduced.

Scour protection damage development tests, such as Test 04, are composed of at least two wave trains. In Section 4, it has been stated that photographs were taken before and after the tests. In Figure 17, the merged photographs are shown for Test 04. The initial state of the scour protection model can be seen in the left panel (Figure 17a), and the final state after 3000 waves on the right panel (Figure 17b). In Figure 17b the displacement of the scour protection material of the inner ring (red stones) can be clearly observed in the direction of the current propagation. Furthermore, deposition of sediment material is seen on top of the scour protection, outside the inner ring region.

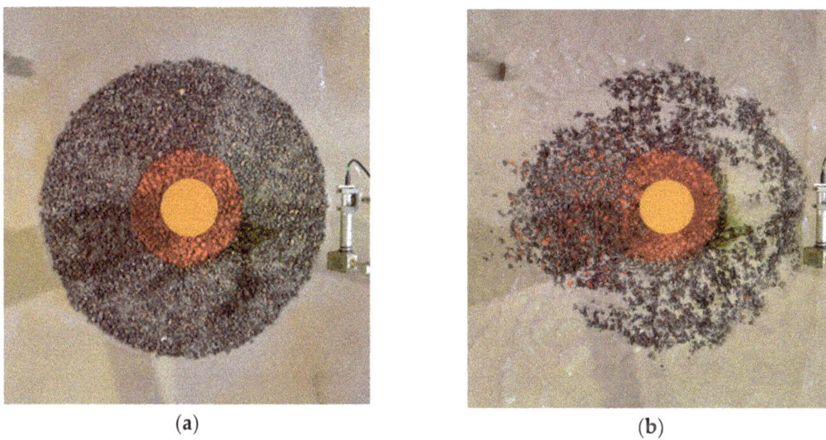

(a)   (b)

Figure 17. Merged picture of the scour protection scale model Tests 04 before (**a**) and after (**b**) the test. The current propagation in this set of figures is from right to left.

This initial visual assessment of the damage of the scour protection is corroborated by the topographic laser scans shown in Figure 18. In Figure 18, the topography of the scour protection material at the initial state, after 1000 waves (Test 04_A) and after 3000 waves (Test 04_B) are shown. Regions with higher elevation are shown in red color, while the lower elevation regions are shown in blue color. Through Test 04, in Figure 18, the development of two symmetrically eroded zones can be observed in the wake of the monopile, in the direction of the current. Upstream, just in front of the monopile in Figure 18, the development of scour is clear and shown by an increasing dark blue region. Furthermore, upstream of the monopile, the sedimentation outside the inner ring is clearly progressing from the scan after 1000 waves (Figure 18b) to the scan after 3000 waves (Figure 18c).

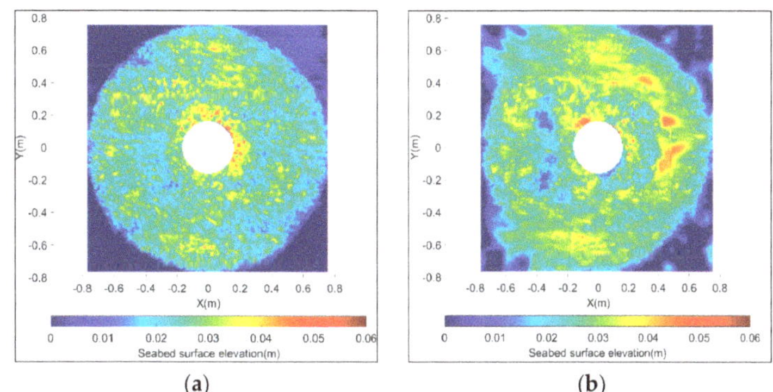

(a)   (b)

Figure 18. *Cont.*

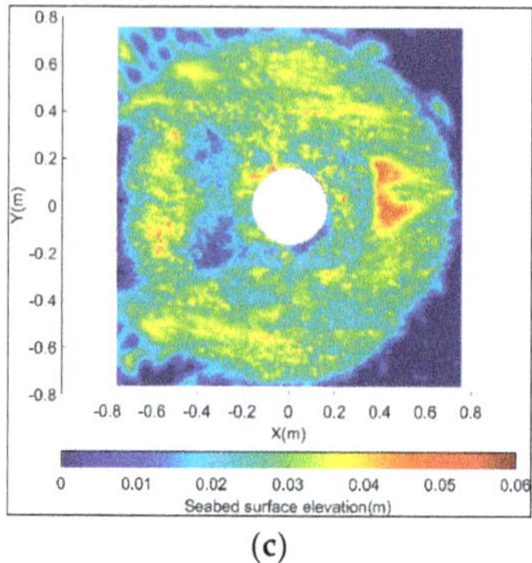

(**c**)

Figure 18. Topography of the scour protection material and of the sand pit measured by the laser scanner before Test 03 (**a**), after the first 1000 waves of Test 04_A (**b**) and after 3000 waves at the end of Test 04_B (**c**). The current propagation in this set of figures is from right to left.

From Figures 17 and 18 it is clear that the scour protection material has undergone damage caused by the hydrodynamic action of the flow. This damage development becomes even clearer when each subsection is considered separately, as in Figure 19.

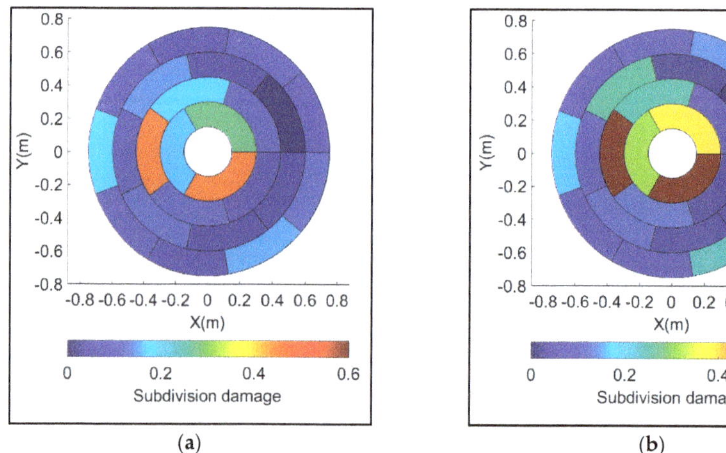

(**a**)　　　　　　　　　　　　　　　(**b**)

Figure 19. Test 04 subdivision damage number: after 1000 waves Test 04_A (**a**) and after 3000 waves Test 04_B (**b**). The current propagation in this set of figures is from right to left.

From the tested hydrodynamic conditions, the measured damage and the predicted damage of the scour protection material are presented in Table 10. The S_{3D} number is the indicator that characterise the scour protection material damage. The predicted damage of the scour protection material is obtained from the damage prediction formula (Equation (11)) presented by [8]:

$$\frac{S_{3D}}{N^{b_0}} = a_0 \frac{U_m^3 T_{m-1,0}^2}{\sqrt{gd}(s-1)^{\frac{3}{2}} D_{n50}^2} + a_1 \left(a_2 + a_3 \frac{\left(\frac{U_c}{w_s}\right)^2 (U_c + a_4 U_m)^2 \sqrt{d}}{g D_{n50}^{3/2}} \right) \tag{11}$$

where, S_{3D} is the damage number; N is the number of waves; U_m is the horizontal orbital wave velocity near the bottom; $T_{m-1,0}$ is the wave energy period; g is the gravitational acceleration; d is the water depth; s is the relative stone density; D_{n50} is the nominal stone diameter; U_c is the current velocity averaged over the water depth and w_s is the fall velocity. a_0, a_2, a_3 and b_0 are non-dimensional parameters determined through fitting and take the value 0.00076, −0.022, 0.0079 and 0.243, respectively. a_1 and a_4 are adimensional parameters as well, they are both dependent on both the current velocity, stone diameter ratio and the current direction:

$$\left. \begin{array}{l} a_1 = 0 \text{ for } \dfrac{U_c}{\sqrt{gD_{n50}}} < 0.92 \text{ and waves following current} \\[3mm] a_1 = 1 \text{ for } \dfrac{U_c}{\sqrt{gD_{n50}}} \geq 0.92 \text{ and waves opposing current} \end{array} \right\} \tag{12}$$

$$\left. \begin{array}{l} a_4 = 1 \text{ waves following current} \\[2mm] a_4 = \dfrac{U_R}{6.4} \text{ waves opposing current} \end{array} \right\} \tag{13}$$

Table 10. Measured and predicted S3D number for Test 04_A and 4_B.

Test No.	Mean Grain Size	Gradation	Number of Waves	Pile Diameter	Water Depth	Significant Wave Height	Peak Period	Horizontal Wave Orbital Velocity	Mean Current Velocity	Predicted Damage Number	Measured Damage Number
	D_{50} (mm)	D_{85}/D_{15} (-)	N (-)	D_p (m)	d (m)	H_s (m)	T_p (m)	U_m (m/s)	U_c (m/s)	Predicted S_{3D}	Measured S_{3D}
Test 04_A	12.5	2.48	1000	0.3	1.2	0.25	2.45	0.177	−0.46	1.834	0.465
Test 04_B	12.5	2.48	2000	0.3	1.2	0.24	2.48	0.166	−0.46	2.141	0.675

More information on Equations (11)–(13) can be found in [8]. A significant deviation in magnitude of the predicted and the measured S_{3D} for the scour protection material can be seen from Table 10. It is important to note that the damage prediction formula, Equation (11), was established for a monopole scale model 1:50 while the monopole model scale of Test 04 is 1:16.7. This deviation could be due to scale effects introduced by the model tests. In Table 10, U_{ADV1} is considered as U_c. The horizontal orbital wave velocity, U_m, is determined from the orbital velocity spectrum as is [2].

$$U_m = \sqrt{2}\,\sigma_U \tag{14}$$

$$\sigma_U^2 = \int_0^\infty S_U(f)df \tag{15}$$

$S_U(f)$ is the power spectrum of the bottom velocity, computed from the spectral analysis of the recordings of the ADVs, in Table 10, U_m is computed from the measurements of ADV1.

7. Conclusions

The PROTEUS experiments performed at the FFF at HR Wallingford within the European Hydralab-PLUS program have yielded a large dataset that provide a benchmark for large scale experiments of scour protection designs around monopiles. The testing program objectives are (i) to compare the performance of single layer wide-graded material used against scouring with current design practices; (ii) to verify the stability of the scour protection designs under extreme flow conditions; (iii) to provide a benchmark dataset for scour protection stability at large scale; and (iv) to investigate the scale effects on scour protection stability. The results will be made available in future studies that will focus in more detail on the impact of specific parameters and methodologies of damage assessment. This first presentation of the dataset obtained highlights the quality of the measurements of hydrodynamic quantities and the scour protection model damage. There is a good agreement of the tested hydrodynamic conditions with respect to the target hydrodynamic conditions. The comparison of the basic analysis of the damage development results and the predicted damage shows that scale effects are not accounted for by the considered prediction formula. Further analysis of the acquired data will provide valuable insight into scale effects and the performance of wide-graded materials. These experiments make use of state of the art optical and acoustic measurement techniques to assess the damage of scour protections under the combined action of waves and currents. These novel PROTEUS tests focus on the study of the grading of the scour protection material as a stabilizing parameter, which has never been done under the combined action of waves and currents at large scale. Scale effects are reduced and thus design uncertainties are minimized. Moreover, the generated data will support the development of future scour protection designs and the validation of numerical models used by researchers worldwide. The target outcomes of the experimental campaign include: (i) study of wide grade material performance with respect to narrow graded materials; (ii) study of scale effects in scour protection around monopiles; (iii) analysis of bed shear stresses in wave-current flows; and (iv) formalization of methodologies for the assessment of the damage of scour protection. These topics will be the basis of our future work within the PROTEUS project. The PROTEUS project will provide unique insight into the behavior of scour protections, improving the design of offshore wind farms, securing the provision of clean and renewable energy, and contributing to deal with climate change challenges.

Author Contributions: Conceptualization, R.W., L.D.V., P.T., L.B. and A.S.; Methodology, L.D.V., L.B., A.S., R.W., P.T. and D.T.; Software, C.E.A.C., M.W. and D.T., Validation, R.W. and L.D.V.; Formal Analysis, M.W. and C.E.A.C.; Investigation, R.W., R.V., L.D.V., L.B., D.K., A.S., M.W., V.S., M.W., C.E.A.C., T.F.-F.; Resources, D.T.; Data Curation, D.T. and V.S.; Writing-Original Draft Preparation, C.E.A.C; Writing-Review & Editing, V.S., A.S., L.D.V., L.B., P.T., T.F.-F, P.R.S., F.T.P, T.S., D.K., M.W., V.S., R.W. and D.T.; Visualization, C.E.A.C. and M.W.; Supervision, D.T., R.W., V.S. and P.T.; Project Administration, D.T. and V.S.; Funding Acquisition, P.T., F.T.P., P.R.S., R.W., D.T., T.S. and A.B.

Funding: The work described in this publication was supported by the European Community's Horizon 2020 Research and Innovation Program through the grant to HYDRALAB-PLUS, Contract no. 654110. The first author would like, in addition, to acknowledge his FWO (Research Foundation-Flanders, project number 3G052716) PhD. funding. Fazeres-Ferradosa was supported by the project POCI-01-0145-FEDER-032170 (ORACLE project), funded by the European Fund for Regional Development (FEDER), through the COMPETE2020, the Programa Operacional Competitividade e Internacionalização (POCI) and FCT/MCTES through national funds (PIDDAC).

Acknowledgments: The authors would like to thank HR Wallingford and the Ghent University technical teams for their availability and help throughout the testing campaign.

References

1. Whitehouse, R.J.S. *Scour at Marine Structures*; Thomas Telford Ltd.: London, UK, 1998.
2. Sumer, B.; Fredsøe, J. *The Mechanics of Scour in the Marine Environment*; Advanced series on ocean engineering; World Scientific: River Edge, NJ, USA, 2002.
3. Chiew, Y.M. Mechanics of Riprap Failure at Bridge Piers. *J. Hydraul. Eng.* **1995**, *121*, 635–643. [CrossRef]
4. Chiew, Y.; Lim, F. Failure behavior of riprap layer at bridge piers under live bed conditions. *J. Hydraul. Eng.* **2000**, *126*, 43–55. [CrossRef]
5. Lauchlan, C.S.; Melville, B.W. Riprap protection at bridge piers. *J. Hydraul. Eng.* **2001**, *127*, 412–418. [CrossRef]
6. Troch, P.; De Rouck, J.; Burcharth, H. Experimental study and numerical modeling of wave induced pore pressure attenuation inside a rubble mound breakwater. In Proceedings of the Coastal Engineering Conference, Cardiff, Wales, UK, 7–12 July 2002; pp. 1607–1619.
7. Galay, V.J.; Yaremko, E.K.; Quazi, M.E. *River Bed Scour and Construction of Stone Riprap Protection*; Sediment Transport in gravel-bed rivers; Thorne, C.R., Bathurst, J.C., Hey, R.D., Eds.; John Wiley & Sons Inc.: New York, NY, USA, 1987; pp. 353–383.
8. De Vos, L.; De Rouck, J.; Troch, P.; Frigaard, P. Empirical design of scour protections around monopile foundations. Part 2: Dynamic approach. *Coast. Eng.* **2012**, *60*, 286–298. [CrossRef]

9.   Loosveldt, N.; Vannieuwenhuyse, K. Experimental Validation of Empirical Design of a Scour Protection around Monopiles under Combined Wave and Current Loading. Master's Thesis, Ghent University, Ghent, Belgium, 2012.

10.   Nielsen, A.W. Scour Protection of Offshore Wind Farms. Ph.D. Thesis, Technical University of Denmark, Lyngby, Denmark, 2011.

11.   Whitehouse, R.; Brown, A.; Audenaert, S.; Bolle, A.; de Schoesitter, P.; Haerens, P.; Baelus, L.; Troch, P.; das Neves, L.; Ferradosa, T.; Pinto, F. Optimising scour protection stability at offshore foundations. In Proceedings of the 7nd International Conference on Scour and Erosion (ICSE-7); CRC Press/Balkema, Leiden, The Netherlands, 2–4 December 2014; pp. 593–600.

12.   Petersen, T.U. Scour around Offshore Wind Turbine Foundations. Ph.D. Thesis, Technical University of Lyngby, Lyngby, Denmark, 2014.

13.   Schendel, A.; Goseberg, N.; Schlurmann, T. Experimental study on the performance of coarse grain materials as scour protection. *Coast. Eng. Proc.* **2014**, *1*, 58. [CrossRef]

14.   Schendel, A.; Goseberg, N.; Schlurmann, T. Erosion Stability of Wide-Graded Quarry-Stone Material under Unidirectional Current. *J. Waterw. Port Coast. Ocean Eng.* **2015**, *142*, 1–19. [CrossRef]

15.   CIRIA, CUR, CETMEF. *The Rock Manual—The Use of Rock in Hydraulic Engineering*, 2nd ed.; C683; CIRIA: London, UK, 2007; ISBN 978-0-86017-683-1.

16.   Petersen, T.U.; Nielsen, A.; Hansen, D.A.; Christensen, E.; Fredsoe, J. Stability of single-graded scour protection around a monopile in current. Scour and Erosion IX. In Proceedings of the 9th International Conference on Scour and Erosion (ICSE 2018), Taipei, Taiwan, 5–8 November 2018.

17.   Fazeres Ferradosa, T.; Taveira-Pinto, F.; Romão, X.; Vanem, E.; Reis, M.T.; Neves, L. Probabilistic design and reliability analysis of scour protections for offshore windfarms. *Eng. Fail. Anal.* **2018**, *91*, 291–305. [CrossRef]

18.   Sutherland, J.; Whitehouse, R.J.S. *Scale Effects in the Physical Modelling of Seabed Scour*; Technical Report TR 64; HR Wallingford Ltd.: Wallingford, Oxfordshire, UK, 1998.

19.   Hughes, S.A. *Physical Models and Laboratory Techniques in Coastal Engineering*; Advanced series on ocean engineering; World Scientific. River Edge, NJ, USA, 1993; Volume 7.

20.   Fazeres-Ferradosa, T.; Taveira-Pinto, F.; Reis, M.T.; das Neves, L. Physical modelling of dynamic scour protections: Analysis of the Damage Number. *Proc. Inst. Civ. Eng. Marit. Eng.* **2018**, *171*, 11–24. [CrossRef]

Article

Experimental Study of a Moored Floating Oscillating Water Column Wave-Energy Converter and of a Moored Cubic Box

Minghao Wu [1,*], **Vasiliki Stratigaki** [1,*], **Peter Troch** [1], **Corrado Altomare** [1], **Tim Verbrugghe** [1], **Alejandro Crespo** [2], **Lorenzo Cappietti** [3], **Matthew Hall** [4] and **Moncho Gómez-Gesteira** [2]

[1] Department of Civil Engineering, Ghent University, Technologiepark 60, B-9052 Zwijnaarde, Belgium; peter.troch@ugent.be (P.T.); corrado.altomare@ugent.be (C.A.); timl.verbrugghe@UGent.be (T.V.)

[2] Environmental Physics Laboratory, Universidade de Vigo, 36310 Pontevedra, Vigo, Spain; alexbexe@uvigo.es (A.C.); mggesteira@uvigo.es (M.G.-G.)

[3] Department of Civil Engineering, Università degli Studi di Firenze - UniFI, Via di Santa Marta 3, 50139 Florence, Italy; lorenzo.cappietti@unifi.it

[4] School of Sustainable Design Engineering, University of Prince Edward Island, Charlottetown, PE C1A 4P3, Canada; mthall@upei.ca

* Correspondence: minghao.wu@ugent.be (M.W.); vicky.stratigaki@ugent.be (V.S.)

Received: 24 April 2019; Accepted: 9 May 2019; Published: 15 May 2019

Abstract: This paper describes experimental research on a floating moored Oscillating Water Column (OWC)-type Wave-Energy Converter (WEC) carried out in the wave flume of the Coastal Engineering Research Group of Ghent University. This research has been introduced to cover the existing data scarcity and knowledge gaps regarding response of moored floating OWC WECs. The obtained data will be available in the future for the validation of nonlinear numerical models. The experiment focuses on the assessment of the nonlinear motion and mooring-line response of a 1:25 floating moored OWC WEC model to regular waves. The OWC WEC model motion has 6 degrees of freedom and is limited by a symmetrical 4-point mooring system. The model is composed of a chamber with an orifice on top of it to simulate the power-take-off (PTO) system and the associated damping of the motion of the OWC WEC model. In the first place, the motion response in waves of the moored floating OWC WEC model is investigated and the water surface elevation in the OWC WEC chamber is measured. Secondly, two different mooring-line materials (iron chains and nylon ropes) are tested and the corresponding OWC WEC model motions and mooring-line tensions are measured. The performance of these two materials is similar in small-amplitude waves but different in large wave-amplitude conditions. Thirdly, the influence of different PTO conditions is investigated by varying the diameter of the top orifice of the OWC WEC model. The results show that the PTO damping does not affect the OWC WEC motion but has an impact on the water surface elevation inside the OWC chamber. In addition, an unbalanced mooring configuration is discussed. Finally, the obtained data for a moored cubic model in waves are presented, which is a benchmarking case for future validation purposes.

Keywords: moored floating wave-energy converter; oscillating water column; wave flume experiment; nonlinear wave condition; 6 degrees of freedom motion; mooring-line tension

---

1. Introduction

The Oscillating Water Column (OWC) is a Wave-Energy Converter (WEC) which mainly consists of a hollow chamber, open below the water level, in which a column of water is forced to oscillate once excited by the external incident waves. The oscillation of water surface inside the chamber introduces an air-pressure variation on the above air volume that drives an air turbine and, in turn, a coaxial electrical generator. Hence, wave energy is firstly converted into pneumatic energy, secondly mechanical, and thirdly into electrical energy; the turbine and generator assembly is called hereafter the power-take-off (PTO) system of the OWC WEC. A common way of installing the OWC WECs is by fixing them to the coastline or to the nearshore seabed. This approach provides convenience in construction, operation, and maintenance. However, the available wave-energy potential decreases due to the energy dissipation of the waves approaching the coast as a result of wave-transformation processes. Consequently, offshore floating OWC WECs are interesting as they are prone to exploit the higher wave-energy resources available at an offshore sea site [1].

A comprehensive review of the history and development of OWC WECs has been given by [2], where several floating OWC WEC concepts such as the Backward Bend Duck Buoy (BBDB) [3], the Spar Buoy [4], and the U-Gen [5] WECs are introduced. Besides the functionality and efficiency of different floating OWC WECs, the hydrodynamic behaviors regarding their motion and mooring system are topics of high interest. Different numerical models have been employed to simulate the dynamics of OWC WECs. Codes based on potential theory, such as WAMIT [6], are widely used for a fast prediction of the motion of different types of floating OWC WECs, for example, the cylinder OWC WEC [7], the BBDB WEC [8,9] and the spar-buoy OWC WEC [10]. Computational Fluid Dynamics (CFD) is another popular methodology to solve the nonlinear air-fluid-mooring-coupling problem. Luo et al. [11] reports simulations of a heave-only floating OWC WEC connected to a spring type of mooring system in a numerical wave tank developed using the Fluent software [12]. Elhanafi et al. [13] thoroughly described a fully 3D numerical investigation of the hydrodynamic behavior of a floating moored OWC WEC by means of the STAR-CCM+ software [14] and validated their results using experimental data. The mooring lines in the employed tests are simulated by four pretensioned springs and are connected vertically to loadcells, and the mooring survivability is investigated in Elhanafi et al. [15]. Besides Eulerian-based methods, Lagrangian-based methods are reported as well in the literature. For example, Crespo et al. [16] presented a numerical model of a floating moored OWC WEC using Smoothed Particle Hydrodynamics (SPH) methods coupled with inelastic catenary theory, and a validation study of this model is presented in [17].

Experimental studies are also seen in the investigation of moored floating OWC WECs, for example, Correia da Fonseca et al. [18] studied the dynamics, energy extraction, and mooring system performance of the spar-buoy OWC WEC. He et al. [19] presented a series of experiments of an OWC WEC integrated to floating box-type breakwaters, primarily with a focus on coastal protection. Although many works have been carried out in the study of floating moored OWC WECs, there exist knowledge gaps in this field. On one hand, the available experimental data of floating moored OWC WECs remains scarce in the reported literature. Most of the studies focus on deep water linear wave condition scenarios, such as in [13,18,20]. On the other hand, much research concerns the heave-only OWC WEC model, such as [19]. Therefore very few studies presented the 6-degrees-of-freedom (6-DOF) motion of a floating moored OWC WEC. Gomes et al. [21] presented the 6-DOF motion response of a very small scale (1:120) slack-moored spar-buoy model, but the mooring-line tension is not investigated. Therefore, it is necessary to carry out more comprehensive studies of the motion and mooring system behaviors of a floating moored OWC WEC, especially, by means of the experiments in wave flume or wave tank as suggested by EMEC [22].

The present paper focuses on an experimental study of a slack-moored floating OWC WEC model in a wave flume. The geometry of this model is originated from the study by Crema et al. [23] who presented a concept of assembling many single units of OWC WECs fixed to a very large floating system power plant to be installed in the Mediterranean Sea. The performance of a single unit fixed OWC WEC is investigated by an experimental study carried out in the wave flume of the University of Florence (LABIMA) [24]. This laboratory scale model has also served as a benchmark test case for the assessment of CFD approaches such as that based on the Lattice Boltzmann Method [25] and the OpenFOAM source code [26]. Moreover, Simonetti et al. [27] have discussed, by means of numerical investigations, the optimization of the main geometric characteristics and the PTO damping.

For the present study of this floating moored OWC WEC model, the primary objective is to investigate the motion behavior and mooring-line tensions of the OWC WEC model subjected to regular waves of different wave periods and wave heights. Two mooring-line materials, iron chain and nylon rope, are used during the experiments for the sake of understanding the impacts of the mooring features on the nonlinear motion of the OWC WEC model. The choice of these two materials has been made to generate numerical validation data for two very different mooring-line materials, with the nylon rope representing a soft and flexible mooring line and with iron chain representing much stiffer mooring line. The PTO damping is simulated by placing an orifice on top of the OWC WEC model, through which the air can exchange between the inside of the OWC chamber and the atmosphere. Different PTO damping characteristics are compared by adjusting the orifice diameter. It should be noted that the study is not aimed at developing a new OWC WEC concept nor improving the energy conversion efficiency. In addition, in our study we include a physical model of a moored floating cubic box with dimensions and mooring system layout similar to that of the OWC WEC model. In the literature, box test cases are often used to validate numerical models (e.g., RANS, SPH, BEM-based solvers). To our knowledge, experimental validation data from tests with boxes are reported only in a few works ([28,29]). However, these test data include only simple free-floating boxes. Therefore, there is a scarcity in test data with moored floating boxes which can serve for validation of numerical models dealing with moored floating structures. As such, in the present study we introduced the box tests to deal with this knowledge gap. The novelty of the work lies on the nonlinear responses of the motion and mooring system of the floating OWC WEC model in nonlinear intermediate depth water wave conditions. Moreover, as the nonlinear numerical models are becoming increasingly popular for the simulation of floating moored OWC WECs, their validation using experimental data is crucial. However, such experimental studies are rarely seen in the literature. Therefore, a second objective of this work is to provide open access experimental data for the validation of numerical models currently under development in the wave-energy research community.

The manuscript starts in Section 1 with an introduction in numerical and experimental floating OWC WEC models and in the objectives of the present study. In Section 2, a description of the experimental setup is provided. In Section 3, validation cases using a simple slack-moored cubic box model are provided. An overview and discussion of the hydrodynamic performance of the laboratory scale OWC WEC model is presented in Section 4. Finally, the main conclusions of this work are summarized in Section 5.

2. Experimental Setup

2.1. Description of the Models

Two models have been employed in the present experimental study. The first one is a simple solid cubic box model (referred to as "BOX" here after), which is used to check the reliability of the installed mooring system, to validate the appropriate installation of all the instruments and of the recorded data, and to provide benchmark experimental data of the BOX motion and mooring-line tensions in nonlinear regular waves. Figure 1a provides an illustration of the BOX model, while all the relevant geometric characteristics are listed in Table 1. The center of gravity is in the geometrical center of the BOX model.

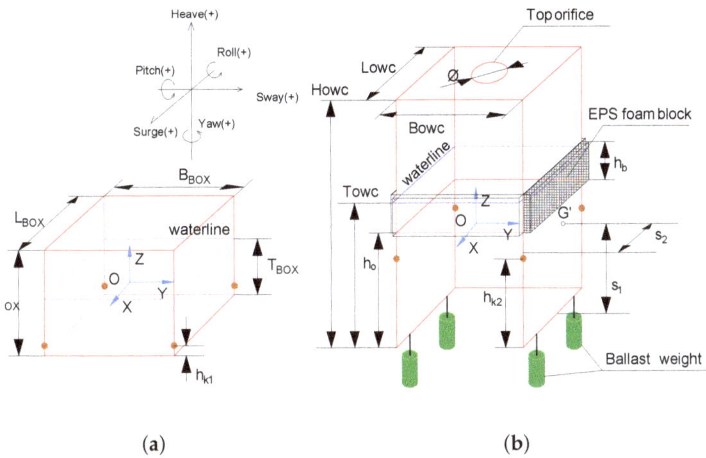

(**a**)   (**b**)

Figure 1. Geometry of floating models: (**a**) BOX; (**b**) OWC WEC.

Table 1. Geometric characteristics of the BOX model.

Symbol	Description	Unit	Value
B_{BOX}	Width of BOX model	cm	20.0
L_{BOX}	Length of BOX model	cm	20.0
H_{BOX}	Height of BOX model	cm	13.2
T_{BOX}	Draft of BOX model	cm	7.9
M_{BOX}	Mass of BOX model	g	3148.0
h_{k1}	Mooring-line fairlead height	cm	0.5
$I_{XX,BOX}$	Moment of inertia around X axis	g·cm^2	1.5×10^5
$I_{YY,BOX}$	Moment of inertia around Y axis	g·cm^2	1.5×10^5
$I_{ZZ,BOX}$	Moment of inertia around Z axis	g·cm^2	2.1×10^5

The second model is a 1:25 scaled OWC WEC shown in Figure 1b. The model is made by light PVC material. The front open side of the OWC WEC model is placed facing the incoming waves. Openings are made on the front and bottom surfaces to obtain higher wave-energy flux. The wall thickness is 10 mm for the four side surfaces and the top surface. Due to this design, the model is asymmetrical regarding its principle axis which results in its destabilization. Therefore, extra ballast weights (green cylinders in Figure 1b) are added at the four bottom corners of the OWC WEC model to lower down the center of gravity and to reach a hydrostatic balance. At the same time, four light expanded polystyrene (EPS) foam blocks (shadowed part in Figure 1b) are attached on the four vertical surfaces near the waterline to provide enough buoyancy and stability. The dimensions of each block are 20.0 cm length, 0.8 cm width and 8.0 cm height. The center of gravity of the OWC WEC model is located on the symmetry plane of the WEC. The distances from the gravity center to the front and the bottom sides are annotated using point **G'** as reference, which is the projection point of the center of gravity on the side surface shown in Figure 1b. To simulate the air turbine PTO damping, an orifice of a diameter of 5.0 cm is made on the top surface of the OWC WEC model, which equals to approximately 6.1% of the top surface area. Table 2 lists the geometric properties of OWC WEC model.

Table 2. Geometric characteristics of the OWC WEC model.

Symbol	Description	Unit	Value
B_{OWC}	Width of OWC model	cm	20.0
L_{OWC}	Length of OWC model	cm	20.0
H_{OWC}	Height of OWC model	cm	44.0
T_{OWC}	Draft of OWC model	cm	26.0
M_{OWC}	Mass of OWC model	g	2593.0
$I_{XX,OWC}$	Moment of inertia around X axis	g·cm^2	7.2×10^5
$I_{YY,OWC}$	Moment of inertia around Y axis	g·cm^2	9.4×10^5
$I_{ZZ,OWC}$	Moment of inertia around Z axis	g·cm^2	5.6×10^5
h_o	Front opening height	cm	19.0
h_{k2}	Mooring-line fairlead height	cm	15.0
h_b	EPS foam block height	cm	8.0
s_1	Vertical height of center of gravity	cm	15.2
s_2	Distance from center of gravity to front surface	cm	9.9
$\varnothing$	Orifice diameter	cm	5.0

2.2. Wave Flume Setup and Instrumentation

The experiments are performed in the 30.0 m long, 1.0 m wide and 1.2 m high wave flume of the Coastal Engineering Research Group at the Department of Civil Engineering of Ghent University. The maximum operating water depth is 80 cm. The global coordinate system $O_0X_0Y_0Z_0$ is defined at the bottom of the wave flume, with the positive O_0X_0 axis pointing to the wave paddle and the positive O_0Z_0 axis with a vertical upward direction. The center of gravity of the OWC WEC model or of the box model is located at the origin point of this global coordinate system. Seven resistive wave gauges (WG1 to WG7) are installed along the wave flume to record wave surface elevations (ζ) at different locations. For the OWC WEC model, one wave gauge (WG8) is installed in the chamber center, which moves with the model and measures the average in-chamber water surface elevation. Figure 2 gives a plan view of the general experimental layout of the wave flume and shows the locations of the employed instrumentation.

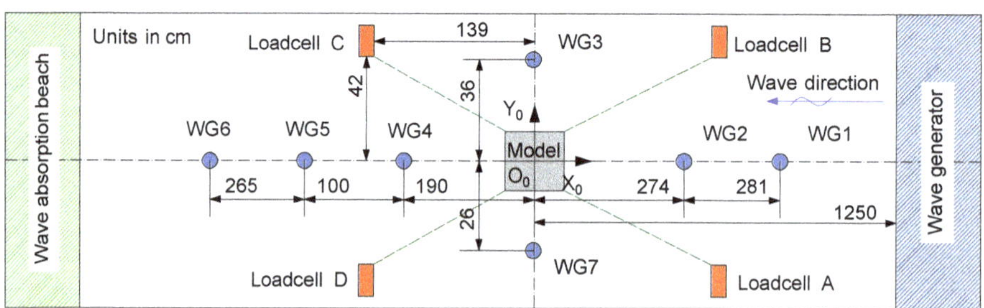

Figure 2. Top view of the experimental setup.

A four-point symmetric slack mooring system is used in the experimental setup. The anchor of each mooring line is connected to a loadcell fixed to the wave flume bottom. The distances between loadcells and models are annotated in Figure 2, while the fairlead of each mooring line is attached to the floating model. Each loadcell measures the horizontal component of the mooring-line tension (F), and has a measurement range of 100 N and a sampling frequency of 1000 Hz. To investigate the effect of the mooring-line material on the model response, iron chains and nylon ropes have been used alternatively. The elasticity properties of the two materials are acquired by performing tension tests at the laboratory of Material Science at Ghent University. The mooring-line parameters are listed in Table 3.

Table 3. Mooring-line parameters.

Material	Symbol	Description	Units	Parameter
	L_C	Total length of chain mooring line	cm	145.5
	k_C	Chain elasticity	N/mm	19.0
Iron chain	l	Length per chain segment	cm	0.8
	w	Chain weight per centimeter	g/cm	0.6
	v	Chain volume per centimeter	cm^3/cm	0.1
Nylon rope	L_R	Total length of rope mooring line	cm	144.0
	k_R	Rope elasticity	N/mm	1.1

The 6-DOF motion of the floating models is captured using an optical tracking motion system, CTrack, developed by Ctech Metrology [30]. CTrack consists of a camera, several marker receivers, and dedicated processing software. The marker receivers are installed on the floating model. Based on the spatial relationship between markers and camera, a local coordinate system OXYZ is created fixed to the center of gravity of each one of the floating models, and the real-time motion is measured. The 6-DOF motion includes translations and rotations along the X, Y, and Z axes, where the three translations are the surge (x), sway (y) and heave (z), and the three rotations are the roll (ϕ), pitch (θ) and yaw (ψ). The local coordinate system and the definition of the 6-DOF motion are visualized in Figure 1a. Figure 3 shows the experimental setup in the wave flume from different perspectives.

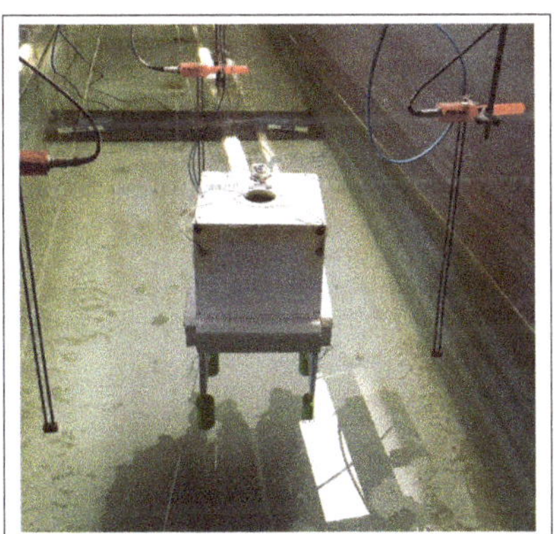

(a)

(b)

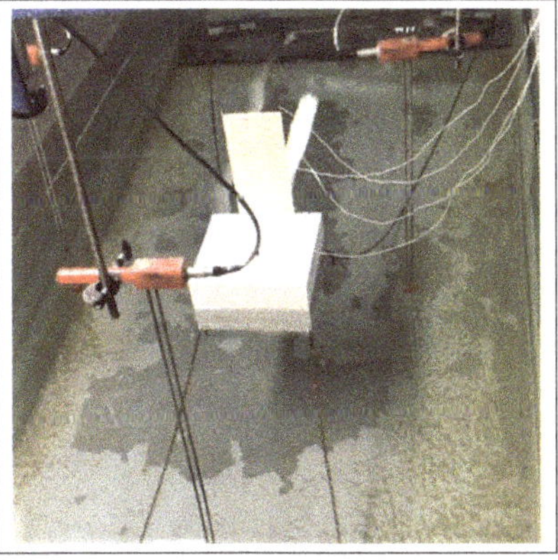

(c)

Figure 3. Moored models in the wave flume: (**a**) front view of the OWC WEC model with rope mooring; (**b**) side and back view of the OWC WEC model with chain mooring (this image of the model "broken" into two pieces is a result of light refraction); (**c**) back view of the BOX model with chain mooring.

2.3. Experimental Program

The present study focuses mainly on the response of the motion and mooring-line tensions of the floating OWC WEC and the BOX model subjected to unidirectional regular waves. State-of-the-art second order wave generation and absorption techniques have been applied [31]. Seven Test groups are conducted. The parameter ranges of the applied wave conditions and water depth in each Test group are listed Table 4. Test group 1 is a validation Test group of the BOX model in the wave conditions of 1.6 s $\leq T \leq$ 2.0 s and 12.0 cm $\leq H \leq$ 15.0 cm, where H is the wave height and T is the wave period. The iron chain mooring lines are applied in this Test group. Test group 2 to 7 investigate the OWC WEC model. Test group 2 studies the motion response of the model in relatively small wave-amplitude conditions (4.0 cm $\leq H \leq$ 8.0 cm) while Test group 3 uses larger wave-amplitude conditions (11.0 cm $\leq H \leq$ 14.0 cm). Test groups 2 and 3 are conducted using the iron chain as the mooring-line material. To compare the impact of different mooring-line materials, in Test groups 4 and 5 the nylon rope is used instead of the iron chain. Test group 4 uses the same wave conditions as in Test group 2 and Test group 5 uses the same wave conditions as in Test group 3. A sensitivity test regarding a scenario of unequal mooring-line lengths, is carried out in Test group 6. This Test group 6 investigates the response of mooring-line tensions due to the variation of the length of one of the four mooring lines, and discusses the effect of this unbalanced mooring system. Moreover, to further investigate the influence of larger orifice PTO damping qualitatively, in Test group 7, a $\varnothing = 1.0$ cm orifice is used and is compared to a $\varnothing = 5.0$ cm one (Test group 3) under the same wave conditions.

For all the wave conditions used in the experimental program, the wave steepness is kept in the range of $H/\lambda < 1/20$ to ensure non-breaking wave conditions, where λ is the wave length. However, as the water depth (d) is limited, the wave conditions are no longer linear. The wave nonlinearity can be assessed by plotting all the applied wave conditions used in each Test group in the adapted Le Méhauté diagram [32], as shown in Figure 4. It is clearly seen that most of the waves used in Test groups 1 to 5 are in the intermediate water depth region and they satisfy the Stokes 2nd order wave theory. Furthermore, for Test group 1, the cases of $H = 15.0$ cm waves are in the Stokes 3rd order wave region. In addition, each wave train contains less than 20 waves and the data acquisition system is stopped before the reflected waves reach the model, so any wave reflection effects from the wave absorption beach located at the opposite end of the wave flume will not be considered. Please note that all presented H and T are target wave condition values. This means that the H and T values resulting from the Fourier transformation analysis are slightly different because all presented results refer to test cases where a BOX or an OWC WEC model is always present in the wave flume. As such wave radiation, diffraction, and reflection induced by the floating objects is included in the presented results.

Table 4. Experimental program.

Test Group	Model	Range of Regular Wave Period, T (s)	Range of Regular Wave Height, H (cm)	Water Depth, d (cm)	Mooring Line Material	Note
1	BOX	1.6–2.0	12.0–15.0	50.0	Chain	Benchmark test
2	OWC	0.7–2.1	4.0–8.0	60.0	Chain	Small wave amplitude
3	OWC	1.5–2.0	11.0–14.0	60.0	Chain	Large wave amplitude
4	OWC	0.7–2.1	4.0–8.0	60.0	Rope	Small wave amplitude
5	OWC	1.5–2.0	11.0–14.0	60.0	Rope	Large wave amplitude
6	OWC	1.7	14.0	60.0	Chain	Unbalanced mooring
7	OWC	1.5–2.0	11.0	60.0	Chain	$\varnothing = 1.0$ cm

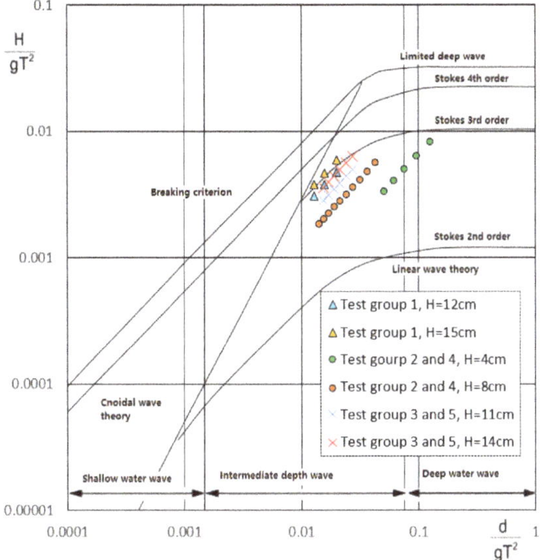

Figure 4. Applied wave conditions plotted in the adapted Le méhauté diagram [32]. All cases in Test group 1 to 5 are plotted. [Adapted with from Le méhauté, B. An Introduction to Hydrodynamics and Water Waves]

2.4. Uncertainty Sources

There are several uncertainty sources related to the obtained experimental data which should be considered. These uncertainty sources may affect the interpretation of the here presented results. The sources of uncertainty are listed as below.

(1) For mooring system, uncertainties are related to the measurement of the mooring-line material elasticity, length, and weight and volume per unit length, the locations of loadcells and fairleads, and the tensions.

(2) For the floating object, uncertainties are related to the measurement of: the geometrical dimensions of the model, the mass, the center of gravity, and momentum of inertia of the model, the spatial position of the model, including the initial position and the 6-DOF motion.

(3) For the wave generation system, uncertainties are related to the measurement of the wave surface elevation and the wave period.

3. BOX model Experimental Results

The BOX model experimental results are obtained from Test group 1. In this paper, we present two datasets with the synchronized information of water surface elevations acquired from WG2, WG3, and WG4 (see Figure 3), motion time series of surge, heave, and pitch obtained from the CTrack system and horizontal components of mooring-line tensions obtained from four loadcells. The case of regular waves of $T = 1.6$ s and $H = 12.0$ cm is displayed in Figure 5, while another test with regular waves of $T = 1.8$ s and $H = 15.0$ cm is shown in Figure 6. The loadcells are initialized when the BOX model is at the equilibrium position. The mooring-line tension data is post-processed via an averaging filter.

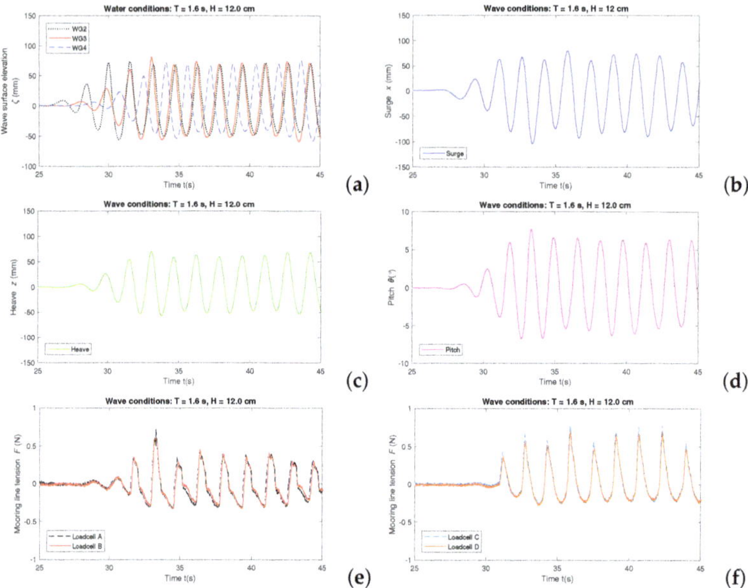

Figure 5. Synchronized data of validation test with BOX model in regular waves of $T = 1.6$ s and $H = 12.0$ cm (target values): (**a**) wave surface elevation; (**b**) surge motion; (**c**) heave motion; (**d**) pitch motion; (**e**) mooring-line tensions measured by Loadcell A and B; (**f**) mooring-line tensions measured by Loadcell C and D.

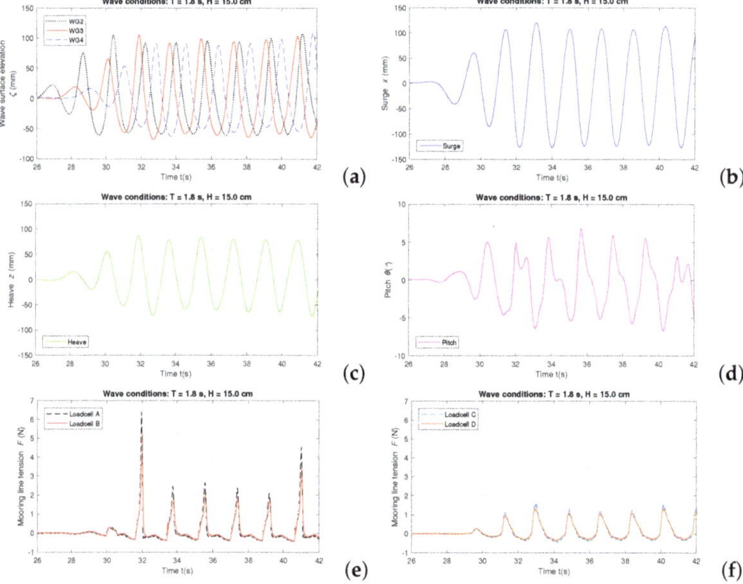

Figure 6. Synchronized data of validation test with BOX model in regular waves of $T = 1.8$ s and $H = 15.0$ cm (target values): (**a**) wave surface elevation; (**b**) surge motion; (**c**) heave motion; (**d**) pitch motion; (**e**) mooring-line tensions measured by Loadcell A and B; (**f**) mooring-line tensions measured by Loadcell C and D.

The results are discussed in terms of wave field modification due to the presence of the BOX model. Firstly, nonlinearity of the incident waves is observed from the recorded water surface elevations. In both Figures 5a and 6a, the recorded data of all the three wave gauges show that the incident wave forms have flat troughs and sharp peaks. Especially in Figure 6a, the average wave trough value obtained by WG2 is $\zeta = -58.7$ mm while the average peak value is $\zeta = 93.4$ mm. The nonlinear wave forms are due to the intermediate water depth, as explained in Figure 4.

Secondly, in both Figures 5a and 6a, the wave heights measured by WG4 are smaller than those measured by WG2 and WG3, since WG4 is located behind the BOX where the incident waves are diffracted after passing the model. Meanwhile, because WG3 is close by the BOX model, the wave heights recorded by WG3 are slightly larger than those recorded by WG2 as both the incident waves and the radiated waves due to the BOX motion are captured.

For the BOX model motion in the regular wave conditions of $T = 1.6$ s and $H = 12.0$ cm, the surge motion of the BOX model (Figure 5b) is not steady regarding the motion amplitude in each wave period. This occurs because the presence of the iron chain mooring lines imposes a low frequency component to the wave induced surge motion. The heave (Figure 5c) and pitch (Figure 5d) motions of the BOX model are regular and the motion amplitudes are steady. The measured mooring-line tensions as plotted in Figure 5e,f show a good balance between the two mooring lines that are connected to Loadcell A and B, and between the two connected to Loadcell C and D, respectively.

When considering the regular wave conditions of $T = 1.8$ s and $H = 15.0$ cm, different motion characteristics of the BOX are observed. As displayed in Figure 6b, the surge motion shows a more regular pattern compared to the surge motion shown in Figure 5b. For the heave motion of the BOX (Figure 6c), although the motion is regular, nonlinear effects are obvious as the average peak value is $z = 80.7$ mm and the average trough value is $z = -61.9$ mm. Moreover, as shown in Figure 6d, the pitch motion of the BOX contains strong nonlinearity and is very irregular. This is due to a combined effect of the nonlinear incident waves and of the mooring lines. For the mooring-line tensions in this test, Figure 6e,f show a good balance between the tensions measured by Loadcell A and B, and between those measured by Loadcell C and D, respectively.

The results of the two cases presented in this section will be employed in the development of fully nonlinear numerical models as a benchmark experiment for numerical validation.

4. OWC WEC Model Experimental Results

4.1. OWC WEC Motion Response

The motion response to regular waves is defined as the motion amplitude of a floating object in regular waves per unit amplitude. By varying the frequency of the incident wave, a motion response curve can be obtained to depict the motion characteristics of the floating system in the frequency domain. For a moored floating OWC WEC, the motion response information reflects its motion amplitude, provides its natural frequency and reveals nonlinear effects due to the mooring system and the incoming waves.

Tests are carried out with both small-amplitude waves (Test groups 2 and 4) and large amplitude waves (Test groups 3 and 5). For Test groups 2 and 4, $H = 4.0$ cm (when $T \leq 1.1$ s) and $H = 8.0$ cm (when $T \geq 1.2$ s) are used; the later one stands for $H = 2.0$ m in prototype scale. For Test groups 3 and 5, the applied wave heights are $H = 11.0$ cm and $H = 14.0$ cm, and the wave period is set within the range of 1.5 s $\leq T \leq 2$ s. Moreover, according to the suggestion given by ITTC guidelines [33], the translation motion response data is expressed by a division of the mean single motion amplitude over an averaged wave amplitude, as x/η_0 or z/η_0, where $\eta_0 = H/2$.

The experimental results of the OWC WEC surge response to small-amplitude waves (Test groups 2 and 4) and large amplitude waves (Test groups 3 and 5) are shown in Figure 7a,b, respectively, and the results of its heave response are shown in Figure 8a,b. In addition, comparisons between the surge and heave response in different wave heights when using one mooring-line material are visualized in Figure 7c (surge response when using only iron chain), Figure 7d (surge response when using only nylon rope) and Figure 8c (heave response when using only iron chain) and Figure 8d (heave response when using only nylon rope).

Firstly, the resonance periods in surge and heave of the OWC WEC model are observed as plotted in Figures 7a and 8a, respectively. For a wave period of $T = 0.9$ s, the OWC WEC surge motion response reaches a minimum value. The heave resonance period ($T_{r,heave}$) for using the nylon rope mooring lines is $T_{r,heave} = 0.9$ s, and when using iron chain $T_{r,heave} = 1.0$ s. This modification of $T_{r,heave}$ shows the effect of the mooring-line material: the floating system has a total mass of both the OWC WEC model and the mooring lines, so the heavy iron chain mooring line increases the resonance period of the OWC WEC heave motion. Secondly, Figures 7a and 8a also show that when 1.0 s $\leq T \leq 1.7$ s, the OWC WEC motion response due to using the chain mooring lines are very similar to the response when using the rope. Obvious differences occur in surge motion when $T \geq 2.0$ s. As the density of nylon rope material is close to the water density, the rope could not provide any stiffness as a free hanging chain unless the mooring lines are fully stretched. As a result, though the rope mooring lines are shorter, they lead to larger surge amplitude than using the chain when $T \geq 2.0$ s. This effect is more obvious in large amplitude waves, as shown Figure 7b. Thirdly, it is clearly seen from Figure 8b that the OWC WEC heave response when using nylon rope material reaches over 1.0 when $T \geq 1.8$ s, and goes up to 1.45 for $T = 2.0$ s.

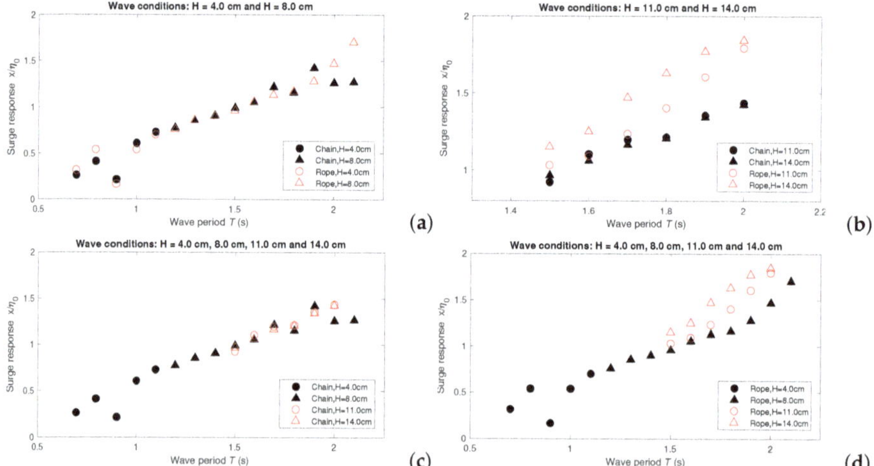

Figure 7. OWC WEC surge motion response for different mooring-line materials: (**a**) wave height $H = 4.0$ cm and $H = 8.0$ cm; (**b**) wave height $H = 11.0$ cm and $H = 14.0$ cm; (**c**) using only iron chain, $H = 4.0$ cm to $H = 14.0$ cm; (**d**) using only nylon rope $H = 4.0$ cm to $H = 14.0$ cm. (All wave conditions are target values).

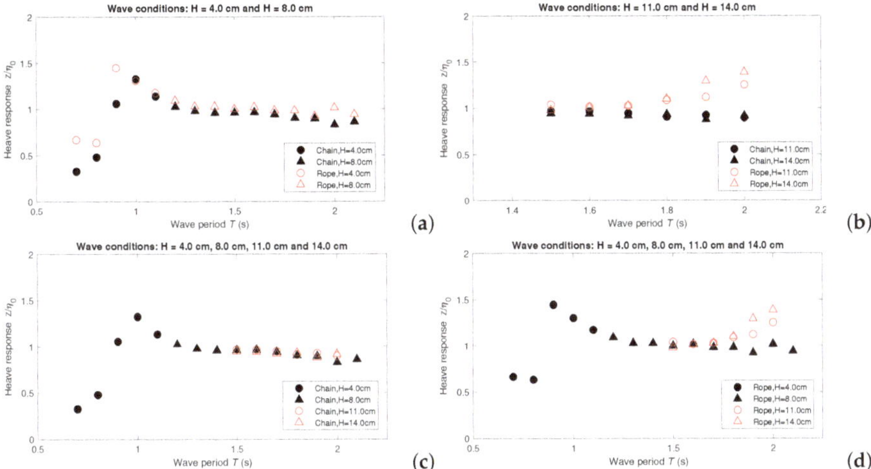

Figure 8. OWC WEC heave motion response for different mooring-line materials: (**a**) wave height $H = 4.0$ cm and $H = 8.0$ cm; (**b**) wave height $H = 11.0$ cm and $H = 14.0$ cm; (**c**) using only iron chain, $H = 4.0$ cm to $H = 14.0$ cm ; (**d**) using only nylon rope $H = 4.0$ cm to $H = 14.0$ cm. (All wave conditions are target values).

To investigate this further, a comparison of the OWC WEC motion response is performed for wave conditions of the same wave period but different wave height, $T = 1.9$ s, $H = 8.0$ cm and $T = 1.9$ s, $H = 14.0$ cm. The motion logs of the OWC WEC model are plotted in Figure 9. It is observed that when $H = 8.0$ cm, the three motions (surge, heave and pitch) are well matched for both of the tested mooring materials. This indicates that the mooring system does not significantly affect the OWC WEC motion at this wave period and when small wave heights are applied. When $H = 14.0$ cm, strong nonlinearity emerges as the incoming waves are close to Stokes 3rd order region (Figure 4). Figure 9b is the surge motion history of the OWC WEC model and shows that the chain mooring line contributes to smaller surge motion than the nylon rope. For the OWC WEC surge motion when using nylon ropes, the average surge peak value is $x = 142.3$ cm and the average trough value is $x = -102.3$ cm, which clearly reflect a nonlinear surge motion behavior. The equilibrium drift, on the contrary to the iron chain case, is to the wave incoming direction. This is caused by the strong shock force from the nylon rope lines. For the OWC WEC heave mode as displayed in Figure 9d, two peaks and troughs appear in one wave period when using the rope mooring lines. The major peak is induced by waves, and the minor peak occurs due to a shock load from the mooring line when the rope is suddenly stretched. The value of this major peak of the heave motion is close to the peak value of the OWC WEC heave motion when using the iron chain. However, when using the nylon rope, the mooring-line shock loads acting on the OWC WEC model lead to lower major heave troughs and amplify the heave motion response up to over 1.0, as shown in Figure 8b. This is primarily due to the low stiffness of the flexible and soft nylon rope material. Moreover, in Figure 9f, nonlinear effect is also obvious for the pitch motion of the OWC WEC model in large waves when using nylon rope, where the average pitch trough is $\theta = -20.6°$ and the average pitch peak is $\theta = 13.8°$. Compared to using iron chain, using nylon rope introduced larger pitch motion amplitude and sharper pitch peaks and pitch troughs, which means the corresponding mooring-line shock loads introduce a more intense rotational acceleration to the OWC WEC model, and hence, result in a possible sloshing effect inside the OWC WEC chamber.

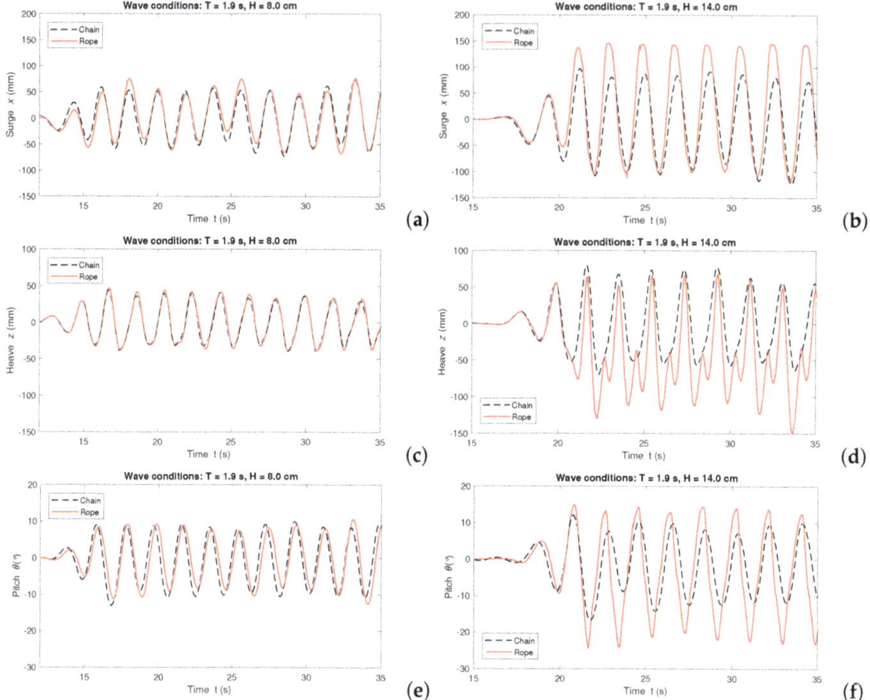

Figure 9. OWC WEC model motions in regular waves of $T = 1.9$ s: (**a**) surge, $H = 8.0$ cm; (**b**) surge, $H = 14.0$ cm; (**c**) heave, $H = 8.0$ cm; (**d**) heave, $H = 14.0$ cm; (**e**) pitch, $H = 8.0$ cm; (**f**) pitch, $H = 14.0$ cm. (All wave conditions are target values).

4.2. Water Surface Elevation Variation Inside the OWC WEC Vhamber

The time averaged energy output of an OWC WEC is given by Equation (1) [13],

$$P_E = \frac{1}{T} \int_0^T \Delta p(t) q(t) dt \tag{1}$$

where P_E is the mean power output, $\Delta p(t)$ is the pressure variation inside the OWC WEC chamber during a wave period, $q(t)$ is the air volume flux and T is the wave period. The mean water surface elevation variation inside the OWC WEC chamber determines the air flux through the orifice at the top of the chamber, and hence, contributes significantly to the potential energy output of the WEC. Based on this, comparisons between the in-chamber water surface elevation and the wave surface elevation outside the OWC WEC model are made for various wave conditions. Figure 10 depicts the water surface elevations (ζ) recorded by WG2 (located in front of the OWC WEC model) and WG8 (inside the OWC WEC chamber). The considered wave conditions are characterized by $T = 0.9$ s and $H = 4.0$ cm in Figure 10a, $T = 1.0$ s and $H = 4.0$ cm in Figure 10b, $T = 1.7$ s and $H = 11.0$ cm in Figure 10c, and $T = 1.7$ s and $H = 14.0$ cm in Figure 10d. The mooring-line material is iron chain for all these cases.

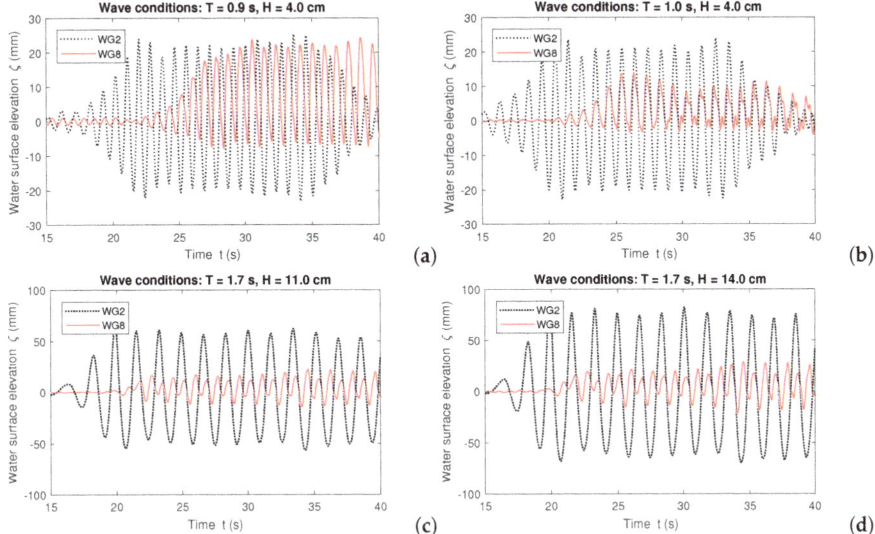

Figure 10. Water surface elevation time series: (**a**) $T = 0.9$ s and $H = 4.0$ cm; (**b**) $T = 1.0$ s and $H = 4.0$ cm; (**c**) $T = 1.7$ s and $H = 11.0$ cm; (**d**) $T = 1.7$ s and $H = 14.0$ cm. (All wave conditions are target values.

Differently from the heave motion resonance period of the OWC WEC $T_{r,heave} = 1.0$ s obtained in Section 4.1, resonance of the in-chamber water surface elevation occurs at $T_{r,chamber} = 0.9$ s. The in-chamber water surface elevation amplitude decreases approximately 50% as the incident wave period increases 0.1 s regarding $T_{r,heave}$. This indicates a narrow frequency band of optimum power output of the studied floating moored OWC WEC, which is similar to the result obtained from the investigation of a fixed detached OWC WEC as described by [23]. For the in-chamber water surface elevation in large wave conditions of $T = 1.7$ s, $H = 11.0$ cm and $H = 14.0$ cm, clearly the resonance does not occur.

4.3. Mooring-Line Tensions

As described in Section 4.1, the mooring-line material plays an important role for the OWC WEC model motion in large amplitude waves. In this section, we show that the mooring-line tension at the anchor location, also differs. Examples are given for the regular wave conditions of $T = 1.7$ s and $H = 14.0$ cm, as presented in Figures 11 and 12. Figure 11 illustrates the horizontal components of the chain mooring-line tensions measured by all four loadcells and Figure 12 illustrates the registered data for the rope mooring lines. It is observed that using rope mooring line has introduced periodic shock loads exceeding 10.0 N, which are significantly larger than those when using chain mooring line for the same wave conditions.

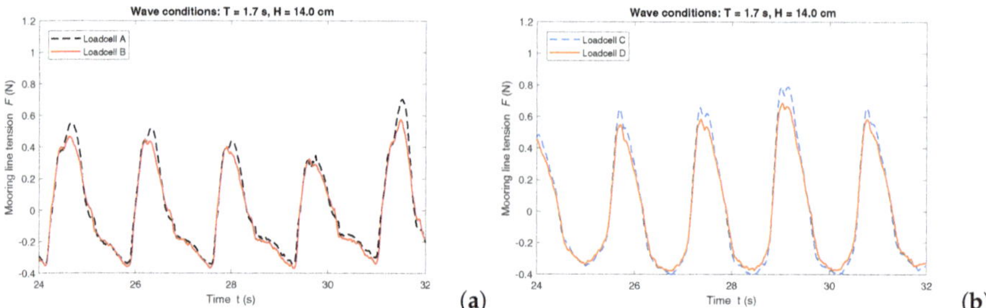

Figure 11. Iron chain mooring-line tensions measured by loadcells in regular waves of $T = 1.7$ s and $H = 14.0$ cm (target values): (**a**) Loadcells A and B; (**b**) Loadcells C and D.

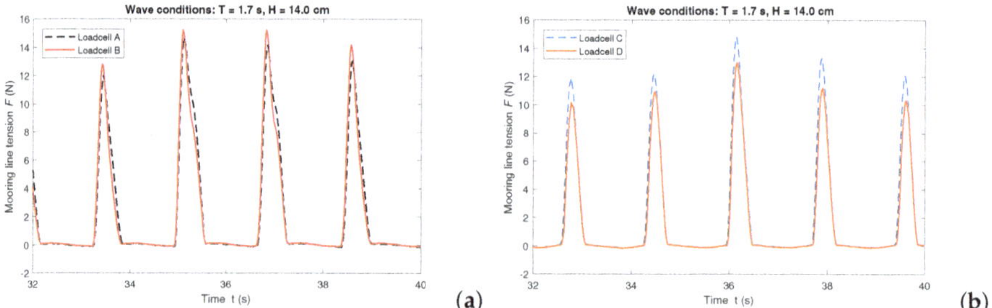

Figure 12. Nylon rope mooring-line tensions measured by loadcells in regular waves of $T = 1.7$ s and $H = 14.0$ cm (target values): (**a**) Loadcells A and B; (**b**) Loadcells C and D.

4.4. Effect of Unequal Mooring-Line Lengths

The mooring-line length affects the performance of a moored floating system. During the tests, a perfect symmetry of the mooring system is difficult to achieve while even slightly unequal lengths of the mooring lines introduce effects on the mooring-line tensions that need to be considered. This section discusses the sensitivity of the mooring-line tensions to the unbalanced mooring-line lengths. For this purpose, Test group 6 (see Table 4) is performed with uniform regular wave conditions of $T = 1.7$ s and $H = 14.0$ cm. The length of the front mooring line connected to Loadcell A, noted as $L_{C,1}$, is adjusted within the range of $\pm1.65\%$ of the initial length L_C (Table 3) . Using $\delta L_{C,1}$ to note the increment of the mooring-line length, the mooring-line length adjustment is within the range of -2.4 cm $\leq \delta L_{C,1} \leq 2.4$ cm, and with a step of 1.6 cm. Given the iron chain mooring-line parameters of Table 3, $\delta L_{C,1} = 0.8$ cm indicates increase of the mooring-line length by adjusting one segment on the chain, while a negative $\delta L_{C,1}$ means decrease of the length. The mooring-line tensions measured by each loadcell when $\delta L_{C,1}$ varies are shown in: Figure 13a for $\delta L_{C,1} = -2.4$ cm (-1.65% of $L_{C,1}$), Figure 13b for $\delta L_{C,1} = -0.8$ cm (-0.55% of $L_{C,1}$), Figure 13c for $\delta L_{C,1} = 0.8$ cm ($+0.55\%$ of $L_{C,1}$) and Figure 13d for $\delta L_{C,1} = 2.4$ cm ($+1.65\%$ of $L_{C,1}$). According to the obtained results, the mooring-line tensions are sensitive to the mooring-line length variation. Shock loads have been observed as the length is shortened by 2.4 cm (or else, when $\delta L_{C,1} = -2.4$ cm, which equals to an adjustment of the chain by 3 segments), when the maximum tension attains to 5 times of the average load of the other mooring lines. This investigation illustrates the necessity of a balanced mooring configuration during the wave flume tests.

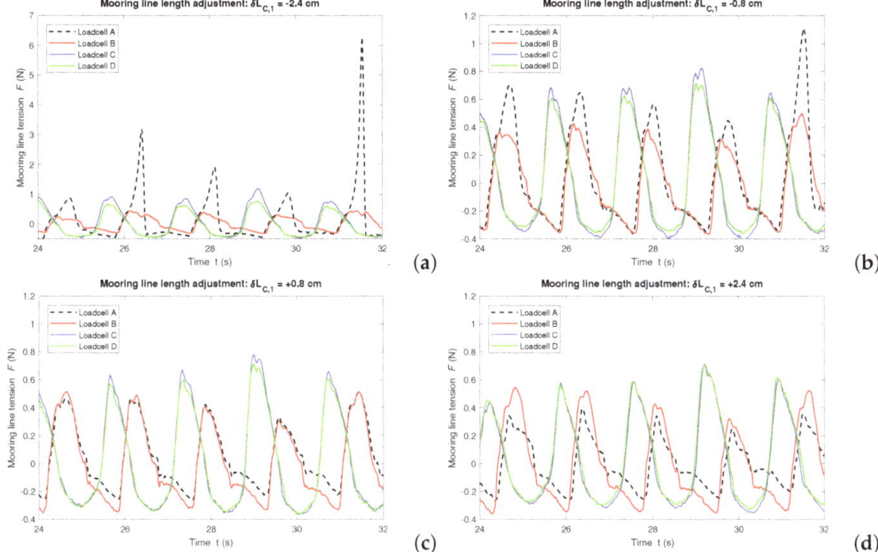

Figure 13. Mooring-line tensions measured by loadcells in regular waves of $T = 1.7$ s and $H = 14.0$ cm (target values) for unequal mooring-line lengths: (**a**) $\delta L_{C,1} = -2.4$ cm $(-1.65\%$ of $L_{C,1})$; (**b**) $\delta L_{C,1} = -0.8$ cm $(-0.55\%$ of $L_{C,1})$; (**c**) $\delta L_{C,1} = 0.8$ cm $(+0.55\%$ of $L_{C,1})$; (**d**) $\delta L_{C,1} = 2.4$ cm $(+1.65\%$ of $L_{C,1})$.

4.5. Effect of the Orifice Diameter at the Top of the OWC WEC Chamber

The size of the orifice diameter on top of the OWC WEC chamber is important for simulating the PTO damping of the OWC WEC since it determines the air-pressure drop inside the chamber [23]. According to the pipe flow theory, a smaller orifice diameter means a higher air-pressure drop and momentum loss rate during the air exchange, which indicates a higher damping of the PTO system. Based on this concept, as described in Table 4, Test group 7 is conducted by replacing the original $\varnothing = 5.0$ cm orifice by a smaller one of $\varnothing = 1.0$ cm diameter. The investigation focuses on the comparison of the OWC WEC motion and of the in-chamber water surface elevation. An example from the regular wave conditions of $T = 1.7$ s and $H = 11.0$ cm is shown in Figure 14. The comparative results for different orifice diameters show almost identical OWC WEC motion time series (see Figure 14a–c). This is because the PTO damping due to the orifice diameter variation is negligible compared to the system's hydrodynamic damping. However, as shown in Figure 14d, the amplitude of in-chamber water surface elevation is reduced to less than 4 mm when the orifice diameter decreases, meaning that there is limited air volume exchange through the $\varnothing = 1.0$ cm orifice.

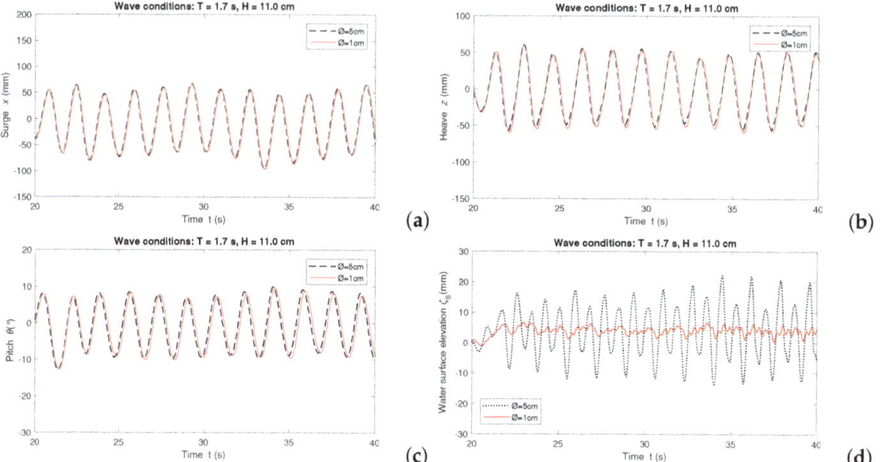

Figure 14. Comparisons of the OWC WEC motion and water surface elevation inside the chamber between different orifice sizes regular waves of $T = 1.7$ s and $H = 11.0$ cm (target values): (**a**) surge; (**b**) heave; (**c**) pitch; (**d**) in-chamber water surface elevation.

5. Conclusions

In this paper, an experimental study for investigating the motion in regular waves and mooring-line tension characteristics of a 1:25 scaled slack-moored floating OWC WEC model is presented. Different wave conditions, PTO damping characteristics and mooring-line materials and lengths are studied.

Firstly, by applying a series of nonlinear regular wave conditions, the motion response of the OWC WEC model is recorded. With using iron chain mooring lines, the heave motion resonance period of the OWC WEC model is $T_{r,heave} = 1.0$ s and the in-chamber water surface elevation resonance is observed when $T_{r,chamber} = 0.9$ s.

Secondly, the effect of different mooring-line materials, nylon rope and iron chain, is investigated in both small and large wave-amplitude conditions. When small wave-amplitude conditions are applied (wave height $H \leq 8.0$ cm), either using iron chain or nylon rope mooring line gives similar surge and heave motion response of the OWC WEC model within the wave period range of 1.0 s $\leq T \leq 1.7$ s. However, the use of nylon rope leads to smaller heave motion resonance period of the OWC WEC, $T_{r,heave} = 0.9$ s, than when using the iron chain, and larger surge motion response amplitude when $T \geq 2.0$ s. On the other hand, when the incident wave heights $H \geq 11.0$ cm, the OWC WEC surge and heave motion of using the nylon rope mooring line is significantly larger than that when using the iron chain mooring line. Meanwhile, strong nonlinear effects in the OWC WEC heave motion occurs when using the nylon rope. Moreover, the use of nylon rope introduces mooring-line shock loads under the regular wave conditions of $H = 14.0$ cm and $T = 1.7$ s.

Thirdly, a scenario of an unbalanced mooring system due to unequal mooring-line lengths is investigated by adjusting the length of one mooring line. The results show that the tensions of the mooring lines of the floating OWC WEC model is sensitive to the variation of the mooring-line length. When decreasing the mooring-line length $L_{C,1}$ (connected to Loadcell A) by 2.4 cm, which is equal to only three segments on the iron chain, severe anchor shock loads have been captured by the loadcell.

Finally, a qualitative study of the PTO damping impact on the motion of the floating moored OWC WEC model is performed through a comparison of the obtained results between two different sized orifices on top of the model. These orifices simulate the effect of the PTO damping. A small orifice represents high PTO damping while a large diameter orifice represents a limited PTO damping.

It is shown that the orifice diameter has a very limited influence on the motion of the OWC WEC model. However, a smaller top orifice introduces a more stable in-chamber water surface elevation.

In addition, two datasets from tests with a generic cubic floating box (BOX) model moored to the wave flume bottom are presented. These data include synchronized wave surface elevations, surge, sway, and pitch motion of the BOX as well as the mooring-line tensions. As a result, of applying nonlinear regular waves and the slack chain mooring system, the nonlinear motion results of the BOX model are obtained and analyzed. The datasets will be further used by researchers as a benchmark case for further development of fully nonlinear numerical models used to simulate the behavior of moored floating objects.

The presented study is novel as it focuses on the nonlinear responses of the motion and mooring system of a floating OWC WEC and BOX model in nonlinear intermediate depth water wave conditions. As nonlinear numerical models are becoming increasingly popular for the simulation of floating moored OWC WEC models, their validation using experimental data is crucial. The present study covers then this existing data gap seen in the literature regarding floating moored OWC WECs, by providing an open access experimental database.

Author Contributions: Formal analysis, M.W.; Investigation, M.W., C.A. and T.V.; Methodology, C.A. and L.C.; Project administration, V.S. and P.T.; Validation, A.C., M.H. and M.G.-G.; Writing—original draft, M.W.; Writing—review and editing, V.S., C.A., A.C. and L.C.

Funding: The present study at UGent is part of the preparatory phase for the MaRINET2 EsfLOWC project which has received funding from the EU H2020 Programme under grant agreement No. 731084. The present experiments at UGent are supported by the Research Foundation Flanders (FWO), Belgium - FWO.OPR.2.0—FWO research project No. 3G029114. The work is also partially financed by the Ministry of Economy and Competitiveness of the Government of Spain under project "WELCOME ENE2016-75074-C2-1-R". The first author would like to acknowledge his PhD funding through a Special Research Fund of UGent, (BOF).

Conflicts of Interest: The authors declare no conflict of interest. The funders had no role in the design of the study; in the collection, analyses, or interpretation of data; in the writing of the manuscript, or in the decision to publish the results.

Abbreviations

The following abbreviations are used in this manuscript:

BBDB	Backward Bend Duck Buoy
BOX	cubic floating box model
CFD	Computational Fluid Dynamics
DOF	Degree of Freedom
EMEC	The European Marine Energy Center
EPS	Expanded Polystyrene
ITTC	The International Towing Tank Conference
OWC	Oscillating Water Column
PTO	power-take-off
SPH	Smoothed Particle Hydrodynamics
WEC	Wave-Energy Converter
WG	wave gauge

References

1. Falcão, A.F.O. Wave energy utilization: A review of the technologies. *Renew. Sustain. Energy Rev.* **2010**, *14*, 899–918. [CrossRef]
2. Falcão, A.F.O.; Henriques, J.C.C. Oscillating-water-column wave energy converters and air turbines. A review. *Renew. Energy* **2016**, *85*, 1391–1424. [CrossRef]
3. Masuda, Y.; Yamazaki, T.; Outa, Y.; McCormick, M.E. Study of back bent duct buoy. *Proc. OCEANS'87* **1987**, *19*, 384–389.
4. McCormick, M.E. Analysis of a wave energy conversion buoy. *J. Hydronaut.* **1974**, *8*, 77–82. [CrossRef]
5. Fonseca, N.; Pessoa, J. Numerical modelling of a wave energy converter based on U-shaped interior oscillating water column. *Appl. Ocean Res.* **2013**, *40*, 60–73. [CrossRef]

6. WAMIT Inc. *WAMIT User Manual*; Version 6.4; WAMIT Inc.: Chestnut Hill, MA, USA, 2019.

7. Sheng, W.; Lewis, A.; Alcorn, R. Numerical studies of a floating cylindrical OWC WEC. In Proceedings of the ASME 2012 31st International Conference on Ocean, Offshore and Arctic Engineering, Rio de Janeiro, Brazil, 1–6 July 2012.

8. Sheng, W.; Lewis, A.; Alcorn, R. Numerical studies on hydrodynamics of a floating oscillating water column. In Proceedings of the ASME 2011 30th International Conference on Ocean, Offshore and Arctic Engineering, Rotterdam, The Netherlands, 19–24 June 2011.

9. Bailey, H.; Robertson, B.R.D.; Buckham, B.J. Wave-to-wire simulation of a floating oscillating water column wave energy converter. *Ocean Eng.* **2016**, *125*, 248–260. [CrossRef]

10. Gomes, R.P.F.; Henriques, J.C.C.; Gato, L.M.C.; Falcão, A.F.O. Wave power extraction of a heaving floating oscillating water column in a wave channel. *Renew. Energy* **2016**, *99*, 1262–1275. [CrossRef]

11. Luo, Y.; Wang, Z.; Peng, G.; Xiao, Y.; Zhai, L.; Liu, X.; Zhang, Q. Numerical simulation of a heave-only floating OWC (oscillating water column) device. *Energy* **2014**, *76*, 799–806. [CrossRef]

12. ANSYS. Ansys Manual, Release12.1. Available online: http://www.afs.enea.it/project/neptunius/docs/fluent/ (accessed on 14 May 2019)

13. Elhanafi, A.; Macfarlane, G.; Fleming, A.; Leong, Z. Experimental and numerical investigations on the hydrodynamic performance of a floating-moored oscillating water column wave energy converter. *Appl. Energy* **2017**, *205*, 369–390. [CrossRef]

14. CD-Adapco. User Guide STAR-CCM+, Version 10.02. Available online: https://mdx.plm.automation.siemens.com/star-ccm-plus (accessed on 14 May 2019).

15. Elhanafi, A.; Macfarlane, G.; Fleming, A.; Leong, Z. Experimental and numerical investigations on the intact and damage survivability of a floating-moored oscillating water column device. *Appl. Ocean Res.* **2017**, *68*, 276–292. [CrossRef]

16. Crespo, A.J.C.; Altomare, C.; Domínguez, J.M.; González-Cao, J.; Gómez-Gesteira, M. Towards simulating floating offshore oscillating water column converters with smoothed particle hydrodynamics. *Coast. Eng.* **2017**, *126*, 11–26. [CrossRef]

17. Crespo, A.J.C.; Matthew, H.; Domínguez, J.M.; Altomare, C.; Wu, M.; Verbrugghe, T.; Stratigaki, V.; Troch, P.; Gómez-Gesteira, M. Floating moored oscillating water column with meshless SPH method. In Proceedings of the ASME 2018 37th International Conference on Ocean, Offshore and Arctic Engineering, Madrid, Spain, 17–22 June 2018.

18. Correia da Fonseca, F.X.; Gomes, R.P.F.; Henriques, J.C.C.; Gato, L.M.C.; Falcão, A.F.O. Model testing of an oscillating water column spar-buoy wave energy converter isolated and in array: Motion and mooring forces. *Energy* **2016**, *112*, 1207–1218. [CrossRef]

19. He, F.; Leng, J.; Zhao, X. An experimental investigation into the wave power extraction of a floating box-type breakwater with dual pneumatic chambers. *Appl. Ocean Res.* **2017**, *67*, 21–30. [CrossRef]

20. Rapaka, E.V.; Natarajan, R.; Neelamani, S. Experimental investigation on the dynamic response of a moored wave energy device under regular sea waves. *Ocean Eng.* **2004**, *31*, 725–743. [CrossRef]

21. Gomes, R.P.F.; Henriques, J.C.C.; Gato, L.M.C.; Falcão, A.F.O. Wave channel tests of a slack-moored floating oscillating water column in regular waves. In Proceedings of the 11th European Wave and Tidal Energy Conference, Nantes, France, 6–11 September 2015.

22. EMEC. Tank testing of wave energy conversion systems. In *Marine Renewable Energy Guides*; The European Marine Energy Center Ltd.: London, UK, 2009.

23. Crema, I.; Simonetti, L.; Cappietti, H.; Oumeraci, H. Laboratory experiments on oscillating water column wave energy converters integrated in a very large floating structure. In Proceedings of the 11th European Wave and Tidal Energy Conference, Nantes, France, 6–11 September 2015.

24. LABIMA. Available online: https://www.labima.unifi.it/ (accessed on 13 May 2019).

25. Thorimbert, Y.; Latt, J.; Cappietti, L.; Chopard, B. Virtual wave flume and oscillating water column modeled by lattice Boltzmann method and comparison with experimental data. *Int. J. Mar. Energy* **2016**, *14*, 41–51. [CrossRef]

26. Simonetti, I.; Cappietti, L.; Elsafti, H.; Oumeraci, H. 3D numerical modelling oscillating water column wave energy conversion devices: Current knowledge and OpenFOAM® implementation. In Proceedings of the 1st International Conference on Renewable Energies Offshore, Lisbon, Portugal, 24–26 Novermber 2014; pp. 497–504.

27. Simonetti, I.; Cappietti, L.; Elsafti, H.; Oumeraci, H. Optimization of the geometry and the turbine induced damping for fixed detached and asymmetric OWC devices: A numerical study. *Energy* **2017**, *139*, 1197–1209. [CrossRef]

28. Ren, B.; He, M.; Dong, P.; Wen, H. Nonlinear simulations of wave-induced motions of a freely floating body using WCSPH method. *Appl. Ocean Res.* **2015**, *50*, 1–12. [CrossRef]

29. Hadžić, I.; Hennig, J.; Perić, C.; Xing-Kaeding, Y. Computation of flow-induced motion of floating bodies. *Appl. Math. Model.* **2005**, *29*, 1196–1210. [CrossRef]

30. Ctech. Available online: http://www.ctechmetrology.com/ (accessed on 13 May 2019).

31. Troch, P. Experimentele Studie en Numerieke Modellering van Golfinteractie Met Stortsteengolfbrekers—Appendix E (in Dutch). Ph.D. Thesis, Ghent University, Ghent, Belgium, 2000.

32. Le Méhauté, B. *An Introduction to Hydrodynamics and Water Waves*; Springer: Berlin, Germany, 1976.

33. ITTC. Seakeeping Experiments, Recommended Procedures and Guidelines; International Towing Tank Conference. Available online: https://ittc.info/ (accessed on 14 May 2019).

Article

Multi-DOF WEC Performance in Variable Bathymetry Regions Using a Hybrid 3D BEM and Optimization

Markos Bonovas [1], Kostas Belibassakis [1] and Eugen Rusu [2,*]

[1] School of Naval Architecture and Marine Engineering, National Technical University of Athens, 15780 Athens, Greece; markosbonovas@hotmail.gr (M.B.); kbel@fluid.mech.ntua.gr (K.B.)

[2] Department of Mechanical Engineering, University Dunarea de Jos of Galati, 800008 Galaţi, Romania

* Correspondence: erusu@ugal.ro; Tel.: +40-740-205-534

Received: 9 April 2019; Accepted: 29 May 2019; Published: 1 June 2019

Abstract: In the present work a hybrid boundary element method is used, in conjunction with a coupled mode model and perfectly matched layer model, for obtaining the solution of the propagation/diffraction/radiation problems of floating bodies in variable bathymetry regions. The implemented methodology is free of mild-slope assumptions and restrictions. The present work extends previous results concerning heaving floaters over a region of general bottom topography in the case of generally shaped wave energy converters (WECs) operating in multiple degrees of freedom. Numerical results concerning the details of the wave field and the power output are presented, and the effects of WEC shape on the optimization of power extraction are discussed. It is demonstrated that consideration of heave in combination with pitch oscillation modes leads to a possible increase of the WEC performance.

Keywords: renewable energy; marine environment; wave energy converters; variable depth effects; multi-DOF WECs; design optimization

---

1. Introduction

Renewable energy from the oceans is increasingly attracting the interest of the scientific and industrial society. Wave energy converters are constantly being deployed in areas characterized by increased potential, and a recent review concerning point absorber wave energy harvesters is presented in [1]. The performance of the devices installed in the nearshore and coastal environment, where the bottom terrain may present significant variations, can be evaluated by formulating and solving interaction problems of free surface gravity waves, floating bodies, and the seafloor; see, e.g., Wehausen [2] and Mei [3]. A thorough presentation of the interaction between waves and oscillating energy systems can be found in Falnes [4]. Models describing coupling methodologies for numerical modelling of near and far-field effects of wave energy converter arrays are presented in various works; see, e.g, [5–7].

The power efficiency and the operation of the WECs is affected by the bottom topography due to the local entrapped modes and of their impact on the wave propagation, with non-negligible results, especially in array layouts; see [5,8,9]. This is also demonstrated in wave propagation over variable seabed topographies or abrupt bathymetries including coastal structures; see, e.g., [10,11].

The numerical method implemented in this work for the treatment of the hydrodynamic problems simulating the WEC operation is a hybrid boundary element method coupled with a perfectly matched layer (BEM-PML) technique, which is used in conjunction with a coupled mode system for the simulation of the propagating waves over general seabed topography, as presented and validated in [12,13]. For the calculation of the propagation wave field over general 3D bottom topographies, including possibly steep parts, the coupled mode model (CMM), developed in [14] and extended for the 3D-domains in [15,16], is applied. The latter method is validated by comparisons against experimental data [17,18] and calculations obtained by the phase-averaged wave model SWAN (simulating waves nearshore) [19]. The boundary element method is then implemented for the calculation of the excitation loads on the floating body, along with the hydrodynamic coefficients of added mass and damping by utilizing the 3D Green's Function, while the perfectly matched layer model is numerically treating the behavior of the outgoing radiating waves at large distances from the floating body [20].

The present method is applied to derive numerical results concerning the details of the wave field and the power output. Different axisymmetric WEC-shapes and power take off (PTO) configurations are examined, and the effects on the optimization of wave energy extraction are discussed. Following previous investigations, as reported in detail by Falnes [4], the consideration of additional degrees of freedom could significantly enhance the performance of a oscillating floating WEC. By using the present hybrid BEM, it is demonstrated that consideration of heave in combination with pitch modes leads to a substantial increase of the WEC performance, up to 300%. What is more important is that the consideration of additional degrees of freedom has an important effect on the determination of the optimal shape of the floater. Finally, the present model supports the application to more complex optimization problems, associated with multi DOF (degree of freedom) WEC performance, which are expected to be excited in variable bathymetry due to general wave incidence, in conjunction with depth inhomogeneity effects.

2. Formulation

2.1. Heaving Cylinder over Variable Bathymetry

We first consider a vertical cylindrical WEC of radius a and draft T, operating in a nearshore environment, characterized by a depth-transition from an incidence-subregion of constant depth $h = h_1$, to a transmission-subregion of constant depth $h = h_3$, with the depth $h_2(\mathbf{x})$ in the middle subdomain exhibiting an arbitrary variation with respect to the horizontal coordinates $\mathbf{x} = (x, y)$. The motion of the floating body is excited by a harmonic wave of angular frequency ω propagating with an incident angle θ. Under the assumption that the wave slope is relatively small, the wave potential and the free-surface elevation are expressed by

$$\Phi(\mathbf{x}, z; t) = \mathrm{Re}\left\{-\frac{igH}{2\omega}\varphi(\mathbf{x}, z; \mu) \cdot \exp(-i\omega t)\right\} \tag{1}$$

$$\eta(\mathbf{x}; t) = \mathrm{Re}\left\{\frac{H}{2}\varphi(\mathbf{x}, z; \mu) \cdot \exp(-i\omega t)\right\} \tag{2}$$

where H is the incident wave height, g is the gravity acceleration, $\mu = \omega^2/g$ is the frequency parameter, and $i = \sqrt{-1}$ is the imaginary unit. According to the hydrodynamic theory of floating bodies (see, e.g., [7]), the complex wave potential is decomposed on several components, namely the propagating wave potential $\varphi_P(\mathbf{x}, z)$, defined without the effect of the body, the diffraction potential $\varphi_D(\mathbf{x}, z)$ due to the presence of the rigid motionless body, and the radiation potential $\varphi_R(\mathbf{x}, z)$, related to the oscillations in the six degrees of freedom of the floater

$$\varphi(\mathbf{x}, z) = \varphi_P(\mathbf{x}, z) + \varphi_D(\mathbf{x}, z) + \frac{2\omega^2}{gH}\varphi_R(\mathbf{x}, z) \tag{3}$$

$$\psi_R(\mathbf{x}, z) = \sum_{\ell=1}^{6} \xi_\ell \varphi_\ell(\mathbf{x}, z) \tag{4}$$

The boundary conditions on the wetted surface of the body are

$$\partial\varphi_D(\mathbf{x}, z)/\partial n = -\partial\varphi_P(\mathbf{x}, z)/\partial n, \quad \partial\varphi_\ell(\mathbf{x}, z)/\partial n = n_\ell, \quad \ell = 1, 2, \ldots 6, \tag{5}$$

where $\mathbf{n} = (n_1, n_2, n_3)$ denotes the normal vector with direction inwards the body, and n_ℓ is the ℓ-component of the generalized normal vector. Moreover, the wave potential (all components) should satisfy the bottom boundary condition on the variable seabed topography

$$\partial\varphi(\mathbf{x}, z)/\partial z + \nabla_2 h(\mathbf{x})\nabla_2\varphi(\mathbf{x}, z) = 0, \quad z = -h(\mathbf{x}) \tag{6}$$

where $\nabla_2 = \left(\partial_x, \partial_y\right)$ denotes the horizontal gradient.

In particular, the heave response of the cylinder is obtained as

$$\xi_3 = (X_P + X_D)/A \tag{7}$$

where X_P and X_D are the Froude–Krylov and diffraction exciting vertical forces due to propagation and diffraction potentials, respectively, defined as follows

$$X_P = \frac{\rho g H}{2} \iint_{\partial D_B} \phi_P n_3 dS \tag{8}$$

$$X_D = \frac{\rho g H}{2} \iint_{\partial D_B} \phi_D n_3 dS \tag{9}$$

The complex coefficient $A(\omega)$ involved in Equation (7) is given by

$$A(\omega) = -\omega^2(M + a_{33}) - i\omega(B_S + b_{33}) + (C_S + c_{33}) \tag{10}$$

where the hydrodynamic coefficients a_{33} and b_{33} (added mass and damping coefficient of the body) are calculated by integrating the heaving radiation potential on the wetted surface of the WEC:

$$a_{33} - \frac{1}{i\omega}b_{33} = \rho \iint_{\partial D_B} \varphi_3 n_3 dS \tag{11}$$

Also, $c_{33} = \rho g A_{WL}$ is the hydrostatic coefficient in heave motion and A_{WL} denotes the waterline surface. The coefficients B_S and C_S are characteristic parameters of the PTO system. Finally, the time-average WEC power output, considering only the ξ_3-heave mode, is calculated by

$$P(\omega, \theta) = \frac{1}{2}\eta_{eff}\omega^2\left|B_S(\xi_3)^2\right| \tag{12}$$

where η_{eff} denotes the efficiency of the PTO. The power output obviously depends on the frequency ω, the direction θ, and the height H of the incident wave, as well as on the characteristics of the PTO installed in the specific environment. Furthermore, the overall performance of the device is dependent on the wave conditions as they are described by the incident directional wave spectrum.

2.2. Propagation Wave Field

The coupled mode model, developed by Athanassoulis and Belibassakis [14] and extended to 3D by Belibassakis et al [15], is appropriate for the efficient numerical simulation of wave propagation problems over a varying sea bottom topography that may contain steep parts, where analytic solutions are not available. The propagation potential over a variable bathymetry, in the absence of the floating body-scatterer, is based on the following local-mode representation

$$\varphi_P(\mathbf{x}, z) = \varphi_{-1}(\mathbf{x})Z_{-1}(z; \mathbf{x}) + \sum_{n=0}^{\infty} \varphi_n(\mathbf{x})Z_n(z; \mathbf{x}) \tag{13}$$

where the vertical functions $Z_n\,(z;\,x)$ are obtained as eigenfunctions of regular Sturm–Liouville problems, formulated at the local depth, and the system is enhanced by appropriate additional terms in order to consistently satisfy the boundary conditions on the sloping seabed. The functions $\varphi_n(x)$ are the complex amplitude of the nth-mode, and are found as the solution of the following coupled mode system (CMS)

$$\sum_{n=-1} \mathbf{A}_{mn}(\mathbf{x})\nabla^2\varphi_n(\mathbf{x}) + \mathbf{B}_{mn}(\mathbf{x})\nabla\varphi_n(\mathbf{x}) + \mathbf{C}_{mn}(\mathbf{x})\varphi_n(\mathbf{x}) = 0 \tag{14}$$

where the matrix coefficients $\mathbf{A}_{mn}$, $\mathbf{B}_{mn}$, and $\mathbf{C}_{mn}$ are defined in terms of the vertical eigenfunctions and are listed in Table 1 of Reference [15]. An important feature of the above CMS is that it can be naturally reduced to well-known simplified models when the environmental parameters permit such simplification. In fact, keeping only the propagating mode ($n = 0$) in Equation (14), the system reduces to a one-equation model, which is exactly the modified mild-slope equation; see [10]. The CMS is also supplemented by appropriate boundary conditions for treating incident, reflection, and transmission phenomena in general bathymetry regions.

2.3. Diffraction and Radiation Potentials

The evaluation for the 3D diffraction and radiation potentials associated with the floating WEC will be treated by the BEM method developed by Belibassakis et al [21] and described in more detail in [13]. In this model, the induced potential and velocity from the collection of the 4-node quadrilateral elements, which are used to discretize all parts of the boundary surface (body, free-surface, seabed surface etc.), is given by

$$\varphi(\mathbf{r}) = \sum_p F_p \Phi_p(\mathbf{r}) \tag{15}$$

$$\nabla\varphi(\mathbf{r}) = \sum_p F_p \mathbf{U}_p(\mathbf{r}) \tag{16}$$

where the summation refers to all panels and Φ_p and U_p denote, respectively, the induced potential and velocity from the pth element with unit singularity distribution to the field point $\mathbf{r} = (x, y, z)$. The induced potential and velocities from each element are obtained by a semi-analytical method, and the discrete solution is finally obtained using the collocation method, used to satisfy the boundary conditions at the centroid of each panel of the geometrical configuration.

2.4. PML Implementation

The domain and the radiating behavior of the diffraction and radiation fields in the far field at large distances from the floating body are numerically simulated by means of an absorbing layer technique, based on a perfectly matched layer (PML) applied all around the borders of the free-surface computational domain; see, e.g., [22]. In the present model, the free-surface boundary condition is expressed by the following formula (see also [12,13,21]):

$$\frac{\partial\Phi}{\partial n} - \mu(\omega)\Phi = 0, \quad r \in \partial D_F \tag{17}$$

where

$$\mu(\omega) = \left\{ \begin{array}{ll} \dfrac{\omega^2}{g}, & R < R_a \\[4mm] \dfrac{\omega^2}{g}\left(1 + \tilde{ic}\dfrac{(R-R_a)^n}{\lambda^n}\right)^2, & R \geq R_a \end{array} \right\} \tag{18}$$

The efficiency of the above technique to damp the outgoing waves with minimal back-scattering is dependent on various parameters of the present PML, including the layer thickness, its activation point R_a, and the coefficient $\tilde{c}$. The latter are optimized by systematic investigation using as objective function the minimization of the error of the numerical solution against analytical results available in the case of floating vertical cylindrical bodies in constant depth; see, e.g., [23]. This procedure is evaluated in detail in [12], and is shown to provide satisfactory results proving the reliability of the present hybrid BEM-PML-CMS numerical scheme.

2.5. Mesh Generation

In computational hydrodynamic problems, mesh generation is an issue of utmost importance. From this perspective, in the present problem, every part of the boundary surface is discretized by distribution of the panels (4-node quadrilateral elements), satisfying perfect junction of various sub-meshes, and ensuring global continuity of geometry. A cylindrical arrangement of panels in all boundary parts is used, as shown in Figure 1, which is found to be suitable for the representation of the radiating behavior of the diffraction and radiation fields; see also [12]. In the example illustrated in Figure 1, the mesh resolution is: 10×88 for the cylindrical WEC (with the first index representing vertical and the second azimuthal discretization). A domain extent of 4 wavelengths on the free surface is discretized into $(4N/\lambda) \times 88$ and 26×88 elements on the bottom surface, respectively, where N denotes the number of elements per wavelength for discetizing the domain. A finer mesh on the floating body corresponding to 18×88 elements is also used for examining the convergence of calculated results.

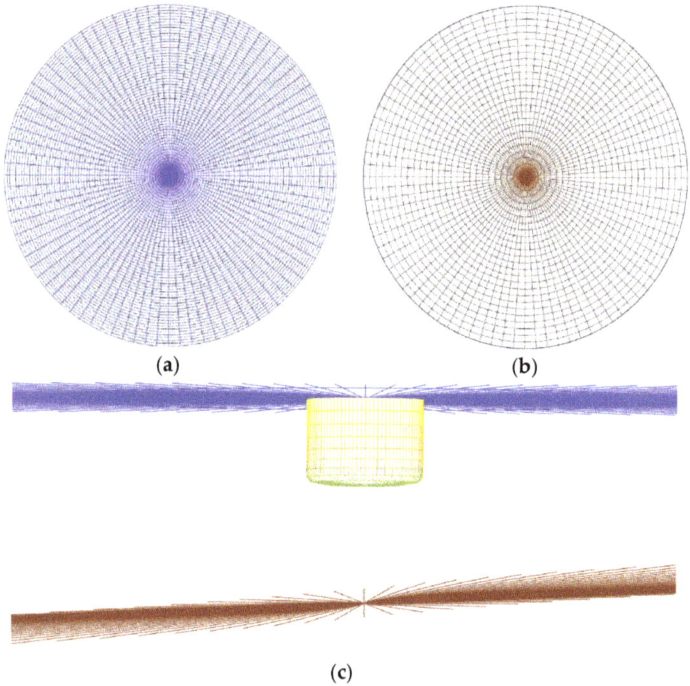

Figure 1. Computational meshes on (**a**) free surface, (**b**) bottom, and (**c**) WEC surface.

2.6. Variable Bathymetry

In this work, the effect of sloping seabeds on the WEC performance is examined by considering an operation over a smooth but steep shoaling region. The seabed profile exhibits a monotonic variation along the x-axis, described as follows

$$h(x) = h_m - 0.5(h_1 - h_3)\tanh(\alpha_{bot}\pi(x - x_{mean})) \tag{19}$$

where the mean depth is: $h_m = 0.5(h_1 + h_3)$ and x_{mean} is the center of the domain span along x-axis, where the WEC is also located. The coefficient α_{bot} controls the bottom slope. The region is characterized by constant depths at infinity, in particular h_1 and h_2.

2.7. 2-DOF WEC Problem Formulation

The idea of combining more than one degree of freedom for harnessing the available power provided by the incident wave is another perspective on improving the WEC efficiency. In this case, the device is able to absorb a higher amount of incident wave power and presents improved grid stability [13,24–26]. Considering the floating WEC operating in two power modes, namely heaving and pitching with amplitudes ξ_{30}, ξ_{50}, respectively, the responses are obtained from the solution of the following system of coupled equations

$$D_{33}\xi_{30} + D_{35}\xi_{50} = X_{P3} + X_{D3} \tag{20}$$

$$D_{53}\xi_{30} + D_{55}\xi_{50} = X_{P5} + X_{D5} \tag{21}$$

where

$$D_{33} = \left(-\omega^2(M + A_{33}) + i\omega(B_{33} + B_{S3}) + C_{33}\right), \quad D_{35} = \left(-\omega^2(A_{35} + I_{35}) + i\omega B_{35} + C_{35}\right) \tag{22}$$

$$D_{53} = \left(-\omega^2(A_{53} + I_{53}) + i\omega B_{53} + C_{53}\right), \quad D_{55} = \left(-\omega^2(A_{55} + I_{55}) + i\omega(B_{55} + B_{S5}) + C_{55}\right) \tag{23}$$

The hydrodynamic coefficients of added mass and damping, as well as the excitation Froude–Krylov and diffraction heave-forces and pitch-moments in the right-hand side of Equations (20) and (21), are calculated by the present BEM solver. As regards the rest of the included coefficients in the above expressions, the fact should be considered that the center of gravity is coincident to the center of the circular waterline of radius a. A typical mass distribution near the surface of the WEC is assumed, corresponding to radius of gyration $R_{yy} = 0.7a$ and thus, $I_{55} = MR_{yy}^2$. Also, $C_{33} = \rho g \pi a^2$, $C_{35} = C_{53} = 0$, and $C_{55} = 0.25\rho g \pi a^4 + M \cdot GB$, where M denotes the mass of the body and GB the vertical distance between the center of buoyancy and the center of gravity. In the present work we have assumed a fixed location of CG coincident the center of buoyancy (GB = 0) for all WEC shapes examined (operating in one or more DOF) in order to provide a first comparative evaluation. It should be mentioned that, even in the latter case, substantial stability concerning the pitch motion is still offered by the available metacentric height. The latter in the case of the cylindrical WEC examined in the paper is 12% of the draft, while for the nailhead, WEC becomes 3.7 times the draft due to increased waterline area. Additional stability is offered by the PTO damping. Based on the above, in conjunction with the fact that in extreme cases a cut-off system is used to ensure safety, the simplified assumption GB = 0 is made and used for the examples considered here in order to illustrate the developed method to calculate the WEC performance and optimization. The consideration of variable mass distributions and different CGs would lead to a substantially more complicated multidimensional optimization problem that is left to be examined in future work. Finally, the coefficients D_{35} and D_{53} are found, for every geometry, to be quite small, compared with D_{33} and D_{55}, and therefore the coupling between the two oscillatory modes is weak.

3. Design Assessment Features

3.1. Geometries Generation

Many different WEC shapes are currently operating in coastal areas all over the world, used both for research purposes and commercial applications. Geometries like the conical, the semi-spherical, and the elliptical, as well as many other shapes, have been examined in a variety of studies (see, e.g., [27–29] and the references cited there). Based on the relevant industry trends and research activities, eight different axisymmetric body shapes have been examined. The geometrical details can be found in [12]. At this point, it should be mentioned that very sophisticated designs and shapes, which are normally accompanied with high R&D (research and development) and manufacturing costs, are not examined. The generation of the above shapes, as well as other axisymmetric geometries, is handled by a parametric model based on a spline representation of the profiles controlled by a set of nodes. As an example, the profile of the nailhead-shaped WEC is shown in Figure 2, and the body surface is obtained by 360 degrees rotation. One significant feature of the present WEC shape model is that it is in compliance with the constant-mass constraint. If the same material is considered for the construction of the considered point absorbers with the same mass distribution, then the radius and the draft of each shape can be calculated in order to maintain the submerged vertical cross section area equal to the area of the reference cylindrical WEC design corresponding to $a/h = 1/3.5$, $T/a = 3/2$, and $d/h = 4/7$, where a is the radius, T is the draft, h represents the water depth, and consequently $d = h\text{-}T$ is the bottom clearance below the WEC. For a more unbiased comparison between designs, the assumption of constant-mass is adopted as a basic reference feature for all the tested geometries.

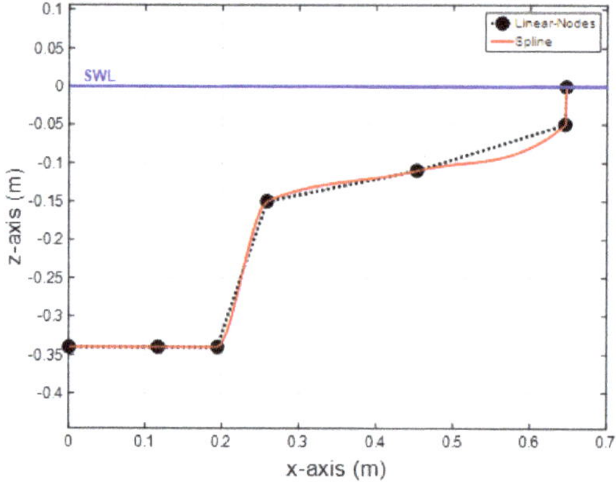

Figure 2. Generating spline of the nailhead-shaped WEC wetted surface.

The above parametric model could be exploited for directly finding the most efficient design. However, in order to reduce the computational cost, a particular set of variants was studied and specific details can be found in Reference [12].

3.2. PTO Damping Configuration

For the evaluation of the absorbed power output by the devices, a typical PTO has been assumed. The heave-PTO coefficient B_{S3} is taken C_{PTO3}-times the mean value of hydrodynamic damping $\bar{b}_{33}$ over the frequency range, $B_{S3} = C_{PTO3}\bar{b}_{33}$, which for the cylindrical WEC considered here is estimated as $2\pi\bar{b}_{33}M\omega_{R3} = 0.12$, and the corresponding resonance frequency is $\omega_{R3}\sqrt{a/g} = 0.7$. A similar PTO assumption is considered for the pitch motion, where the PTO damping is taken C_{PTO5}-times the mean value of hydrodynamic damping $\bar{b}_{55}$ over frequency, which is estimated as $2\pi\bar{b}_{55}M\omega_{R5} = 0.01$, where the corresponding resonance frequency is $\omega_{R5}\sqrt{a/g} = 0.45$. The values for the coefficients C_{PTO3} and C_{PTO5} are selected over a wide range, according to the geometry, contextually, and more detailed investigation of their operational restrictions. Very low values of the PTO coefficients, both for heave and pitch mode, are not considered either because they are incompatible with the current industry standards, or because they lead to high resonance amplitudes which are not permitted due to operational limits and survivability restrictions. These limits are decided after a careful examination of the relevant responses and performances. The damping coefficient of the power take off system is a decisive parameter for the performance of the devices, and after fine tuning and optimization is expected to provide significant improvements concerning the overall WEC performance.

3.3. Performance Index Definition

The averaged power-P output of the device, normalized with respect to the incident wave power, is plotted in Figure 3 in the case of the cylindrical-shaped WEC. According to the selection of PTO damping, there are cases of concentrated power maximization in the near-resonance frequency bandwidth, or cases of lower power levels, corresponding to responses covering wider frequency bandwidth. The curve with the largest area below defines the optimum PTO damping value. The latter behavior concerning the extraction of incident wave energy is exploited in the definition of an index quantifying the performance of different designs, which is based on the area below the "most-efficient-curve" of normalized power output. The result will be used, in normalized form with respect to incident power, to quantify the overall performance of the WEC and will be called performance index (PI%).

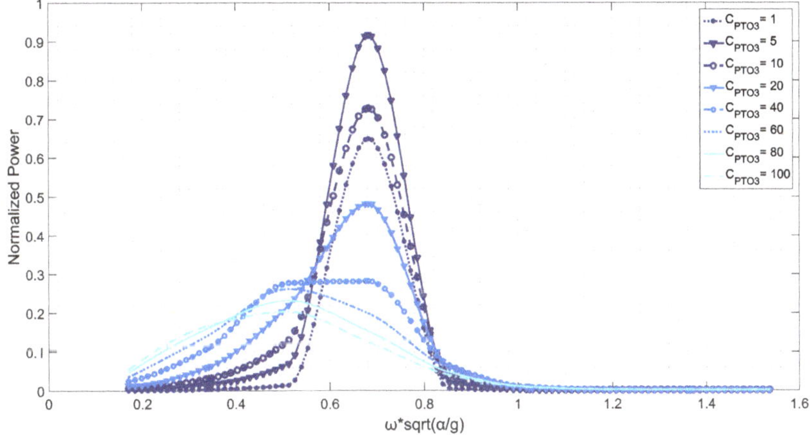

Figure 3. Normalized power output by the reference cylindrical heaving WEC (a/h = 1/3.5, T/a = 3/2, d/h = 4/7) over the variable bottom (h_1 = 1.2 m, h_3 = 0.8 m, α_{bot} = 0.5) for different PTO damping configurations. Indicative values of the $C_{PTO3} = B_{S3}/\bar{b}_{33}$ considered here are C_{PTO3} = [1, 5, 10, 20, 40, 80, 100].

In more detail, each WEC design corresponds to a different waterline radius due to mass-constraint implementation. Therefore, the operational bandwidth of non-dimension angular frequency varies among the WECs. In order to obtain a reliable index for assessing their performance, the P.I. is defined as the ratio of the area below the power curve for the optimum PTO damping to the area of the rectangle [0,1] and [0.2, $\max(\omega \sqrt{a/g})$], representing the total available power for absorption by the system. It should be considered that the fact for this rectangle is that the horizontal axis span differs to the efficiency frequency bandwidth, commonly used in WEC evaluations, a fact that can be observed with a closer look of Figure 3, where this difference is apparent. It should also not be overlooked that even if this index takes small values, almost insignificant compared with the capture width ratio (CWR) of WEC devices, however, as its definition explains, is a totally different index and every alteration of its value, caused for example by a variable-depth seafloor, is countable. For the extension of the latter definition to the 2-DOF WEC, the total power output, calculated as the summation of the contribution by each mode individually, will be used as the performance index.

4. Numerical Results and Discussion

4.1. Heaving Cylindrical WEC over Variable Bathymetry

The investigation of the heaving response of the reference cylindrical WEC over a variable seabed is discussed in this section. As an example, the cylindrical WEC with $a/h = 1/3.5$, $T/a = 3/2$, and $d/h = 4/7$ operates over a variable bottom topography described by Equation (19), with $h_1 = 1.5$ m, $h_3 = 0.5$ m, and $\alpha_{bot} = 0.5$. The propagation field, evaluated by the CMS, is illustrated in Figure 4 in the case of normally incident waves of frequency $\omega \sqrt{a/g} = 0.7$ (resonance frequency). The corresponding bathymetric non-dimensional frequency of the waves is $\omega \sqrt{h/g} = 1.3$. It is clearly observed that the propagating field is diffracted and reflected, and the bottom boundary condition is consistently satisfied, by the fact that the equipotential lines intersect perpendicularly in the seabed profile. In Figure 4, the illustration of the free-surface elevation is also indicated by using solid black line.

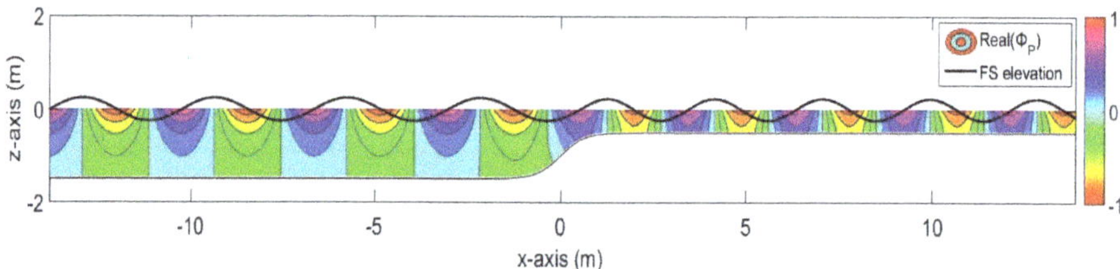

Figure 4. Propagating field (real part) over a smooth shoal on the vertical plane as calculated by the CMS for normally incident waves $\omega \sqrt{h/g} = 1.3$ $\theta = 0$ degrees. Variable bottom ($h_1 = 1.5$ m, $h_3 = 0.5$ m, $\alpha_{bot} = 0.5$).

Using the data concerning the propagating wave field and its normal derivative over the motionless WEC-scatterer, the diffraction field is calculated by the present BEM-PML solver which is illustrated in Figure 5, where the bottom contours are also plotted using dashed lines and the body's position is indicated with the white disk in the center of the domain. The heave radiation field on the free surface for WEC frequency $\omega \sqrt{a/g} = 0.7$ over the variable bottom topography ($h_1 = 1.5$ m, $h_3 = 0.5$ m, $\alpha_{bot} = 0.5$), is shown in Figure 6 as calculated by the present method. The bottom contours are indicated by using dashed lines. Furthermore, the calculated radiation potentials related to the rest oscillation modes (except of yaw, which is not excited) are presented in Figure 7 for the same frequency as before. The details of the radiated wave pattern are clearly observed in these plots, as well as the effects of variable seabed topography, resulting in an amplification of wave amplitudes in the shallower region due to shoaling effects. The effectiveness of the present PML model is clear in these plots by noticing the damping of the outgoing waves after the PML-activation point, which in the cases considered is set at a radius of three wavelengths away from the floating body.

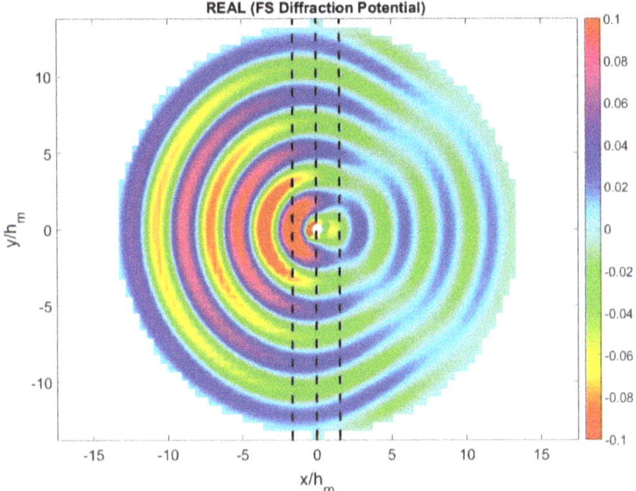

Figure 5. Diffraction field (real part) on the free surface for normally incident waves $\omega\sqrt{a/g} = 0.7$ $\theta = 0$ degrees as calculated by the present hybrid BEM over the variable bottom topography ($h_1 = 1.5$ m, $h_3 = 0.5$ m, $\alpha_{bot} = 0.5$). The bottom contours are indicated by using dashed lines.

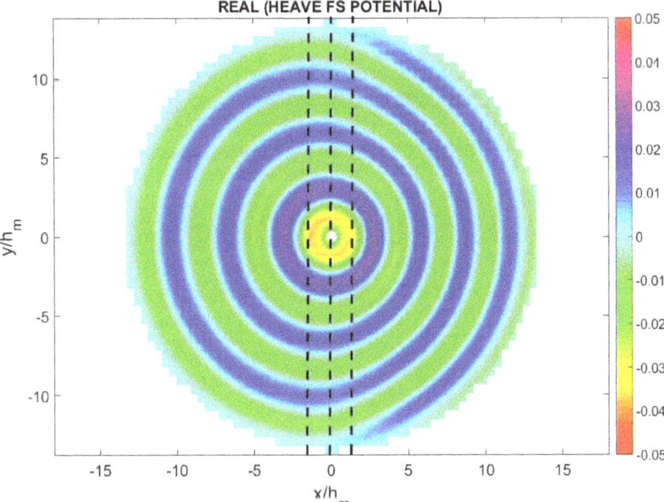

Figure 6. Heave radiation field on the free surface for WEC frequency $\omega\sqrt{a/g} = 0.7$ over the variable bottom topography ($h_1 = 1.5$ m, $h_3 = 0.5$ m, $\alpha_{bot} = 0.5$), as calculated by the present method. The bottom contours are indicated by using dashed lines.

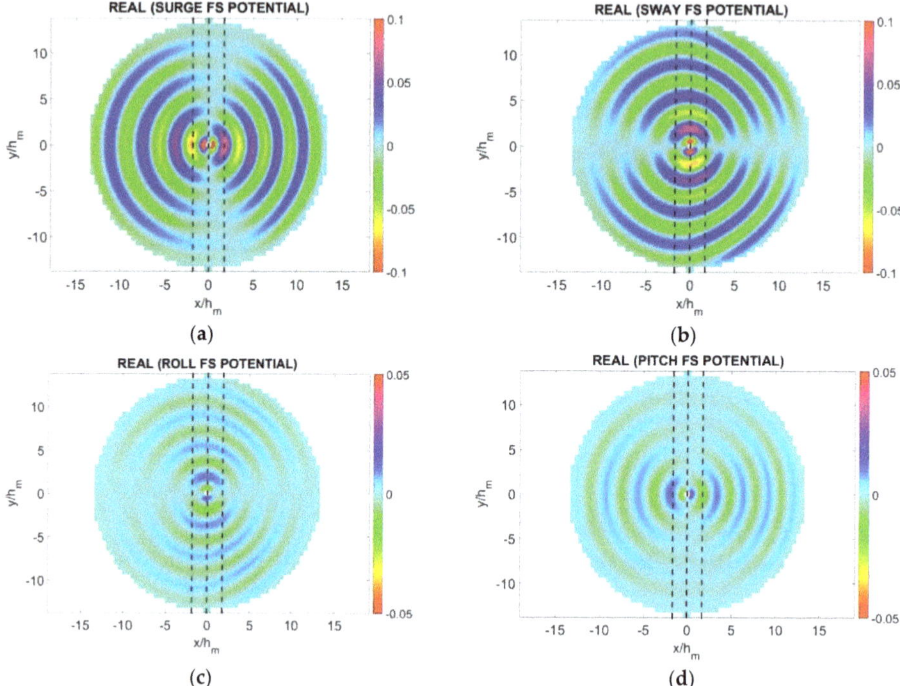

Figure 7. Radiation fields on the free surface for WEC frequency $\omega \sqrt{a/g} = 0.7$ over the variable bottom topography ($h_1 = 1.5$ m, $h_3 = 0.5$ m, $\alpha_{bot} = 0.5$), as calculated by the present method. (**a**) surge, (**b**) sway, (**c**) roll, (**d**) pitch.

4.2. Heaving (1-DOF) WECs over Variable Bathymetry

In order to examine the effect of variable seabed topography on the WEC performance over shoaling bottom topography, the responses are evaluated for the eight different WEC shapes which have been presented and discussed in detail in Reference [12]. Except for the steep bottom profile considered in the previous section, a second less steep bathymetric profile was initially examined, which is also described by Equation (19), for $h_1 = 1.2$ m, $h_3 = 0.8$ m, and $\alpha_{bot} = 0.5$. Both seabed geometries have the same mean depth, $h_m = 1$ m, but the maximum bottom slope decreases from 0.7 to 0.3.

The main aim of the present work is the assessment of the performance of the devices by means of the achieved performance index. The damping of the PTO is set initially, for the assessment of the designs, as: $C_{PTO3} = [1, 5, 10, 50, 100, 250, 500, 750, 1000]$. The latter configuration will be optimized in the sequel in order to estimate the value of the exact PTO damping coefficient providing the best performance. Numerical results concerning the performance index are presented in Table 1 for each WEC design over a flat bottom and variable bottom topography. We observe that the conical WEC is selected as the design with the best performance, among all the investigated alternative designs in the case of an 1-DOF device and this holds true for both of the examined depth configurations. Furthermore, comparison between the two most efficient shapes, i.e. the conical and the semi-spherical, is also performed in [28] and concludes again in the superiority of the conical WEC against the semi-spherical one.

Moreover, the shoaling sea bottom topography is shown to affect the performance of every WEC-design, and more specifically it causes an increase of the device's ability for harnessing wave energy. The highest upgrade is detected for the disk-shaped WEC, being equal to almost +3.3%.

However, it should be taken under consideration the fact that the performance index (P.I.) used in the present work to quantify the overall WEC performance, accounts for the whole frequency bandwidth and not only the efficient bandwidth of each WEC operation. For example, in the case of the cylindrical WEC, it is clearly observed in Figure 3 that the efficient bandwidth is $0.2 < \omega \sqrt{a/g} < 1$ while the total bandwidth extends in the range $0 < \omega \sqrt{a/g} < 1.8$. The latter is used because the efficient frequency bandwidth of other WEC shapes is different but it is included in the extended range. The above has the effect that changes of the calculated PI due to bathymetry remain rather small.

Results like this are very promising and may be a first indication for installations of WECs over a deliberately sloping seabed, with man-made constructions and interventions that could be combined with water wave lenses producing the focusing of the wave energy in the location of the WEC or WEC-array and contributing to achieve higher power absorption by the devices. Further studies of this effect and optimization will be subjects of future work.

Table 1. 1-DOF WEC shapes and PI for flat and variable bottom topographies ($h_1 = 1.2$ m, $h_3 = 0.8$ m, $\alpha_{bot} = 0.5$).

WEC Design	Max {Performance Index} Flat Bottom	Max {Performance Index} Variable Bottom
Cylindrical	11.29%	11.41%
Nailhead-shaped	15.38%	15.74%
Disk-shaped	13.66%	14.11%
Elliptical	14.73%	15.11%
Egg-shaped	10.63%	10.82%
Conical	17.70%	17.92%
Floater-shaped	9.87%	10.10%
Semi-spherical	13.30%	13.39%

Proceeding to the determination of the optimal PTO damping values for the conical WEC, it is found by finer discretization of the parameters that a performance index of 17.95% is achieved for a PTO coefficient equal to $C_{PTO3} = 77$ in the case of the flat bottom. The latter index has further improved to 18.10% with $C_{PTO3} = 80$ over the smooth shoal, Equation (19). This result is more proof of the significance of the effects by a varying topography in the design-task of the WECs.

4.3. 2-DOF WECs over Variable Bathymetry

The accomplished performance index by the 1-DOF WEC design may be appreciably upgraded by considering the device operating in 2-DOFs (heave and pitch). These modes of operation are normally coupled with surge oscillation mode and all those three degrees of freedom determine the maximum possible energy absorption by a wave device, according to [4]. However, in this study, surge power mode is omitted, with the WEC assumed to be properly moored. Moreover, surge motion is strongly affected by more complex phenomena such as wave drift forces and thus this assumption is in compliance with the complexity of the developed hydrodynamic theory. The same procedure for the selection of the most efficient shape and the optimal PTO damping coefficient is followed, using the subsequent values of PTOs coefficients $C_{PTO3} = [1, 5, 10, 20:10:340]$ and $C_{PTO5} = [50:10:400]$, resulting in the data presented in Table 2.

In this case, the two qualified designs are the cylindrical WEC and the nailhead-shape WEC, which are expected to be the best performing designs in the case of heaving-pitching body due to its larger waterline area. The results summarizing the performance over flat and sloping seabed topography are listed in Table 3. It is observed that the effect induced by the variable seafloor is sometimes constructive, for example, as for the Disk-shaped device (+0.27%), and other times destructive, as it happens in the case of the cylindrical device (−0.43%).

As stated previously, the impact of the seabed profile is now not so simple and affects the performance of the device in a more complicated way. The PTOs used for harnessing power from the sea environment are exhibiting different damping values. Thus, the optimization of the PTO coefficients is investigated with respect to three principal design parameters: (i) the geometry of the WEC, (ii) the heave-PTO damping coefficient, and (iii) the pitch-PTO damping coefficient. The calculated values of the performance index for the cylindrical and the nailhead-shape WECs operating over a flat bottom topography are plotted in Figure 8, allowing the prompt comparison and the determination of the optimal PTO damping coefficients.

Table 2. 2-DOFs WEC shapes and PI for flat and variable bottom topographies ($h_1 = 1.2$ m, $h_3 = 0.8$ m, $\alpha_{bot} = 0.5$).

WEC Design	Max {Performance Index (Heave+Pitch)} Flat Bottom	Max {Performance Index (Heave+Pitch)} Variable Bottom
Cylindrical	32.43%	32.00%
Nailhead-shaped	29.00%	29.08%
Disk-shaped	22.04%	22.31%
Elliptical	27.12%	27.26%
Egg-shaped	26.25%	25.84%
Conical	21.99%	22.07%
Floater-shaped	19.65%	19.59%
Semi-spherical	13.44%	13.60%

Table 3. Optimum WEC designs, P.I., and PTO damping for flat and variable bottom topography in the case of smooth shoal ($h_1 = 1.2$ m, $h_3 = 0.8$m, $\alpha_{bot} = 0.5$).

Design & Performance		Cylindrical WEC	Nailhead-Shaped WEC
Max{Performance Index (Heave+Pitch)}	Flat	32.47%	29.00%
	Variable	32.04%	29.08%
PTO damping coefficient: $[C_{PTO3}, C_{PTO5}]$	Flat	[4, 50]	[261, 214]
	Variable	[4, 50]	[276, 213]

Finally, the bathymetry also affects the optimal values of PTO damping for each device, and this can be studied considering it as an additional design constraint. The effects on the WEC performances by the variable bathymetry in the case of the smooth shoal defined by Equation (19) are presented in Figure 9, with the strongest influence detected for the pitch power mode of the cylindrical WEC and the heave power mode of the nailhead-shaped WEC.

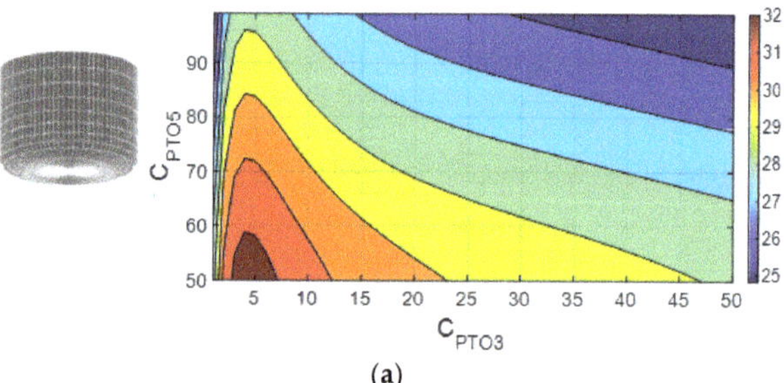

Figure 8. *Cont.*

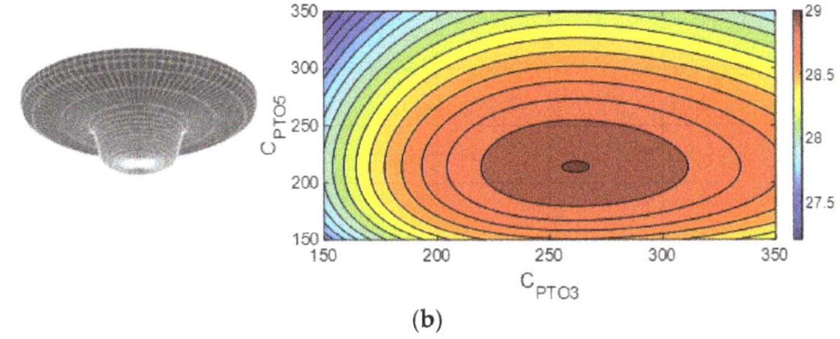

(b)

Figure 8. Isoperformance curves of (**a**) Cylindrical WEC and (**b**) Nailhead-shaped WEC over flat bottom.

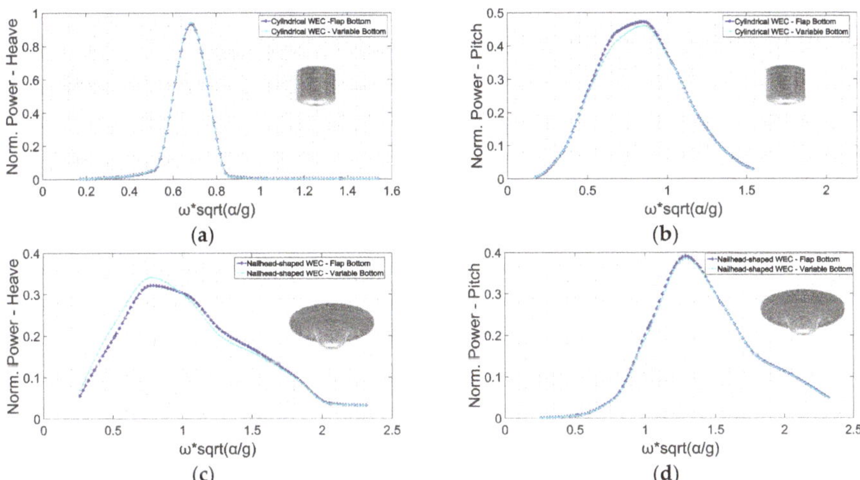

(a)   (b)

(c)   (d)

Figure 9. Normalized power output over flat and variable bottom topography ($h_1 = 1.2$ m, $h_3 = 0.8$ m, $\alpha_{bot} = 0.5$) using the optimized PTO. Cylindrical WEC in (**a**) heave mode and (**b**) pitch mode, and nailhead-shaped WEC in (**c**) heave mode and (**d**) pitch mode.

5. Sloping Seabed Effect

In order to better illusrate the effect of the sloping seabed on the performance of WECs, a second, steeper, bottom topography is considered, with bathymetric profile defined by Equation (19) for $h_1 = 1.5$ m, $h_3 = 0.5$ m, and $\alpha_{bot} = 0.5$; see Figure 4. The results, presented in Table 4 for the optimum heaving-WECs and 2-DOF WECs, prove that the effects by the seabottom cannot be predicted a priori. For the heaving WEC, a slight increase of PI is observed, while for the 2-DOF systems, an increase for the cylindrical WEC is obtained. The nailhead-shaped design performance index seems almost unaffected by the bottom topography. Again, the small changes of P.I. values are due to the bandwidth extent that has been considered for the definition of the index in order to enable comparison between various WEC shapes characterized by different resonant frequencies.

Table 4. Optimum WEC designs, P.I. over flat and sloping seabeds with increased steepness.

Type	Geometry	P.I. % Flat	δ P.I. % Steep vs Flat	δ P.I. % Steep vs Variable
Heaving WECs	Cylindrical	11.3%	6.2%	5.2%
	Conical	17.7%	3.1%	0.8%
2-DOF WECs	Cylindrical	32.4%	0.3%	1.6%
	Nailhead-shaped	29.0%	0.1%	0.4%

6. Conclusions

In the present work, several aspects related to the performance of wave energy converters of the type of point absorbers are studied, using a boundary element method for solving the associated hydrodynamic problems. In particular, a hybrid BEM coupled with a perfectly matched layer model is used for calculating the diffraction and radiation fields, based on information concerning the wave conditions around the floating bodies derived from a coupled mode model. The present method is shown to provide useful information, being able to treat general body geometry of the floating body operating in various oscillation modes over flat or sloping seabeds. Subsequently, it is systematically applied to numerically simulate the WEC performance in variable bathymetry regions. For the assessment and comparison of various designs, a specific performance index was defined as a useful indicator for the estimation of the power absorbing capacity. Regarding the PTO damping, which is exhibiting various constraints, this was found to be very important concerning the optimum performance of a WEC. From the results of the present study, it is concluded that the conical floater appears to be the most efficient design in the case of a heaving WEC, while the cylindrical and the nailhead-shaped forms are the ones exhibiting the highest performance operating as 2-DOF devices in coupled heaving and pitching modes. It is demonstrated that multiple DOF systems could substantially increase the levels of the extracted wave energy, and that the sloping seabed could also have an effect on the overall behavior of the devices and it should be taken into account. Future extensions of the present work include the examination of the performance and optimization of WEC arrays in more than two power modes in variable bathymetry, as well as the investigation of viscosity effects.

Author Contributions: The main idea of this work belongs to K.B. and M.B., while the main draft of the text and numerical simulations were handled by M.B.; E.R. participated in verification of the results, operated various corrections and acted as corresponding author. All authors agreed with the final form of this article.

Funding: The work of the corresponding author was carried out in the framework of the research project REMARC (Renewable Energy extraction in MARine environment and its Coastal impact), supported by the Romanian Executive Agency for Higher Education, Research, Development and Innovation Funding – UEFISCDI, grant number PN-III-P4-IDPCE-2016-0017.

Conflicts of Interest: The authors declare no conflict of interest.

Abbreviations

BC	Boundary Condition
BEM	Boundary Element Method
CMM	Coupled-Mode Model
CMS	Coupled-Mode System
DOF	Degree Of Freedom
FS	Free Surface
PI	Performance Index
PML	Perfectly Matched Layer
PTO	Power Take Off
RAO	Response Amplitude Operator
R&D	Research & Development
SWAN	Simulating Waves Nearshore
SWL	Sea Water Level
WEC	Wave Energy Converter

Nomenclature

a	Waterline radius
A	Hydrodynamic matrix
a_{bot}	Bottom slope coefficient
a_{ij}	Added mass coefficient (i, j=1, … ,6)
b_{ij}	Hydr. damping coefficient (i, j=1, … ,6)
B_{mn}	CMS coefficient
B_s	PTO damping coefficient
c_{ij}	Hydrostatic coefficient (i, j=1, … ,6)
C_{mn}	CMS coefficient
C_{PTOi}	PTO mean-damping multiplier
C_s	PTO hydrostatic coefficient
d	Bottom clearance
D_{ij}	Coupling coefficient of i-j modes
F_p	Source strength of pth element
g	Gravity acceleration
GB	Center of buoyancy
H	Wave height
h	Depth
i	Imaginary unit
M	Mass
N	Number of elements per wavelength
n_i	Normal vector of i-mode (i=1, … ,6)
P	Power
$P.I.$	Performance Index
R	Radial distance
r	Position vector
R_a	PML activation radius
R_{yy}	Radius of gyration
t	Time
T	Draft
U_p	Induced velocity by p^{th} element
x	x-Coordinate
X_D	Diffraction excitation forces
X_P	Froude Krylov excitation forces
y	y-Coordinate
z	z-Coordinate
Z_n	Vertical function of nth-mode (n=-1, 0, 1, …)
A_{mn}	CMS coefficient
η	Free surface elevation
η_{eff}	PTO efficiency
λ	Wavelength
μ	Frequency parameter
ξ_i	Response amplitude of i-mode (i=1, … ,6)
ρ	Water density
Φ	Wave potential
φ_d	Diffraction potential
φ_i	Radiation potential of i-mode (i=1, … ,6)
φ_n	Amplitude of nth-mode (n=-1, 0, 1, …)
φ_p	Propagation potential
Φ_P	Induced potential by p^{th} element
φ_R	Radiation potential
ω	Angular frequency
ω_R	Resonance angular frequency

References

1. Al Shami, E.; Zhang, R.; Wang, X. Point Absorber Wave Energy Harvesters: A Review of Recent Developments. *Energies* **2019**, *12*, 47. [CrossRef]
2. Wehausen, J.V. The Motion of Floating Bodies. *Annu. Rev. Fluid Mech.* **1971**, *3*, 237–268. [CrossRef]
3. Mei, C.C. *The Applied Dynamics of Ocean Surface Waves*; World Scientific: Singapore, 1989.
4. Falnes, J. *Ocean Waves and Oscillating Systems: Linear Interactions including Wave Energy Extraction*; Cambridge University Press: Cambridge, UK, 2004.
5. Verao Fernandez, G.; Balitsky, P.; Stratigaki, V.; Troch, P. Coupling Methodology for Studying the Far Field Effects of Wave Energy Converter Arrays over a Varying Bathymetry. *Energies* **2018**, *11*, 2899. [CrossRef]
6. Balitsky, P.; Quartier, N.; Verao Fernandez, G.; Stratigaki, V.; Troch, P. Analyzing the Near-Field Effects and the Power Production of an Array of Heaving Cylindrical WECs and OSWECs Using a Coupled Hydrodynamic-PTO Model. *Energies* **2018**, *11*, 3489. [CrossRef]
7. Fernández, G.V.; Stratigaki, V.; Troch, P. Irregular Wave Validation of a Coupling Methodology for Numerical Modelling of Near and Far Field Effects of Wave Energy Converter Arrays. *Energies* **2019**, *12*, 538. [CrossRef]
8. Charrayre, F.; Peyrard, C.; Benoit, M.; Babarit, A. A Coupled Methodology for Wave-Body Interactions at the Scale of a Farm of Wave Energy Converters Including Irregular Bathymetry. In Proceedings of the 33rd International Conference on Ocean, Offshore and Arctic Engineering (ASME 2014), San Francisco, CA, USA, 8–13 June 2014.
9. McCallum, P.; Forehand, D.; Sykes, R. On the Performance of an Array of Floating Wave Energy Converters for Different Water Depths. In Proceedings of the 33rd International Conference on Ocean, Offshore and Arctic Engineering (ASME 2014), San Francisco, CA, USA, 8–13 June 2014.
10. Massel, S.R. Extended refraction-diffraction equation for surface waves. *Coast. Eng.* **1993**, *19*, 97–126. [CrossRef]
11. Touboul, J.; Rey, V. Bottom pressure distribution due to wave scattering near a submerged obstacle. *J. Fluid Mech.* **2012**, *702*, 444–459. [CrossRef]
12. Belibassakis, K.; Bonovas, M.; Rusu, E. A Novel Method for Estimating Wave Energy Converter Performance in Variable Bathymetry Regions and Applications. *Energies* **2018**, *11*, 2092. [CrossRef]
13. Bonovas, M. WECs over General Bathymetry: A Novel Approach for Performance Evaluation and Optimization. Master's Thesis, National Technical University of Athens, Athens, Greece, 2019.
14. Athanassoulis, G.A.; Belibassakis, K.A. A consistent coupled-mode theory for the propagation of small-amplitude water waves over variable bathymetry regions. *J. Fluid Mech.* **1999**, *389*, 275–301. [CrossRef]
15. Belibassakis, K.A.; Athanassoulis, G.A.; Gerostathis, T.P. A coupled-mode model for the refraction-diffraction of linear waves over steep three-dimensional bathymetry. *Appl. Ocean Res.* **2001**, *23*, 319–336. [CrossRef]
16. Belibassakis, K.A.; Gerosthathis, T.P.; Athanassoulis, G.A. A Coupled-Mode Model for the Transformation of Wave Systems Over Inhomogeneous Sea/Coastal Environment. In Proceedings of the 29th International Conference on Offshore Mechanics and Arctic Engineering (OMAE2010), Shanghai, China, 6–11 June 2010; pp. 471–478. [CrossRef]
17. Berkhoff, J.C.W.; Booij, N.; Radder, A.C. Verification of numerical wave propagation models for simple harmonic linear water waves. *Coast. Eng.* **1982**, *6*, 255–279. [CrossRef]
18. Vincent, C.L.; Briggs, M.J. Refraction–diffraction of irregular waves over a mound. *J. Waterw. Port Coast. Ocean Eng.* **1989**, *115*, 269–284. [CrossRef]
19. Booij, N.; Ris, R.C.; Holthuijsen, L.H. A third-generation wave model for coastal regions: 1. Model description and validation. *J. Geoph. Res.* **1999**, *104*, 7649–7666. [CrossRef]
20. Ryszard, M.S. *Ocean Surface Waves: Their Physics and Prediction*; World Scientific: Singapore, 1996.
21. Belibassakis, K.A.; Gerostathis, T.P.; Athanassoulis, G.A. A 3D-BEM coupled-mode method for WEC arrays in variable bathymetry. In *Progress in Renewable Energies Offshore, Proceedings of the 2nd International Conference on Renewable Energies Offshore (RENEW2016), Lisbon, Portugal, 24–26 October 2016*; CRC Press: Boca Raton, FL, USA, 2016; p. 365.
22. Turkel, E.; Yefet, A. Absorbing PML boundary layers for wave-like equations. *Appl. Numer. Math.* **1998**, *27*, 533–557. [CrossRef]
23. Yeung, R. Added mass and damping of a vertical cylinder in finite depth waves. *Appl. Ocean Res.* **1981**, *3*, 119–133. [CrossRef]

24. Brooke, J. *Wave Energy Conversion*; Elsevier Science: Amsterdam, The Netherlands, 2003.
25. Davis, A.F.; Thomson, J.; Mundon, T.R.; Fabien, B.C. Modeling and Analysis of a Multi Degree of Freedom Point Absorber Wave Energy Converter. In Proceedings of the 33rd International Conference on Ocean, Offshore and Arctic Engineering (ASME 2014), San Francisco, CA, USA, 8–13 June 2014. [CrossRef]
26. Sergiienko, N.Y. Three-Tether Wave Energy Converter: Hydrodynamic Modelling, Performance Assessment and Control. Ph.D. Thesis, University of Adelaide, Adelaide, Australia, 2018.
27. Backer, G. Hydrodynamic Design Optimization of Wave Energy Converters Consisting of Heaving Point Absorbers. Ph.D. Thesis, University of Gent, Ghent, Belgium, 2009.
28. Blommaert, C. Composite Floating Point Absorbers for Wave Energy Converters: Survivability Design, Production Method and Large-Scale Testing. Ph.D. Thesis, University of Gent, Ghent, Belgium, 2018.
29. Franzitta, V.; Curto, D.; Rao, D.; Viola, A. Hydrogen Production from Sea Wave for Alternative Energy Vehicles for Public Transport in Trapani (Italy). *Energies* **2016**, *9*, 850. [CrossRef]

Article

Sealing Performance Analysis of an End Fitting for Marine Unbonded Flexible Pipes Based on Hydraulic-Thermal Finite Element Modeling

Liping Tang [1], Wei He [1], Xiaohua Zhu [1,*] and Yunlai Zhou [2]

[1] School of Mechatronic Engineering, Southwest Petroleum University, Chengdu 610500, China; lipingtang@swpu.edu.cn (L.T.); hewei@stu.swpu.edu.cn (W.H.)
[2] Department of Civil and Environmental Engineering, The Hong Kong Polytechnic University, Hong Kong 00852, SAR, China; YunLai.zhou@alumnos.upm.es
* Correspondence: zhuxh@swpu.edu.cn

Received: 11 May 2019; Accepted: 30 May 2019; Published: 10 June 2019

Abstract: End fittings are essential components in marine flexible pipe systems, performing the two main functions of connecting and sealing. To investigate the sealing principle and the influence of the temperature on the sealing performance, a hydraulic-thermal finite element (FE) model for the end fitting sealing structure was developed. The sealing mechanism of the end fitting was revealed by simulating the sealing behavior under the pressure penetration criteria. To investigate the effect of temperature, the sealing behavior of the sealing ring under different temperature fields was analyzed and discussed. The results showed that the contact pressure of path 1 (i.e., metal-to-polymer seal) was 31.7 MPa, which was much lower than that of path 2 (metal-to-metal seal) at 195.6 MPa. It was indicated that the sealing capacities were different for the two leak paths, and that the sealing performance of the metal-to-polymer interface had more complicated characteristics. Results also showed that the finite element analysis can be used in conjunction with pressure penetration criteria to evaluate the sealing capacity. According to the model, when the fluid pressures are 20 and 30 MPa, no leakage occurs in the sealing structure, while the sealing structure fails at the fluid pressure of 40 MPa. In addition, it was shown that temperature plays a significant role in the thermal deformation of a sealing structure under a temperature field and that an appropriately high temperature can increase the sealing capacity.

Keywords: end fitting; unbonded flexible pipe; sealing performance; pressure penetration; temperature

---

1. Introduction

The development of offshore resources has traditionally relied on floating production systems, such as floating production storage and offloading (FPSO) units and semi-submerged ships [1]. The hydrocarbons produced by an FPSO or from nearby subsea templates are transported through a pipeline or offloaded onto a tanker [2,3]. In terms of offshore pipes, submarine pipelines, which are buried in a trench or laid on the seabed, are commonly used [4,5]. Compared with conventional steel pipes, flexible pipe systems have the characteristics of higher flexibility, greater applicability, and enhanced recyclability [6]. Flexible pipes can be classified into two primary types: bonded flexible pipes and unbonded flexible pipes [7,8]. An unbonded flexible pipe usually comprises an outer polymeric layer, helical tensile armor, anti-wear layers, pressure armor layers, and an inner carcass layer [9], as shown in Figure 1. With the rapid development of techniques for exploiting deep-water resources, unbonded flexible pipes now play a significant role in transferring oil and gas resources from offshore platforms to onshore facilities.

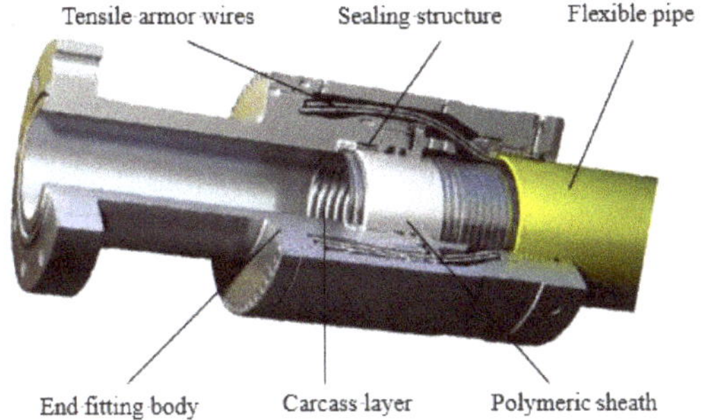

Figure 1. Typical cross-section of flexible pipeline end fitting [9].

Connecting the subsea infrastructure to surface facilities and transporting hydrocarbon products are the major applications of flexible pipes in the offshore oil and gas industry [10]. However, the harsh deep-sea environment imposes significant challenges on flexible pipes, necessitating higher mechanical response and performance characteristics [11]. According to the American Petroleum Institute (API), the terminations of a flexible pipe are defined as end fittings (as shown in Figure 2), of which the functions are: (1) to provide a transition between the pipe body and the connecting component and (2) to transmit the loads acting on the pipe without allowing the pipe to fail [12]. The widespread use of flexible pipes under more demanding operational conditions makes the safety performance of end fittings particularly important [13].

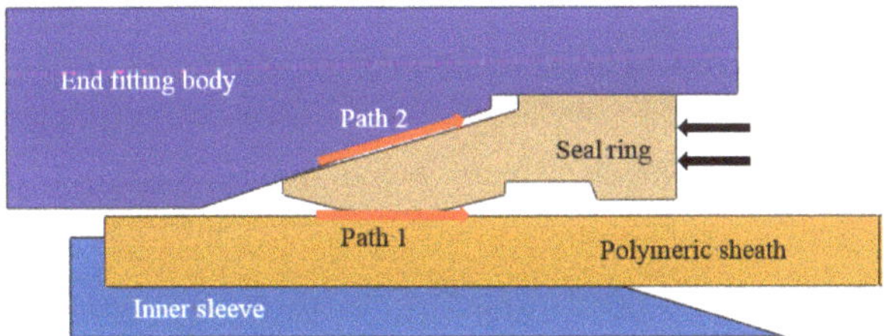

Figure 2. Schematic illustration of the sealing structure.

Experiences in offshore environments have shown that the end fitting of a flexible pipe may be the weakest point [13]. In service, end fittings will be subjected to similar environmental loads and conditions as the pipe, such as axial tension, inner pressure, and external hydrostatic pressure [14]. Apart from mechanical load, the end fitting has to offer thermal insulation and be leak-proof [13]. Therefore, if the sealing capacity of the end fitting is insufficient, there will be a risk of oil and gas leakage, which can have serious consequences. Because the composite structure of flexible pipe consists of many independent concentric metallic and polymeric layers, the structure of the end fitting is also multifaceted and complex [15]. To ensure that the end fitting has adequate sealing performance, it is necessary to investigate its sealing capacity.

In general, existing studies related to the sealing performance of a structure have concentrated on the sealing rings [16]. Typical examples are "O" rings, although these are different from the structure of an end fitting sealing assembly. In addition, a number of studies have focused on the mechanical behavior of flexible pipe, such as the instability of the armor wire [17], the collapse of the carcass layers [18], and fatigue reliability analysis [19]. Although the sealing behavior of the end fitting is unlike the mechanical behavior of the flexible pipe body or the layers inside the end fitting, which have been extensively investigated [8,20], there has been relatively little research on the sealing performance of the end fittings themselves.

The composite materials used in flexible pipe have different properties to metallic materials in terms of anisotropy, thermal expansion coefficient, thermal conductivity, and stiffness. Indeed, the structural properties of composite materials are more complex than those of metallic materials. This may lead to interface failure, such as when the composite material separates from the metallic material, thus losing the ability to maintain leak-tight integrity. Hatton et al. [15] studied the design of sealing assemblies in different types of end fittings using finite element (FE) analysis and laboratory testing.

To understand the sealing performance of a mechanical connector in a subsea pipeline, Wang et al. [21] investigated the critical condition of the sealing structure and created a new method to analyze the contact pressure of the sealing surface by examining the static metal sealing mechanism. An optimized design for a new subsea pipeline mechanical connector was proposed and an approach for determining the contact pressure of various dimensions was provided.

For an end fitting in a high-pressure pipe, it is challenging to create the necessary sealing performance. Fernando et al. [22] developed an FE model of a flexible pipe end fitting and presented a method of evaluating the sealing performance of the sealing assembly and the design requirements for the sealing assembly of the end fitting. In their work, FE analysis was conducted using specially established leak criteria. In addition, an ultrasonic technique was used to measure the contact pressure at the metal-to-metal interface, which showed that their method had significant promise.

Li et al. [23] considered the sealing performance of the sealing assembly in a deep-water flexible pipe end fitting and established an FE model using the ABAQUS software (6.11). They studied the key parameters under different conditions, providing further references for research on flexible pipe end fittings. By summarizing the general sealing criteria, Zhang [24] introduced the concept of "contact pressure amplification factor" to evaluate the sealing capability of end fittings, while Marion et al. [25] investigated the suitability of end fittings for high-temperature thermal cycling conditions using specially designed pipe samples and facilities that satisfy the API specifications.

Previous studies have analyzed and optimized the sealing criteria and the geometric parameters of the sealing assembly. However, there have been few studies related to the sealing behavior. In general, research on the sealing performance of the end fittings is not comprehensive. In this study, FE methods were used to develop a two-dimensional axisymmetric numerical model of the sealing structure of an end fitting, including the temperature field. The pressure penetration criteria were applied to this model to analyze the performance of the sealing structure.

2. Sealing Analysis

2.1. Sealing Structure

According to the API SPEC 17J and 17B standards [12,26], the sealing structure of a typical flanged unbonded flexible pipe end fitting is as illustrated in Figure 2. As can be seen in the figure, the functional structure of the sealing system is mainly composed of four parts: the inner sleeve, polymeric sheath, sealing ring, and end fitting body. The inner sleeve is the innermost component of the end fitting. This plays a supporting role in the whole structure and is used to bear the radial force while in service. The polymeric sheath of the flexible pipe is pressed on the outer side of the inner sleeve, and is one of the most important parts of the sealing structure. The end fitting body is the outermost part of the sealing structure, and is used to protect all components in the end fitting. The wedge sealing ring provides the critical sealing capacity through axial extrusion.

The sealing ring is designed in advance according to the specifications of the flexible pipe and end fitting. The sealing assembly in an end fitting is usually formed by swaging a metallic sealing ring into the area between the polymeric layer and the end fitting body. During assembly, the contact surface between the sealing ring and the end fitting body and between the sealing ring and the polymeric sheath creates two leakage paths [14,22]. Path 2 is the metal-to-metal microscopic gap between the end fitting body and the sealing ring, where a higher contact pressure ensures better sealing capacity. Path 1 refers to the metal-to-polymer contact interface between the sealing ring and the polymeric sheath of the flexible pipe. This path involves complex interactions such as elastic–plastic deformation and nonlinear contact. The contact pressure is lower than in path 1, so leaks are more likely to occur through this path.

2.2. Sealing Criteria

Sealing can be either dynamic or static. The contact sealing between the sealing ring and the polymeric material in the inner layer of the flexible pipe is a type of static sealing [27]. In engineering applications, the performance of this type of sealing is evaluated by comparing the contact pressure and the length of the two contact surfaces. To obtain good sealing capacity, it is necessary to achieve a relatively large contact pressure, and so the length of the contact surface should be as long as possible. Of course, the premise is that the physical properties of the material itself cannot be destroyed.

For the sealing to remain valid, the contact pressure of the sealing path must be greater than the critical failure pressure. However, calculating the critical failure pressure is complicated, and the influence of the material and the medium should be considered. In a previous study [22], the nominal critical failure pressure was expressed as

$$p_c = \alpha p_f + (1 - \alpha)\sigma_y \tag{1}$$

where p_c is the critical failure pressure and p_f is the fluid pressure in the sealing system, σ_Y is the smaller yield stress of the two materials in contact, and α is the ratio of the length of the fluid infiltrating into the contact surface to the length of the contact path. However, in practical applications, the geometry and roughness of the contact surface exert significant influences, so this formula is not exact.

In this study, to simulate the real process of fluid intrusion, a pressure penetration module was employed in the FE software [28,29]. The loading principle of the pressure penetration criteria is illustrated in Figure 3. On the two contact surfaces, the surface with elements 3 and 4 is defined as the master surface, and that with elements 1 and 2 is defined as the slave surface. Nodes 11 and 12 belong to element 1 on the side of the contact surface.

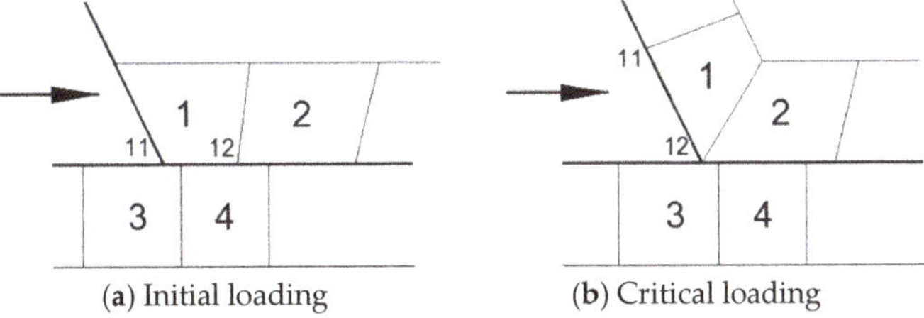

(a) Initial loading   (b) Critical loading

Figure 3. Pressure penetration criteria diagram.

As shown by the arrows in this figure, when the fluid penetrates from left to right, the fluid pressure load will be applied normal to the surface from left to right. Node 11 is the first node exposed to the fluid pressure. If the contact pressure of node 11 is higher than the applied fluid pressure, the pressure penetration stops and node 11 becomes the critical node for the contact pressure (Figure 3a). If the contact pressure of node 11 is less than the fluid pressure, the pressure penetration will continue to load along the contact surface until a new critical contact pressure node is reached (Figure 3b). In addition, if no critical node exists, the contact path cannot provide the sealing capacity under this fluid pressure. The application of pressure penetration can identify the critical node dynamically along the path and determine whether the sealing capacity is sufficient [30].

2.3. Thermal Sealing Analysis

High temperatures will lead to thermal expansion and material deformation, which will change the contact behavior and affect the sealing performance [31]. However, in previous studies on the sealing performance of end fittings, temperature has seldom been taken into account. Over recent years, operating temperatures and pressures have risen as water depths have continued to increase, making the design, manufacture, and installation of flexible pipe a complex challenge. Therefore, the thermal sealing performance at different temperatures is investigated in this paper. Changes in temperature should follow the basic thermal conduction equation. According to the law of conservation of energy, this can be calculated as follows [32]:

$$\rho c \frac{\partial \theta}{\partial t} = \frac{\partial}{\partial x}(k \frac{\partial \theta}{\partial x}) + q(x, t) \tag{2}$$

At the same time, the FE method has the ability to model heat transfer with convection. Based on the work of Yu and Heinrich [33,34], the formulation can be obtained by the following expression:

$$\int \delta\theta \left[\rho c \left\{ \frac{\partial \theta}{\partial t} + V \cdot \frac{\partial \theta}{\partial x} \right\} - \frac{\partial}{\partial x} \cdot \left(k \cdot \frac{\partial z}{\partial x} \right) - q \right] dV + \int_{S_q} \partial\theta \left[n \cdot k \cdot \frac{\partial \theta}{\partial x} - q_s \right] dS = 0 \tag{3}$$

where $c(\theta)$ is the specific heat capacity of the fluid, $\rho(\theta)$ is the fluid density, $\theta(x, t)$ is the temperature at a spatial position x at time t, $k(\theta)$ is the conductivity of the fluid, $q(x, t)$ is the heat added per unit volume from external sources, $\delta\theta(x, t)$ is an arbitrary variational field, q_s is the heat flowing into the volume across the surface on which the temperature is not prescribed (S_q), and n is the outward normal to the surface.

The boundary condition of this thermal equilibrium equation is that $\theta(x)$ is prescribed over some part of the surface S_θ, and that the heat flux per unit area entering the domain across the rest of the surface, $q_s(x)$, is defined by convection or radiation conditions. In the conditions considered in this study, the boundary term in the equation defines

$$q_s = -n \cdot k \cdot \frac{\partial \theta}{\partial x} \tag{4}$$

This implies that q_s is the flux associated with conduction across the surface only; any convection of energy across the surface is not included in q_s.

2.4. Mechanical Analysis

Based on the analysis of the flexible pipe end fitting sealing structure in the preceding section, the polymeric sheath is squeezed between the inner sleeve and the end fitting body when assembled. Thus, in the FE model, the axial degrees of freedom at both ends of the polymeric sheath are restrained to prevent axial motions, but radial free expansion and contraction are not affected. When variations in temperature lead to thermal deformation of the polymeric sheath, the expansion can only happen in the radial direction. In previous studies on the deformation of polymeric material [27], the radial deformation of the seal can be calculated by converting the deformation in the axial direction to a deformation in the radial direction. Based on the principle of volume invariance, the deformation relation between the radial direction and the axial direction for a cylinder specimen can be described as

$$\frac{\pi d^2}{4}(l + \Delta l) = \frac{\pi (d + \Delta d)^2}{4} l \tag{5}$$

where d is the diameter of the cylinder specimen before the deformation, l is the length of the cylinder specimen before deformation, and Δd and Δl are the radial and axial increments, respectively, of the cylinder specimen after deformation.

3. FE Modeling Procedures

Considering the geometry and axisymmetric load characteristics of the sealing system, a two-dimensional plane axisymmetric solid model is employed to predict the seal performance. This section describes a thermal coupling sealing structure model of a flexible pipe with a design pressure of 20 MPa. This model was developed using FE with the ABAQUS software. Note that the factory acceptance test pressure is 1.5 times the design pressure, so the critical pressure acting on the end fitting is 30 MPa in service [12]. The model comprises the inner sleeve, polymeric sheath, sealing ring, and end fitting body, from inside to outside. The inner and outer diameters of the flexible pipe are 139.7 mm and 209.5 mm, respectively, and the thickness of the polymeric sheath is 5 mm. The basic physical properties of each component are listed in Table 1.

Table 1. Materials and properties for each part of the model.

Component	Young's Modulus (GPa)	Poisson's Ratio	Yield Strength (MPa)
End fitting body	210	0.3	355
inner sleeve	210	0.3	355
Sealing ring	191	0.3	758
Polymeric sheath	0.571	0.45	20.74

In addition, the polymeric sheath of the flexible pipe is usually made from high-molecular-weight polymeric materials, such as high-density polyethylene (HDPE) [35]. In this study, the polymeric material parameters were taken from the work of Malta and Martins [36], and the elastic–plastic properties are illustrated in Figure 4.

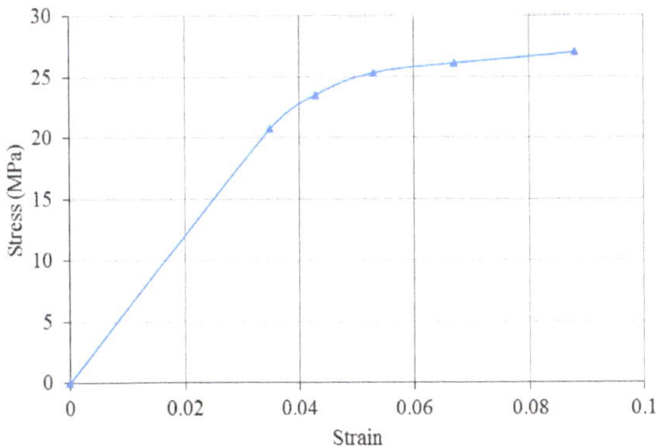

Figure 4. Stress versus strain curve for the polymeric material [36].

To model the incompressible or quasi-incompressible characteristics of these materials, the planar axisymmetric hybrid element CAX4H is selected. Structured and sweep meshing techniques are used in each part of the model, and the mesh is refined around the contact area to improve the accuracy of the simulation. Because of the nonlinear contact characteristics of metallic and polymeric materials, the Mohr-Coulomb friction criterion is employed to describe the contact relationship (i.e., normal contact is "hard" and tangential contact incurs a penalty under a friction coefficient of 0.1) [37].

To simulate the sealing process of the sealing ring, full constraints are applied to the inner sleeve, end fitting body, and polymeric sheath of the flexible pipe, while the sealing ring is free to undergo axial displacement. Pressure penetration is then applied to predict the effectiveness of the sealing. When analyzing the parameter sensitivity of the sealing structure, a temperature field is applied to the model. In addition, an implicit solver is used to obtain improved solution convergence and performance. The FE model of the sealing structure is illustrated in Figure 5.

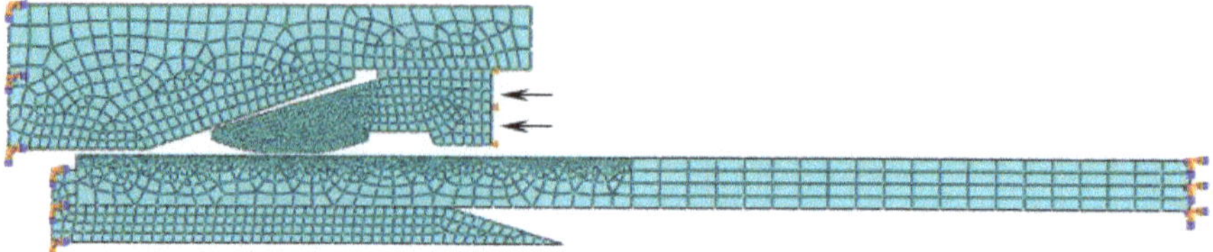

Figure 5. Description of the finite element (FE) model.

4. Results and Discussion

4.1. Simulation of Sealing Principle

According to the design specifications and assembly requirements of the end fitting, axial displacement is applied to the sealing ring to achieve an interference fit. In this section, the von Mises stress and the contact pressure of the model are investigated to analyze the sealing performance of the sealing structure. The von Mises stress distribution of the sealing structure after assembly at 20 °C is shown in Figure 6.

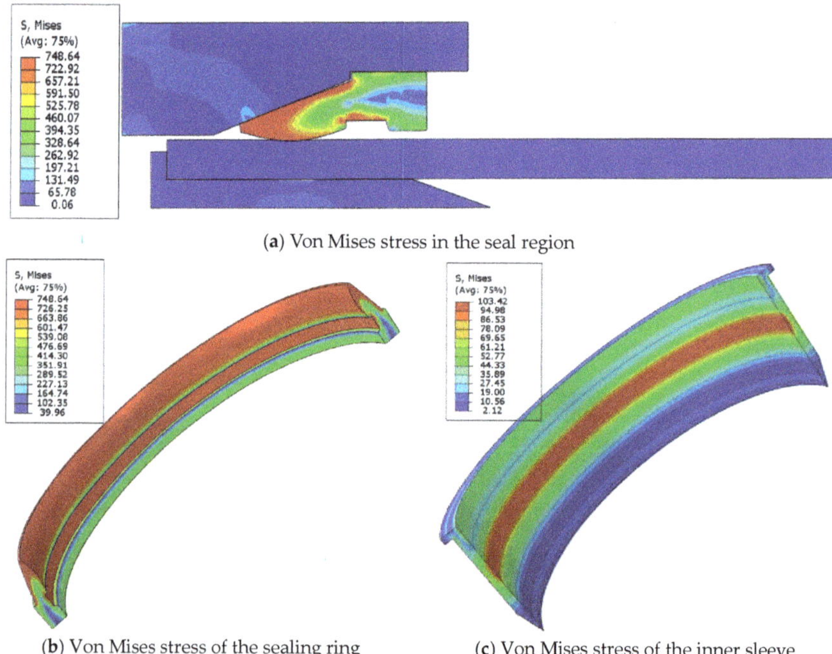

(**a**) Von Mises stress in the seal region

(**b**) Von Mises stress of the sealing ring

(**c**) Von Mises stress of the inner sleeve

Figure 6. Von Mises stress distribution of the FE model.

As can be seen from Figure 6a, the von Mises stress of the model is mainly concentrated on the wedge sealing ring, and so the stress of the inner sleeve and the end fitting body is relatively small. The closer to the tip of the sealing ring, the greater the von Mises stress, because the tip is subjected to a greater pressure. A three-dimensional von Mises stress contour distribution of the sealing ring, obtained by rotating the planar sealing ring, is shown in Figure 6b. The maximum von Mises stress of the wedge sealing ring is 748.6 MPa in the FE model, which does not exceed the yield stress of the material. This shows that no plastic deformation occurs in the sealing ring and the property of the material itself is not destroyed. In addition, some stress concentration occurs in the contact area of the inner sleeve (as shown in Figure 6c), but the maximum stress is still less than the yield stress of the material.

To clarify the characteristics of the contact pressure distribution in the sealing structure, the three-dimensional contact pressure contour distribution of the FE model is presented in Figure 7. On the polymeric sheath of the flexible pipe, some contact pressure occurs on the outer side of the contact region, although the contact pressure on the inner side is lower. For the sealing ring, the contact pressure is higher on the outer interface close to the end fitting body (i.e., path 2), whereas the contact pressure is lower on the inner interface close to the polymeric sheath (i.e., path 1). This may be caused by the property of the materials along the contact paths.

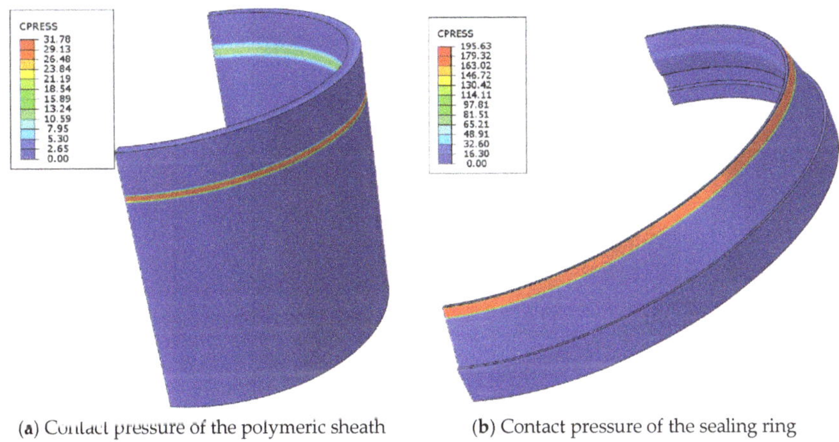

(**a**) Contact pressure of the polymeric sheath   (**b**) Contact pressure of the sealing ring

Figure 7. Contact analysis of the sealing structure.

Based on the FE model of the sealing structure, the contact pressure distributions on the nodes along the two paths were obtained. They are illustrated in Figure 8. The maximum contact pressure along path 1 is 31.74 MPa, which is slightly greater than the test pressure of 30 MPa. The maximum contact pressure along path 2 is 195.6 MPa, which is much larger than the design pressure and test pressure of the end fitting. This indicates that there will be no seal leakage along this path, which is consistent with the results of previous works [22,23]. Hence, we focus on path 1 in the following analysis.

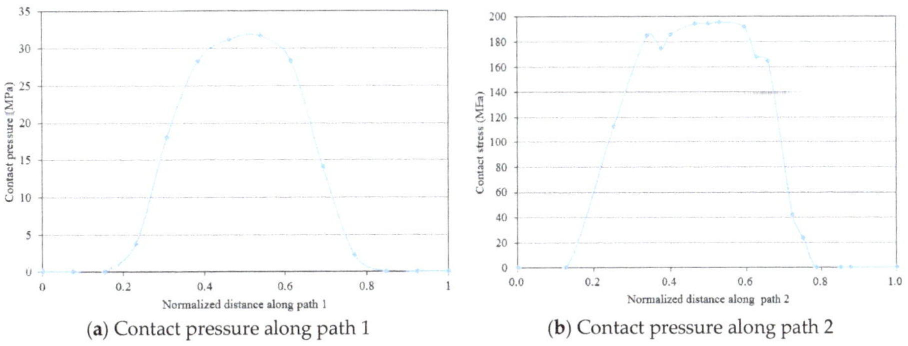

(**a**) Contact pressure along path 1   (**b**) Contact pressure along path 2

Figure 8. Comparison of the sealing performance along the two paths.

4.2. Analysis of Pressure Penetration

The pressure penetration criteria can be applied to evaluate the sealing capacity of the sealing assembly. According to the design pressure and the maximum test pressure of the end fitting, the sealing performance of the sealing structure along path 1 was analyzed at fluid pressures of 20, 30, and 40 MPa at a temperature of 20 °C. The contact pressure on the polymeric sheath reflects the variation in the pressure penetration, and Figure 9 shows the contact pressure of the polymeric sheath under the different fluid pressures. With an increase in fluid pressure, the contact pressure on one side of the inner surface of the polymeric sheath increases, whereas the pressure on the outer surface gradually decreases. This is because when the fluid acts along path 1, the contact surface between the sealing ring and the polymeric sheath is continuously penetrated by the pressure, causing these components to separate from each other. On the contrary, the contact surface between the sheath and the inner sleeve is squeezed and shrunken along the radial direction.

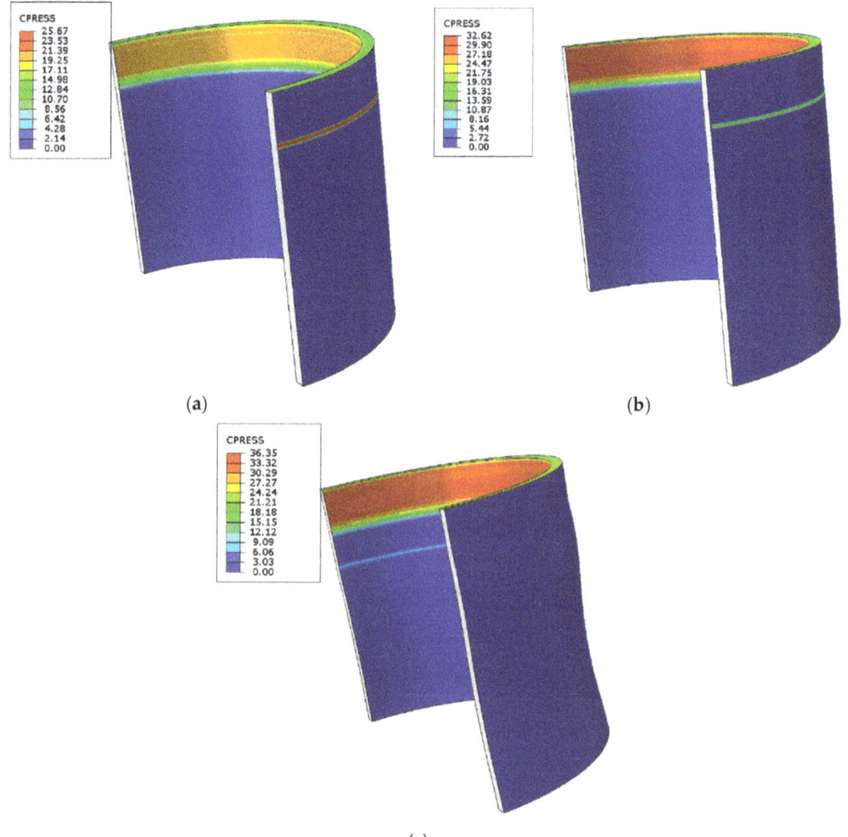

Figure 9. Contact pressure of polymeric sheath under different fluid pressures: (**a**) 20 MPa; (**b**) 30 MPa; (**c**) 40 MPa.

Figure 10 shows the contact pressure distribution of the nodes along path 1. When no pressure penetration is applied, the maximum contact pressure is 31.7 MPa. At fluid pressures of 20 and 30 MPa, the maximum contact pressure and contact length decrease, and the pressure distribution moves to the right. At this time, the existence of the contact pressure indicates that no leakage will occur in the sealing structure.

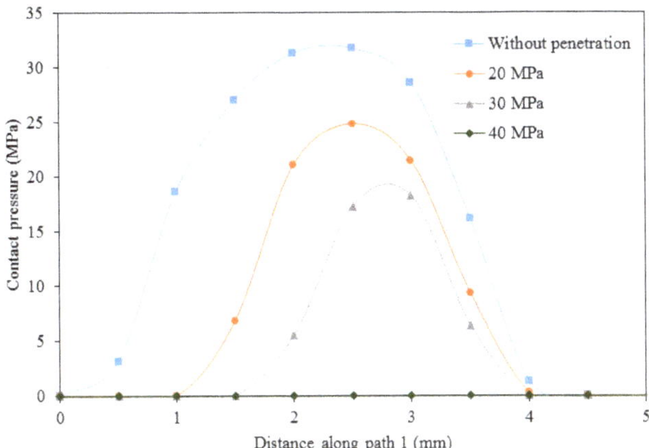

Figure 10. Contact pressure under different fluid pressure.

When the fluid pressure is 40 MPa, the contact pressure along path 1 becomes 0 MPa, which indicates that the fluid pressure is too high, and the sealing structure fails. As shown in Figure 9, the fluid leakage through this path flows around the annulus of the flexible pipe, acting to compress the polymeric sheath and make it collapse. In addition, leakage accidents may occur if the sealing structure fails. It should be emphasized that, after many simulation calculations and predictions, the sealing structure was found to leak at a fluid pressure of 35.5 MPa. In other words, the critical fluid pressure of the sealing structure in this study is 35.5 MPa.

4.3. Analysis of the Thermal Sealing

Under the actual working conditions of flexible pipe end fittings, the hydrocarbon products transported along the pipe can reach high temperatures, so thermal effects in the sealing structure cannot be ignored. This section reports the changes in sealing performance of the sealing structure under temperature fields of 20, 50, 60, 70, and 80 °C in the model [38].

Compared with metallic materials, the polymeric materials in the flexible pipe will be more strongly influenced by temperature. Therefore, after applying different temperature fields to the FE model of the sealing structure, the contact pressure distributions of the polymeric sheath were recorded (as shown in Figure 11). From the perspective of deformation, an increase in temperature deforms the polymeric sheath and causes the flexible pipe to expand. Regarding the contact pressure along path 1, as the temperature increases from 20–80 °C, the maximum contact pressure increases from 31.7–41.3 MPa—a rise of 30% (as shown in Figure 12). In addition, the contact length increases because the material expands at higher temperatures, increasing the number of contactable nodes.

From the results in Table 2, the maximum von Mises stress of the model exhibits a continuous increase, becoming close to the yield strength of the material at 80 °C. The contact pressures along the two paths gradually increase, which indicates that the sealing capacity is also increasing. In summary, thermal effects can enhance the sealing performance, but the temperature should not exceed the rated temperature of the material, otherwise failure will occur.

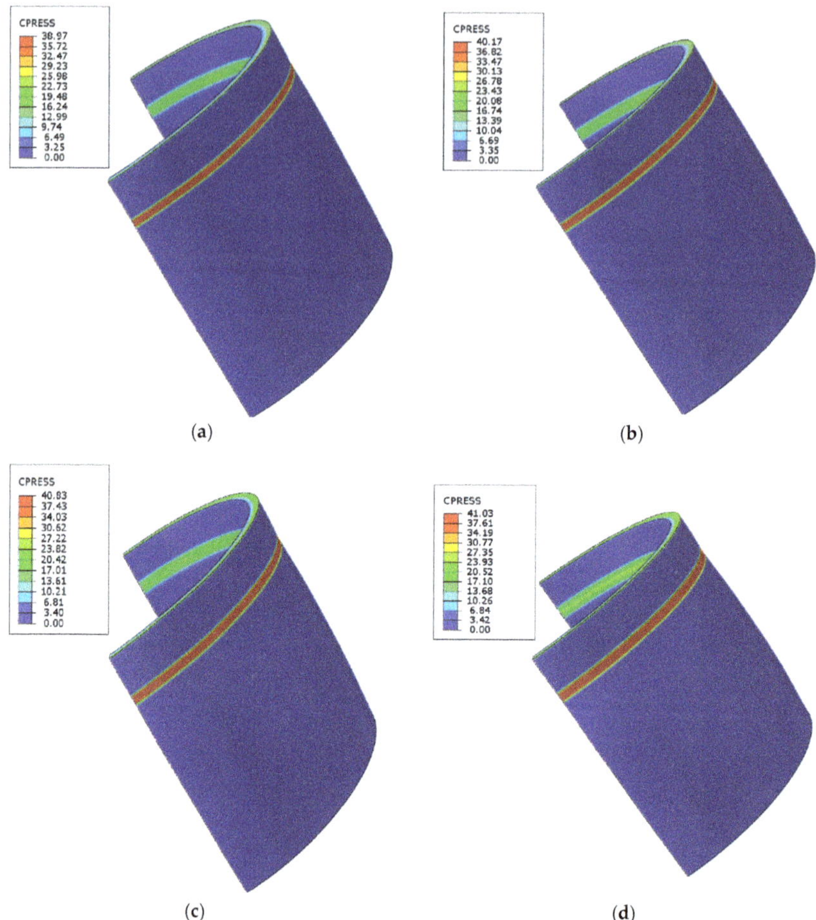

Figure 11. Contact pressure distribution of the polymeric sheath: (**a**) 50 °C; (**b**) 60 °C; (**c**) 70 °C; (**d**) 80 °C.

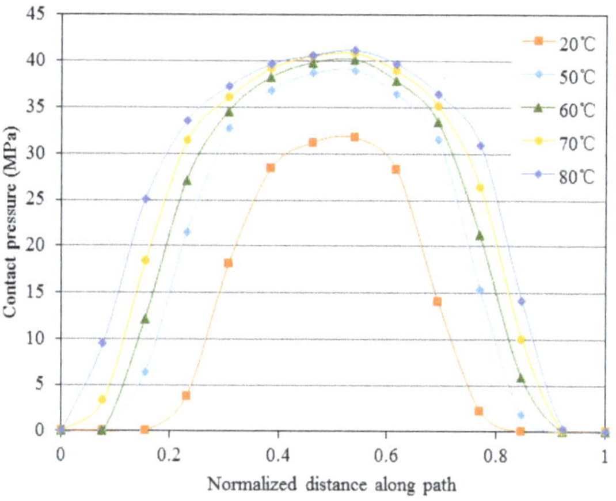

Figure 12. Contact pressure along path 1.

Table 2. Results of the FE model at different temperatures.

Temperature (°C)	20	50	60	70	80
Von Mises stress (MPa)	748.6	750.1	751.6	753.7	756.8
Contact pressure in path 1 (MPa)	31.7	38.9	40.1	40.8	41.3
Contact pressure in path 2 (MPa)	195.6	199.7	227.3	229.4	234.5

5. Conclusions

A classical unbonded flexible pipe is a combination of polymeric and metallic layers. An end fitting with reliable sealing properties is a precondition of a successful flexible pipe application. In this study, a hydraulic-thermal FE model was developed to investigate the sealing performance of a flexible pipe end fitting. The FE model employed the pressure penetration criterion and considered the temperature field, which is suitable for real applications. Using this model, the sealing principle was simulated and the influence of thermal effects on the sealing capacity was investigated.

The results showed that the maximum von Mises stress occurs on the sealing ring during the sealing process, whereas the stresses on the other components are relatively small. In terms of the contact pressure distribution, the maximum value appears in the sealing region, and is higher along path 2. By introducing pressure penetration, the sealing performance could be predicted and the dynamical pressure critical node was identified. In the model described in this paper, the critical fluid pressure of the end fitting is 35.5 MPa, which means that leakage occurs when the working pressure exceeds this value. In previous studies, thermal effects were usually omitted. The results in this paper, however, show that temperature is an important factor in the sealing performance of the sealing assembly, and should not be neglected. Thermal effects cause the components of the sealing structure to deform and expand. By increasing the contact length and contact pressure, the sealing ability of the sealing structure can be improved. Of course, very high temperatures are not appropriate, because the strain on the sealing ring should be considered in actual applications.

Author Contributions: L.T. contributed significantly to analysis and manuscript writing; W.H. performed the numerical simulation and data analyses; X.Z. contributed to the conception of the investigation; Y.Z. helped perform the analysis with constructive discussions.

Funding: This research was funded by the China Postdoctoral Science Foundation (43XB3793XB), National Key Research and Development Program (2018YFC0310204), National Natural Science Foundation of China (No. 51674214), and Scientific Innovation Group for Youths of Sichuan Province (No. 2019JDTD0017). And The APC was funded by National Key Research and Development Program (2018YFC0310204).

Conflicts of Interest: This article does not involve conflicts of interest.

References

1. Moan, T.; Amdahl, J.; Ersdal, G. Assessment of ship impact risk to offshore structures-New NORSOK N-003 guidelines. *Mar. Struct.* **2019**, *63*, 480–494. [CrossRef]

2. Yang, C.K.; Kim, M.H. Numerical assessment of the global performance of spar and FPSO connected by horizontal pipeline bundle. *Ocean Eng.* **2018**, *159*, 150–164. [CrossRef]

3. Zhang, J.M.; Li, X.S.; Chen, Z.Y.; Zhang, Y.; Li, G.; Yan, K.F.; Lv, T. Gas-Lifting characteristics of methane-water mixture and its potential application for self-eruption production of marine natural gas hydrates. *Energies* **2018**, *11*, 240. [CrossRef]

4. Guha, I.; White, D.J.; Randolph, M.F. Subsea pipeline walking with velocity dependent seabed friction. *Appl. Ocean Res.* **2019**, *82*, 296–308. [CrossRef]

5. Zhang, P.; Huang, Y.F.; Wu, Y. Springback coefficient research of API X60 pipe with dent defect. *Energies* **2018**, *11*, 3213. [CrossRef]

6. Paiva, L.F.; Vaz, M.A. An empirical model for flexible pipe armor wire lateral buckling failure load. *Appl. Ocean Res.* **2017**, *66*, 46–54. [CrossRef]

7. Anderson, T.A.; Vermiyea, M.E.; Dodds, V.J.N.; Finch, D.; Latto, J.R. Qualification of flexible fiber-reinforced pipe for 10,000-foot water depths. In Proceedings of the Offshore Technology Conference, Houston, TX, USA, 6–9 May 2013.

8. Pham, D.C.; Sridhar, N.; Qian, X.; Sobey, A.J.; Achintha, M. A review on design, manufacture and mechanics of composite risers. *Ocean Eng.* **2016**, *112*, 82–96. [CrossRef]

9. Dahl, C.S.; Andersen, B.; Groenne, M. Developments in managing flexible risers and pipelines, a suppliers perspective. In Proceedings of the Offshore Technology Conference, Houston, TA, USA, 2–5 May 2011.

10. Ebrahimi, A.; Kenny, S.; Hussein, A. Radial buckling of tensile armor wires in subsea flexible pipe—Numerical assessment of key factors. *ASME J. Offshore Mech. Arct. Eng.* **2016**, *138*, 031701. [CrossRef]

11. Drumond, G.P.; Geovana, P.; Pasqualino, I.P.; Pinheiro, B.C.; Estefen, S.F. Pipelines, risers and umbilicals failures: A literature review. *Ocean Eng.* **2017**, *148*, 412–425. [CrossRef]

12. API 17B. *Recommended Practice for Flexible Pipe*, 5th ed.; American Petroleum Institute: Washington, DC, USA, 2014.

13. Campello, G.C.; Sousa, J.R.M.; Vardaro, E. An analytical approach to predict the fatigue life of flexible pipes inside end fittings. In Proceedings of the Offshore Technology Conference, Houston, TX, USA, 2–5 May 2016.

14. Fernando, U.S.; Nott, P.; Graham, G.; Roberts, A.E.; Sheldrake, T. Experimental evaluation of the metal-to-metal seal design for high-pressure flexible pipes. In Proceedings of the Offshore Technology Conference, Houston, TX, USA, 30 April–3 May 2012.

15. Hatton, S.; Rumsey, L.; Biragoni, P.; Roberts, D. Development and qualification of end fittings for composite riser pipe. In Proceedings of the Offshore Technology Conference, Houston, TX, USA, 6–9 May 2013.

16. Peng, G.; Zhang, Z.; Li, W. Computer vision algorithm for measurement and inspection of O-rings. *Measurement* **2016**, *94*, 828–836. [CrossRef]

17. Liu, X.; Li, G.; Yue, Q.; Oberlies, R. Acceleration-oriented design optimization of ice-resistant jacket platforms in the Bohai Gulf. *Ocean Eng.* **2009**, *36*, 1295–1302. [CrossRef]

18. Tang, M.; Lu, Q.; Yan, J.; Yue, Q. Buckling collapse study for the carcass layer of flexible pipes using a strain energy equivalence method. *Ocean Eng.* **2016**, *111*, 209–217. [CrossRef]

19. Cour, D.; Kristensen, C.; Nielsen, N.J.R. Managing fatigue in deepwater flexible risers. In Proceedings of the Offshore Technology Conference, Houston, TX, USA, 5–8 May 2008.

20. Li, X.; Jiang, X.; Hopman, H. A review on predicting critical collapse pressure of flexible risers for ultra-deep oil and gas production. *Appl. Ocean Res.* **2018**, *80*, 1–10. [CrossRef]

21. Wang, L.Q.; Wei, Z.L.; Yao, S.M.; Guan, Y.; Li, S.K. Sealing performance and optimization of a subsea pipeline mechanical connector. *Chin. J. Mech. Eng.* **2018**, *31*, 1–14. [CrossRef]

22. Fernando, U.S.; Karabelas, G. Analysis of end fitting barrier seal performance in high pressure un-bonded flexible pipes. In Proceedings of the 33rd ASME International Conference on Ocean, Offshore and Arctic Engineering, San Francisco, CA, USA, 8–13 June 2014.

23. Li, X.; Du, X.; Wan, J.; Xiao, H. Structure analysis of flexible pipe end fitting seal system. In Proceedings of the 34th ASME International Conference on Ocean, Offshore and Arctic Engineering, St John's, NL, Canada, 31 May–5 June 2015.

24. Zhang, L.; Yang, Z.; Lu, Q.; Yan, J.; Chen, J.; Yue, Q. Numerical simulation on the sealing performance of serrated teeth inside the wedgy sealing ring of end fitting of marine flexible pipeline. *Oil Gas Storage Transp.* **2017**, *37*, 108–115. (In Chinese)

25. Marion, A.; Rigaud, J.; Werth, M.; Martin, J. γ-Flex®: A new material for high temperature flexible pipes. In Proceedings of the Offshore Technology Conference, Houston, TX, USA, 6–9 May 2002.

26. API 17J. *Specification for Un-Bonded Flexible Pipe*, 4th ed.; American Petroleum Institute: Washington, DC, USA, 2014.

27. Yan, H.; Zhao, Y.; Liu, J.; Jiang, H. Analyses toward factors influencing sealing clearance of a metal rubber seal and derivation of a calculation formula. *Chin. J. Aeronaut.* **2016**, *29*, 292–296. [CrossRef]

28. Gorash, Y.; Dempster, W.; Nicholls, W.D.; Hamilton, R. Fluid pressure penetration for advanced FEA of metal-to-metal seals. *Proc. Appl. Math. Mech.* **2015**, *15*, 197–198. [CrossRef]

29. Zhao, B.; Zhao, Y.; Wu, X.; Xiong, H. Sealing performance analysis of P-shape seal with fluid pressure penetration loading method. *IOP Conf. Ser. Mater. Sci. Eng.* **2018**, *397*, 012126. [CrossRef]

30. Slee, A.J.; Stobbart, J.; Gethin, D.T.; Hardy, S.J. Case study on a complex seal design for a high pressure vessel application. In Proceedings of the ASME Pressure Vessels and Piping Conference, Anaheim, CA, USA, 20–24 July 2014.
31. Wang, B.; Peng, X.; Meng, X. A thermo-elastohydrodynamic lubrication model for hydraulic rod O-ring seals under mixed lubrication conditions. *Tribol. Int.* **2019**, *129*, 442–458. [CrossRef]
32. Deng, D.; Zhang, C.; Pu, X.; Liang, W. Influence of material model on prediction accuracy of welding residual stress in an austenitic stainless steel multi-pass buttwelded joint. *J. Mater. Eng. Perform.* **2017**, *26*, 1494–1505. [CrossRef]
33. Yu, C.; Heinrich, J. Petrov-Galerkin methods for the time-dependent convective transport equation. *Int. J. Numer. Meth. Eng.* **1986**, *23*, 883–902. [CrossRef]
34. Yu, C.; Heinrich, J. Petrov-Galerkin method for multidimensional, time-dependent, Convective-Diffusive equations. *Int. J. Numer. Meth. Eng.* **1987**, *24*, 2201–2215. [CrossRef]
35. Yu, R.; Yuan, P. Structure and research focus of marine un-bonded flexible pipes. *Oil Gas Storage Transp.* **2016**, *35*, 1255–1260. (In Chinese)
36. Malta, E.; Martins, C. Finite element analysis of flexible pipes under axial compression: Influence of the sample length. *ASME J. Offshore Mech. Arct. Eng.* **2017**, *139*, 011701. [CrossRef]
37. Cuamatzi-Melendez, R.; Castillo-Hernandez, O.; Vazquez-Hernandez, A.O.; Vaz, M.A. Finite element and theoretical analyses of bisymmetric collapses in flexible risers for deepwaters developments. *Ocean Eng.* **2017**, *140*, 195–208. [CrossRef]
38. Pethrick, R.A.; Banks, W.M.; Brodesser, M. Ageing of thermoplastic umbilical hose materials used in a marine environment 1-Polyethylene. *Proc. Inst. Mech. Eng. Part L J. Mater. Des. Appl.* **2014**, *228*, 45–62. [CrossRef]

Article

An Assessment of Wind Energy Potential in the Caspian Sea

Florin Onea and Eugen Rusu *

Department of Mechanical Engineering, Faculty of Engineering, Dunarea de Jos University of Galati,
47 Domneasca Street, 800008 Galati, Romania
* Correspondence: eugen.rusu@ugal.ro; Tel.: +40-740-205-534

Received: 10 June 2019; Accepted: 25 June 2019; Published: 1 July 2019

Abstract: At this time, there are plans to develop offshore wind projects in the Caspian Sea. The aim of the present work was to estimate the possible benefits coming from such a project. As a first step, the wind profile of this region was established by considering reanalysis data coming from the ERA-Interim project, the time interval covered being between January 1999 and December 2018. According to these results, significant resources have been noticed in the northern part where the wind speed frequently reached 8 m/s, being identified also as a hot-spot south of Olya site. In the second part, the performances of some offshore wind turbines were established. These were defined by rated capacities ranging from 3 MW to 8.8 MW. The downtime period of some generators can reach 90% in the central and southern sectors, while for the capacity factor, the authors expected a maximum of 33.07% for a turbine rated at 4.2 MW. From a financial point of view, the values of the levelized cost of energy (LCOE) indicated that the sites from the north and central parts of the Caspian Sea have been defined by an average LCOE of 0.25 USD/kWh. Thus, they can represent viable locations for wind farm projects.

Keywords: Caspian Sea; wind speed; offshore turbines; capacity factor; LCOE

---

1. Introduction

The wind energy market is a dynamic sector that is continuously expanding. Nevertheless, this evolution will finally reach saturation, taking into account that the range of suitable sites for such projects will reduce. As an alternative, one of the best locations to develop a wind project is near the coastal areas, which can easily support the development of either onshore and offshore projects. At this time, there is growing interest to develop systems capable of operating in deep water areas, the floating platforms being the most viable solution [1,2]. A significant percentage of the projects have operated in Europe (almost 409 projects), being estimated that in 2018, only 18 projects were added to the grid. Per total, the operating projects are defined by a capacity of 18.5 GW, this value being supported by 11 countries, such as UK, Germany, Denmark, Norway or Portugal, as being more relevant. One of the largest wind turbines connected at this moment is the Haliade 150-6 MW (Merkur project, Germany), with the mention that some other projects currently under construction may include systems defined by comparable capacity, or even higher [3].

Most of the research is focused on the coastal areas facing the ocean environment [4–6], since the wind resources are more consistent in such areas. During recent years, the enclosed sea basins were also considered for investigation [7–9]. More advanced progress is related to the Mediterranean Sea, where a pilot farm of 24.8 MW will be developed off the coast of Gruissan in the Aude region, France. It is expected that through four wind turbines, it is possible to generate approximately 100 million kWh/year, which should be enough to cover the annual electricity consumption of more than 50,000 inhabitants [10]. Other sources indicate that a 30 MW wind project will be installed in front of the Taranto Harbour, Puglia, Southern Italy, which will consider ten wind turbines produced by Senvion [11]. The French coastal area located in this region seems to present important wind resources, and the implementation of several floating projects is expected in the near future [12].

As for the Caspian Sea wind resources, there is some research focused in this aspect. In the research by Amirinia et al [13], the distribution of the wind and wave resources associated to the southern area was evaluated. By considering various databases (including in-situ measurements) and the characteristics of a 3 MW wind turbine, it was found that more promising results have been reported in the eastern part of this region. In Rusu and Onea [8], the wind and wave regime from this region were also assessed, and more energetic conditions in the centre and northern regions were observed, with an average wind speed of 6 m/s (at 80 m height). In addition, observations of the bathymetric map of the Caspian Sea highlight that the entire northern area is defined by a lower water depth, which is very suitable for the development of the wind projects. In Onea and Rusu [14], the performances of some wind turbines that may operate in the coastal areas of the Black and Caspian seas were discussed. The turbines were reported to an operating hub height of 80 m, while the power output of the particular systems were estimated using a Betz coefficient of 0.5. By applying these simplifications, it appears that the performances of some wind turbines have been overestimated, resulting in capacity factors of 70%, compared to the usual capacity factor values between 20 and 40%. An important aspect, which is highlighted in this work, is related to the diurnal and nocturnal fluctuations of the wind conditions. Kerimov et al [15] analysed the long term distribution of wind resources, also proposed some suitable sites, and evaluated the associated costs required to couple some wind generators to the electrical grid. According to these results, one of the best sites to implement an offshore wind project is located in the northern part of Absheron peninsula (Azerbaijan). Here, a suitable connection grid was also identified. Although, this region is an enclosed sea defined by moderate conditions, there is some research focused on the local wave resources that brings into discussion the viability of a commercial wave farm in this environment, especially considering hybrid wind-wave approaches [16–18].

On the other hand, it is well known that the Caspian Sea region is defined by important oil reserves and therefore, at this time, it seems that the interest for a renewable project is not very high [19]. Nevertheless, taking into account that there are problems with the oil pollution in this region [20,21], some changes are expected to occur. Thus, at this moment there are plans to develop a 200 MW offshore wind farm in the coastal areas of Azerbaijan, which will significantly increase the existing wind capacity (66.7 MW) of this country. It is estimated that the project will cost 392 million euro. This is one of the multiple wind parks proposed for this area, and more likely will be developed close to the capital, Baku [22,23].

In this context, the objective of the present work is to provide a more complete picture of the wind resources in the Caspian Sea in order to identify some suitable sites for the development of a wind project, and also to establish the performances of some state-of-the-art wind turbines that may operate in the vicinity of some major cities from this region.

2. Materials and Methods

2.1. The Target Area

Figure 1 presents a first perspective of the Caspian Sea. Thus, Figure 1a,b presented the distribution of the water depth, in general, and for the reference sites considered for evaluation. Regarding the northern sector, this is defined by lower water depths that do not exceed 50 m, compared to the southern side, where the water depth values easily reach 500 m, and even higher, if the Nowshahr site as a reference is considered. Several distances from the shoreline are taken into account (5 km, 25 km and 50 km) in order to identify the variations from the nearshore to offshore. Furthermore, in Table 1 the characteristics of the sites located near the shore are presented. A total of 10 reference sites are considered for evaluation, the main selection criteria being related to the fact that they are important port cities capable of easily offering the infrastructure and technical support for the development of a wind project. The sites located at 5 km, are defined by lower water depths (<50 m), but going towards the 50 km limit, these values significantly increase. This is the case of the Iranian sites of Babolsar, Nowshahr, Anzali and Astara, where the depths can easily exceed 400 m. The 50 km limit was not arbitrarily selected, and represents an acceptable threshold at which a renewable project can still be competitive [24].

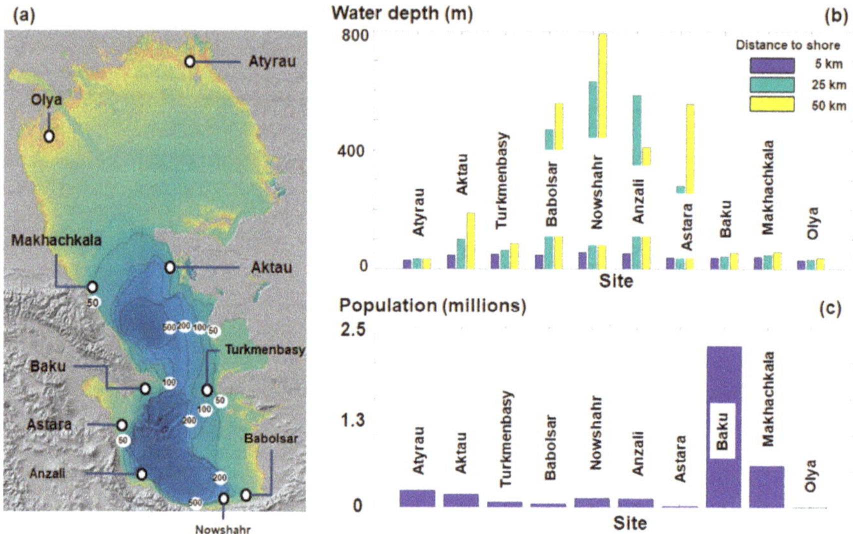

Figure 1. The Caspian Sea and the reference sites considered for assessment, where: (**a**) bathymetry map [25]; (**b**) water depths corresponding to each reference site [25]; (**c**) number of inhabitants corresponding to the reference area considered [26].

Table 1. The main characteristics of the considered sites located at a distance of 5 km from shore.

No.	Site	Country	Long (°)	Lat (°)	Water Depth (m)
P1	Atyrau	Kazakhstan	51.745	46.680	30
P2	Aktau	Kazakhstan	51.101	43.615	46
P3	Türkmenbasy	Turkmenistan	52.714	39.994	50
P4	Babolsar	Iran	52.643	36.759	48
P5	Nowshahr	Iran	51.492	36.705	56
P6	Anzali	Iran	49.512	37.519	52
P7	Astara	Iran	48.938	38.425	38
P8	Baku	Azerbaijan	49.889	40.324	37
P9	Makhachkala	Russia	47.584	42.997	41
P10	Olya	Russia	48.136	45.269	30

The port of Baku is the capital of Azerbaijan and one of the largest cities in the Caucasus, this aspect being reflected by the number of its inhabitants (Figure 1c), of almost 2.5 million, being followed by the Makhachkala city with 0.6 million inhabitants and Atyrau with 0.24 million.

2.2. Wind Dataset

ERA-Interim Reanalysis Data

The ERA-Interim wind product has been considered for assessment, this being a project maintained by the European Center for Medium-Range Weather Forecasts (ECMWF). This global model covers the time interval from 1979 to the present, being defined by a temporal resolution of 6h per day

(00-06-12-18 UTC). In the near future, this model may be replaced by the ERA5 product defined by 1h resolution and some others significant improvements [27,28]. For the present work, only the data covering the 20-year time interval from January 1999 to December 2018 are used, considering a spatial resolution of $0.125° \times 0.125°$ (the highest one). The U and V components of the wind speed are reported to a 10 m height above the sea level, and in this case, the wind speed is denoted as *U10*. Apart from the assessment of the wind energy potential, another objective of the present work is to estimate the performance of some offshore wind turbines, and therefore the *U10* values need to be adjusted to the hub height of a particular system. This is indicated as [29,30]:

$$U_{turbine} = U10 \frac{\ln(z_{turbine}) - \ln(z_{10})}{\ln(z_{10}) - \ln(z_0)} \tag{1}$$

where, $U_{turbine}$—wind speed at hub height, *U10*—initial wind speed (at 10 m), z_0—roughness factor (calm sea surface −0.0002 m), z_{10} and $z_{turbine}$—reference heights.

2.3. Wind Turbines

Several offshore wind turbines are considered for evaluation, their characteristics being presented in Table 2. The rated power of the generators is in the range of 3–8.8 MW, with the systems rated below 6 MW currently in operation, compared to the others that are the next generation turbines and are being implemented at this moment. In general, the cut-in values are approximately 3–3.5 m/s. The system, Adwen AD 5-135, has a lower rated speed value of 11.4 m/s, which means that this turbine will reach his rated capacity quicker. Each turbine is defined by a particular hub height, some of them being specifically indicated by the manufacturers, but in most of the cases these are mentioned as site specific, being possible to develop in this way multiple scenarios. A minimum of 65 m corresponds to the system Vestas V90-3.0 MW, while a maximum value of 140 m corresponds to Vestas V164-8.0 MW.

Table 2. The main characteristics of the wind turbines considered [31].

Turbine	Rated Power (MW)	Cut-in Speed (m/s)	Rated Speed (m/s)	Cut-out Speed (m/s)	Diameter (m)	Hub Height (m)
Vestas V90-3.0	3	3.5	15	25	90	65–105
Vestas V112-3.45	3.45	3	13	25	112	84
Siemens SWT-3.6-120	3.6	3.5	12	25	120	90
Siemens SWT-3.6-107	3.6	4	13.5	25	107	80 or site specific
Siemens SWT-4.0-130	4	5	12	25	130	89.5
Envision EN 136-4.2	4.2	3	10.5	25	136	80/90/110
Areva M5000-116	5	4	12.5	25	116	site specific
Adwen AD 5-135	5.05	3.5	11.4	30	135	site specific
Siemens SWT-6.0-154	6	4	13	25	154	site specific
Senvion 6.2M126	6.15	3.5	14	30	126	85/95
Siemens SG 7.0-154	7	3	13	25	154	site specific
Siemens SG 8.0-167	8	3	12	25	167	site specific
Vestas V164-8.0	8	4	13	25	164	105/140
Vestas V164-8.8	8.8	4	13	25	164	site specific

The annual electricity production (*AEP*) indicator is frequently used to quantify the electricity output expected in a particular system. In the present work, the following expression was considered for this indicator [32]:

$$AEP = T \cdot \int_{cut-in}^{cut-out} f(u)P(u)du \tag{2}$$

where, AEP—in MWh, T—average hours per year (8760 hr/year), $f(u)$—Weibull probability density function, $P(u)$—power curve of a turbine, cut-in and cut-out values of a turbine.

One way to estimate the overall performances of a particular system is through the capacity factor (C_f), that can be defined as [32]:

$$C_f = \frac{P_E}{R_P} \cdot 100,\tag{3}$$

where: P_E is the electric power expected to be generated and R_P represents the rated power of the system.

3. Results

3.1. Verification of the Reanalysis Data

The ERA-Interim data were obtained through numerical simulations and therefore the accuracy of these data for this enclosed sea area need to be discussed. One way is to use the satellite measurements provided by the AVISO (Archiving, Validation and Interpretation of Satellite Oceanographic Data) project that include daily multi-mission measurements [33]. This dataset was defined by only one measurement per day ($U10$ values), and for the current work the values corresponding to the interval January 2010 and December 2017 were processed. Figure 2 presents a direct comparison between AVISO and ERA-Interim data, including a quality check of the satellite measurements. A common problem associated to the altimeter data, has been represented by the accuracy of these systems to measure the marine resources at the land-water interface. As can be seen from Figure 1a, a maximum of 50.46% corresponded to the site of Atyrau, compared to a minimum of 20.76% indicated for Baku, while a constant distribution of 41% was noticed close to the sites Babolsar, Nowshahr and Anzali, respectively.

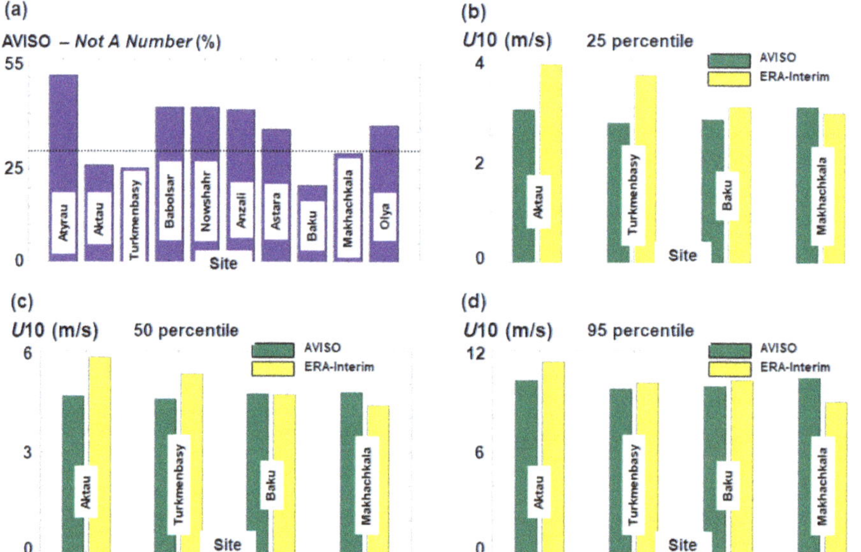

Figure 2. The comparisons between the AVISO measurements and the ERA–Interim reanalysis data for the time interval 2010–2017. The results are structured in: (**a**) NaN corresponding to the AVISO measurements; (**b**) 25 percentile; (**c**) 50 percentile; (**d**) 95 percentile.

In order to compare the two datasets, only the sites defined by much lower NaN (not a number) occurrences were considered for evaluation, and for this work, the ones chosen did not exceed 30% (indicated by dotted line). From the percentile analysis (25, 50 and 95), the ERA-Interim indicated higher values than the satellite measurements, this aspect being valid only for the sites Aktau, Türkmenbasy and Baku located in the eastern and southern sectors. The Aktau site was defined by more significant variations, and a difference of 0.9 m/s (25 percentile), 1.16 m/s (50 percentile) and 1.12 m/s (95 percentile) was observed. A reverse pattern corresponded to the Makhachkala site (west coast–north of Baku), where ERA-Interim slightly underestimated the wind resources, with a maximum difference of 1.62 m/s for the 95 percentile.

3.2. Analysis of the Wind Speed

A first perspective of the wind distribution is presented in Figure 3, including the full time distribution (denoted as the total time) and the four main seasons that include: spring (March-April-May), summer (June-July-August), autumn (September-October-November) and winter (December-January-February). Regardless the period taken into account, it is clear that the northern part of the Caspian Sea is defined by more important wind resources. For the total time, a maximum value of 8.5 m/s was noticed in the southern regions of Olya site (north-west), and on an axis that crosses the northern area of this basin. The values significantly decreased towards the centre of this target area, the eastern part being defined by more important resources that can reach 6 m/s. The areas located in the centre-west and south seem to be defined by hot-spot regions where the wind conditions frequently indicated values of approximately 2 m/s, which meant that they are not very attractive for a wind project.

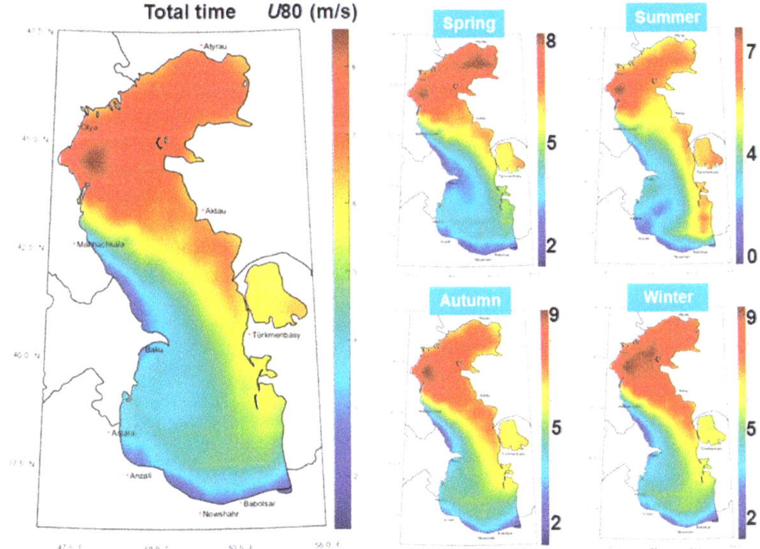

Figure 3. Spatial distribution of the *U80* parameter (average values) considering the 20-year time interval (January 1999 to December 2018) of ERA-Interim data.

During the springtime, the spatial distribution was similar to the total time, where the maximum values can increase to 8 m/s. As for the summer season, moderate conditions were noticed and the hot-spot area located near the Olya site was identified, where the wind speed reached a maximum value of 7.3 m/s. For the rest of the regions, the conditions decreased below 6 m/s, with a significant percentage of the southern part did not exceed 4 m/s. During autumn and winter, the wind strength significantly increased, and more frequent wind speeds of 9 m/s were observed.

Figure 4 presents a slightly better representation of the seasonal variations, where the average values corresponding to each season are reported to the total time distribution. Various patterns were noticed. During spring, in the north, wind conditions increased approximately 10% and a decrease of the resources to almost 20% was observed close to the Absheron Peninsula (Baku site). During summer, the entire region indicated a decrease of the values to almost 30%. There is a hot spot close to the Babolsar site (south-east) where the wind speed may increase to 30%. During autumn, a wind farm located close to the Absheron Peninsula may have indicated better performances with an expected increase of the wind speed to almost 20%. The northern and centre regions indicated, in general, a slight increase of the values that may reach a maximum of 10%. During winter, the wind conditions presented a constant increase in intensity ($\approx$15%), and in some sites, the balance was close to zero. In this case, the eastern area of Babolsar indicated a 15% decrease of the values.

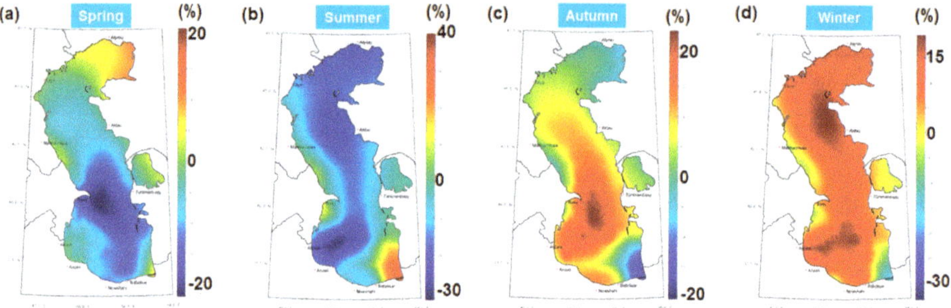

Figure 4. The spatial differences (in %) between the total time data and the four main seasons. The results are reported for the $U80$ parameter (average values) considering the 20-year time interval of ERA-Interim data, where: (**a**) spring; (**b**) summer; (**c**) autumn and (**d**) winter.

In Figure 5a, the evolution of the $U80$ parameter is presented for the reference sites indicated in Table 1. The average values (dotted line) reported to all the sites were approximately 5.2 m/s, and above this threshold, this study found: Atyrau (7.12 m/s), Olya (6.96 m/s), Aktau (6.68 m/s), Turkmenbasy (6.36 m/s) and Baku (5.87 m/s). A minimum wind speed value of 3.12 m/s corresponded to the site Anzali (south-west).

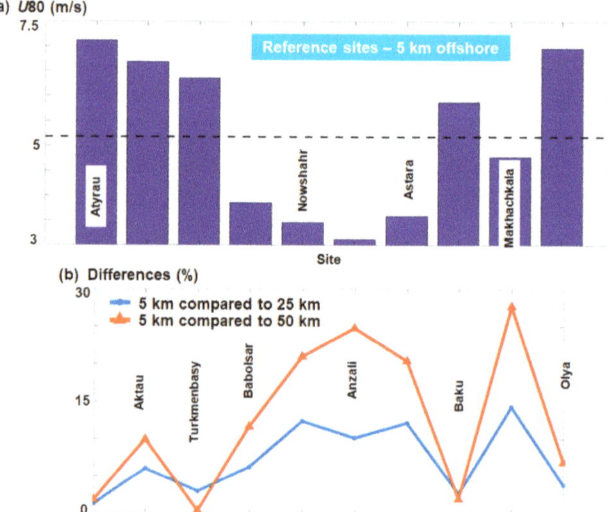

Figure 5. The distribution of the $U80$ parameter (average values) per reference sites considering the 20-year time interval of ERA-Interim data, where: (**a**) $U80$ values reported by the sites located at 5 km nearshore; (**b**) differences (in %) between the sites located at 5 km and the ones located at 25 km and 50 km, respectively.

The variation of the wind speed between various distances from the shoreline (5 km-25 km-50 km) is presented in Figure 5b. In general, the sites defined by important wind resources have not indicated important variations, this being the case of Atyrau, Baku or Olya. This suggests that a wind project located close to the shoreline will be a suitable solution. For the Turkmenbasy site, it was observed that the wind speed at 50 km offshore was lower than the one at 25 km, the differences being approximately 2.61%. It is clear that, in general, the wind speed increased heading towards offshore, this being the situation of Babolsar (25 km—6.2%; 50 km—11.62%), Anzali (10.1%; 24.79%) or Makhachkala (14.27%; 27.67%). In this case, the site Makhachkala (50 km) with 6.1 m/s exceeded the value of 6 m/s, which corresponded to the Baku site.

Figure 6 presents the monthly evolution of the $U80$ parameter taking into account all the available wind data. As expected, the northern part was dominated by more important conditions for the entire time interval considered., In July, it was observed that some wind energy hot-spots close to Olya site (in the south) or nearby Balbosar (in the north), reached a maximum value of 7 m/s. During the interval January–April, the consistency of the wind resources was more present in the entire northern region, with wind speed values observed at 9 m/s. In general, the sites located close to Olya (in the south) constantly indicated a higher wind resource, which makes them suitable candidates for the development of a wind project, if there is any energy demand in this region.

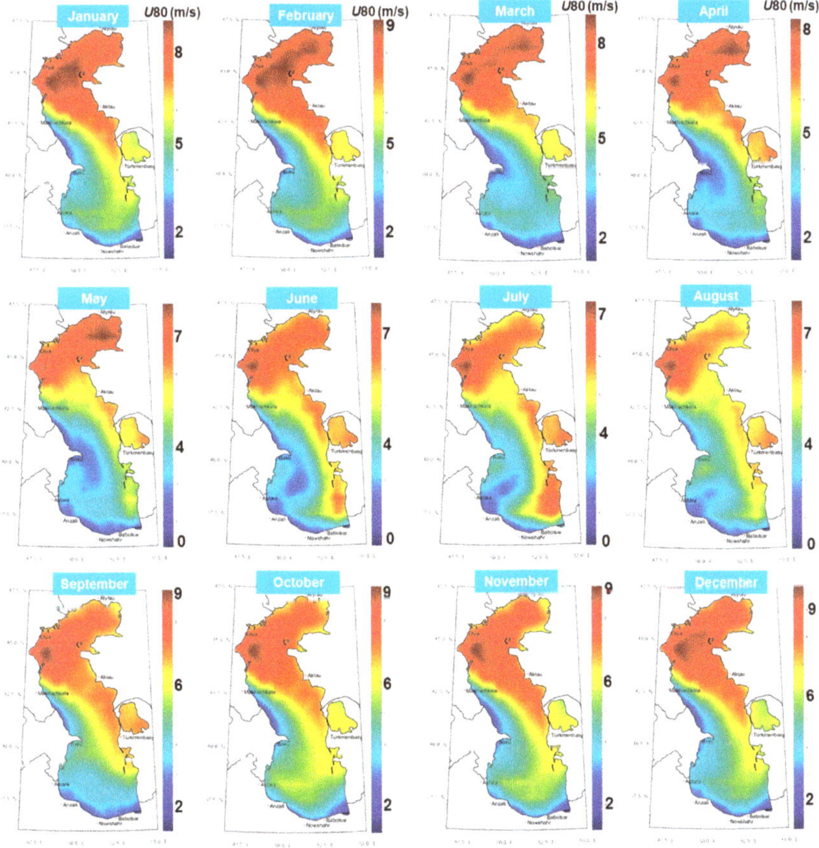

Figure 6. Monthly distribution of the $U80$ parameter (average values) reported by the ERA-Interim wind data for the 20-year time interval from January 1999 to December 2018.

3.3. Evaluation of the Wind Turbines

The downtime interval represents the inactivity period, during which the turbine will be shut down since the wind conditions are not suitable for the electricity production. In Figure 7, some case studies are presented. As the wind distribution located below the cut-in value and a dotted line that marks a distance of 50 km from the shore, a wind project can be usually implemented. In the case of the value 3 m/s, a minimum of 20% may be expected for the northern region, this value gradually increasing to 50% for the sites located in the central part of the target area, while a maximum of 70% may be expected in some isolated regions from the west and south. Small differences were noticed between the values 3 m/s and 3.5 m/s, the more notable values corresponded to the 5 m/s scenario (close to 90%). This indicates that a system as Siemens SWT-4.0 is not a suitable option for this enclosed basin.

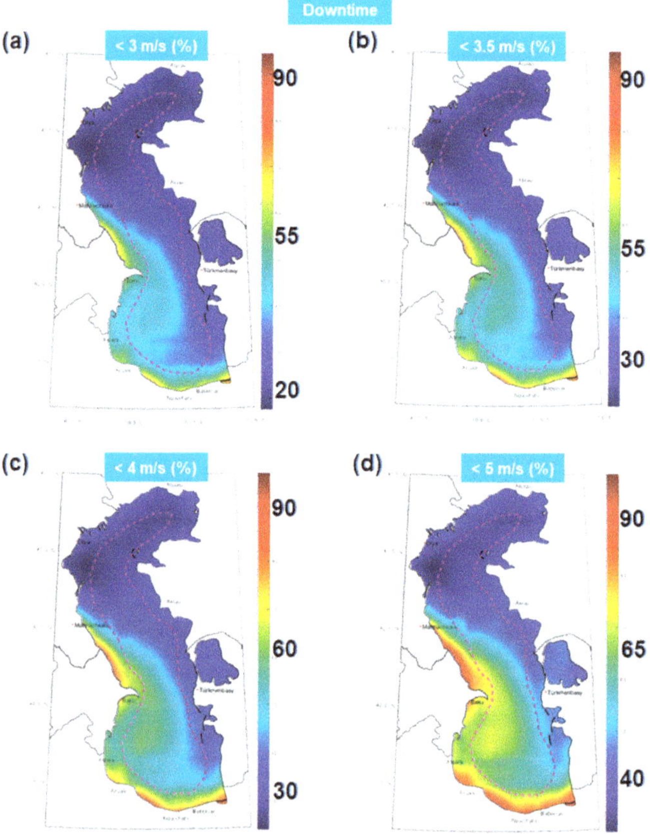

Figure 7. The downtime interval (%) corresponding to different cut-in values, where: (**a**) 3 m/s; (**b**) 3.5 m/s; (**c**) 4 m/s; (**d**) 5 m/s. The dotted line is located at approximately 50 km from the shoreline.

A top five downtime is presented in Table 3, taking into account as a reference the 3 m/s value. The lowest values were accounted by the same sites that include Atyrau, Olya, Aktau, Türkmenbasy and Makhachkala, regardless of the distance to the shore considered for evaluation. For the 5 km limit, a minimum of 20.76% was expected close to Atyrau and a maximum of 36.01% near Makhachkala. The same pattern was repeated for the 25 km, with a minimum of 20.31%, while in the case of 50 km, there was a change of places between the fourth and fifth positions, for which a maximum of 34.4% was noticed.

Table 3. The top five downtimes sorted in an ascending order (corresponding to a 3 m/s cut-in value).

Distance to Shore	Site (%)				
5 km	Atyrau (20.76)	Olya (21.28)	Aktau (25.97)	Türkmenbasy (29.47)	Makhachkala (36.01)
25 km	Atyrau (20.31)	Olya (20.49)	Aktau (24.92)	Türkmenbasy (30.62)	Makhachkala (30.75)
50 km	Olya (19.89)	Atyrau (20.11)	Aktau (24.56)	Makhachkala (27.32)	Türkmenbasy (34.40)

Through the rated capacity, the percentage of time during which a particular turbine will operate at a full capacity can be identified. This is done by taking into account the wind speed located between the rated wind speed and the cut-out values. More details are presented in Figure 8. Only one spatial map was represented, since the colour distribution remained the same, and only the scale of the map was represented for different wind speed values. As expected, the values significantly decreased from 10.5 m/s to 15 m/s, It was observed that the energy hot spot located close to the Olya site did not exceed the 50 km limit (dotted line), where the best results were reported. The maximum values, oscillated between 15% and 30%, depending on the considered rated speed. This distribution closely followed the evolution of the wind speed with better results expected in north. In the southern region, it is possible that the considered wind turbines from Table 2 reported a negative performance (0%).

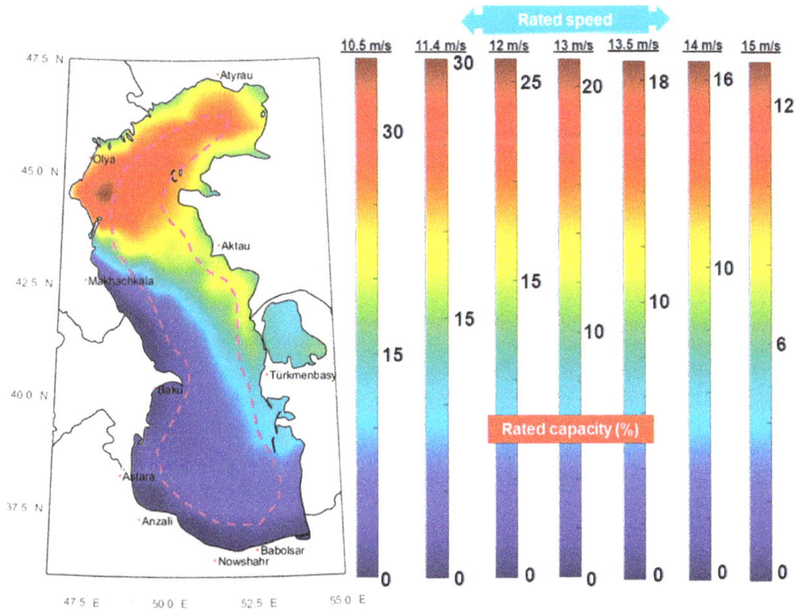

Figure 8. The rated capacity (%) reported for different rated wind speeds.

A more concise evaluation of the rated capacity is presented in Table 4, with the rated speed of 13 m/s as a reference. From the selected sites, Olya presented the best performances that varied between 14.7% (5 km) to 17.81% (50 km), compared to a minimum value of 3.96% (50 km) reported by Türkmenbasy, that was included in this classification.

Table 4. The top five rated capacities sorted in a descending order (reported to a 13 m/s rated speed).

Distance to Shore	Site (%)				
5 km	Olya (14.70)	Atyrau (13.17)	Aktau (7.61)	Türkmenbasy (5.30)	Makhachkala (1.34)
25 km	Olya (16.65)	Atyrau (14.09)	Aktau (9.59)	Türkmenbasy (5.30)	Makhachkala (3.77)
50 km	Olya (17.81)	Atyrau (14.50)	Aktau (10.38)	Makhachkala (6.25)	Türkmenbasy (3.96)

As a next step, the annual electricity production (AEP) was assessed by considering two reference sites, namely Atyrau and Baku. The selection of the Baku site was made by taking into account the interest for development of an offshore project in this region, the large population that lives in this region, and that it has some wind resources considering the results presented in Figure 5a. Figure 9 is focused on this evaluation which includes all the selected wind turbines. In the case of the systems where the hub height was indicated as site specific, different values were considered for assessment.

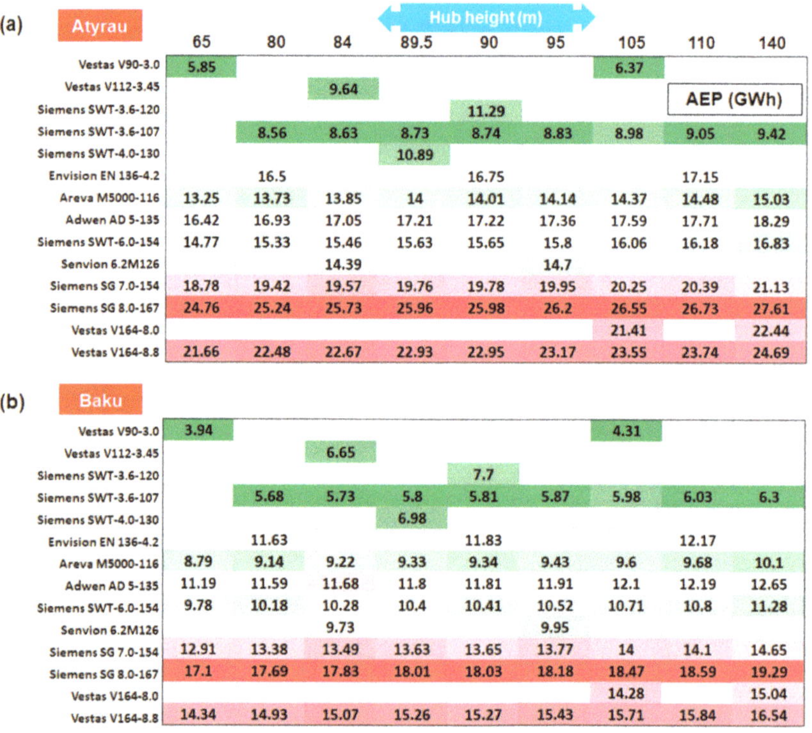

(a) Atyrau — AEP (GWh), Hub height (m)

Turbine	65	80	84	89.5	90	95	105	110	140
Vestas V90-3.0	5.85						6.37		
Vestas V112-3.45			9.64						
Siemens SWT-3.6-120					11.29				
Siemens SWT-3.6-107		8.56	8.63	8.73	8.74	8.83	8.98	9.05	9.42
Siemens SWT-4.0-130				10.89					
Envision EN 136-4.2		16.5			16.75			17.15	
Areva M5000-116	13.25	13.73	13.85	14	14.01	14.14	14.37	14.48	15.03
Adwen AD 5-135	16.42	16.93	17.05	17.21	17.22	17.36	17.59	17.71	18.29
Siemens SWT-6.0-154	14.77	15.33	15.46	15.63	15.65	15.8	16.06	16.18	16.83
Senvion 6.2M126			14.39			14.7			
Siemens SG 7.0-154	18.78	19.42	19.57	19.76	19.78	19.95	20.25	20.39	21.13
Siemens SG 8.0-167	24.76	25.24	25.73	25.96	25.98	26.2	26.55	26.73	27.61
Vestas V164-8.0							21.41		22.44
Vestas V164-8.8	21.66	22.48	22.67	22.93	22.95	23.17	23.55	23.74	24.69

(b) Baku

Turbine	65	80	84	89.5	90	95	105	110	140
Vestas V90-3.0	3.94						4.31		
Vestas V112-3.45			6.65						
Siemens SWT-3.6-120					7.7				
Siemens SWT-3.6-107		5.68	5.73	5.8	5.81	5.87	5.98	6.03	6.3
Siemens SWT-4.0-130				6.98					
Envision EN 136-4.2		11.63			11.83			12.17	
Areva M5000-116	8.79	9.14	9.22	9.33	9.34	9.43	9.6	9.68	10.1
Adwen AD 5-135	11.19	11.59	11.68	11.8	11.81	11.91	12.1	12.19	12.65
Siemens SWT-6.0-154	9.78	10.18	10.28	10.4	10.41	10.52	10.71	10.8	11.28
Senvion 6.2M126			9.73			9.95			
Siemens SG 7.0-154	12.91	13.38	13.49	13.63	13.65	13.77	14	14.1	14.65
Siemens SG 8.0-167	17.1	17.69	17.83	18.01	18.03	18.18	18.47	18.59	19.29
Vestas V164-8.0							14.28		15.04
Vestas V164-8.8	14.34	14.93	15.07	15.26	15.27	15.43	15.71	15.84	16.54

Figure 9. Annual electricity production (GWh) reported for the full time distribution, considering the sites: (**a**) Atyrau; (**b**) Baku. The points are located at a distance of 5 km from the shoreline.

It is clear that the site Atyrau (5 km from shore) indicated better performances than Baku (5 km from shore), regardless of the selected wind turbine, the lower and higher values being highlighted as a reference to the individual hub heights. As expected, the electricity production was related to the rated capacity and at the bottom, the systems rated below 4 MW were found. In addition, it is important to mention that the system, Siemens SG 8.0 MW, presented better results than the other two Vestas systems rated at 8 MW and 8.8 MW, making it a suitable candidate for this region. For the Atyrau site, the authors expected a maximum value of 5.85 GWH in the case of Vestas V90-3.0 MW. These systems operate at a hub height of 65 m, while a maximum of 27.761 GWh is indicated by the Siemens SG 8.0 MW, if the rotor of this system works at a height of 140 m. In the case of the Baku site, a similar pattern was noticed, with the values oscillating between 3.94 GWh and 19.29 GWh. Also, it seems that some wind turbines presented lower performances compared to others, although they were defined by a higher rated capacity. This was the case of Vestas V112-3.45 MW compared to Siemens SWT-3.6-107, or for the generator, Adwen AD5-135, compared to Siemens SWT-6.0-154 and Senvion 6.2M126.

A similar analysis was performed in Figure 10, taking into account the capacity factor index. According to these values, Envision 4.2 MW stood out with more important values, being followed by Adwen AD5-135 and Siemens SG 8.0 MW. Regarding the Atyrau site, by observing the 80 m hub height, the values in the range of 27.13–44.86%, which were specific to some turbines rated at 3.6 MW and 4.2 MW, respectively, were noticed. Compared to these, the turbine rated at 8.8 MW indicated a value of approximately 29.16%. This is close to the lower value and similar to the one reported by Siemens SWT-6.0-154. In the case of the turbines, where the hub height was indicated as site specific, the authors obtained the following differences from 65 m to 140 m: Areva Multibrid M5000—4.07%; Adwen AD 5-135—4.23%; Siemens SWT-6.0-154—3.93%; Siemens Gamesa 7 MW—3.82% or Vestas V164-8.8 MW—3.93%. For the Baku site, lower values were expected from the turbines Vestas V90-3.0, Siemens SWT-3.6-107 and Senvion 6.2M126, while the turbine Envision 4.2MW seemed to be a suitable solution for this site. The turbines rated above 4.2 MW exceeded, in general, by 20%, and reached a maximum of 33.07% in the case of Envison 4.2 MW (110 m height), this turbine exceeded, in general, by 30%.

(a) Atyrau — Hub height (m), Cf (%)

	65	80	84	89.5	90	95	105	110	140
Vestas V90-3.0	22.24						24.24		
Vestas V112-3.45			31.91						
Siemens SWT-3.6-120					35.8				
Siemens SWT-3.6-107		27.13	27.38	27.69	27.72	28	28.47	28.7	29.89
Siemens SWT-4.0-130				31.09					
Envision EN 136-4.2		44.86			45.52			46.6	
Areva M5000-116	30.25	31.36	31.62	31.96	31.98	32.29	32.8	33.05	34.32
Adwen AD 5-135	37.11	38.27	38.54	38.9	38.93	39.24	39.77	40.03	41.34
Siemens SWT-6.0-154	28.1	29.16	29.41	29.74	29.77	30.06	30.55	30.79	32.03
Senvion 6.2M126			26.71			27.29			
Siemens SG 7.0-154	30.63	31.66	31.91	32.23	32.26	32.54	33.02	33.25	34.45
Siemens SG 8.0-167	35.34	36.45	36.71	37.05	37.08	37.38	37.89	38.14	39.4
Vestas V164-8.0							30.55		32.03
Vestas V164-8.8	28.1	29.16	29.41	29.74	29.77	30.06	30.55	30.79	32.03

(b) Baku — Cf (%)

	65	80	84	89.5	90	95	105	110	140
Vestas V90-3.0	14.98						16.41		
Vestas V112-3.45			22						
Siemens SWT-3.6-120					24.42				
Siemens SWT-3.6-107		18.01	18.18	18.4	18.42	18.61	18.96	19.12	19.99
Siemens SWT-4.0-130				19.93					
Envision EN 136-4.2		31.62			32.16			33.07	
Areva M5000-116	20.06	20.87	21.06	21.31	21.33	21.54	21.93	22.1	23.06
Adwen AD 5-135	25.29	26.19	26.4	26.67	26.7	26.93	27.36	27.55	28.6
Siemens SWT-6.0-154	18.61	19.37	19.55	19.79	19.81	20.01	20.38	20.55	21.46
Senvion 6.2M126			18.06			18.47			
Siemens SG 7.0-154	21.06	21.82	22	22.23	22.26	22.45	22.82	22.99	23.89
Siemens SG 8.0-167	24.4	25.25	25.44	25.7	25.73	25.95	26.35	26.53	27.52
Vestas V164-8.0							20.38		21.46
Vestas V164-8.8	18.61	19.37	19.55	19.79	19.81	20.01	20.38	20.55	21.46

Figure 10. The capacity factor (%) reported for the full time distribution, considering the sites: (**a**) Atyrau; (**b**) Baku. The points are located at a distance of 5 km from the shoreline.

4. Discussions and Conclusions

One objective of the present work is to identify some suitable areas for the development of a wind project. From the literature review, it seems that the offshore sites that exceeded an average wind speed of 6 m/s (reported at 10 m height) are recommended [34]. Figure 11 presents this analysis, where the total time interval can be noticed, with the more promising areas being located in the northern sector. As expected, the region located south of the Olya site concentrated most of the wind energy. Some acceptable areas were noticed heading towards the north-east. During the winter time (December-January-February), the suitable areas were significantly extended, reaching the northern and eastern coast of the Caspian Sea. During this season, some promising results are expected from the sites located in the central part of this sea, more precisely, close to the Aktau site (east). Per total, according to this criterion ($U10 < 6$ m/s), a significant part of the Caspian Sea is not suitable for the development of an offshore wind project, this being also the case of the Baku location, where some offshore wind farms may occur in the near future.

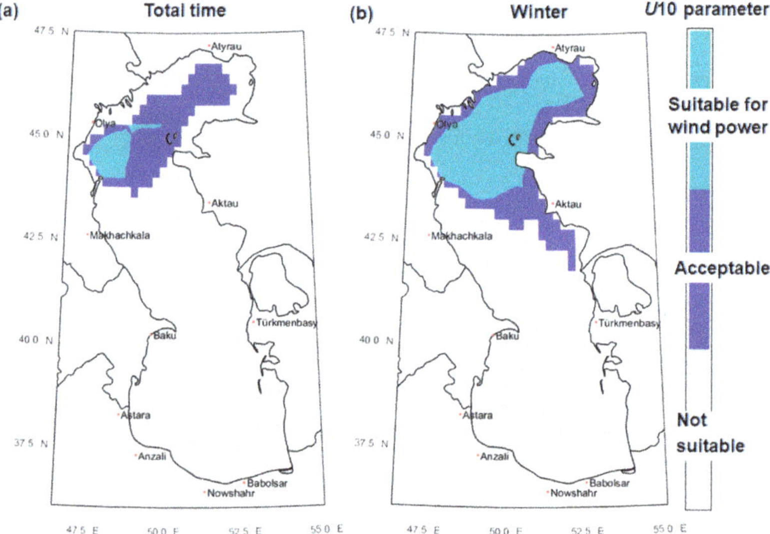

Figure 11. The suitability map for wind power with an average wind speed of 6 m/s as a reference. The results are indicated for: (**a**) full time distribution; (**b**) winter time.

Until now, the performances of the wind turbines were discussed in terms of their rated power. Another important parameter that needs to be taken into account is the turbine diameter, and this analysis is provided in Figure 12. The wind turbines presented in Table 2 were grouped in three main categories (small-medium-large), and the results were presented only for the sites Atyrau and Baku. These results showed a similar distribution for both sites, with the Atyrau site presenting slightly higher values.

From the analysis of the wind turbines included in the first category (small), the system, Envision EN 136-4.2 (hub height 80-90-110 m), seems to be more suitable for this location compared to Siemens SWT-4.0-130 that has a comparable rated capacity and operates at a height of 89.5 m. As for the second group (medium), significant results were noticed between Areva M500-116 and Siemens SWT-6.0-154, although similar hub heights were taken into account. From the systems rated above 7 MW, the better performances were expected from the Siemens SG 8.0-167, compared to Vestas V164-8.0 and Siemens SG 7.0-154, which were defined by a much lower rated capacity.

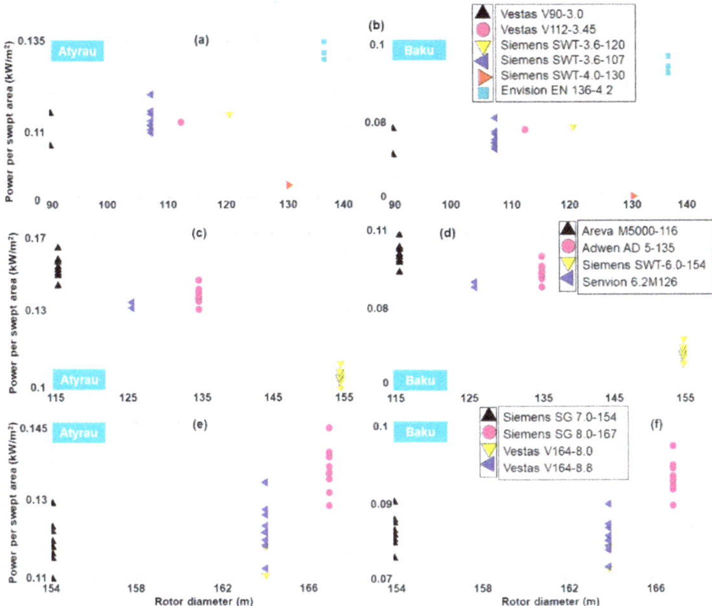

Figure 12. The classification of the wind turbines selected for the sites Atyrau and Baku, taking into account the power output per swept area. The results are grouped as follows: (**a**,**b**) small (3–4.2 MW); (**c**,**d**) medium (5–6.2 MW); (**e**,**f**) large (7–8.8 MW).

In the present work, a more complete description of the wind conditions corresponding to the Caspian Sea was provided by considering spatial maps and specific sites located near the major harbour cities of this region. According to the bathymetric data, significantly lower water depths are noticed in the northern part of this region. This means that they are suitable for the development of mono pile wind projects, taking into account that this solution can be used for water depths of 50 m [35]. For the rest of the regions, a floating wind farm has been indicated [36]. According to the wind data coming from the ERA-Interim project, the northern part of this sea was also defined by important wind resources, that seemed to be very suitable for a wind project. For this region, an average wind speed of 8–8.5 m/s (at 80 m height) were noticed, with a significant increase in the wind power which corresponded to autumn and winter. This environment was defined by important seasonal variations, which can reach a maximum of 40% in the southeast (in summer) or an increase with 15% of the wind speed for the entire basin (in winter).

When discussing the viability of an offshore wind project, an important issue to take into account are the financial aspects, especially in the case of the Caspian Sea. A common way to do this is to consider the levelized cost of energy (LCOE) that can be defined as [24]:

$$LCOE = \frac{CAPEX + \sum_{t=1}^{\eta} \frac{OPEX_t}{(1+r)^t}}{\sum_{t=1}^{\eta} \frac{AEP_t}{(1+r)^t}} \tag{4}$$

where: CAPEX—represents Capital Expenditure; $OPEX_t$—Operational expenditures reported for year t; AEP_t—Annual Energy Production corresponding to year t; r—the discount rate; η—lifetime of the project; t—year from the start of project.

Considering the methodology presented in Rusu and Onea [24], the following set-up was used: (a) Discount rate = 10%; (b) inflation = 2%; (c) ageing of the turbines = 1.6%; (d) CAPEX cost = 4500 USD/kW; (e) OPEX = 0.048 USD/kW; (f) project lifetime = 20 years (1999–2018). For simplicity, the hub height of the turbines was fixed to 80 m height and all the reference sites located at 5 km were taken into account. However from the results presented in Figure 2a, it is possible that some of them need to be defined by a lower quality of the wind data.

Figure 13 presents such an analysis, where a similar spatial pattern of the LCOE is indicated by the three groups of wind turbines. The better results corresponded to the sites Atyrau, Aktau, Turkmenbasy, Baku and Olya, where the values gradually decreased below 0.5 USD/kWh. A maximum LCOE value of 10 USD/kWh was reported by the site Anzali, this value gradually decreasing to 4.5 USD/kWh and 3 USD/kWh for turbines defined by higher rated capacity. Taking into account that the target of the European Union for the year 2025 is to obtain an LCOE of 0.11 USD/kWh [24], this study can conclude that the sites that exceed at this moment 0.5 USD/kWh cannot be taken into account for an offshore project. In Figure 13 (subplots b1 and c1), the turbines rated between 7 and 8.8 MW indicate better performances, reporting in some cases LCOE values below 0.25 USD/kWh. There is room for improvement and better performances can be obtained.

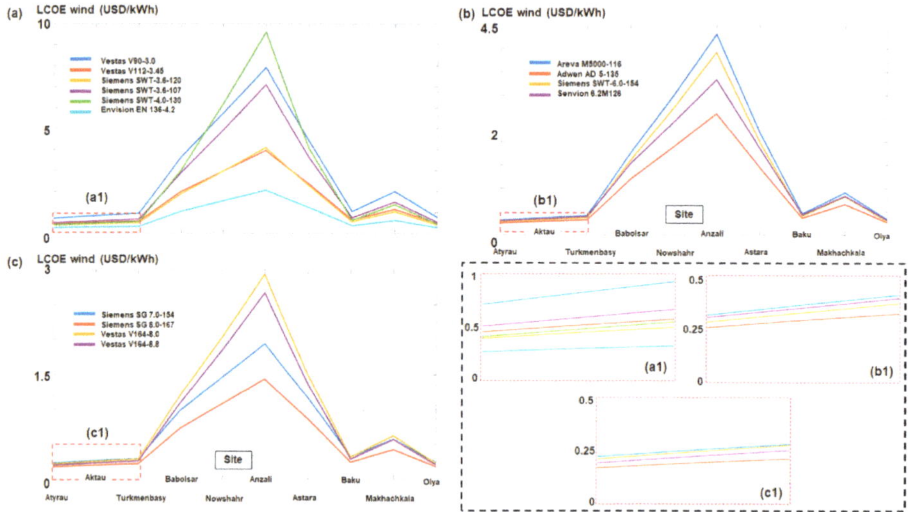

Figure 13. LCOE assessment considering the reference sites located at 5 km from the shore. The results were grouped in small, medium and large systems (**a**–**c**) including a detailed view of the values reported by the sites Atyrau, Aktau and Turkmenbasy (**a**1–**c**1). All the turbines were evaluated for an 80 m hub height.

The best sites to develop wind projects are close to Atyrau (north) and Olya (northwest), with the mention that the spatial maps indicate a hot-spot south of Olya (close to 44° latitude north) that presents more important wind resources. At this moment, there are plans to develop offshore wind projects in the vicinity of the Baku region (Absheron peninsula). By looking at the wind map, it can be noticed that this region is located in an area with moderate wind resources, where the wind speed conditions do not increase heading towards the offshore region. This means that a wind project located close to the shore will be preferable as the energy output will not significantly increase. Also to be considered is the development of a project along the shoreline in order to avoid the problems associated with marine areas, such as weather windows [37].

As in most of the regions located in the northern hemisphere, the wintertime presents more energetic wind resources [38,39], this being also the case for the Caspian Sea. During December, January and February, the entire northern area has been defined by important wind resources, which constantly indicated average wind speed values of 8.5–9 m/s. During the summer time, the areas suitable for a renewable project are significantly reduced, being possible to register a maximum of 7 m/s only in the southern part of Olya site.

As for the wind turbine performances, different generators were considered for assessment using as a reference the sites of Atyrau and Baku. The annual electricity production is directly related to the rated capacity of the turbines, but even so, an 8 MW system will perform much better than an 8.8 MW generator. By considering different hub heights, the electricity output for different configurations was established, being reported a maximum production of 27.61 GWh and 19.29 GWh for the Atyrau and Baku site, respectively. To obtain a higher capacity factor, probably a system rated at 4.2 MW or 5 MW will be more indicated, having reported values close to 50% for the Atyrau sites, and between 25% and 33% for the Baku location.

Finally, the Caspian Sea is an important area in terms of energy resources, being well known for the hydrocarbon extractions [40]. Nevertheless, there is interest for natural resources, and the development of a wind farm which will represent one-step forward for the development of a renewable portfolio, as it can be designed according to the electricity demand of the coastal communities from this region.

Author Contributions: F.O. performed the literature review, processed the wind data, carried out the statistical analysis and interpreted the results. E.R. guided this research, wrote the manuscript and drawn the conclusions. The final manuscript has been approved by all authors.

Funding: This work was carried out in the framework of the research project REMARC (Renewable Energy extraction in MARine environment and its Coastal impact), supported by the Romanian Executive Agency for Higher Education, Research, Development and Innovation Funding—UEFISCDI, grant number PN–III–P4–IDPCE–2016–0017.

Acknowledgments: ECMWF ERA-Interim data used in this study have been obtained from the ECMWF data server. The altimeter products were generated and distributed by Aviso (http://www.aviso.altimetry.fr/) as part of the SSALTO ground processing segment.

Conflicts of Interest: The authors declare no conflicts of interest.

Nomenclature

T	Avera–year (8760 hr/year)
r	discount rate
$U10$	wind speed reported for a 10 m height above sea level
$P(u)$	power curve of a wind turbine
P_E	electric power expected to be generated
R_P	rated power of the system
OPEXt	Operational expenditures reported for year t
CAPEX	Capital Expenditure
LCOE	Levelized Cost of Energy
AEP	Annual Electricity Production
NaN	Not A Number
$f(u)$	Weibull probability density function
ECMWF	European Centre for Medium-Range Weather Forecasts
Cf	Capacity factor
AVISO	Archiving, Validation and Interpretation of Satellite Oceanographic Data
$z_{10}; z_{80}$	reference heights
z_0	roughness of the sea surface

References

1. Yang, W.; Tian, W.; Hvalbye, O.; Peng, Z.; Wei, K.; Tian, X. Experimental research for stabilizing offshore floating wind turbines. *Energies* **2019**, *12*, 1947. [CrossRef]
2. Ishie, J.; Wang, K.; Ong, M. Structural dynamic analysis of semi-submersible floating vertical axis wind turbines. *Energies* **2016**, *9*, 1047. [CrossRef]
3. Offshore Wind in Europe—Key Trends and Statistics 2018. Available online: https://windeurope.org/about-wind/statistics/offshore/european-offshore-wind-industry-key-trends-statistics-2018/ (accessed on 8 March 2019).
4. Klinge Jacobsen, H.; Hevia-Koch, P.; Wolter, C. Nearshore and offshore wind development: Costs and competitive advantage exemplified by nearshore wind in Denmark. *Energy Sustain. Dev.* **2019**, *50*, 91–100. [CrossRef]
5. Vieira, M.; Henriques, E.; Amaral, M.; Arantes-Oliveira, N.; Reis, L. Path discussion for offshore wind in Portugal up to 2030. *Mar. Policy* **2019**, *100*, 122–131. [CrossRef]
6. Rusu, E.; Onea, F. Joint evaluation of the wave and offshore wind energy resources in the developing countries. *Energies* **2017**, *10*, 1866. [CrossRef]
7. Ganea, D.; Mereuta, E.; Rusu, E. An evaluation of the wind and wave dynamics along the european coasts. *JMSE* **2019**, *7*, 43. [CrossRef]
8. Rusu, E.; Onea, F. Evaluation of the wind and wave energy along the Caspian Sea. *Energy* **2013**, *50*, 1–14. [CrossRef]
9. Zountouridou, E.I.; Kiokes, G.C.; Chakalis, S.; Georgilakis, P.S.; Hatziargyriou, N.D. Offshore floating wind parks in the deep waters of Mediterranean Sea. *Renew. Sustain. Energy Rev.* **2015**, *51*, 433–448. [CrossRef]
10. EolMed—The First Offshore Wind Farm in the Mediterranean Sea. Available online: http://www.eolmed.fr/en/ (accessed on 29 May 2019).
11. Italy on Pole in Race for First Mediterranean Offshore Wind | Recharge. Available online: https://www.rechargenews.com/wind/1712851/italy-on-pole-in-race-for-first-mediterranean-offshore-wind (accessed on 29 May 2019).
12. Onea, F.; Deleanu, L.; Rusu, L.; Georgescu, C. Evaluation of the wind energy potential along the Mediterranean Sea coasts. *Energy Explor. Exploit.* **2016**, *34*, 766–792. [CrossRef]
13. Amirinia, G.; Kamranzad, B.; Mafi, S. Wind and wave energy potential in southern Caspian Sea using uncertainty analysis. *Energy* **2017**, *120*, 332–345. [CrossRef]
14. Onea, F.; Rusu, E. Efficiency assessments for some state of the art wind turbines in the coastal environments of the Black and the Caspian seas. *Energy Explor. Exploit.* **2016**, *34*, 217–234. [CrossRef]
15. Kerimov, R.; Ismailova, Z.; Rahmanov, N.R. Modeling of wind power producing in Caspian Sea conditions. *Int. J. Tech. Phys. Problems Eng.* **2013**, *15*, 136–142.
16. Alamian, R.; Shafaghat, R.; Miri, S.J.; Yazdanshenas, N.; Shakeri, M. Evaluation of technologies for harvesting wave energy in Caspian Sea. *Renew. Sustain. Energy Rev.* **2014**, *32*, 468–476. [CrossRef]
17. Kamranzad, B.; Etemad-Shahidi, A.; Chegini, V. Sustainability of wave energy resources in southern Caspian Sea. *Energy* **2016**, *97*, 549–559. [CrossRef]
18. Alamian, R.; Shafaghat, R.; Safaei, M.R. Multi-objective optimization of a pitch point absorber wave energy converter. *Water* **2019**, *11*, 969. [CrossRef]
19. Tavana, M.; Behzadian, M.; Pirdashti, M.; Pirdashti, H. A PROMETHEE-GDSS for oil and gas pipeline planning in the Caspian Sea basin. *Energy Econ.* **2013**, *36*, 716–728. [CrossRef]
20. Mityagina, M.; Lavrova, O. Satellite survey of Inner Seas: Oil pollution in the Black and Caspian Seas. *Remote Sens.* **2016**, *8*, 875. [CrossRef]
21. Efendiyeva, I.M. Ecological problems of oil exploitation in the Caspian Sea area. *J. Pet. Sci. Eng.* **2000**, *28*, 227–231. [CrossRef]
22. INTERVIEW—Azerbaijan's 200MW Offshore Wind Plan May be Tweaked After Feasibility Study. Available online: /news/interview-azerbaijans-200mw-offshore-wind-plan-may-be-tweaked-after-feasibility-study-503906/ (accessed on 30 May 2019).
23. Wind Power in Azerbaijan. REVE, Archives: Wind energy in Azerbaijan. Available online: https://www.evwind.es/tags/wind-energy-in-azerbaijan (accessed on 30 May 2019).

24. Rusu, E.; Onea, F. An assessment of the wind and wave power potential in the island environment. *Energy* **2019**, *175*, 830–846. [CrossRef]
25. Bathymetric Data Viewer. Available online: https://maps.ngdc.noaa.gov/viewers/bathymetry/ (accessed on 31 May 2019).
26. Home. United Nations. Available online: https://www.un.org/en/ (accessed on 31 May 2019).
27. Poli, P. Browse Reanalysis Datasets. Available online: https://www.ecmwf.int/en/forecasts/datasets/browse-reanalysis-datasets (accessed on 31 May 2019).
28. Wan, Y.; Fan, C.; Dai, Y.; Li, L.; Sun, W.; Zhou, P.; Qu, X. Assessment of the joint development potential of wave and wind energy in the South China Sea. *Energies* **2018**, *11*, 398. [CrossRef]
29. Onea, F.; Rusu, L. Evaluation of some state-of-the-art wind technologies in the nearshore of the Black Sea. *Energies* **2018**, *11*, 2452. [CrossRef]
30. Onea, F.; Rusu, L. A Study on the wind energy potential in the Romanian coastal environment. *J. Mar. Sci. Eng.* **2019**, *7*, 142. [CrossRef]
31. Welcome to Wind-Turbine-Models.com. Available online: https://en.wind-turbine-models.com/ (accessed on 20 May 2018).
32. Rusu, E.; Onea, F. A parallel evaluation of the wind and wave energy resources along the Latin American and European coastal environments. *Renew. Energy* **2019**, *143*, 1594–1607. [CrossRef]
33. MSWH/MWind: Aviso+. Available online: https://www.aviso.altimetry.fr/en/data/products/windwave-products/mswhmwind.html (accessed on 18 August 2018).
34. Vagiona, D.G.; Kamilakis, M. Sustainable site selection for offshore wind farms in the South Aegean—Greece. *Sustainability* **2018**, *10*, 749. [CrossRef]
35. Njomo-Wandji, W.; Natarajan, A.; Dimitrov, N. Influence of model parameters on the design of large diameter monopiles for multi-megawatt offshore wind turbines at 50-m water depths. *Wind Energy* **2019**, *22*, 794–812. [CrossRef]
36. Lerch, M.; De-Prada-Gil, M.; Molins, C. The influence of different wind and wave conditions on the energy yield and downtime of a Spar-buoy floating wind turbine. *Renew. Energy* **2019**, *136*, 1–14. [CrossRef]
37. O'Connor, M.; Lewis, T.; Dalton, G. Weather window analysis of Irish west coast wave data with relevance to operations 82 maintenance of marine renewables. *Renew. Energy* **2013**, *52*, 57–66. [CrossRef]
38. Onea, F.; Rusu, E. Wind energy assessments along the Black Sea basin. *Meteorol. Appl.* **2014**, *21*, 316–329. [CrossRef]
39. Onea, F.; Raileanu, A.; Rusu, E. Evaluation of the wind energy potential in the coastal environment of two enclosed seas. *Adv. Meteorol.* **2015**. [CrossRef]
40. Clem, R.S. Energy in the Caspian region: Present and future. *Eurasian Geogr. Econ.* **2002**, *43*, 661–662. [CrossRef]

 *energies*

Article

Volume-Based Assessment of Erosion Patterns around a Hydrodynamic Transparent Offshore Structure

Mario Welzel *, Alexander Schendel, Torsten Schlurmann and Arndt Hildebrandt

Ludwig-Franzius-Institute for Hydraulic, Estuarine and Coastal Engineering, Leibniz Universität Hannover, 30167 Hannover, Germany
* Correspondence: welzel@lufi.uni-hannover.de

Received: 1 July 2019; Accepted: 6 August 2019; Published: 10 August 2019

Abstract: The present article presents results of a laboratory study on the assessment of erosion patterns around a hydrodynamic transparent offshore foundation exposed to combined waves and currents. The model tests were conducted under irregular, long-crested waves in a scale of 1:30 in a wave-current basin. A terrestrial 3D laser scanner was used to acquire data of the sediment surface around the foundation structure. Tests have been conducted systematically varying from wave- to current-dominated conditions. Different volume analyzing methods are introduced, which can be related for any offshore or coastal structure to disclose physical processes in complex erosion patterns. Empirical formulations are proposed for the quantification of spatially eroded sediment volumes and scour depths in the near-field and vicinity of the structure. Findings from the present study agree well with in-situ data stemming from the field. Contrasting spatial erosion development between experimental and in-situ data determines a stable maximum of erosion intensity at a distance of 1.25 A, 1.25 times the structure's footprint A, as well as a global scour extent of 2.1–2.7 A within the present study and about 2.7–2.8 A from the field. By this means, a structure-induced environmental footprint as a measure for erosion of sediment affecting marine habitat is quantified.

Keywords: offshore wind farm; jacket; scour; wave-current interaction; spatial resolution; erosion patterns; sediment transport; laboratory tests

---

1. Introduction

To meet the rising demand for renewable energy, the expansion of offshore wind energy converters (OWECs) in coastal waters is progressing steadily. Due to continuing technological development, upcoming offshore wind parks will not only utilize larger turbines with a capacity of 10 MW and beyond [1], but also create opportunities to open new available space in larger water depths. As the average water depth increases in projected wind parks globally, different construction types are adopted that are more complex and have a larger footprint than commonly used monopiles. However, the installation and operation of those structures, especially if several are closely aligned next to each other, may lead to impacts on the formerly unaffected marine environment in the near- but also in the far-field. Potential impacts [2–4] include large scale morphological changes and entrainment of large quantities of sediment in the water body due to interaction of the structure with ocean currents and waves [5–7]. Of course, the scouring processes might also affect the sustainability of the structure itself over time. Unfortunately, only a limited understanding of environmental impacts and the impairment of the structure's stability over its lifetime due to scouring processes around complex foundation structures exist. This is why for some structures that are affected by scouring, e.g., gravity-based foundations (GBF), the installation of a scour protection system became mandatory. The protection of those structures against the degradation due to scour is often designed following a conservative, and thus, inefficient approach that is based on monopiles. Yet, this evident mismatch may also lead to incorrect prediction of scour depths and unreliable design of scour protection (see Rudolph et al. [8]). This in turn might also impose an effect of superimposing global scouring processes, that possibly contribute to the subsidence of the seabed, in particular around complex structures see, e.g., Rudolph et al. [8] and Baelus et al. [9]. While the equilibrium scour depth around monopile foundations has been investigated and published extensively over the last decades, limited understanding exists for jacket-type foundations, see [8–11]. Even though the bed topography was measured, these studies were more focused on the local scour development for specific conditions rather than on the spatial scour development on a global extent. A literature study on model tests and field studies related to jacket-type foundations can be found in Welzel et al. [12].

Research conducted for groups of circular cylinders represents the basis of knowledge to understand the initiation and development of local and global scour around a hydrodynamic transparent structure like a jacket. For groups of circular cylinders, several studies outlined a dependency between the distance of piles (gap ratio) and the local as well as global scour development [13–15]. Furthermore, it is reported that hydrodynamic interactions between individual circular piles are small if the distance between them exceeds six times the piles' diameter (see e.g. [16–18]). Bolle et al. [10] transferred this knowledge to jacket structures, arguing that the distance was clearly above 6 D in their study and thus global scour does not have to be considered. To gain insights into potential effects of global scouring processes on the marine environment and the structures' stability, spatial seabed changes in the vicinity of the structure need to be measured. Although different measurement techniques and analysis methods were already applied in previous studies, they are rarely used to provide information beyond the calculation of volumes of displaced sediment. Porter [19] used a photogrammetric-based measurement system to analyze the scour hole, depth and shape in tidal currents around a monopile. Margheritini et al. [20] conducted physical model tests for the scour development around monopile foundations in unidirectional and tidal currents. They analyzed the scour volume by means of a laser probe bottom profiler. Stahlmann and Schlurmann [21] conducted small scale, 1:40, as well as large scale, 1:12, physical model tests for a tripod foundation in regular and irregular wave conditions and evaluated the scour development by using either a laser distance bottom profiler (for 1:40) or a multi-beam echo sounder (for 1:12). Hartvig et al. [22] investigated the scour and backfilling processes around a monopile foundation due to steady currents and combined wave-current load. A laser probe bottom profiler was used to obtain bed topography measurements for several time steps. Insights about scour processes and results on scour depth, scour volume and a scour shape factor are derived as a function of time and space. As the spatial investigation of erosion volumes around offshore structures so far has attracted little research interest (also for technical reasons of measurement instruments), few studies exist which may provide a systematic analysis. Studies of Margheritini et al. [20] and Hartvig et al. [22] systematically analyzed erosion processes and introduced a dimensionless erosion volume (normalized with a structural volume). Nevertheless, the lack of a spatial reference (e.g., the related interrogation area of the erosion volume) to the information of eroded sediment volume seems to be an important point missing for a further normalization.

However, several aspects regarding scouring processes around complex offshore structures remain (so far) disregarded and demand a more systematic investigation of erosion patterns. Consequently, the objective of the present paper addresses a systematic volume-based analysis of erosion processes to evaluate the degree and extent of the local and global scour development around a jacket structure. This enables the sediment redistribution footprint of the offshore structure to be deduced in the transition between the near- and the far-field. Therefore, hydraulic model tests have been carried out in the wave and current basin of the Ludwig-Franzius-Institute to conduct a systematic study of erosion processes around a jacket-type offshore foundation under waves, combined waves and current as well as steady current conditions. Different volume-analyzing concepts and calculation methods are introduced, which can be adapted, generally, for any offshore structure or coastal structure to reveal physical processes in complex erosion patterns.

The objectives of this paper are:

(1) The systematic study of global scour patterns in combined waves and current conditions around a jacket foundation.
(2) Gaining further insights into the spatial scouring process around jacket structures with detailed 3D laser scan measurements.
(3) The introduction and application of a novel method to analyze volume-based erosion processes with a spatial reference.
(4) The improvement of prediction methods to account for local and global erosion volumes/scour depths and the extent of global and local scour around jacket type offshore structures.
(5) The quantification of eroded sediment volume, and the determination of areas, which exhibit an increased erosion rate, and therefore, have an impact on the natural dynamics of the ocean floor.

It should be noted that the present physical model tests were part of a fundamental study previously described in [12]. While Welzel et al. [12] focused on local scour depths, measured for a wide range of wave, wave-current and steady current conditions, the present study concentrates on the volume-based assessment of spatial erosion processes in the near-field and vicinity of the structure.

2. Experimental Setup

The physical model tests have been conducted using a jacket-type model in the 3D wave and current basin of the Ludwig-Franzius-Institute, Leibniz Universität Hannover, Germany. The wave basin has a maximum water depth of 1 m, a total length of 40 m and a width of 24 m (see Figure 1). Passive wave absorbers are installed at three sides of the wave basin, resulting in an effective usable length of 30 to 15 m (see Figure 1). An integrated active wave absorption system further reduces reflections. The snake type wave machine consists of 72 wave paddles, allowing generation of regular and irregular waves at angles of 45° between 135° degrees. For the present study a perpendicular wave direction was set to 90° to the current coming from 0° (see Figure 1).

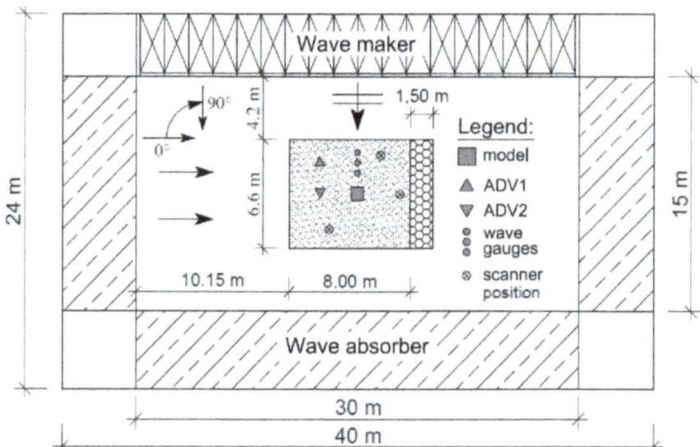

Figure 1. Sketch of the wave basin, plan view including the test setup, unidirectional current is coming from left to right with 0°, waves are propagating perpendicular in 90°.

This superposition angle was chosen to enable a better comparability to other studies as well as to investigate the influence of waves approaching perpendicular to a current on the global scour development. As shown in Figure 2b, the water depth of 0.67 m, representing a water level of 20 m in prototype scale according to the model scale of 1:30, was kept constant during the study. The sediment pit is located in the center of the basin, providing an additional depth of 1.2 m, a length of 8 m and a width of 6.6 m. A sediment trap with a length of 1.5 m at the downstream side of the pit was installed to prevent large amounts of sediment from being transported as bed load into the pump sump. A jacket structure was assembled to physically mimic a generic structure of a jacket-type foundation without considering the influence of post piles or mud-mats. The physical model was constructed using 3D printed parts, which were glued together, sanded down and painted with filler and lacquer in multiple layers to achieve a smooth coating of the surface. The jacket structure has been built in a quadratic cross section consisting of four main piles with a diameter of 4 cm each and with a distance of 0.55 m between them. The piles below the lowest nodes were made out of aluminum and were connected to the bottom of the wave basin. The model was installed in the middle of the sediment pit. As the jacket structure was not rotated during the laboratory experiments, only one orientation of the model with respect to the current and wave direction was investigated in the present study. The model was installed and constructed in a way that the lowest node of the structure was positioned in a distance of one pile diameter D (with D = 4 cm) above the sediment bed, as shown in Figure 2. Consequently, a substantial influence of the jacket structure on the flow and thus on the sediment bed was intended. The sediment pit was filled with sand with a median diameter of $d_{50} = 0.19$ mm. To achieve a good compaction, without entrapped air, the sand was installed in wet condition and levelled with aluminum bars.

Two Acoustic Doppler Velocimeters (ADVs, Vectrino+, Nortek AS, Rud, Norway) measured wave and current induced flow velocities. One was placed 2.5 m upstream (in current direction) from the model (ADV2), and the other one (ADV1) was positioned in line with the wave gauge array.

Both ADVs were installed in a distance of 2.5 D (10 cm) over the sediment surface, vertically (looking down). Additional preliminary tests were carried out to measure the undisturbed current velocity U_c. Furthermore, vertical velocity profiles of horizontal flow components were measured at the location of ADV2 to calculate the undisturbed and depth-averaged current velocity $\overline{U}$. The undisturbed orbital velocity U was measured during previously conducted tests with ADV1 at a distance of 10 cm above the sand level. A terrestrial 3D laser scanner (Focus 3D, FARO, Lake Mary, FL, USA) was used to measure the surface elevation around the model. The FARO Focus 3D laser scanner offers the advantage of high-resolution measurements of up to 70 million data points per scan. In comparison to photogrammetric [20], echo sounder based measurement techniques [23] as well as measurements with laser probe bottom profiler [22], the 3D laser scanning method provides higher resolution and therefore a better accuracy to measure the scour patterns around the structure. The high accuracy of up to ±1 mm is especially important for volume analyzes around small/thin objects (e.g., the present structure with diameters D = 4 cm).

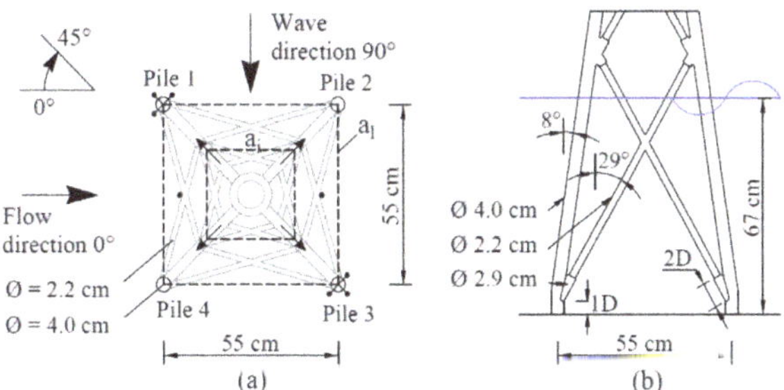

Figure 2. Schematic view of the jacket model (**a**) plan view on the model, including dimensions, the mounted echo sounding transducer, the reference structure footprint a_1 and the increasing interrogation area a_i to compute the erosion volume (**b**) side view of the model with dimensions, angles and water level, D = 4 cm.

As mentioned previously, the present tests were part of a larger investigation, in which the time dependent development of the scour around the jacket at different locations was measured and analyzed (cp. [12]). For this, measurements were carried out by means of echo sounding transducers with a diameter of ~1cm around pile 1 and pile 3 as well as in between the piles. These echo sounders are additionally illustrated in Figure 2. For more detailed information about the experimental setup and procedure refer to Welzel et al. [12].

Experimental Procedure and Test Conditions

Before each test, the sediment bed was smoothed under wet conditions (water depth kept at the level of sediment bed) by making use of aluminum bars. Subsequently the wave basin was carefully filled overnight. No general starvation of the bed (general seabed lowering) has been observed in the present study. Therefore, the measured erosion patterns can be attributed to the presence of the structure. JONSWAP wave spectra were generated until a maximum amount of 6500 waves (or 7 h for test 5) were reached in one test. Studies related on scour around OWECs in combined waves and current exhibit test durations in ranges of 1 to 3 hours (see, e.g., [11,24–26]). Some of those studies did not reach an equilibrium stage of the related scour depth. Therefore, a maximum amount of 6500 waves was chosen in the present study (test durations of 3.7 between 8.3 hours) to ensure the erosion process to reach its equilibrium stage. The test procedure can be summarized as follows:

(1)   Smoothing the sand level and carefully filling the basin to avoid disturbances of the adjusted sand level.
(2)   Running the desired test until the scour process has attained an (almost) equilibrium stage.
(3)   Emptying the wave basin and carefully draining the sand pit to avoid further influence on the scour pattern.
(4)   3D scans of the global sediment surface around the structure.

The maximum value of the undisturbed orbital velocity U_m (see Equation (1)) is calculated with U_{rms}, the root-mean-square value (RMS) of the orbital velocity U at the bottom in the direction of the waves defined as $U_{rms}^2 = \int_0^\infty S(f)df$, with $S(f)$ = power spectrum of U which corresponds to the wave component, with f = frequency. As investigated in [27], the Keulegan-Carpenter number is defined as $KC = U_m T_p / D$, in which T_p = peak wave period and D = pile diameter. The parameter U_{cw} (see Equation (2)) represents a wave current velocity ratio introduced by Sumer and Fredsøe [27] to assess the ratio of undisturbed current U_c to undisturbed wave generated flow velocity U_m:

$$U_m = \sqrt{2} U_{rms} \tag{1}$$

$$U_{cw} = \frac{U_c}{(U_c + U_m)} \tag{2}$$

The Shields parameter, which is based on the current velocity U_c and the orbital velocity U_m, was determined by the approach of Soulsby [28]. The test program consisted of a wave only test, three tests in which different wave and current loads were combined, and a final test with current-only conditions. Current velocities and wave parameters were selected to cover a wide range of U_{cw} and KC numbers. A critical Shields parameter of $\theta_{cr} = 0.049$ was calculated for the sediment of the present study. Considering the Shields parameters given in Table 1, all tests have thus been conducted under live bed conditions.

Table 1. Test conditions/Measured values (waves are propagating in 90° to the current).

Test	H_s [m]	T_p [s]	Bed Orbital Velocity U_m [cm/s]	Depth Averaged Current Velocity u [cm/s]	Current Velocity 10 cm Above Bed U_c [cm/s]	KC [-]	U_{cw} [-]	Shields Parameter θ [-]	Global Eroded Volume: for an Area of 1.25 A V_D [-]	Local Eroded Volume: Diameter of 6 D V_D [-]
1	0.165	4.5	20.8	-	-	23.4	0.00	0.080	−0.49	−0.49
2	0.165	4.5	20.8	11.4	10.1	23.4	0.33	0.085	−14.14	−10.62
3	0.158	3.4	17.5	24.3	22.5	14.9	0.56	0.087	−27.26	−18.50
4	0.147	2.0	13.3	41.7	38.8	6.7	0.75	0.123	−43.61	−23.82
5	-	-	-	41.7	38.8	-	1.00	0.084	−55.52	−27.19

* Test 1–5 are related (in the same order) to test 3, 10, 8, 6 and 13 in Welzel et al. [12].

3. Calculation of Erosion Volumes

The determination of dimensionless erosion volumes analyzed over increasing interrogation areas (a_i), enables the spatial assessment of eroded sediment quantities (eroded sediment per surface area) around offshore structures. To enable an improved analysis of erosion processes, in particular of the related erosion patterns, only erosion volumes of sediment are considered within each interrogation area without the additional influence of deposited sediment volume as the deposited sediment patterns would affect the analysis of the considered erosion. Different techniques for the analysis of these volumes are introduced in the following. The application of those techniques is not limited to jacket-type structures as used in this study but can also be adapted for other complex structures to reveal physical processes behind complex erosion patterns. To avoid optically shadowed interrogation areas in close proximity of the structure, scans were taken from three different angles and subsequently merged into a single 3D point cloud, providing a high-resolution bed topography in a point density of 100 data points per cm^2 in the carefully drained wave basin around and beneath the structure. No changes in the sediment bed have been observed while slowly draining the sediment pit. To ensure the same spatial reference for every laser scan taken, 6 reference spheres were set in preparation of every measurement. Each merged measurement was pre-processed to reduce the number of outliers as well as to cut unnecessary data points. The pre-processed point cloud data were exported and further processed with MATLAB®. The irregular spaced point cloud data were interpolated by a cubic Delaunay triangulation to calculate the Z value of each data point and converted into a regular 3D mesh grid in a 2 mm resolution. Similar as proposed by Vosselman [29] a slope-based filtering method was implemented to reduce remaining outliers. The x- and y- coordinate system was centered to the middle of the structure. Erosion volumes $V_{erosion}(a_i)$ located in a considered interrogation area a_i, were calculated by subtracting the digital elevation model (DEM) from the reference level at the beginning of a test. As stated by Raudkivi and Ettema [30] for the case of monopiles, the extent and shape of a scour hole is related to the pile diameter, but can also be expressed by a dimensionless volume (see Margheritini et al. [20], Hartvig et al. [22]). Accordingly, the dimensionless eroded volume within a certain area a_i is defined in this study as:

$$V_{D,i} = \frac{V_{erosion}(a_i)}{n\,D^3} \tag{3}$$

in which n is the number of piles, a_i a rectangular area (see Figure 3) and D the pile diameter. The investigated erosion patterns are a result of the complex flow and scouring mechanism, which are briefly described in Section 4.1. In the present study the global erosion volume is related with the four main piles of the jacket ($n = 4$) for global erosion calculations and $n = 1$ for local erosion analyses in the near-field of an individual single pile.

In contrast to the erosion volume $V_{D,i}$, the cumulative erosion volume $V_{A,i}$ represents a volume which is related to a normalized ratio of the area a_i/a_1, with a_1 being the area of the structure's footprint. By considering monotonously increasing areas around the structure, the development of the erosion volume with increasing distance from the center of the structure can be evaluated. This approach enables both an insight into the spatial extent and a quantitative measure how the erosion process translates from local to global patterns. By this means also the distance to which the structure has a quantifiable influence on the mobile seabed can be projected. The cumulative erosion volume is defined as:

$$V_{A,i} = \frac{V_{D,i}}{a_i/a_1} \tag{4}$$

$V_{A,i}$ and $V_{I,i}$ both provide a quantity of an erosion volume in reference to a specific area (see Figure 3). The incremental erosion volume $V_{I,i}$ is representing the net gradient volume ($V_{D,i} - V_{D,i-1}$) related to a corresponding ratio of areas $a_i/a_1 - a_{i-1}/a_1$. In addition to $V_{A,i}$, the incremental erosion volume $V_{I,i}$ provides information on the variation of erosion volumes between two individual areas a_i and a_{i-1}. Thereby, it is possible to assess and analyse the volumetric change of sediment with the erosion intensity as eroded volume per area with an increasing distance from the center of the structure. The incremental erosion volume is given by:

$$V_{I,i} = \frac{V_{D,i} - V_{D,i-1}}{a_i/a_1 - a_{i-1}/a_1}$$

(5)

The incremental erosion volume $V_{I,i}$ and the incremental erosion depth $D_{V,i}$ are representing both a net gradient volume, relating to an area between two adjacent interrogation areas a_i and a_{i-1}. The incremental erosion volume $V_{I,i}$ Equation (5) is directly related to Equations (3) and (4) and represents an erosion volume in relation to an area. On the other hand, the incremental parameter $D_{V,i}$ depicts an erosion depth for an interrogation area, calculated by normalisation with the pile diameter. This enables a comparison between scour depths and bathymetric surface data as well as a quantification of local and global scour extends for the design and prediction of the foundation structure:

$$D_{V,i} = \left(\frac{V_i - V_{i-1}}{a_i - a_{i-1}}\right)/D$$

(6)

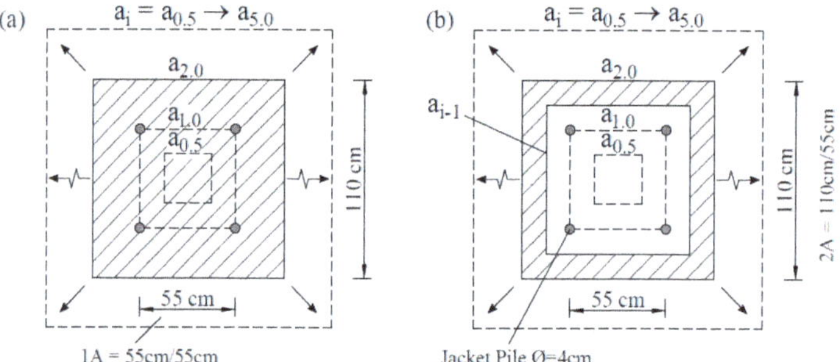

Figure 3. Schematic sketch of increasing rectangular interrogation areas a_i around the present model, (**a**) illustrated exemplarily for an area $a_i = a_2$, related to Equation (4) ($a_2 = 1.1 \times 1.1$ m, shaded area); (**b**) as well as increasing incremental areas, related to Equation (5). Interrogation areas a_i are centered to the structure, starting at $a_{0.5}$ up to a_5.

4. Results

4.1. Changes in Bed Topography

It is reasonable to assume from current understanding of scouring and scour extent, that the near bed flow acceleration around a jacket structure is focused along the individual piles. This in turn leads to a mobilization and transport of the sediment bed. Furthermore, it can be assumed that structural elements as braces which are close to the seabed are generally causing additional vortex shedding and streamline contraction, also leading to a potential increase of bed shear stresses, and thus, sediment mobilization. The interaction of structural elements, flow and sediment bed is particularly pronounced if the structural elements are located close to the seabed, as shown in Welzel et al. [31]. To visualize the impact of the structure on the spatial erosion and deposition of sediment, Figure 4 presents an exemplary photo of the model setup and the final scour pattern after test 4 (additional photos of tests 2–5 are provided and described in Welzel et al. [12]).

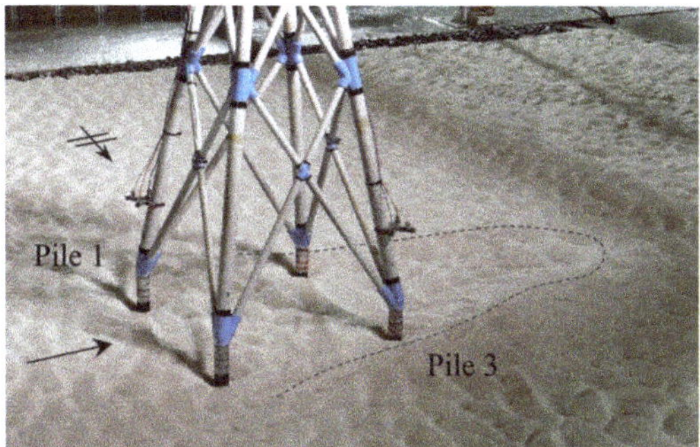

Figure 4. Exemplary photo of the model setup and scour pattern after test 4; the dashed line illustrates the extent of the global scour, current is coming from left to right with 0°, waves are propagating perpendicular in 90°.

In the following, the measured bed topography of tests 1–5 are described and graphically illustrated. Figure 5 is divided in to two color representations, color-coding on the left is optimized to differentiate between global (green) and local erosion (blue) as well as deposition (red). Whereas the colorbar on the right side is optimized regarding different elevations, using an HSV colourmap with an additional computed light source. Distances along the x and y axis are given as a dimensionless value times the structure footprint spacing (see Figure 5), defined in the following as A (A = x or y distance/structure footprint reference distance; e.g., 0.5 A = 0.275 m/0.55 m). Test 1 was conducted under wave only conditions ($U_{cw} = 0$), with waves approaching from 90°. For this condition, an overall deposition of sediment around and beneath the structure was observed (Figure 5a). However, reasonable magnitudes of local scour depths were measured around the piles. Similar to those observed in the local scour development (cp. Welzel et al. [12]), measurements of the present study in general confirm a low erosion, in particular for globally affected areas.

Figure 5b illustrates results stemming from test 2, which is conducted under wave dominated flow conditions ($U_m = 20.8$ cm/s, KC = 23.4, $U_{cw} = 0.33$) with a superimposed current of $U_c = 10.1$ cm/s. The 3D scan reveals short-crested sand ripple migration in the wave direction with slightly longer crests compared to test 1. As the same wave spectra were used as in test 1, the change in bed topography indicates that the superimposed steady current of $U_c = 10.1$ cm/s has a significant impact on the erosion of sediment around the structure. In contrast to test 1, full coverage of the spatial erosion processes is now visible, which is confined to the area beneath the structure. On the other hand, a large deposition of sediment formed at the lee side of the structure (in terms of current flow direction). The extent of the sediment deposition might be a result of the waves, which prevent the sediment from settling down and instead re-distribute the sediment over a relatively large area. It is assumed that this erosion and deposition pattern is a direct consequence of the sediment being picked-up and entrained by the waves and then being transported downstream by the current. These processes cumulate in an increase in the global scour, which reached a maximum global scour depth of about $D_V = -0.3$ D in a distance of 1.2–1.25 A.

The 3D scan of test 3 (Figure 5c) reveals a long-crested sand ripple migration into the direction of the progressing wave. The test was conducted with a slightly smaller orbital velocity ($U_m = 17.5$ cm/s, KC = 14.9) and an increased current velocity ($U_c = 22.5$ cm/s) compared to the tests before. This led to a wave current velocity ratio of $U_{cw} = 0.56$. This particular pattern reveals that erosion of sediment was taking place mainly in the direction of wave propagation. This might be a result of the combination of flow contraction stemming from the current component and vortices induced by waves that dominate the local sediment transport process. Nevertheless, in relation to the current direction, deposition of sediment was found on the lee side and at a small distance from the structure, confined to a long-crested dune-like sediment accumulation. For this test, an erosion depth up to $D_V = -0.54$ D was measured for a distance of approx. 1.2–1.25 A.

Test 4 (Figure 5d) was conducted under current-dominated conditions ($U_{cw} = 0.75$). Here, sand ripples with comparable long crests and smaller heights (as compared to steady current conditions) migrated in the current direction. Measurements for test 4 indicate a more globally affected erosion process in particular in between the structure's footprint area. Laterally distributed erosion areas, emerging from the upstream located piles (pile 1 and 4), indicate that the bed topography is influenced by structure-induced near-bed vortices that arise from the front piles and might be influenced by perpendicular approaching orbital wave motion. Sediment is deposited behind the structure, elongated in the current direction. The global scour in between the structure footprint (areas ≤ 1 A) is increased significantly ($D_V = -0.6$ D) in comparison to tests 1–3. The incremental erosion depth value increases on up to $D_V = -0.71$ D over a distance of approximately 1.2–1.25 A.

Figure 5e illustrates test 5, conducted under steady current conditions without the presence of waves for a flow velocity of $U_C = 38.8$ cm/s ($U_{cw} = 1.0$). The absence of the superposed orbital wave motion led to shorter crested sand ripples migrating in the current direction. Ripples are generally higher and longer than those under combined wave-current load (see Figure 5d). A comparison with studies of [32] and [33] shows a similar bedform of ripples under current only conditions, indicating a fully developed ripple length and height. The 3D scan, depicted in Figure 5d reveals a globally affected erosion pattern with comparable high scour depths ($D_V = -0.75$ D) in between the structure footprint (areas ≤ 1 A) and a maximum global scour depth of -0.84 D (for 1.2–1.25 A), which is further increased in comparison to test 4. The eroded sediment is deposited behind the structure elongated over a longer distance than shown in test 4.

Present results show that the extent and distribution of the spatial scouring process depends on the hydraulic conditions, i.e., whether the flow is current, or wave dominated. For wave dominated conditions the oscillating flow induced by irregular waves is leading to backfilling of once eroded areas, and thus also to a considerably lower erosion rate, while a more current dominated flow is leading to a more constant bed load and suspended load transport in downstream direction. Therefore, wave dominated conditions of the present study lead to smaller global scour depths, whose pattern seems to align with the direction of wave propagation. Current dominated conditions cause a deeper global scour, whose largest intensity can be found between the individual piles of the jacket structure.

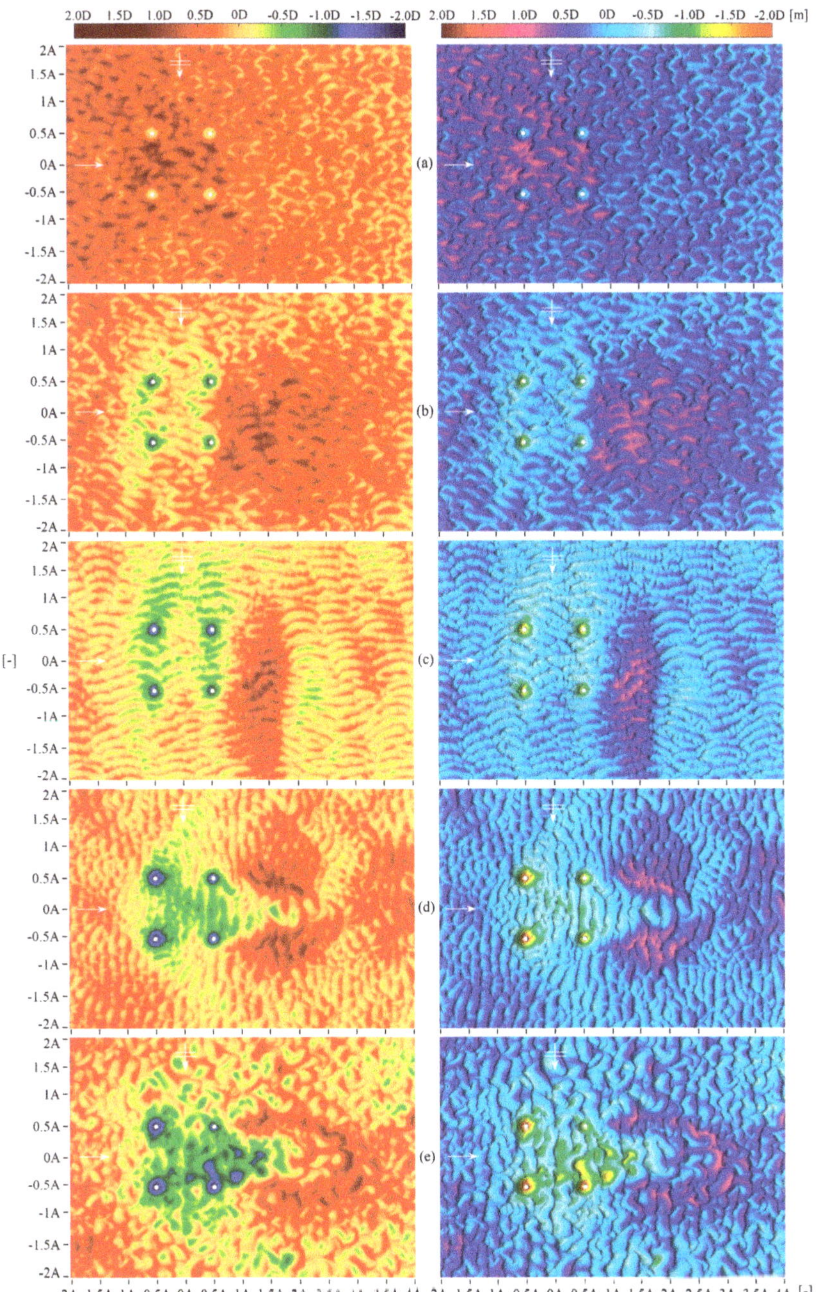

Figure 5. Bed topography measured after test 1–5 (**a–e**), top view on erosion and deposition around the jacket structure, current is coming from left to right, waves are propagating in 90° to the current, x and y distances given times the structure footprint length of 0.55 m, 1 A = 0.55 m/0.55 m [-]; (**left**) optimized illustration to differ between global (green) and local (blue) erosion depths as well as deposition (red); (**right**) HSV colourmap to differentiate between different elevations.

4.2. Analysis of Global Erosion Volumes

Figure 6a illustrates the dimensionless eroded sediment volume $V_{D,i}$ with increasing distance from the center of the structure. A general trend of increasing erosion volumes depending on the wave current velocity ratio U_{cw} of each test is depicted. This correlation is in agreement with the observations on the bed topography around the structure as illustrated in Figure 5. The volume of eroded sediment is increased (independently of U_{cw}) with increasing distance from the structure as expected. However, the rate of volume growth with increasing distance is not constant see Figure 6a. Instead, the volume of eroded sediment is increased rapidly in areas in close proximity to the structure and loses its influence with distance from the jacket. Again, this might be expected, as the amount of eroded volume near the structure is mainly controlled by the local scouring process around the piles. With growing distance from the structure, the influence of the global scouring process on the erosion volume diminishes. Thus, the inconsistent development of $V_{D,i}$ is an indication for different erosion processes taking place, the change in the dominance between those processes and thus also an indication for different erosion intensities (eroded sediment volume per area) around the structure.

In addition, to elucidate the dependency of the erosion process on the distance from the structure, Figure 6b illustrates the cumulative erosion volume $V_{A,i}$ as a function of A. Test 1–3 have been conducted under wave current velocity ratios of $U_{cw} = 0$–0.56. In tests 1–3, wave spectra with higher orbital velocities were studied, leading to a more wave dominated erosion pattern with lower values of $V_{A,i}$. In comparison, test 4–5 have been conducted under current dominated conditions, leading to higher magnitudes of $V_{A,i}$. To allow a comparison between each test, whether a more global or more local erosion process dominated the morphodynamic regime, values around the main piles for approx. 1–1.25 A are compared with magnitudes in between the structure footprint 0.5–1 A, which is more affected by global erosion processes. Therefore, the representation of the cumulative erosion volume $V_{A,i}$ reveals a more locally pronounced erosion of sediment around the main piles for test 1–3, as values in between the structure 0.5–1 A are significantly smaller than the maximum value observed in a distance of 1.25 A around the main piles. While measurements for tests 4 and 5 are indicating a more globally affected erosion process as values in between the structure footprint (0.5–1 A) are exhibiting magnitudes of a similar values than the maximum erosion value (1.25 A). However, Figure 6b shows a stable peak at about ~1.25A for each test. Here, the local scour around each pile is superimposed with the global erosion pattern around the foundation structure. Areas in between the structure footprint (A < 1) show larger fluctuations as they are further away from the more stable local scour and normalized over smaller areas. Subsequently, Figure 6b indicates a similar decrease of the cumulative erosion volume $V_{A,i}$ for gradients beyond 1.25 A.

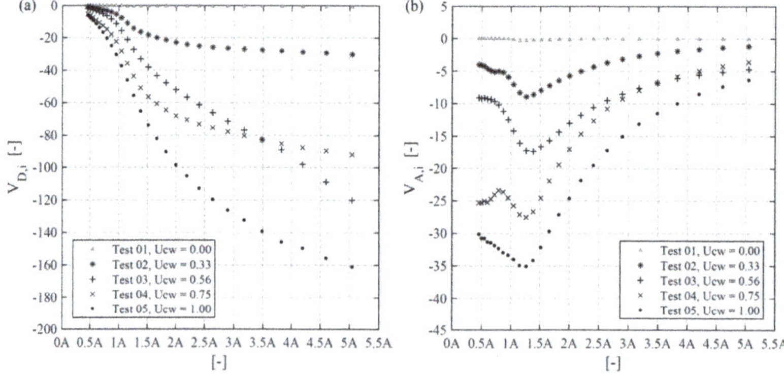

Figure 6. (a) Dimensionless eroded sediment volume $V_{D,i}$ (see Equation (3)) as function of the distance from the center of the structure A; (b) cumulative eroded volume $V_{A,i}$ (see Equation (4)) in dependency to the dimensionless distance A given times the structure footprint of 0.55 m, 1 A = 0.55 m/0.55 m [-].

As deposition of sediment is not considered, the decrease of eroded volume per area might be explained with a general decrease of flow contraction and disturbances due to vortices. The aforementioned peak of the cumulative eroded volume $V_{A,i}$ at around 1.25 A is compared to U_{cw} in Figure 7. Obviously, a slight current superimposed on waves causes the erosion volume to increase significantly. With further increasing values of U_{cw} the maximum eroded sediment volume increased as well. Furthermore, the comparison of values measured in test 4 and 5 shows the addition of irregular waves reduce the amount of eroded sediment. Overall, Figure 7 reveals a rather strong dependency of the maximum eroded volume on U_{cw}. The maximum of eroded sediment $V_{A,max}$ can be described as a function of the wave current velocity ratio as follows:

$$V_{A,max} = -1.3(0.1 + \exp(-4.6\,U_{cw}))^{-1.5} \tag{7}$$

Equation (7) enables the calculation and prediction of the maximum cumulative erosion volume $V_{A,max}$, which appears to be found in a distance of about 1.25 times the structure footprint. Thus, $V_{A,max}$ quantifies the maximum erosion volume depending on U_{cw} and is convertible into the parameters introduced in Section 3 (Equations (3)–(6)). The knowledge of this distance, as well as of the values of the eroded sediment at this point, might be of practical use for the prediction of scour and the design of a scour protection system in a graduated intensity starting at $V_{A,max}$. By this means, it might be possible to adapt areas with different erosion intensities to different scour protection areas (e.g., different stone sizes) with a reference on the maximum erosion intensity at $V_{A,max}$. Due to the limited number of tests, the influence of different KC numbers on the transition between wave and current dominated hydrodynamic conditions could not be studied as of now. However, it is expected, that a wide range of U_{cw} and KC values would lead to an array of non-dimensional erosion volume curves comparable to the transition of wave and current dominated scour depth at a jacket structure, see Welzel et al. [12] as well as originally developed for cylindrical piles (see e.g., [14,26,27,34–36]).

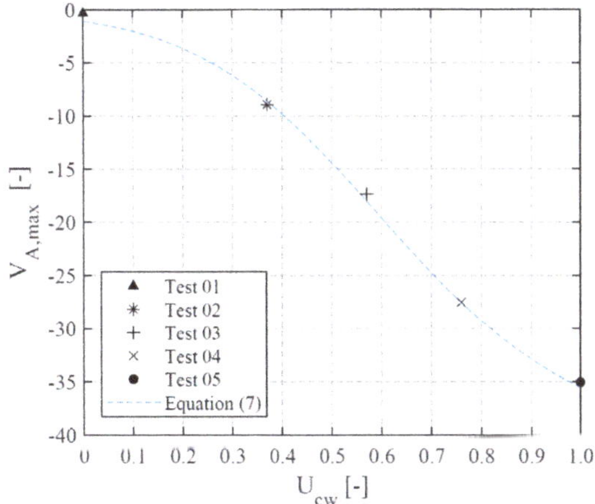

Figure 7. Maximum cumulative erosion volume $V_{A,max}$ (see Equation (4)) of test 1–5, depicted over the wave current velocity ratio U_{cw} (see Equation (2)) in comparison to Equation (7) (see also Figure 6b; ~1.25 A).

Figure 8 illustrates the cumulative erosion volume $V_{A,i}$ (see also Figure 6b) normalized with the maximum value $V_{A,max}$ for each test to further elaborate the differences in the development of erosion volumes over the distance from the center of the structure. Starting from a distance of 1.25 A, this representation reveals a rapidly increasing dependency on hydraulic conditions with decreasing distance. Therefore, values corresponding to test 1–3 reveal lower magnitudes in between the structure footprint (0.5–1 A), while values of test 4–5 show magnitudes (0.5–1 A) similar to $V_{A,max}$ around the main piles. On the other hand, for distances larger than 1.25 A, the dependency of erosion volumes on the hydraulic condition is much less pronounced. Values for distances > 1.25 A reveal a similar decreasing trend for gradients of eroded sediment. Figure 8 also indicates that the global scouring process is more affected by a change of hydraulic conditions and the interaction of the structure than by the local erosion process, which is limited to approximately 1–1.25 A. An exception is test 1, where large amounts of sediment were deposited near the structure. Test 1 was considered as an outlier for the derivation of Equations (9) and (10) as the test was influenced due to sediment deposition around the structure (see Section 4.1). A combination of Equation (7) and additional terms, that relate to the increase and decrease of erosion volume with change in U_{cw} and distance to structure, leads to a general description of non-dimensional areal erosion volumes for the present study:

$$V_{A,i} = V_{area} - 1.3(0.1 + \exp(-4.6\ U_{cw}))^{-1.5} \tag{8}$$

with:

$$V_{area\ A\ <\ 1.25A} = (0.1 + \exp(-10\ A + 8))^{-0.9} B - C \tag{9}$$

$$V_{area\ A\ >\ 1.25A} = -2.2\ \exp(-0.7\ A) - 0.11 \tag{10}$$

In which $V_{A,i}$ is calculated for the development over the wave current velocity ratio U_{cw}, multiplicated with a factor V_{area} in relation to the size of the area to consider for areal volume differences. Equation (9) accounts for the development of distances < 1.25 A, with B = $(-5.2\ U_{cw} + 6.9)10^{-2}$ and C = $(3.8\ U_{cw} + 4.9)10^{-1}$ to account for different wave current velocity ratios (U_{cw}), while Equation (10) describes the decreasing trend of dimensionless erosion volumes for areas > 1.25 A in dependency to the size of the considered area, which is given with the dimensionless distance A (structure footprint = 1 A).

Equation (8) (approx. R^2 = 0.87) allows the calculation of cumulative erosion volumes $V_{A,i}$ which makes it possible to quantify the intensity of erosion in reference to the spatial extent. Values calculated with Equation (8) can be converted into parameters introduced in Section 3, Equations (3)–(6). In consequence, it is possible to estimate sediment volumes or scour depths in relation to the hydraulic condition or distance from the center of the structure. The calculated erosion volumes can give important information about a possible impact of a structure or a wind park on the natural dynamics of the ocean floor environment. The knowledge of the value of eroded sediment around and within the foundation structure, thus is also of practical use for the prediction of scour depths or eroded sediment volumes, in example for the design of a structure or a scour protection system around such complex foundation structures.

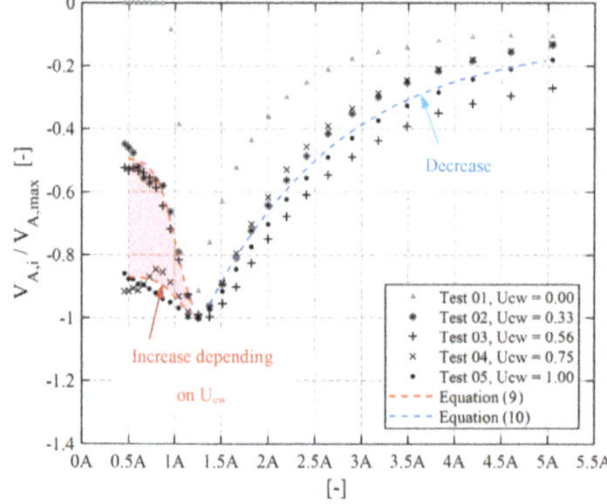

Figure 8. Cumulative erosion volume $V_{A,i}$ (see Equation (4)), normalized with the maximum value $V_{A,max}$ as a function of the distance A to the center of the structure, compared with predicted erosion volumes relating to Equation (9), A < 1.25 A and Equation (10) for A > 1.25 A, R^2 = 0.87 for Equation (8) including Equations (9) and (10).

Figure 9 is illustrating cumulative erosion volumes $V_{A,i}$, compared with calculated values of Equation (8) in relation to the distance of each interrogation area to the center of the structure. Measured areal erosion volumes generally agree with the calculated values relating to Equation (8). Nevertheless, some minor differences can be recognized for the areal development (regarding the representation of Equations (9) and (10)) of test 3 and 4.

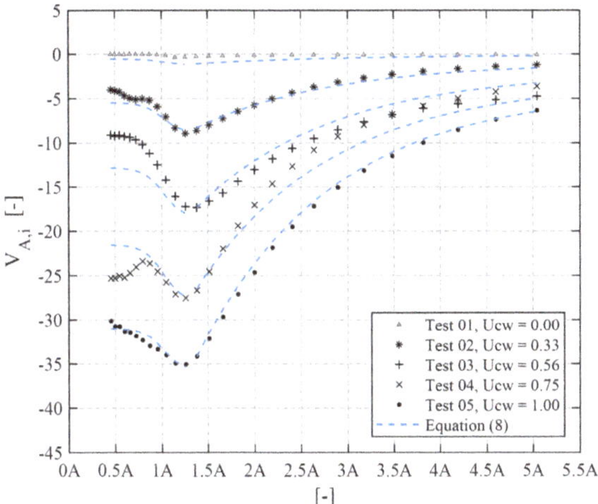

Figure 9. Cumulative erosion volume $V_{A,i}$ (see Equation (4)) as a function of the distance A to the center of the structure, compared with calculated values after Equation (8).

A comparison of erosion volumes for test 3 (see Figure 9) shows lower eroded sediment volumes within areas of 0.5 A up to 1.25 A as well as slightly higher erosion volumes for areas > 1.25 A. In contrast, test 4 reveals a slightly higher erosion intensity (erosion per area) for the areal development within 0.5–1.25 A as well as a lower intensity for areas > 1.25 A for measured compared to predicted values, see Figure 9. Additionally, Figure 10 illustrates the non-dimensional incremental erosion volumes around the jacket structure and compares them with predicted values obtained with Equation (8). The comparison with Equation (8) shows that predicted values (calculated with Equation (8) and converted from $V_{A,i}$ in to $V_{I,i}$) can correctly describe measured values of $V_{I,i}$ of the present study. In contrast to $V_{A,i}$, the incremental erosion volumes $V_{I,i}$ exhibit an enhanced fluctuation near the pile which can be explained with the calculation method related to Equation (5). For a distance from the center of the structure > 1.2 A, erosion rates decrease significantly towards an equilibrium stage at a distance of about 3.5 A. The convergence of erosion volume within all tests might thus be used to define the maximum extent of the global scour around the structure. Due to a natural ripple migration under live bed conditions, a certain erosion rate still remained for areas which were unaffected by the immediate structure's influence on the flow. Therefore, the global scour extent is defined with a threshold of less than 10% of the peak value (at 1.2 A), in reference to the value of an interrogation area, located in a distance of 5 A. For all tests this was the case at a distance between 2.1–2.7 A. As a consequence, the maximum erosion intensity (erosion per area) is reduced by about 90% in this distance. Contrary to approaches which were partly applied to predict global scour development in the past (see e.g. Bolle et al. [10] and Sumer and Fredsøe, [14]), the present study reveals a significant impact of the complex structure on the morphodynamic regime, which extends up to a distance of 2.7 A from the structure.

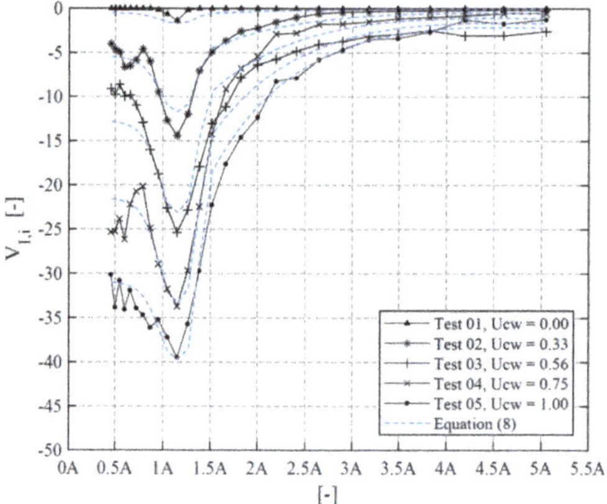

Figure 10. Incremental erosion volume $V_{I,i}$ (see Equation (5)) depicted over the structure footprint length A and compared with calculated values stemming from Equation (8).

To provide a comparison to data obtained in the field, Figure 11 juxtaposes data measured in this study with in-situ field data from Rudolph et al. [8], Baelus et al. [9] and Bolle et al. [10]. As erosion volumes are not available for the field datasets, the data is given in terms of the incremental erosion depth $D_{V,i}$ in Figure 11. Bolle et al. [10] and Baelus et al. [9] presented field data around the same jacket structure (with D_{pile} approx. = 2 m) in the Thornton Bank offshore wind farm located in the southern North Sea. The authors analyzed several scans of the bed topography at various points in time over a period of three years. Rudolph et al. [8] analyzed bathymetric surveys after three years around a wellhead jacket production platform which was founded by additional post piles and near bed braces. A comparison between common prediction approaches for single piles and the scour development observed in [8] led to a factor of 3–4 times the predicted local scour depths, which might be explained by the disturbing effect of additional structural elements close to the seabed. In consequence, the disturbance impact of the foundation structure on the scour development could not be represented by the pile diameter $D_{pile} = 1.2$ m alone. Instead, the influence of additional post piles as well as the horizontal and diagonal braces have to be considered since additional effects increase the local contraction of the flow and influence the vortex system. Therefore, data of Rudolph et al. [8] are compared in Figure 11 with the post pile diameter = 1.2 m, as well as with an artificially increased diameter = 2 m, similar to the one in [9] and [10]. Furthermore, the data shown in Figure 11 are calculated in reference to the post dredging survey given in [9] and [10] as well as a reference value given in [8]. Whereas a direct comparison of magnitudes of erosion depths between the present study and the field measurements is hindered by missing information on hydraulic conditions, the development of erosion depth with changing distance from the center of the structure can be compared. Therefore, it is revealed that the in-situ field data are generally in agreement with the areal distribution of the incremental erosion depth measured in the present study, given in Figure 11 with the dependence to the dimensionless distance A. The comparison shows in particular that the bathymetric surveys, similarly, yield a peak of erosion depth at a distance of around 1.2–1.25 A. In addition, all datasets exhibit a similar development of erosion depths before and after this distance, i.e., a rapid increase of depths towards the structure and a slower decrease with increasing distance.

As the bed topography, especially shown in Baelus et al. [9] was illustrated over an area of A > 3, the global scour extent is analyzed related to the previously introduced definition. The bathymetric surveys of [8] and [9] indicate a reduction of the maximum erosion rate by 90% in a distance of 2.7–2.8 times the structure footprint length (2.7–2.8 A), which is in line with the distance obtained for the data in this study. In this respect, the bathymetric surveys of Bolle et al. [10] show a time dependent context of the global scour development around the jacket foundation as those surveys were conducted 3 and 5 months after the installation. On the other hand, the bathymetric surveys of Rudolph et al. [8] and Baelus et al. [9] were measured three years after the installation of the jacket structure. In addition to the varying and unknown hydraulic conditions, differences between the field data and the test results might stem from a difference in the time dependent stage of the global scour development as well as differences in the structural design.

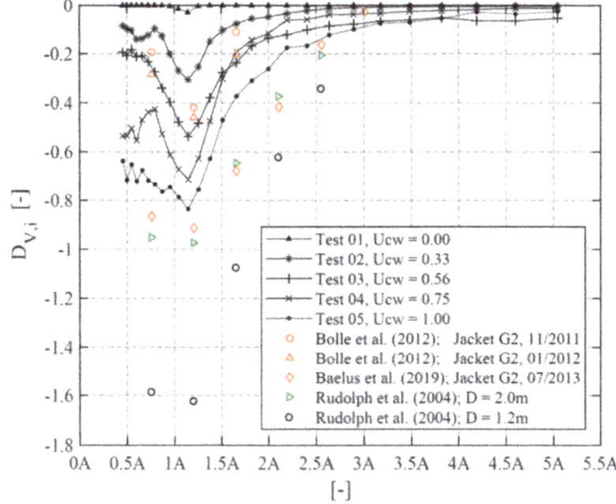

Figure 11. Global erosion of the dimensionless incremental erosion depth $D_{V,i}$ (see Equation (6)) around the jacket structure plotted over the structure footprint A, compared with in-situ field measurements of Rudolph et al. [8], Baelus et al. [9] and Bolle et al. [10].

The previously described comparison illustrates that findings stemming from the present laboratory study, generally agree with data from field studies (surveys of Rudolph et al. [8]; Baelus et al., [9] and Bolle et al. [10]). Furthermore, it is shown that predicted values (calculated with Equation (8)) generally agree with measurements from the present study. It is shown that the introduced equations and methods generally account for the spatial extent of global erosion volumes and scour depths around the compared jacket structures, also for jacket structures which have a non-symmetric footprint as shown with the compared field data of Rudolph et al. [8]. Additionally, it is revealed that it is possible to determine areas which exhibit or exceed a certain erosion rate. A similar areal distribution of eroded sediment volume as well as a global scour extent in a range of the present study (2.1–2.7 A) is found with a comparison of in-situ data (2.7–2.8 A). The comparison between the present study and in-situ field measurements reveals a significant impact on the near field (A < 1) but also on the morphodynamic regime close to the structure (2.7 > A > 1). Furthermore, the analysis proves, that jacket structures, known as "hydrodynamic transparent" can also cause global scouring under certain hydrodynamic conditions.

4.3. Local Scour around Individual Piles

In this section the local scour around each pile is analyzed thoroughly and compared with the global erosion on the near-field bed topography around the structure. For this, local and global erosion volumes are analyzed and defined. In contrast to definitions given in previous paragraphs, related interrogation areas a_i are arranged in a circular pattern and are related to a distance times the pile diameter (see Figure 12).

Figure 13 shows a top view of the measured bed topography over an area of 7 D around each pile. X and Y coordinates refer to the center of each individual pile and are given in multiples of the diameter. This allows a direct comparison between the bed topography and the analyzed erosion depths $D_{V,i}$. A shadowing effect of the jacket piles, which could not be eliminated with the present filtering method, seemed to have only marginal influence on the computed surface elevation near the piles. This is observed for example in Figure 13e around pile 1. The influences were found to be negligible for the calculated erosion volumes.

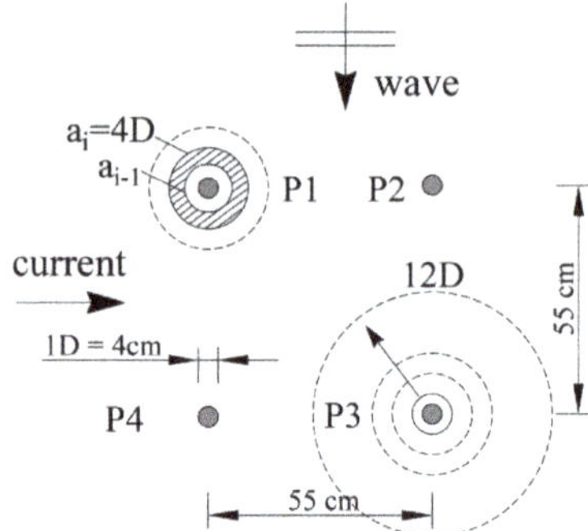

Figure 12. Schematic showing the increasing circle interrogation areas a_i around pile (P1–P4), illustrated exemplarily for a circle area a_i with a diameter of 4 D for pile 1, related to Equation (6).

Similar to what was found in the global erosion analysis, the local bed topography generally confirms the trend of deeper scour with an increasing wave current velocity ratio. The bed topography of test 1 (Figure 13b), on the one hand, illustrates the deposited sediment volume as an influence on the measurement, but on the other hand, also shows a reasonable local erosion depth around each pile. Furthermore, the bed topography for tests 2–5 (Figure 13c–f) shows an increase in scouring on the upstream side of each pile, relative to the current direction, as well as higher erosion rates for pile 1 and 4 (upstream located, in current direction) in comparison to pile 2 and 3 (downstream located). The bed topography measured for tests 4 and 5 around pile 1 and 4 (Figure 13e,f) shows a similar magnitude of erosion. As shown by Welzel et al. [12], for current dominated flow conditions with values of $U_{cw} >$ 0.7 the local scour depth approach values similar to that in current only conditions, whereas the bed topography measured after test 5 (Figure 13f) shows a significant deeper scour around pile 3 and a less pronounced scour depth around pile 2 than around the equivalent piles after test 4. Nevertheless, a clear explanation despite the fluctuations due to erosion and backfilling is not found yet for the difference between pile 2 and 3 at the end of test 5. Furthermore, especially test 4 and 5 reveal a slightly higher erosion of sediment in an area under the diagonal braces (crossing the current direction), which is an indicator for increased bed shear stresses below these diagonal braces. A similar erosion pattern with increased scour depths below the braces was found by Welzel et al. [31], who conducted tests for the same hydraulic conditions and the same jacket structure, only with the difference that the lowest nodes were closer to the seabed.

Figure 14 compares the development of the erosion depth $D_{v,i}$ for each pile over increasing circle areas. Independent of the hydraulic condition and the position of each pile, the maximum erosion depth was always found at a distance from the pile of around 2 D. At the inner most points measurements show that the bed elevation appears to decrease again. As shown in Figure 14 measurements, related to areas > 2 D reveal a similar decreasing trend of erosion, in relation to the slope of the scour hole. As sand is being re-distributed, the local slope angle exceeds the internal friction angle, hence the inner frictional forces of sediment grains are not able to withstand the gravity acting on the grains and thus sediment sliding occurs. Sediment slides from higher to lower locations of the bed and erodes again. A comparison of the incremental erosion depth $D_{V,i}$ of global areas (cp. Figure 11) with local erosion depth values (Figure 14) illustrates a significantly higher erosion per surface area for eroded sediment close around each pile. With distance from the pile, the erosion depth resembles that of the global erosion value. However, it remains strongly dependent on U_{cw} even for larger distances to the pile. The global scour depths for different values of U_{cw} clearly converged with increasing distance from the structure, indicating a boundary of influence of the structure's interaction with the flow. The influence is not as clear for the local scouring processes around the main piles, illustrated in Figure 14. The global erosion causes a subsidence of the seabed simultaneously to the deepening of the local scour hole. However, Figure 14 reveals a boundary of influence at a distance from the pile at which the gradient of the erosion depth clearly decreases. Based on this definition, the size of the local scour hole is found to be about 5–6 D for the present study and 3 D as an outlier for test 1. Thus, the maximum extent of the local scour hole is defined as six times the pile diameter. The distance of about 6 D, at which the gradient of erosion clearly decreases, thus might be interpreted as the boundary of influence of the local scour processes on the global ones. The knowledge of this distance, as well as of the values of eroded sediment, is of practical use for the design of a scour protection system and the required spatial extent.

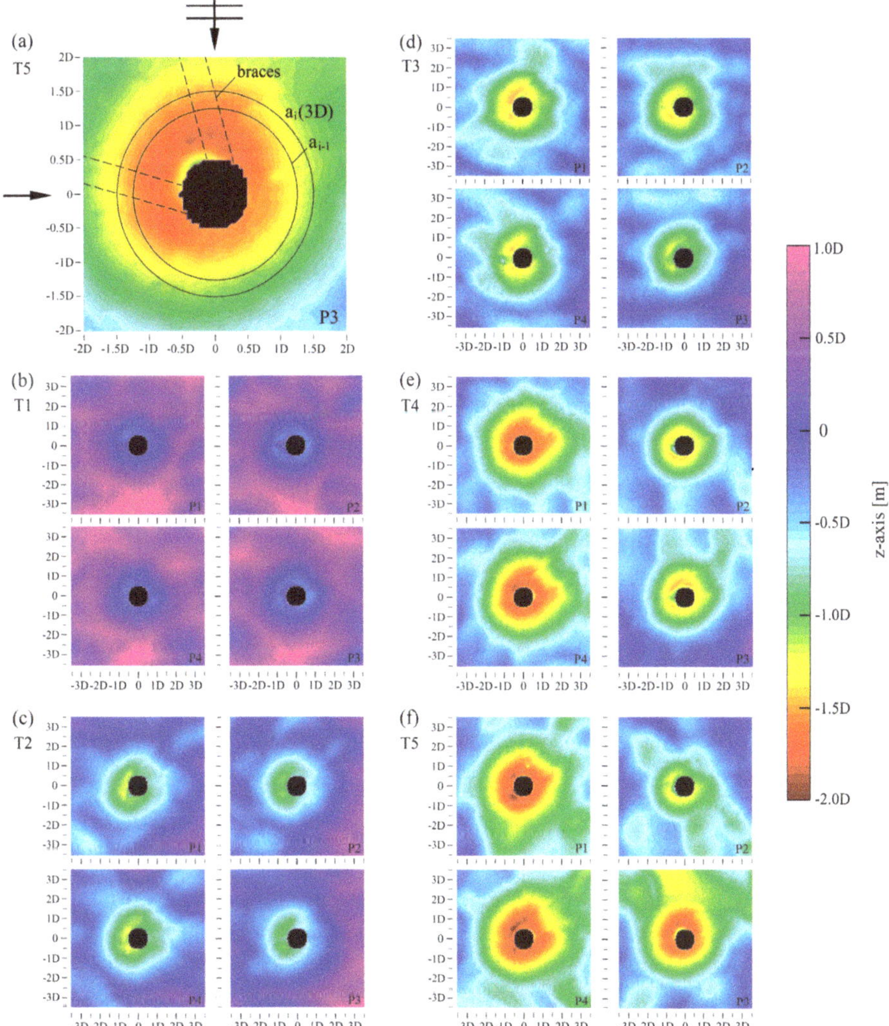

Figure 13. (**a**) Local bed topography, illustrated as an example for test 5, pile 3 with a sketch of the diagonal braces, including the direction of the incoming current, which is coming from left to right, waves are perpendicular to the current as well as exemplary circles related to an incremental interrogation area of 3 D and the directions of the diagonal braces; (**b–f**) local bed topography related to test 1–5; (**b–f**) in an area of 7 D around each main pile of the jacket structure.

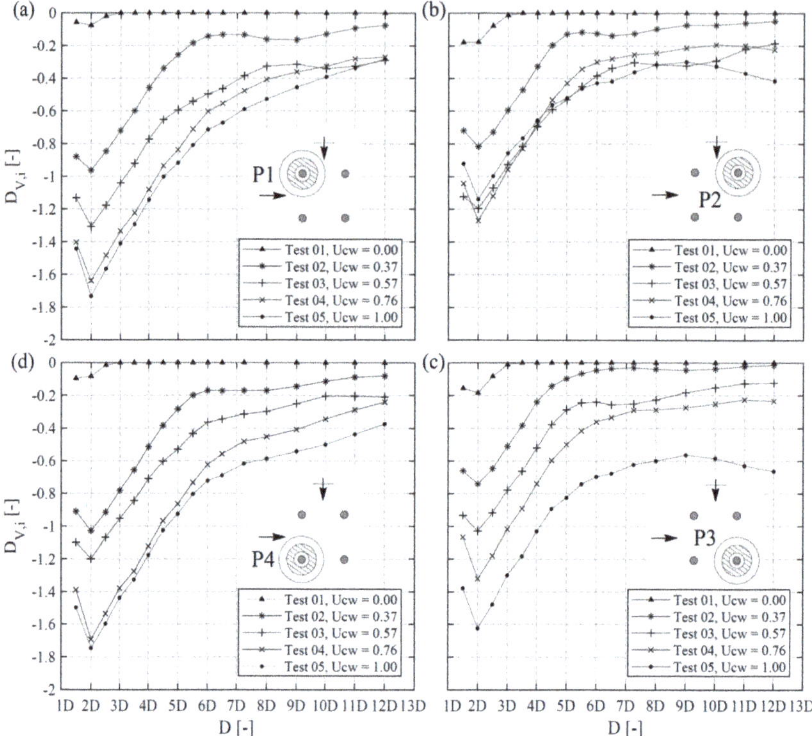

Figure 14. Local, nondimensional erosion depth D_V (see Equation (6)) plotted over increasing diameters D around pile 1–4 (**a–d**), inner graphical subset of figure a–d is showing the related pile location, current (0° from left to right) and wave (90° perpendicular to the current) direction.

The literature reports that hydrodynamic interactions are relatively small for distances > 6 D around circular cylinders [16–18], which is partly transferred as a border of influence for global scouring processes [10]. The present jacket model has a spacing of ~14 D between the pile centers, which is far above a previously described critical distance of 6 D. The volume analysis presented in paragraph 3 proves that there is a significant influence on the global erosion for areas > 6 D for the present structure. Nevertheless, the illustrated erosion depth shown in Figure 14 also confirms that hydrodynamic interactions are significantly smaller for distances > 6 D.

To quantify and compare local and global erosion processes on the overall eroded sediment volume, global erosion volumes are calculated for a threshold of 1.25 A as this distance was found to mark the maximum erosion intensity around the structure. The local eroded sediment volume is considered to be limited to an area of approximately 6 D, as discussed previously, and is summed up for all four piles. Figure 15 illustrates the difference between local and global dimensionless erosion volumes V_D as a function of U_{cw}. For values smaller than $U_{cw} = 0.56$ large amounts of the global erosion volume can be attributed to local scouring processes around the individual piles. In particular, for $U_{cw} = 0.56$, 68% of the erosion volumes can be referred to locally eroded sediment and 32% to globally eroded sediment volumes. However, Figure 15 also reveals that the share of global erosion processes is significantly increased in current dominated hydrodynamic conditions. Measurements of the present study show an increase to about twice the amount of locally eroded sediment volume under current dominated conditions, see Figure 15.

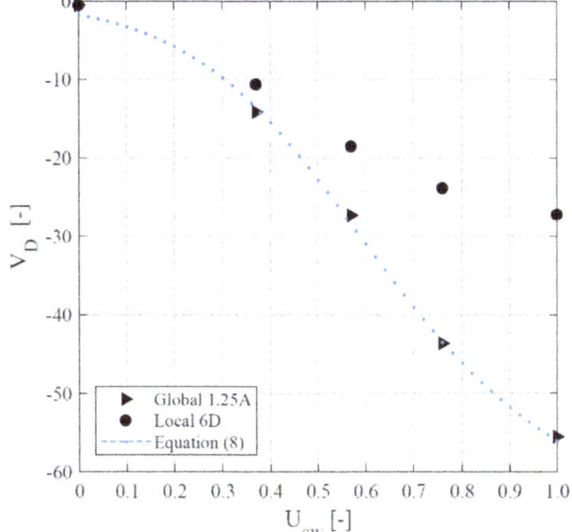

Figure 15. Comparison between non-dimensional local erosion volumes V_D (see Equation (3)) of pile 1–4, depicted over the wave current velocity ratio U_{cw} (see Equation (2)) for circular areas of 6 D and global erosion volumes for rectangular areas of 1.25 times the structure footprint length (1.25 A).

5. Remarks Regarding Practical Application and Scale Effects

Remark 1: Uncertainties regarding various scale effects may exist and must be considered when results of small-scale experiments are extrapolated to prototype conditions. Scale effects in laboratory experiments with a movable bed, e.g., the erosion of sediment, are attributed to the well-known difficulties in geometrically scaling sediment [37,38]. To avoid cohesive behaviour, the sediment was not geometrically scaled in accordance with the model length scale. Instead, to ensure some form of similitude of sediment mobility, the flow velocities were scaled related to the desired flow intensities, i.e. to reach shields parameters (calculated after Soulsby and Clarke [39]) close to the critical Shields parameter. In addition, wave parameters were selected to achieve a certain range of Keulegan-Carpenter numbers and wave current velocity ratios U_{cw}. As a result of the disproportional scaled sediment, i.e., relatively large grain sizes compared to field conditions with comparable small velocities in the present tests, it is expected that bedload transport is thus overrepresented in relation to suspended sediment transport. In consequence, a complete similitude in sediment transport and sediment pick-up rate is not reached. A possible scale effect in consequence of the underestimated ration of suspended load might be that sediment is transported over shorter distances as in prototype scale, presumably leading to different deposition patterns of sediment.

Remark 2: As a consequence of the disproportionally scaled sediment, the bed forms, e.g., ripples are also larger in the laboratory experiments than in the field. The dynamic flow field over ripples causes form drag and turbulence associated with erosion on the stoss side (upstream side) and deposition of sediment on the lee side (downstream side) of the ripples. As an effect of the increased ripple size the boundary layer is also affected and might be increased in thickness, thus leading to a larger horseshoe vortex, influencing the scour development around the piles [14]. Furthermore, according to Sutherland and Whitehouse [40] there is an increased sediment transport due to ripple migration in a model with non-linear flows with proportional larger ripples than in a prototype scale with equally non-linear flows.

Remark 3: In order to compare the present results with field measurements, it is also important to know which temporal stage of the scouring process was reached during the tests, i.e., whether or not an equilibrium stage was achieved. As tests of the present study have been part of Welzel et al. [12], in which the development of scour around the jacket at different locations over time was analyzed, the test duration was generally chosen to enable the scour process to reach the equilibrium stage. By an extrapolation of the expected equilibrium scour depths, Welzel et al. [12] concluded that around 90% of the equilibrium depths were reached for the local scouring processes at the end of the tests. In contrast, the study did not show a clear attainment of an equilibrium stage for measurements related to the global scour depth, particularly under higher current velocities. With a scale factor of 1:30, present test durations correspond to a storm duration of ~20–45 hours, depending on the wave period. This is in the range of typical storm durations in the North Sea [41], but presumably not enough to reach a global scour equilibrium stage (also under laboratory conditions). The local scour depth data and combined bed topography data of Baelus et al. [9] and Bolle et al. [10] indicate a similar trend of a faster developing local scour and a slower developing global scour. Results of [9,10] and [12] therefore indicate that local and global scouring rates are controlled by scouring processes on different time scales. However, the results also indicate that local and global scouring processes are affected by a characteristic depth ratio but are correlated to each other by means of entwined feedback mechanisms, presumably leading to an influence of the global scour on the local scour development and vice versa (see [12]). In particular, the timescale of the global scour development as well as the impact on the local scour development seems to be an important research question in this context, which remains unsolved.

Remark 4: Furthermore, it should be noted that prediction approaches as Equation (8) and Equation (7) are derived for tests based on the present study. Therefore, caution must be exercised when these equations are applied and extrapolated to prototype conditions. To better estimate differences to prototype conditions, scale effects have been discussed and measurements of the present study were compared to available field studies on the scour development around jacket structures (see Figure 11).

6. Summary and Conclusions

Only a few studies exist which provide an approach to analyze complex erosion patterns around offshore foundations. Therefore, hydraulic model tests were carried out, investigating the spatial erosion process in the near-field and vicinity of the hydrodynamic transparent jacket-structure in combined wave and current conditions. The main conclusions can be summarized as follows:

- Different volume analyzing methods and dimensionless parameters are introduced which can be generally adapted for any other offshore structure or coastal structure to reveal physical processes in complex erosion patterns. Therefore, eroded sediment volumes are normalized in relation to a structural volume $V_{D,i}$ as well as in relation to the considered erosion area, $V_{A,i}$, $V_{I,i}$ and the structural diameter $D_{V,i}$, given in Equations (3)–(6).

- A comparison between locally (circle area of 6 D around each pile) and globally (area of 1.25 times the structure footprint) attributed erosion volumes revealed that wave dominated hydrodynamic conditions with $U_{cw} \leq 0.57$ led to scour patterns which were dominated by local erosion around the piles (68% locally, 32% globally, for $U_{cw} = 0.57$). Furthermore, it is shown that the share of globally eroded sediment volume is significantly increased in current dominated conditions $U_{cw} \geq 0.75$ (33% locally, 67% globally, for $U_{cw} = 1.0$).

- The literature reports that hydrodynamic interactions between groups of circular piles are small if the distance between them is larger than six times the piles' diameter [14–16]. In the past, this was partly interpreted as a border beyond which global scour around jacket-type foundations may not occur [10]. In contrast to this, insights from the present study illustrate that the area of the seafloor affected by a supposedly transparent hydraulic structure is considerably larger than expected and is estimated to be 2.1–2.7 times the structure's footprint for the present study.

- A comparison reveals that findings stemming from the present study generally agree well with in-situ data from field studies [8–10]. Similar areal distributions of eroded sediment volume with a stable maximum of the erosion intensity at 1.25 A (1.25 times the structure's footprint) as well as a global scour extent in a similar range to the present study (2.1–2.7 A) is found from a comparison of in-situ data (2.7–2.8 A).

- To improve the prediction of global scour around jacket-type offshore foundations, empirical expressions (Equations (7) and (8)) are proposed to account for the areal development and extent of global erosion volumes and scour depths in the near-field and vicinity of the foundation structure. The analysis and derivation is explained stepwise and is based on insights of the introduced methods. Furthermore, the knowledge of the extent of erosion patterns in relation to the erosion intensity, as well as of the value of the eroded sediment at different points, is useful for the design of a scour protection system around such complex foundation structures. While the former might be used to determine the required spatial extent of a scour protection, the latter helps determining the volume, which has to be refilled by a subsequently installed scour protection.

- Results allow a prediction of areas which exceed a certain erosion rate as well as a quantification of spatially eroded sediment in the near-field and vicinity of the foundation structure. By this means a structure-induced environmental footprint as a measure of eroded sediments partially affecting marine habitat can be exposed. Once eroded sediment is entrained into the water column it deposes behind the structure with the effect of burying marine habitats and can be transported over long distances due to long lasting vortices and an increased turbulence and mixing [5–7]. As a consequence, not only areas in the vicinity but also in the far-field of the structure can be affected, with potential impacts [2–4] on the marine wildlife and the ocean seabed environment in general. These potential impacts to the marine environment might represent an important hurdle for the future of wind technology in general.

Author Contributions: For this research article the specifying of the individual contributions of the authors is as follows: idea and concept of the article, M.W.; methodology, M.W.; investigation and volume calculations, M.W.; resources, T.S.; writing—original draft preparation, M.W.; writing—review and editing, M.W., A.S., T.S. and A.H.; supervision, T.S. and A.H.; project administration, T.S. and A.H.; funding acquisition, T.S.

Funding: The present study is part of the research project "HyConCast—Hybrid substructure of high strength concrete and ductile iron castings for offshore wind turbines" (BMWI: 0325651A). The authors gratefully acknowledge the support of the German Federal Ministry for Economic Affairs and Energy within the funded project. The publication of this article was funded by the Open Access fund of Leibniz Universität Hannover.

Acknowledgments: The authors gratefully acknowledge the support of T. Kreklow, and F. Faltin for their support in conducting the laboratory experiments and pre-processing of the 3D scan data. The authors thank also the FZK and in particular M. Miranda-Lange for support related to the 3D scanner.

Conflicts of Interest: The authors declare no conflict of interest.

Notations

	Reference distance times the structure footprint length;
A	A given times the structure footprint distance, for this structure 0.55 m in both directions, 1 A = 0.55 m/0.55 m, A = x or y distance / structure footprint distance in x or y direction
a_1	Structure footprint area; for the present study $a_1 = 0.55\,m \times 0.55\,m$
a_i	Interrogation area a_i in dependence to i
B	Additional term of equation (9); $B = 10^{-2}(-5.2\,U_{cw} + 6.9)$
C	Additional term of equation (9); $C = 10^{-1}(3.8\,U_{cw} + 4.9)$
D	Pile Diameter of the main struts of the jacket structure D or D_{pile}
D_{sleeve}	Diameter of the pile sleeve of the jacket structure
D_{leg}	Diameter of the legs of the jacket structure
$D_{V,i}$	Incremental erosion depth; representing an erosion depth of the related interrogation area times the pile diameter
d_{50}	Grain size for which 50% of the material by weight is finer
f	Frequency
g	Gravitational acceleration
H_s	Significant wave height
KC	Keulegan-Carpenter number
n	Number of piles

$S(f)$	Velocity frequency spectrum
U	Orbital velocity at the bed in direction of the waves
U_c	Undisturbed current velocity at 2.5 D from bed
$\overline{U}$	Mean current velocity of the vertical profile
U_{cw}	Wave-current velocity ratio $U_{cw} = U_c/(U_c + U_m)$
U_m	Undisturbed maximum orbital velocity at 2.5 D from bed
U_{rms}	Root-mean-square (RMS) value U of at the seabed
$V_{A,i}$	Cumulative erosion volume; $V_{D,i}$ in relation to each normalized area a_i/a_1
V_{area}	Additional term in equation (8) to account for the areal development of volumes
$V_{D,i}$	Dimensionless erosion volume; $V_{erosion}$ of an interrogation area a_i in relation to the structural reference volume
$V_{I,i}$	Incremental erosion volume; the net gradient volume $\left(V_{D,i} - V_{D,i-1}\right)$ in relation to each corresponding area $a_i/a_1 - a_{i-1}/a_1$.
$V_{erosion}$	Eroded sediment volume in m^3 below a reference value based on the pre-scans
θ	Shields parameter
θ_{cr}	Critical value of the Shields parameter

References

1. International Renewable Energy Agency (IRENA). *Offshore Innovation Widens Renewable Energy Options: Opportunities, Challenges and the Vital Role of International Co-Operation to Spur the Global Energy Transformation (Brief to G7 Policy Makers)*; IRENA: Abu Dhabi, UAE, 2018.
2. Carpenter, J.R.; Merckelbach, L.; Callies, U.; Clark, S.; Gaslikova, L.; Baschek, B. Potential Impacts of Offshore Wind Farms on North Sea Stratification. *PLoS ONE* **2016**, *11*, e0160830. [CrossRef] [PubMed]
3. Shields, M.A.; Woolf, D.K.; Grist, E.P.M.; Kerr, S.A.; Jackson, A.C.; Harris, R.E.; Bell, M.C.; Beharie, R.; Want, A.; Osalusi, E.; et al. Marine renewable energy: The ecological implications of altering the hydrodynamics of the marine environment. *Ocean Coast. Manag.* **2011**, *54*, 2–9. [CrossRef]
4. Miller, R.G.; Hutchison, Z.L.; Macleod, A.K.; Burrows, M.T.; Cook, E.J.; Last, K.S.; Wilson, B. Marine renewable energy development: Assessing the Benthic Footprint at multiple scales. *Front. Ecol. Environ.* **2013**, *11*, 433–440. [CrossRef]
5. Grashorn, S.; Stanev, E.V. Kármán vortex and turbulent wake generation by wind park piles. *Ocean Dyn.* **2016**, *66*, 1543–1557. [CrossRef]
6. Vanhellemont, Q.; Ruddick, K. Turbid wakes associated with offshore wind turbines observed with Landsat 8. *Remote Sens. Environ.* **2014**, *145*, 105–115. [CrossRef]
7. Vanhellemont, Q.; Ruddick, K. Landsat-8 as a Precursor to Sentinel-2: Observations of Human Impacts in Coastal Waters. Presented at the 2014 European Space Agency Sentinel-2 for Science Workshop, Frascati, Italy, 20–23 May 2014. ESA Special Publication SP-726.
8. Rudolph, D.; Bos, K.J.; Luijendijk, A.P.; Rietema, K.; Out, J.M.M. Scour Around Offshore structures—Analysis of Field Measurements. In Proceedings of the Second International Conference on Scour and Erosion, ICSE 2, Singapore, Singapore, 14–17 November 2004.
9. Baelus, L.; Bolle, A.; Szengel, V. Long term scour monitoring around offshore jacket foundations on a sandy seabed. In Proceedings of the Ninth International Conference on Scour and Erosion, ICSE 9, Taipei, Taiwan, 5–8 November 2018.
10. Bolle, A.; de Winter, J.; Goossens, W.; Haerens, P.; Dewaele, G. Scour monitoring around offshore jackets and gravity based foundations. In Proceedings of the Sixth International Conference on Scour and Erosion, ICSE 6, Paris, France, 27–31 August 2012.
11. Chen, H.H.; Yang, R.Y.; Hwung, H.H. Study of Hard and Soft Countermeasures for Protection of the Jacket-Type Offshore Wind Turbine Foundation. *J. Mar. Sci. Eng.* **2014**, *2*, 551–567. [CrossRef]
12. Welzel, M.; Schendel, A.; Hildebrandt, A.; Schlurmann, T. Scour development around a jacket structure in combined waves and current conditions compared to monopile foundations. *Coast. Eng.* **2019**, *152*, 103515. [CrossRef]
13. Sumer, B.M.; Fredsøe, J. Wave scour around group of vertical piles. *J. Waterw. Port Coast. Ocean Eng.* **1998**, *124*, 248–256. [CrossRef]

14. Sumer, B.M.; Fredsøe, J. *The Mechanics of Scour in the Marine Environment*; World Scientific: Hackensack, NJ, USA; Singapore; London, UK; Hong Kong, China, 2002.
15. Yagci, O.; Yildirim, I.; Celik, M.F.; Kitsikoudis, V.; Duran, Z.; Kirca, V.S.O. Clear water scour around a finite array of cylinders. *Appl. Ocean Res.* **2017**, *68*, 114–129. [CrossRef]
16. Breuers, H.N.C. *Local Scour Near Offshore Structures*; Delft Hydraulics Publication: Delft, The Netherlands, 1972.
17. Hirai, S.; Kuruta, K. *Scour around multiple- and submerged circular cylinders. Memoirs Faculty of Engineering*; Osaka City University: Osaka, Japan, 1982; Volume 23, pp. 183–190.
18. Hildebrandt, A.; Sparboom, U.; Oumeraci, H. Wave forces on groups of slender cylinders in comparison to an isolated cylinder due to non-breaking waves. In Proceedings of the International Conference on Coastal Engineering, NO. 31, Hamburg, Germany, 31 August 2008.
19. Porter, K.E. Seabed Scour Around Marine Structures in Mixed and Layered Sediments. Ph.D. Thesis, University College London (UCL), London, UK, 2016.
20. Margheritini, L.; Frigaard, P.; Martinelli, L.; Lamberti, A. Scour around monopile foundations for offshore wind turbines. In Proceedings of the First International Conference on the Application of Physical Modelling to Port and Coastal Protection (CoastLab06), Porto, Portugal, 8–10 May 2006. Faculty of Engineering, University of Porto.
21. Stahlmann, A.; Schlurmann, T. Kolkbildung an komplexen Gründungsstrukturen für Offshore-Windenergieanlagen—Untersuchungen zu Tripod-Gründungen in der Nordsee. *Bautechnik* **2012**, *89*, 293–300. [CrossRef]
22. Hartvig, P.A.; Thomsen, J.M.; Frigaard, P.; Andersen, T.L. Experimental Study of the development of scour and backfilling. *Coast. Eng.* **2010**, *52*, 157–194. [CrossRef]
23. Porter, K.; Simons, R.; Harris, J. Comparison of three techniques for scour depth measurement: Photogrammetry, Echosounder profiling and a calibrated pile. In Proceedings of the International Conference on Coastal Engineering, No. 34, Seoul, Korea, 15–20 June 2014.
24. Raaijmakers, T.; Rudolph, D. Time-dependent scour development under combined current and waves conditions—laboratory experiments with online monitoring technique. In Proceedings of the Fourth International Conference on Scour and Erosion, ICSE 4, Tokyo, Japan, 5–7 November 2008.
25. Petersen, T.U.; Sumer, B.M.; Fredsøe, J. Time scale of scour around pile in combined waves and current. In Proceedings of the Sixth International Conference on Scour and Erosion, ICSE 6, Paris, France, 27–31 August 2012.
26. Qi, W.G.; Gao, F.P. Physical modeling of local scour development around a large-diameter monopile in combined waves and current. *Coast. Eng.* **2014**, *83*, 72–81. [CrossRef]
27. Sumer, B.M.; Fredsøe, J. Scour around pile in combined waves and current. *J. Hydraul. Eng.* **2001**, *127*, 403–411. [CrossRef]
28. Soulsby, R. *Dynamics of Marine Sands: A Manual for Practical Applications*; Thomas Telford: London, UK, 1997.
29. Vosselman, G. Slope Based Filtering of Laser Altimetry Data. *Int. Soc. Photogramm. Remote Sens. Congr. Amst.* **2000**, *33*, 935–942.
30. Raudkivi, A.J.; Ettema, R. Clear-water scour at cylindrical piers. *J. Hydraul. Eng.* **1983**, *109*, 338–350. [CrossRef]
31. Welzel, M.; Schlurmann, T.; Hildebrandt, A. Local scour development and global sediment redistribution around a jacket-structure in combined waves and current. In Proceedings of the Ninth International Conference on Scour and Erosion, ICSE 9, Taipei, Taiwan, 5–8 November 2018.
32. Coleman, S.E.; Melville, B.W. Bed-Form Development. *J. Hydraul. Eng.* **1994**, *120*, 544–560. [CrossRef]
33. Flemming, B.W. Zur Klassifikation subaquatischer, strömungstransversaler Transportkörper. *Boch. Geol. Geotech. Arb.* **1988**, *29*, 93–97.
34. Rudolph, D.; Bos, K.J. Scour around a monopile under combined wave current conditions and low KC-numbers. In Proceedings of the Third International Conference on Scour and Erosion, ICSE 3, Amsterdam, The Netherlands, 1–3 November 2006.
35. Zanke, U.C.E.; Hsu, T.W.; Roland, A.; Oscar, L.; Reda, D. Equilibrium scour depths around piles in noncohesive sediments under currents and waves. *Coast. Eng.* **2011**, *58*, 986–991. [CrossRef]
36. Sumer, B.M.; Petersen, T.U.; Locatelli, L.; Fredsøe, J.; Musumeci, R.E.; Foti, E. Backfilling of a Scour Hole around a Pile in Waves and Current. *J. Waterw. Port Coast. Ocean Eng.* **2013**, *139*, 9–23. [CrossRef]

37. Ettema, R.; Melville, B.W.; Barkdoll, B. Scale effects in pier-scour experiments. *J. Hydraul. Eng.* **1998**, *124*, 639–642. [CrossRef]

38. Hughes, S.A. *Physical Models and Laboratory Techniques in Coastal Engineering*; World Scientific Publishing Co. Pte. Ltd: Singapore; London, UK, 1993.

39. Soulsby, R.; Clarke, S. *Bed Shear-Stresses under Combined Waves and Currents on Smooth and Rough Beds*; Report TR 137 Rev 1.0; HR Wallingford: Wallingford, UK, 2005.

40. Sutherland, J.; Whitehouse, R.J.S. *Scale Effects in the Physical Modelling of Seabed Scour*; Report TR 64; HR Wallingford: Wallingford, UK, 1998.

41. Hildebrandt, A.; Schmidt, B.; Marx, S. Wind-wave misalignment and a combination method for direction-dependent extreme incidents. *Ocean Eng.* **2019**, *180*, 10–22. [CrossRef]

 *energies*

Review

Control Strategies Applied to Wave Energy Converters: State of the Art

Aleix Maria-Arenas [1,*], Aitor J. Garrido [2], Eugen Rusu [3] and Izaskun Garrido [2]

[1] Department of Engineering, Wedge Global S.L., 35017 Las palmas de Gran Canaria, Spain
[2] Automatic Control Group—ACG, Department of Automatic Control and Systems Engineering,
Engineering School of Bilbao, University of the Basque Country (UPV/EHU), 48012 Bilbao, Spain
[3] Department of Applied Mechanics, University Dunarea de Jos of Galati, Galati 800008, Romania
* Correspondence: aarenas@wedgeglobal.com

Received: 29 June 2019; Accepted: 6 August 2019; Published: 14 August 2019

Abstract: Wave energy's path towards commercialization requires maximizing reliability, survivability, an improvement in energy harvested from the wave and efficiency of the wave to wire conversion. In this sense, control strategies directly impact the survivability and safe operation of the device, as well as the ability to harness the energy from the wave. For example, tuning the device's natural frequency to the incoming wave allows resonance mode operation and amplifies the velocity, which has a quadratic proportionality to the extracted energy. In this article, a review of the main control strategies applied in wave energy conversion is presented along their corresponding power take-off (PTO) systems.

Keywords: ocean energy; marine energy; wave energy; renewable energy; wave energy converter; control system

---

1. Introduction

Marine Renewable Energy (MRE) is one of the least tapped renewable energy resources. Despite decades of development efforts, more than 90% of the 529 MW of the MRE operating capacity at the end of 2017 was represented mainly by two tidal barrage facilities: 254 MW from Sihwa plant in the Republic of Korea (completed in 2011) and the 240 MW La Rance tidal power station in France (built in 1966) [1]. Main barriers for MRE causing this slow development are analyzed in [2] with special interest in the Mediterranean Sea (MS), although the results can be applied to a great extent for any other geographical areas. Remarkable conclusions about MRE barriers include:

- Bathymetry and distance to shore: Near-shore facilities can have a direct affection on nearby coastal areas, maritime routes, fishing areas or visual impacts. Narrow and step near-shore continental shelfs can also have a negative impact for economic reasons, mainly installation (including grid connectivity) and maintenance.
- Electricity infrastructure: On the contrary, if we move large distances from the shore to minimize the previously cited negative impacts over coastal areas, it can significantly increase the relevant cost of cabling and substations, especially for areas with large depths.
- Potential environmental impacts: Underwater noise, sediment dispersal, increased turbidity, electromagnetic field effects (EMF), wave radiation and diffraction alteration can lead to significant changes to coastal morphology; fixed structures can also generate artificial reef effect. It is concluded that environmental impacts for MRE are hardly clear and not sufficiently quantified, hence, more research is necessary about this topic.
- Economics: The great diversity of MRE technologies and the early development status result in a wide range of levelized cost of energy (LCOE) [3], in the particular case of wave energy

ranging from 108 €/MWh to 530 €/MWh. This level of economic uncertainty creates a less favorable environment for investments.

- Legal and regulatory framework: There is a lot of uncertainty in this regard for MRE, and it is not adequately addressed by the relevant national/international entities. Many key parameters affecting directly any MRE installation are interrelated with several other aspects such as the environmental impact assessment, rights and ownership, international law, management of ocean space, etc.

Geographical or energy islands remain as the most interesting technological enablers for MRE research and deployment, as highlighted in [4]. For remote areas or small islands, the electrical energy production is still based on outdated, polluting and expensive technologies, powered by fossil fuels. Many islands around the world are working on important projects in order to achieve energetic independence based on MRE. Hybrid solutions such as solar-wave [5] or wind-wave [6] can also favor the development of MRE, combining more stablished technologies such as offshore wind and PV in the same offshore device (e.g., spar sharing) or energy farm.

MRE technologies have seen new capacity come online over the recent years; in particular, wave energy development in Spain [7] holds some notorious projects, such as the so called MARMOK-A-5 being the first operational point absorber connected to the grid in Spain, developed by OCEANTEC (acquired by IDOM in September 2018), installed at BIMEP test site in 2016. Mutriku Wave Power Plant, operates since July 2011 connected to the grid, being the first oscillating water column (OWC) multi-turbine facility acting as a test rig for different technology developers; the Portuguese company Kymaner finished testing of its bi-radial air turbine last August 2018. WEP+ demonstration project, based on the industrial scale W1 point absorber developed by Wedge Global, accumulated roughly 5 years of testing at PLOCAN test site on the Canary Islands. LifeDemoWave demonstration project deployed a 25-kW prototype in June 2018 in the Galician coast for a no grid connected test. European highlights from 2018 resulted from H2020 funded projects, including the fabrication of the second Penguin Wave Energy Converter (WEC) at EMEC as part of the CEFOW project, the deployment of the Corpower WEC at EMEC, the installation of the new turbine on the Marmok wave device and the deployment of Minesto's Deepgreen500 device.

The wave energy sector is still in a very early development stage when compared to other renewable energies, especially wind or solar PV. There is a low level of consensus among the WEC technology developers that are still, at best, in a prototype demonstration phase, and there is a lot of discussion among said developers about the best WEC topology and/or PTO configurations. It is not so common to discuss at this level about control strategies given the difficulties in effectively implementing most of the control strategies listed in Section 3 of this review. Most of the existing industry scale prototypes in the water have the simplest control strategy resulting in low energy absorption.

The main purpose of this article is to establish a framework for the control strategies being discussed over the recent years (2017–2019) for WECs. References older than 2017 will not be discussed (unless for introductory purposes) to avoid redundancies with older reviews with a similar approach. A brief review will be provided about the analytical formulation of the mean power absorption and optimal control [8] particularized for heaving WECs, which is the most extended WEC technology in the form of floating point absorbers [9]; however, the same approach is extensible to any other oscillatory mode (sway, surge, pitch, roll or yaw). A point absorber is a floating buoy moored (2 bodies) or fixed (1 body) to the seabed. The incoming waves induce into the floating body a synchronous oscillation mainly in heave motion. It is this movement that is converted into an energy vector (e.g., hydrogen [10], desalination [11] or electricity) through a power take-off located inside the buoy.

Great effort has been made in the last decades about control strategies for WEC, as we can see from older state-of-the-art reviews such as [12–14]. The main objective in wave energy conversion for control is the maximization of power absorption, aiming for resonance operation. The best control approach to achieve said resonance is known as complex-conjugate control, based on solving the impedance matching problem. As we will explain later, this control always achieves optimal power absorption since it regulates the system reactance and resistance simultaneously. However, this control strategy is not practical for real world applications because of large motions and loads. Hence, under-optimal control solutions, it needs to be analyzed and implemented considering physical constrains in motion, force and power rating of the WEC.

In Section 2 a brief introduction about wave energy technology is presented with the purpose of helping the uninitiated reader understand the different technologies and parts (oscillator body, PTO) of WECs. In Section 3, we will start with a comprehensive introduction for WEC modelling and optimal control, as originally introduced in [15], to establish a reference framework, continuing with classification, introduction and discussion for each control strategy found in recent WEC control publications. Finally, conclusion and further research end the article in Section 4.

2. Wave Energy Technology

In the same way that a wind turbine transforms the kinetic energy of the wind into mechanical energy of rotation through the blades, and this one into electrical energy through the electrical generator, a WEC transforms the energy from the waves into mechanical energy through the oscillating body and this one into an energy vector through the PTO.

In the wind energy sector, there is a predominant type over the rest, the three-bladed horizontal axis turbine; it is not so for wave energy, where there is a wide spectrum of options for both oscillator body and PTO without a clear predominant type. Over 1000 different wave energy conversion techniques were patented in Japan, North America and Europe [16] just in 2004, and it is expected this number has greatly increased since 2004 based on the general renewables energies patent numbers shared recently by EMW Law [17].

2.1. Wave Energy Converters

Despite the great number of different technologies for harvesting wave energy, all of them can generally be categorized on the basis of three criteria [18–21].

2.1.1. Location

This classification is according to the relative distance between the device, coast and seabed depth. This classification is somehow qualitative because of the many differences between continental shelfs around the world. Nevertheless, this classification is not used in practical discussions and remains as a first qualitative approach to WEC technologies.

- Onshore: Located in coast proximity, commonly affected by swallow waters ($h/\lambda < 1/25$), where h is the water depth and λ is the wavelength. These converters are usually integrated in a breakwater, dam, fixed to a cliff or rest on the seabed. The distinctive characteristics for these converters are easy maintenance and installation. The drawbacks are that coastline waves have less energy than deep-water waves along with a potential coastline reshape.
- Nearshore: They are installed close to the shore, commonly affected by swallow or intermediate waters ($1/25 < h/\lambda < 1/2$). Their deployment and maintenance expenses are limited since they do not need mooring systems as they are usually fixed or rest on the seabed.
- Offshore: They are placed in deep waters ($h/\lambda > 1/2$), far from the shore. They are able to harvest energy from the most energetic places, but installation and maintenance can be much more expensive because of the required mooring systems (high depth), long underwater cabling, underwater substations and offshore maintenance.

2.1.2. Dimensions of the Prime Mover and Orientation with Respect to the Wave

This classification is according to the orientation of the wave energy device with respect to the wave propagation front (Figure 1). This classification along with the working principle are used to clearly differentiate any WEC.

- Attenuators: The length of the device is of the same order of magnitude (or larger) than the wavelength; these devices are oriented in such a way that they are parallel to the incident wave.
- Terminators: Similar in dimensions to attenuators but placed perpendicular to the incident wave.
- Point Absorbers: Axisymmetric devices capable of harvesting waves from any direction, known as antenna effect, their dimensions are usually an order of magnitude lower than the wavelength.

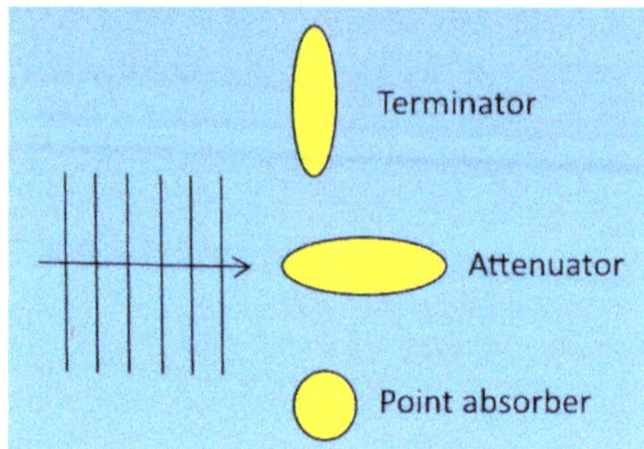

Figure 1. Classification according to device dimensions and orientation.

2.1.3. Working Principle

Oscillating water column: This type of technology builds, among its main elements, an air chamber. It is this air, subjected to oscillating pressure by the action of the waves, which ascends or descends moving a conventional air turbine linked to an electrical generator (Figure 2). Hence, the air turbine can take advantage of the complete oscillation cycle of the wave. It is generally installed in a breakwater, but there are shore-based and floating models. The main benefits for oscillating water column concept commonly accepted are its simplicity and robustness [22]. Common examples are Oceanlinx device deployed in 2005 designed to sit in shallow water, approximately 21 m wide and 24 m long, and Mutriku wave energy plant, located in the Bay of Biscay and commissioned in July 2011, which is one of the few wave energy plants still in operation.

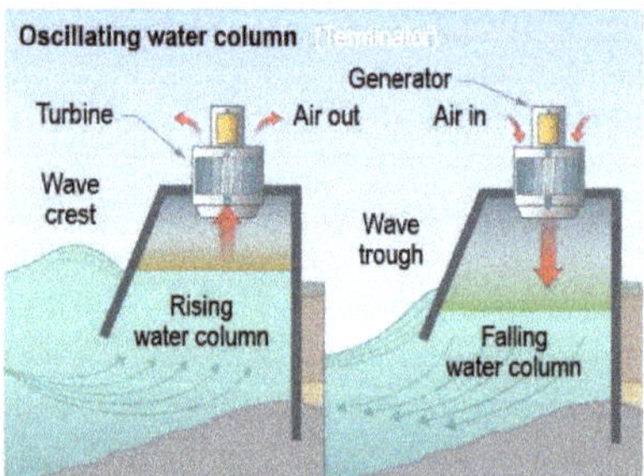

Figure 2. Oscillating water column working mode of operation [22].

Floating structures: Unibody or multibody structures moving in heave, pitch, roll or in any combination of the three (Figure 3) when affected by a wave. The relative movement between different parts of the device allows converting it into electricity. These kinds of devices are rarely named as floating structures but using the dimensions with respect to the wave: attenuators or floating-point absorbers. Multiple examples can be found for this kind of technology. Pelamis was an attenuator floating structure deployed during 2007. The machine is composed by a number of semi-submerged, linked sections. These sections move relatively when the waves pass along the length of the machine. W1 is a point absorber floating technology deployed in 2014. The machine has two main bodies linked without restrictions in heave motion, which allows the relative movement between them.

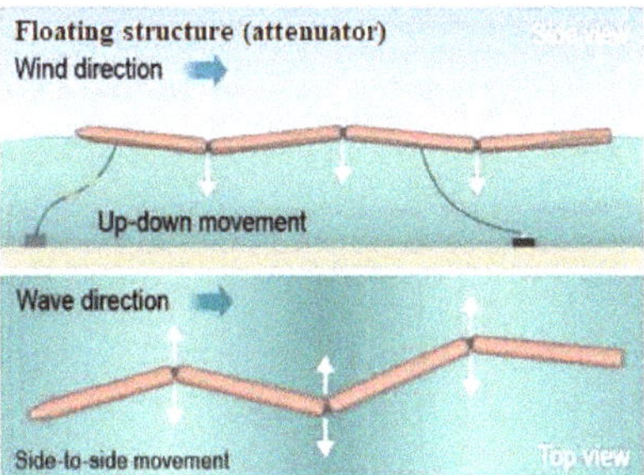

Figure 3. Floating structure with multiples bodies mode of operation. Reprinted from [22].

Pressure differential: Typically located nearshore, this kind of technology can be explained as a combination of two technologies working together—oscillating water column and floating point-absorbers. This is because it uses the working principle for both: difference of pressure and relative heave/pitch/roll displacement between parts, the fixed air chamber in the seabed and the moveable upper body (Figure 4). When a wave crest passes over the device, the water pressure above the device compresses the air within the cylinder, moving the upper cylinder down creating a relative movement in the same way as in punctual absorbers, this happens in the opposite way when a trough passes over. Potential advantages of these devices include: Better survivability, they are not exposed to splash zone corrosion nor the various hazards that could take effect when floating on surface and reduced/negligible visual impact. A major drawback for pressure differential technology is the required underwater maintenance. A good example of evolution with pressure differential technology is Carnegie Clean Energy device (CETO). Deployed in 2015, CETO 5 served the purpose of delivering pressured water for reverse osmosis membranes in the desalination plant, but CETO 6 (still in development) will include electrical generation onboard.

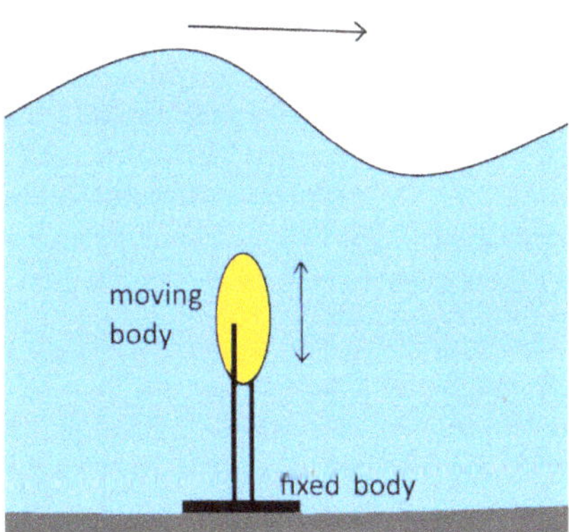

Figure 4. Pressure differential mode of operation.

Overtopping devices: These devices collect the water from the incident waves into a reservoir in order to move one or more reduced jump hydraulic turbines, usually Kaplan turbines. They take advantage of the potential energy of the waves to convert it, through synchronous generators, into electrical energy (Figure 5). Within this type of device, we can distinguish between converters with a fixed structure located on the coast (onshore) and those with a floating structure far away from it (nearshore–offshore). A common example of these kind of devices is the Wave Dragon [23], which is characterized by having a reflector that directs the incident waves towards a ramp to the reservoir above sea level.

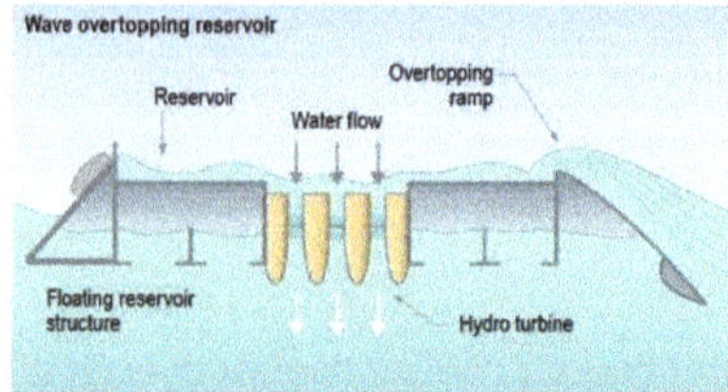

Figure 5. Overtopping mode of operation [22].

Oscillating wave surge: These devices typically have one end attached to a fixed structure or the bottom of the sea while the other end is free to move. A hinged deflector, this part is positioned perpendicular to the wave direction, terminator (Figure 6). The axis of the deflector (or paddle) oscillates like a pendulum mounted on a pivoting joint in response to the impact of the horizontal movement of the wave particle. They often come in the form of floats, fins or membranes. This working principle could be associated to the unique Japanese 'Pendulor' system [24]; but these devices do not take advantage of any harbor resonance. An example for this kind of technology is the Aquamarine Power Oyster, a nearshore device, where the top of the deflector is above the water surface and is hinged from the sea bed.

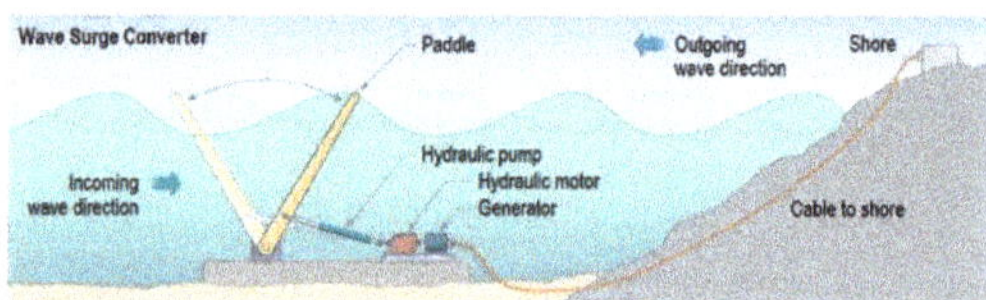

Figure 6. Oscillating wave surge mode of operation [22].

2.2. Power Take-Off Systems

In this section, a brief introduction for each of the different PTO technologies, as presented in [18,25–27], is given. Three main technology paths can be applied to obtain electricity from the wave power conversion chain (PCC), converting the energy being carried by the wave into fluid capture, linear motion or rotary motion. Many different rotary electrical solutions can be applied, but these technologies imply a lot of intermediate steps: Pistons, accumulators, air chambers or mechanical gear systems [25]. The number of intermediate steps is critical for the wave energy conversion efficiency and reliability:

- Efficiency: The larger the number of intermediate steps, the greater are the mechanical and transformation losses that we obtain as a result of the PCC. This causes a reduction in the annual energy production (AEP), which in turn affects the levelized cost of the electricity (LCOE), increasing it.
- Reliability: The offshore equipment undergoes an accelerated degradation in comparison with the same equipment implemented within a ground installation due to the high salinity of the maritime environment where it is implemented. This fact makes it desirable to minimize the amount of equipment to monitor and maintain while the equipment is in operation.

2.2.1. Air Turbines

Commonly used in OWC devices, air turbines require an air chamber to convert wave energy into mechanical power (Section 2.1.3). The basic principle is to drive the air turbine with the oscillating air pressure in the air chamber as a consequence of the oscillating water level. As a result, this PTO solution presents a challenge coming from the bidirectional nature of the flow. A possible solution for this challenge includes non-returning valves combined with a conventional turbine. However, due to complexity, size and high maintenance costs, this configuration is not considered as a viable option [25]. A better solution involves a self-rectifying air turbine that converts an alternating air flow into a unidirectional rotation.

2.2.2. Hydraulic Systems

These are typically used in attenuators, point absorbers and wave surge devices (Section 2.1.3), in which the energy conversion system is based on taking advantage of the linear movement generated by the interaction of the body (or bodies) with the waves. Conventional rotary electrical solutions may not be directly compatible [25]. Therefore, a suitable conversion interface is required between the linear energy capture and the electrical generator, capable of operating with high forces at low frequencies, such as hydraulic systems that operate reversed with respect to their traditional counterpart, that is, the movement of the body feeds the energy of the hydraulic motor which in turn feeds an electric generator.

2.2.3. Hydro Turbines

Used for overtopping devices [25], hydraulic turbines [28] take advantage of the potential energy of the water stored in the accumulation chamber of the device, which is converted to mechanical power using low-head turbines and rotary electrical generators.

2.2.4. Direct Mechanical Drive Systems

This form of PTO solution requires additional mechanical systems driving a rotary electrical generator [25]. It can comprise pulleys, cables, gear boxes or energy storage systems, such as Flywheels (for rotation-based systems) in order to accumulate and release energy, if needed, for reactive operation or to smooth any power variation.

2.2.5. Direct Electrical Drive Systems

Direct electrical drive PTO directly couples the moving part of the electrical generator with the moving body of the WEC [25]. A direct electrical drive PTO system presents two main parts: (i) a translator coupled to the moving body of the WEC, which can be equipped with permanent magnets (conventional solution) or magnetic steel (switched reluctance) and (ii) the stator equipped with coils. The waves induce a heave motion in the moving body coupled with the translator, generating a relative displacement of the translator within the stator, inducing electrical current.

Critical added value of direct linear drive systems is the ability to move instantly in any of the 4-quadrant modes of operation, commonly called 4-quadrant control, allowing instant swap from motor to generator mode at any given moment of the wave oscillation cycle (upwards or downwards) to handle the required reactive power for some of the control strategies we will list in Section 3.

In the first (I) and third (III) quadrants, the electric machine delivers positive power, clockwise or counterclockwise, supplying mechanical energy (motor). On the other hand, in the second (II) and fourth (IV) quadrants, the electric machine delivers negative power, supplying electrical power (generator). Applying a single cycle of regular waves to a WEC, for example, the period between two consecutive wave peaks, the required operation for each quadrant will be as follows: During the downward movement the electric machine will work in downward generator mode (quadrant II) once it reaches the valley, and the consequent upward movement the electric machine will start working in upward generator mode (quadrant IV). If the WEC is operating with a control strategy that requires to brake or accelerate the machine within the same cycle to achieve resonance, as we will see in the next section, the instantaneous swap between the quadrants II-I-III or IV-I-III will be required.

Additionally, direct drive electrical systems have less components to maintain, avoiding intermediate steps while providing simpler/cheaper construction and better reliability. Thus, direct drive is the preferable technology for WECs and offshore facilities where reliability and efficiency are key parameters.

3. Control Strategies

The control problem for the wave energy sector does not fit the classic description of control for other industries where control strategies involve the use of feedback (open loop, closed loop and set-point tracking) and forcing the system variables to a constant value. Instead, WEC control aims for maximization of captured energy while relying on feedforward control to generate optimal device velocity or PTO force setpoints (Figure 7).

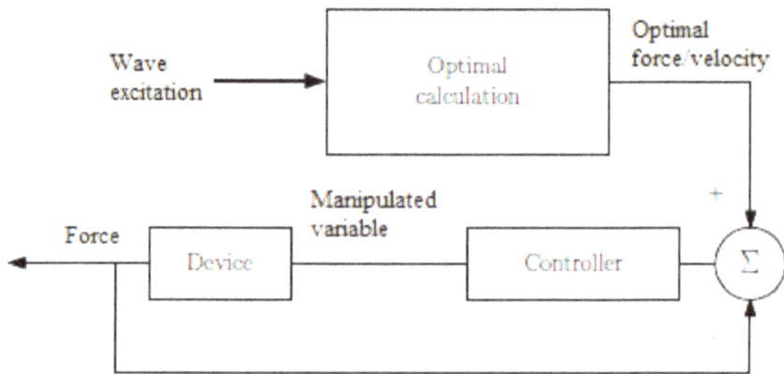

Figure 7. Hierarchical control structure. Manipulated variable depends on the PTO: Bypass valves, swashplate angle, excitation current or conduction angle. Optimal force/velocity calculated as setpoint for the feedforward control.

Optimal calculation involves the performance function of the form:

$$J = \int_0^T v(t) f_{PTO}(t) dt \tag{1}$$

where $v(t)$ is the device velocity, and $f_{PTO}(t)$ is the exerted PTO force. To ease the understanding of how control maximizes this captured energy, we will start in Section 3.1 with a simple analytical revision of the mean absorbed power and optimal control as originally defined in [15] and further discussed in [8], concluding with a discussion of why suboptimal control approaches are required before starting with the main topic for this paper, recent studies about different control strategies for wave energy converters.

A good qualitative first approach to understand how to maximize the absorbed power is the concept of resonance. A system being excited at its natural frequency is described as resonant. When operating in resonance, the response amplitude is highest. Resonance does not usually occur naturally for wave energy converters that have a natural frequency higher (Figure 8) than the power-rich frequency components of a typical wave spectrum, so we have to trick the system into resonance tuning the PTO damping and stiffness as needed, solving the impedance matching problem as we will explain later.

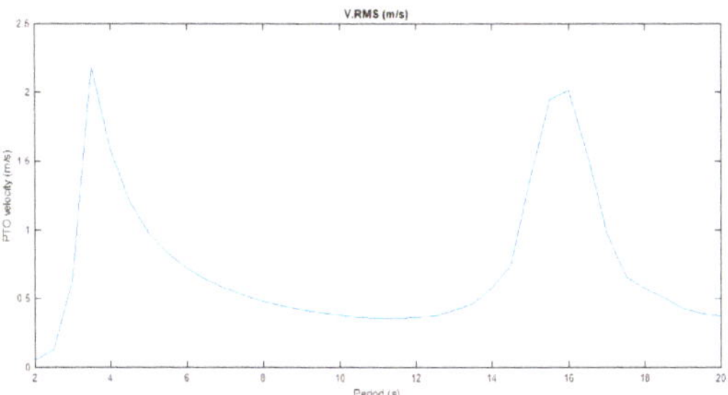

Figure 8. Variation of WEC oscillator velocity (no PTO) for a set of regular wave frequencies with Reference Model 3 [29] geometry with minor variations; simulations performed in WECSIM [30]. Natural resonance operation found for periods of 3.8 s and 16.4 s.

3.1. Numerical Modeling

For any WEC, the inertial force is balanced by the whole forces acting on the WEC. These forces are usually split into external loads, WEC-wave interaction (hydrostatic force, excitation load and radiation force) and reaction forces (caused by PTO, mooring or end-stop mechanism). Interaction between WECs (i.e., floater) and ocean waves is a high-order nonlinear process that can be simplified to linear equations for waves and small-amplitude device oscillation motions, which is acceptable throughout the device's operational regime. This means that the superposition principle applies [31].

The PTO system results in a complex nonlinear dynamic behavior. To keep the superposition principle valid, the PTO forces must be linearized. In this linear form, the PTO force is composed of two contributions [32]: A force proportional to velocity (damper) and a force proportional to the displacement (spring). Mooring systems are often represented by a linear function of the captor displacement and the mooring spring stiffness. End-stop mechanism and other constrains (velocity or PTO force operational limits) are abrupt nonlinear forces which are usually not considered, given the complexity of a nonlinear approach for wave energy conversion. Instead, the optimum method of achieving an acceptable displacement amplitude is to increase the PTO damping until the body has the maximum allowance displacement [33].

In [8], M. Alves obtains the mean absorbed power assuming linearity and sinusoidal waves for a heave motion wave energy converter as:

$$P_a = \frac{1}{2} \frac{B_{pto}\omega^2 |\hat{F}_e|^2}{\left[-\omega^2(m+A) + G + K_{pto} + K_m\right]^2 - \omega^2(R + B_{pto})^2} = \frac{1}{2} \frac{B_{pto}\omega^2 |\hat{F}_e|^2}{|Z_i + Z_{pto}|^2} \tag{2}$$

where ω is the wave frequency, $\hat{F}_e$ is the excitation force, m is the total inertia of the captor, A is the added mass, G is the hydrostatic spring stiffness, K_{pto} is the PTO mechanical spring, K_m is the mooring spring stiffness, R is the radiation damping, B_{pto} is the PTO damping, Z_i is the intrinsic impedance and Z_{pto} is the PTO impedance.

An alternative, yet equivalent, formulation considers the force-to-velocity model of a WEC in the frequency domain [15] as,

$$\frac{V(\omega)}{F_{ex}(\omega) + F_u(\omega)} = \frac{1}{Z_i(\omega)} \tag{3}$$

where $V(\omega)$, $F_{ex}(\omega)$, and $F_u(\omega)$ represent the Fourier transform of the velocity v(t), excitation force f_{ex}(t) and control force f_{pto}(t), respectively. $Z_i(\omega)$ is the intrinsic impedance in the frequency domain of the system as

$$Z_i(\omega) = B_r(\omega) + \omega[M + M_a(\omega) - \frac{K_b}{\omega^2}] \tag{4}$$

where $B_r(\omega)$ is the radiation damping (real and even) and $M_a(\omega)$ is the frequency-dependent added mass, often replaced by its high-frequency asymptote m∞.

The model in (4) allows the derivation of conditions for optimal energy absorption assuming a linear approach, and the intuitive design of the energy-maximizing controller in the frequency domain [15] as

$$Z_{PTO}(\omega) = Z_i^*(\omega) \tag{5}$$

The choice of Z_{PTO} as in (5) is referred to as optimal, reactive or complex conjugate control which is the solution to the so-called impedance-matching problem. Technically, reactive control refers only to the fact that the PTO reactance must cancel the inherent reactance. However, the PTO resistance and the hydrodynamic resistance must also be equal. Thus, complex-conjugate control is a more accurate description since it refers to the fact that the optimum PTO impedance equals the complex conjugate of the intrinsic impedance.

The result in (5) has a number of relevant implications [34]:

- The result is frequency dependent, implying a great optimization difficulty for irregular seas containing a mixture of frequencies.
- Future knowledge of the excitation force may be required. While this knowledge is straightforward for regular waves, it is more complex for irregular seas.
- Since force and velocity can have opposite signs, the PTO may need to supply power for some parts of the sinusoidal cycle.
- The optimal control takes no constrains into consideration; it is more than likely a real system will have velocity and displacement constrains.

Nevertheless, delivering optimal control may be infeasible due to the associated excessive motions and loads in extreme waves. Hence, alternative suboptimal control schemes have been implemented, which include physical constraints on the motions, forces and power rating of the device. While a lot of discussion and different approaches can be found over the recent years for sub-optimal control solutions, we have classified most of them according to the nomenclature that most commonly appears—damping, reactive, latching and model predictive control.

3.2. Damping Control

A widely studied approach to avoid the difficulties in the implementation of the feedback control of the WECs is known in the literature as linear damping of the PTO, also called passive loading [35] or resistive [36], a suboptimal approach where the instantaneous value of the PTO force is linearly proportional to the oscillating body speed, that is to say

$$f_{pto}(t) = -B_{pto}v(t) \tag{6}$$

where $B_{pto} > 0$ is the PTO damping coefficient. This methodology does not require a prediction of the excitation force, thus making it a simple strategy to implement. In fact, it is the one we can usually find in the demonstrators or pre-commercial devices currently deployed around the world. Conventionally, it only requires knowing the instantaneous value of the PTO velocity, for which measurement instruments are usually available in the market.

Figure 9 shows a simulation example in WECSIM [30] for Reference Model 3 using damping control; the electric power (Pe) is always negative, Pe < 0, so the machine does not need to return energy at any point of the oscillating cycle to maximize the energy output in resonance operation.

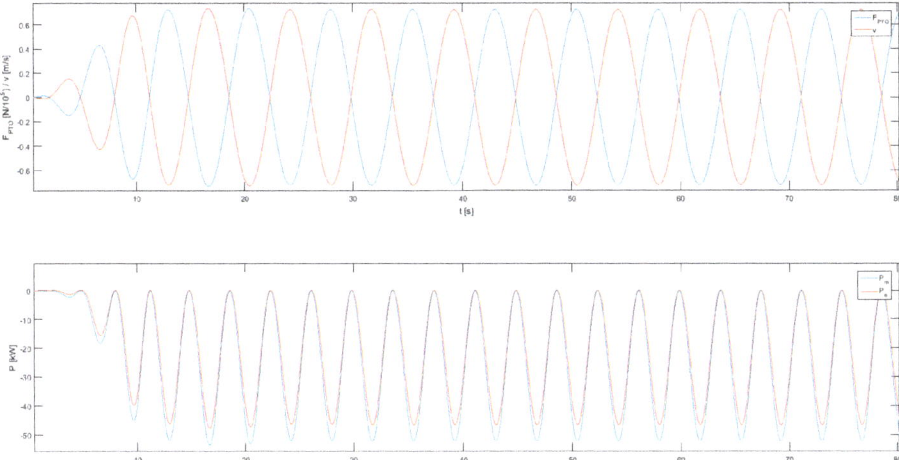

Figure 9. Linear damping–WEC simulation for regular waves. In the upper graphic PTO force (blue) is compared with PTO velocity (red). The lower graphic represents ideal power output (blue) and power output considering electrical losses (red); negative values for power means the WEC is delivering energy to the grid.

Damping control, however, provides a much smaller amount of power absorbed when compared to other strategies such as reactive control [37], as we will see in the next section, and the linear relationship between the speed and the force of the PTO, when it is a straightforward relation, may not be easy to implement without using any feedback control. In addition, the optimal value of the PTO damping, which is the value of B_{pto} that maximizes the instantaneous power absorbed, can be easily calculated for regular waves. However, in practice, where the incident wave is irregular (defined by the wave spectrum), B_{pto} is more difficult to calculate because of the changes in the spectral components of the incident wave which are not constant over time, so a real time feedback control for a time-varying damping value is required.

Therefore, we can distinguish between a real time-varying damping control and a constant (or passive) damping control. First generation WEC control is based on damping strategies with constant values for B_{pto}. This particular strategy is still very common in recent WEC prototypes by technology developers (given the simplicity of implementation).

3.2.1. Constant Damping Control

The work [38] presents a PTO force via constant damping coefficient applied to compare the power conversion performances of three WEC devices modelled in a computational fluid dynamic software (CFD) model based on a 1/50 scale heaving point absorber WEC. Results from this article quantify crucial hydrodynamic parameters for the three devices, revealing a prominent affection of the device amplitude response in free motion without PTO. When PTO is included under effect of regular and irregular waves, the joint effects of geometry and PTO damping on the power absorption are very significant.

Experimental evidence with CECO device (a floating point absorber) with different linear damping coefficients is shown in [39] with the following conclusions: (a) optimal PTO damping coefficients for low-energy irregular waves are higher than for high-energy regular waves, and (b) wave conditions affect significantly the optimal damping coefficients.

3.2.2. Time-Varying Damping Control

Passive damping control is analyzed and compared in [35] with a real-time passive control (PC) based on the Hilbert–Huang transform (HHT). For this solution the damping coefficient is time-varying and tuned instantaneously, based on the frequency of the excitation force. This solution adds a grade of complexity to damping control, since it is required that excitation force be known. The results of this study prove that the proposed solution with real-time calibration of the damping coefficient improves from 21% to 65% the results that a conventional damping control strategy can obtain.

An experimental solution to calibrate the optimum damping coefficients has been presented in [40], based on tank testing experiments on the power performance of a bottom hinged oscillating wave surge converter (OWSC) for regular and irregular waves with damping control, testing different damping coefficients for different wave conditions. The best damping coefficient based on performance was obtained. In this study it is concluded that there are no differences between linear or non-linear strategies in relation to the amount of energy absorbed, but nonlinear strategies have better stability and a broader damping range.

Damping control is electronically implemented in a solid-state relay (SSR) with pulse-width modulation (PWM) in [38]. The objective for this analysis is to mimic analog current flow and compare it with a nonlinear model predictive control (NMPC). It is concluded that peak values of absorbed power and the capture width greatly improve, compared with passive damping strategy.

3.3. Reactive Control

Reactive control is often misleading in the literature and can be confused with complex conjugate control. As the differences between these definitions were already explained in Section 3.1, for clarity reasons, we will keep "reactive control" as it can be usually found in the literature, but a new term such as "sub-optimal reactive control" should be used, as it is done in [41]. These control strategies usually involve the tuning of both PTO resistance and reactance (B_{pto} and K_{pto}), taking into account constrains such as PTO power rating or displacement limits, adjusting the resistance of the PTO to avoid non-linear approaches [33]. Therefore, we will need to consider the generic approach to a PTO characterization as explained in Section 3.1.

$$f_{pto}(t) = -B_{pto}v(t) - K_{pto}x(t) \tag{7}$$

where K_{pto} is the stiffness coefficient, and $x(t)$ is the displacement in the PTO. This kind of control when implemented in demo prototypes usually employs a tabular approach to alleviate the computational constrains required to calculate optimum values in real time. Hence, sub-optimal values for damping and stiffness coefficients are pre-calculated with an optimization algorithm to be stored in tables. For this reason, this particular technique is prone to modelling errors requiring a reanalysis of the constant values after a certain testing period.

Figure 10 shows a simulation in WECSIM for Reference Model 3 [29] using reactive control. The electric power (Pe) varies between positive and negative values, so the PTO needs to switch from motor to generator mode and vice versa at least two times for every oscillation cycle. This kind of mode switching is commonly called "4-quadrant control" and is not obtainable within the time constrains for all of the available typologies of PTOs in the market. Reactive power is a back and forth exchange between the PTO and the oscillating body and does not contribute to the facility energy production. This energy may be supplied by any hydraulic, compressed-air, thermal, chemical, kinetic, electrostatic, electromagnetic storage source [42] or the electrical grid. The biggest disadvantage of reactive strategies comes from the reactive energy exchange process. This process does not suppose an electrical energy gain, but it is subject to dissipative energy loss processes. The magnitude of these losses can negatively affect the overall efficiency of the device.

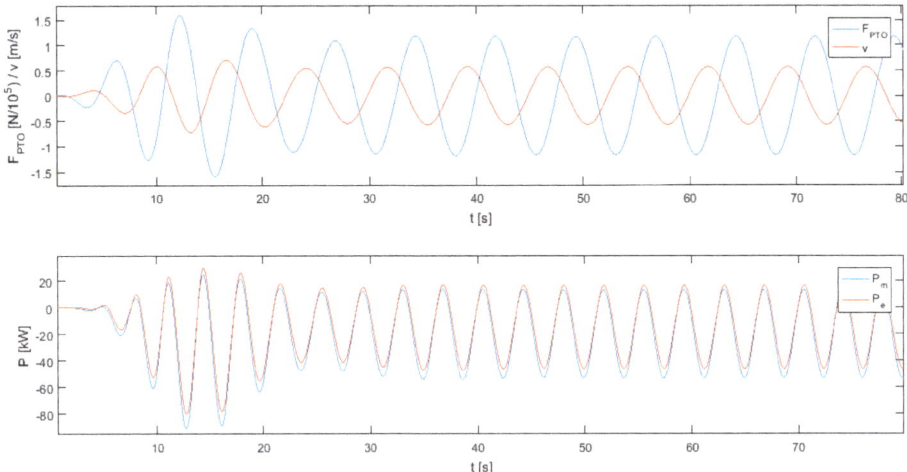

Figure 10. Reactive control–WEC simulation for regular waves.

Energy storage requirements for the reactive power are analyzed in [41] based on a time-domain approach. These storage systems facilitate the exchange of reactive energy and can help to decrease the associated losses, so they are a critical element of the system to maximize the power absorption.

The performance of a floating heaving-only point absorber is analyzed in [43]. The objective is to maximize the wave energy absorption by actively controlling damping and stiffness parameters on the basis of a linear model in the frequency domain. The study concludes with a comparison of the results with similarly validated studies.

Reinforcement learning methodologies are studied in [44]. Calculating the optimum reactive control variables by means of a Q-learning algorithm, the model is able to maximize the energy absorbed for each sea state.

3.4. Latching/Unlatching

Firstly suggested in [45], the latching control is based on achieving the resonance of the WEC through a clamping system, fixing the device during a certain part of the wave oscillation cycle [46]. When the device is released, the control of the device is usually governed by a linear damping as in Section 3.2. This way, the device presents resonance operation without need of reactive power control. However, some energy needs to be drawn from an external source in order to activate the clamping system when the device velocity is null. The critical point for this control strategy is the calculation of the latching-unlatching time periods. Latching control avoids the two-way energy transfer and the associated energy dissipation that characterize reactive control, so a wider spectrum of PTO systems operating only in generator mode can be used under this control strategy.

Setting as base case scenario the passive damping control strategy in [47], the performance improvement when latching control strategy is applied was quantified. The results show that the capture width increases by 70% and the optimal damping coefficient decreases by 60%.

An economic approach was made with different latching control strategies assessed for the WEC in [48], including an interesting comparison with passive damping control. Results are based on the simulated performance of the WEC using regular monochromatic waves, revealing similar annual energy production for constant damping when compared with suboptimal latching, 201 and 197 MWh/yr, respectively. Optimal latching shows the best results with a 45% increase over the annual energy production, 286 MWh/yr.

3.5. Model Predictive Control

Due to its ability to deal with linear and non-linear models, together with the system constrains and real time evaluation of future behavior, model predictive control (MPC) is a widely used and analyzed strategy in the industry [49], and it should not be different for WECs. MPC solutions can handle the physical constraints present for any WEC technology and the non-causal optimal control solution.

However, the problem of maximizing WEC energy requires an important modification over the regular approach in the objective function of the MPC, resulting in a potentially non-convex optimization problem. Given the benefits and growing understanding of these algorithms, this strategy has become the most common control research topic in recent years. MPC maximizes the energy absorption, applying at each time step the optimum force to achieve resonance over a future time horizon, as firstly defined in [50].

As a starting point, [51] presents results of a comparison between MPC control and classical (complex-conjugate control) methods for a Linear Permanent Magnet (LPMG) PTO controlled by a machine side back-to-back actuator. It is concluded that complex-conjugate control when applied to real world solutions shows to be inefficient in maximizing the power absorption from the ocean waves.

Presented as an improvement to reactive control, a predictive strategy is analyzed in [52] where a neural network trained with machine learning is used to predict future waves (height and period), affecting the WEC and optimizing in real time the relevant parameters for the wave energy absorption (PTO stiffness coefficient and PTO damping coefficient). The algorithm does not present any improvements over similar state-of-the-art reactive control solutions in relation with the absorbed power, but it solves associated control inaccuracies from laboratory calibration and enables the controller to be adaptive to variations in the machine response caused by ageing.

In a similar way for a heaving point-absorber, a neural network is employed to forecast the short-term wave height and period in [53] to implement real-time adaptive latching control. This work presents some results comparing the differences of absorbed power for a particular wave scenario with and without control.

An innovative MPC solution is proposed in [54]. Named as robust model predictive control (R-MPC), it combines a predictive controller considering PTO constrains, ensuring maximum power absorption while being realistic, and an innovative model to solve some parametric uncertainties and model mismatches.

An interesting approach to control strategies for 3-degree of freedom WECs is presented in [55] and compared with classical heave-only WECs. Presenting a parametric MPC, it optimizes independently the pitch-surge and heave motion. Numerical algorithms are employed to find the optimal conditions and results. In this work, several numerical tests are conducted for regular and irregular waves. The presented results reveal a great improvement in absorbed power over heavy-only WECs. Contrary to these results, in [56] A. Korde states that near-optimal control for pitch-surge motions are not significant for wave energy absorption when compared to heave motion which is presented as the dominant contributor to power absorption.

A hybrid MPC strategy is presented in [57], constrains are applied to PTO damping and damping force for a two-body WEC. A Mixed-integer Quadratic Programming (MIQP) problem is proposed to obtain the maximum power absorption. Results from this problem are compared with other MPC solutions and classical models for an irregular wave scenario.

Future wave frequency prediction is used in [58] using a Fuzzy Logic controller to determine the optimum PTO damping and stiffness coefficients in real time. The proposed solution combines some regular tuning techniques with an innovative slow tuning methodology.

Fatigue, reliability and survivability controlled by MPC are analyzed in [36]. The results show a trade-off between maximized electrical power and the necessary dimensions for the WEC to resist large loads and fatigue periods. These results are also compared with conventional reactive control, where MPC improves the average annual energy production by 29%.

3.6. Others

This category includes any mixed or innovative control strategies that do not clearly fit into any of the categories presented above.

A genetic algorithm is used to optimize truncated power series along with the geometry for nonlinear WECs in [59]. It enables higher energy harvesting without large motions and less dependence of reactive power as a result.

A so-called Adaptive Parameter Estimation (APE) is proposed in [60]. The algorithm updates in real time several WEC model parameters such as the radiation and excitation force coefficients, combining the benefits associated with optimal control (maximum energy output) and APE dealing with any of the model parameter variation.

In [61] a new power take-off technique is proposed for oscillating wave surge WECs. The main innovation is to avoid any kind of braking system while keeping the amplitude within the specified range. Then, the results are compared with constant damping control, showing the benefits of the new proposed control system.

A new controller which is a variation for the complex conjugate through impedance matching in the time-domain is proposed in [62,63]. The main benefit for the proposed control lies in that it does not need a wave prediction or measurement. It is novel in that it is a feedback strategy with a multi-resonant generator strategy, decomposing the control problem into multiple sub-problems with independent single-frequency controllers. The solution is based on the spectral decomposition of the measurement signal which is employed to construct the optimal solution.

Stochastic control derived from optimal control for heave-only point absorbers considering force constrains is analyzed in [64]. Results indicate performance close to optimal in terms of mean absorbed power.

A crosscutting solution can be found for a cabin-suspended catamaran with a motion control system in [65]. The main objective is to minimize the heave velocity in the cabin, but a secondary measured result of interest for this review is the power absorption from incoming waves which can then be used as an energy vector for different applications, such as feeding auxiliary systems or driving the main engines.

4. Conclusions and Further Research

Since the mean absorbed power for any WEC is frequency dependent, maximum power absorption is achieved in resonance operation when the natural frequency of the WEC matches the wave frequency, causing the excitation force. We can force the WEC into resonance with different control strategies tuning the PTO damping and stiffens constants (B_{pto} and K_{pto}). In this article, we have classified different wave energy technologies based on different criteria commonly used in the literature. Optimal control strategy (complex conjugate control) based on solving the impedance matching problem is impractical for implementation, given the need for future knowledge of the excitation force in irregular waves and the absence of constrains in force and speed for the PTO. Hence, suboptimal control techniques are required, such as damping, reactive (misleading definition which should be revised to suboptimal reactive), latching, MPC and other novelty control ideas.

Wave Energy Technologies are still far from the commercialization point. Only a few successful demonstration projects can be found all over the world and even less when we try to find grid connected projects. Several regulatory, social, economic, environmental and technological barriers need to be addressed from different stakeholders at the same time to perceive an effective pulling action. Control strategies is one of the main technological topics to be discussed. Great efforts have been made over the years in developing effective suboptimal solutions for WECs.

Damping control, usually constant damping control, has been (and still is) the best approach for technology developers willing to test an industrial scale WEC device, given the simplicity of implementation and the safety of operation. Safety is not a minor issue for industrial scale marine devices. Large forces and motions provided by optimal control and top suboptimal approaches could exceed the operational limits of prototype devices fabricated by shipyards and/or associated industries still unfamiliar with WEC technology.

Reactive control is the natural evolution from damping control, presenting an affordable tabular approach for any WEC prototype being deployed for a demonstration phase. Calibration of PTO constants based on simulations or previous experiences do not represent a challenge to the current state of art, although this kind of control strategy requires a PTO capable of switching from motor to generator mode multiple times in a single wave oscillation.

Latching control solves the PTO limitation of reactive control, but it requires an additional clamping mechanism to be installed and energized in the WEC, generating extra costs while also lowering the reliability. Offshore equipment, especially mechanical pieces, are prone to failure because of the extreme salinity ambient. Even well protected (marinization) pieces require a yearly basis maintenance to avoid failures.

MPC solutions represent the best approach to optimal control. Enabling excitation force prediction applies at each time step the optimum PTO force for maximum energy absorption while still considering constrains (non-linear models). MPC has been found to be the most interesting topic among the scientific community (Figure 11) over the recent years, given the good results presented in different articles. This is due to the growing experience in simulated and tank-testing environments, the incremental available advances in computational capacity and the improved expertise with environments based on neural networks. Nevertheless, the complexity of implementation and absence of industry scale demonstration projects disfavored MPC solutions for WEC technology developers. MPC strategies can become a WEC technology enabler in the near future. Maximizing the energy reliability while maintaining equipment costs will result in overall reduction of LCOE, along with the support from different stakeholders. Caring about the other WEC barriers previously presented will result in a market competitive renewable energy technology.

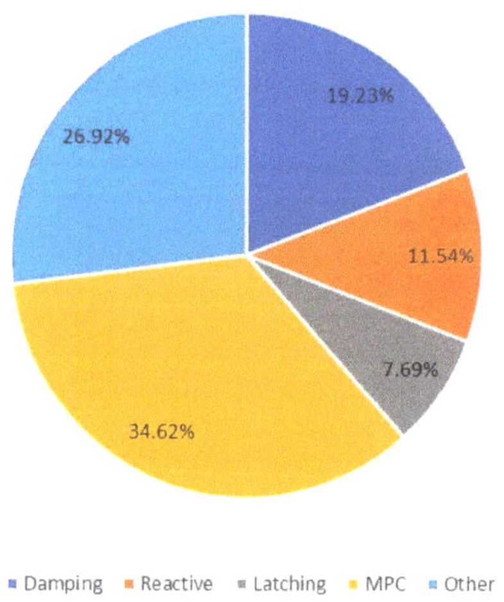

Figure 11. Recent studies for different WEC control strategies, 2017–2019.

Author Contributions: A.M.-A. conceived of the presented idea and took the lead in writing the manuscript. A.J.G., E.R. and I.G. reviewed and supervised the manuscript. All authors discussed the results and contributed to the final manuscript.

Funding: This work was supported in part by the Basque Government through project IT1207-19 and by the MCIU through the Research Project RTI2018-094902-B-C22 (MCIU/AEI/FEDER, UE).

Acknowledgments: The authors gratefully acknowledge Wedge Global for the helpful discussion and support.

Conflicts of Interest: The authors declare no conflict of interest.

References

1. REN21. *Renewables 2018 Global Status Report*; REN21: Paris, France, 2018.
2. Soukissian, T.H.; Denaxa, D.; Karathanasi, F.; Prospathopoulos, A.; Sarantakos, K.; Iona, A.; Georgantas, K.; Mavrakos, S. Marine Renewable Energy in the Mediterranean Sea: Status and Perspectives. *Energies* **2017**, *10*, 1512. [CrossRef]
3. Doe Office of Indian Energy. US Department of Energy. Available online: https://www.energy.gov/sites/prod/files/2015/08/f25/LCOE.pdf (accessed on 14 July 2019).
4. Franzitta, V.; Curto, D.; Rao, D. Energetic Sustainability Using Renewable Energies in the Mediterranean Sea. *Sustainability* **2016**, *8*, 1164. [CrossRef]
5. Franzitta, V.; Catrini, P.; Curto, D. Wave Energy Assessment along Sicilian Coastline, Based on DEIM Point Absorber. *Energies* **2017**, *10*, 376. [CrossRef]
6. Karimirad, M.; Koushan, K. WindWEC: Combining wind and wave energy inspired by hywind and wavestar. In Proceedings of the 2016 IEEE International Conference on Renewable Energy Research and Applications (ICRERA), Birmingham, UK, 20–23 November 2016.
7. OES. Ocean Energy Systems. IEA. Available online: https://www.ocean-energy-systems.org/ocean-energy-in-the-world/ (accessed on 12 July 2019).
8. Alves, M.; Causon, D.; Child, B.; Davidson, J.; Elsäßer, B.; Ferreira, C.; Fitzgerald, C.; Folley, M.; Forehand, D.; Giorgi, S.; et al. *Numerical Modelling of Wave Energy Converters*; Elsevier: London, UK, 2016.
9. Magagna, D.; Monfardini, R.; Uihlein, A. *JRC Ocean Energy Status Report*; European Commission: Brussels, Belgium, 2016.
10. Orecchini, F. The era of energy vectors. *Int. J. Hydrogen Energy* **2006**, *31*, 1951–1954. [CrossRef]
11. Franzitta, V.; Curto, D.; Milone, D.; Viola, A. The Desalination Process Driven by Wave Energy: A Challenge for the Future. *Energies* **2016**, *9*, 1032. [CrossRef]
12. Wilson, D.; Bacelli, G.; Coe, R.G.; Bull, D.L.; Abdelkhalik, O.; Korde, U.A.; Robinett, R.D., III. *A Comparison of WEC Control Strategies*; Sandia National Laboratories: Albuquerque, NM, USA, 2015.
13. Özkop, E.; Altas, I.H. Control, power and electrical components in wave energy conversion systems: A review of the technologies. *Renew. Sustain. Energy Rev.* **2017**, *67*, 106–115. [CrossRef]
14. Wang, L.; Isberg, J.; Tedeschi, E. Review of control strategies for wave energy conversion systems and their validation: The wave-to-wire approach. *Renew. Sustain. Energy Rev.* **2018**, *81*, 366–379. [CrossRef]
15. Falnes, J. *Ocean Waves and Oscillating Systems*; Cambridge University Press: Trondheim, Norway, 2002.
16. Duckers, L. Wave energy. In *Renewable Energy*, 2nd ed.; Boyle, G., Ed.; Oxford University Press: Oxford, UK, 2004; p. 8.
17. Bioenergy International. Global Green Energy Patent Filings 2017 Jump 43% Compared to 2016. Available online: https://bioenergyinternational.com/research-development/global-green-energy-patent-filings-2017-jump-43-compared-to-2016 (accessed on 13 July 2019).
18. Ochs, M.E.; Bull, D.L.; Laird, D.L.; Jepsen, R.A.; Boren, B. *Technological Cost-Reduction Pathways for Point Absorber Wave Energy Converters in the Marine Hydrokinetic Environment*; Sandia Report 7204; Sandia National Laboratories: Albuquerque, NM, USA, 2013.
19. Drew, B.; Plummer, A.R.; Sahinkaya, M.N. A review of wave energy converter technology. *Proc. Inst. Mech. Eng. Part A J. Power Energy* **2009**, *223*, 887–902. [CrossRef]
20. Clément, A.; McCullen, P.; Falcao, A.; Fiorentino, A.; Gardner, F.; Hammarlund, K.; Lemonis, G.; Lewis, T.; Nielsen, K.; Petroncini, S.; et al. Wave energy in Europe: Current status and perspectives. *Renew. Sustain. Energy Rev.* **2002**, *6*, 405–431. [CrossRef]
21. O'Sullivan, D.; Blavette, A.; Mollaghan, D.; Alcorn, R. *Dynamic Characteristics of Wave and Tidal Energy Converters and a Recommended Structure for Development of a Generic Model for Grid Connection*; A Report Prepared by HMRC-UCC for OES-IA under ANNEX III, Document No. T0321. Cork, Ireland, 2010. Available online: https://hal.archives-ouvertes.fr/hal-01265981/document (accessed on 14 August 2019).
22. Laboratory, N.R.E. Marine and Hydrokinetic Technology Glossary. Available online: https://openei.org/wiki/Marine_and_Hydrokinetic_Technology_Glossary (accessed on 22 May 2019).
23. Wave Dragon. Available online: http://www.wavedragon.net/ (accessed on 22 May 2019).
24. Matt, F.; Trevor, W.; Max, O. The Oscillating Wave Surge Converter. In Proceedings of the International Offshore and Polar Engineering Conference, Toulon, France, 23–28 May 2004.

25. Pecher, A.; Kofoed, J.P. *Handbook of Ocean Wave Energy. Ocean Engineering & Oceanography*; No. Power Take-Off Systems for WECs; Springer: Berlin/Heidelberg, Germany, 2017; Volume 7.
26. Schwartz, D.; Mentzer, A. *Feasibility of Linear Induction Wave Power Generation*. Available online: https://linearinductionwavepower.weebly.com/ (accessed on 22 May 2019).
27. So, R.; Casey, S.; Kanner, S.; Simmons, T.K.A.B.A. PTO-Sim: Development of a Power Take Off Modeling Tool for Ocean Wave Energy Conversion. 06 2014. Available online: https://energy.sandia.gov/wp-content/uploads/2014/06/2015-IEEE-PES_PTO-Sim_Nak.pdf (accessed on 22 May 2019).
28. Basic Principles of Turbomachines. IIT 2016. Available online: http://nptel.ac.in/courses/112104117/chapter_7/ (accessed on 19 July 2019).
29. Sandia. Reference Model Project. Available online: https://energy.sandia.gov/energy/renewable-energy/water-power/technology-development/reference-model-project-rmp/ (accessed on 21 June 2019).
30. NREL&Sandia. WECSIM. Available online: https://wec-sim.github.io/WEC-Sim/ (accessed on 22 May 2019).
31. Denis, M.S. Some Cautions on the Employment of the Spectral Technique to Describe the Waves of the Sea and the Response Thereto of Oceanic Systems. In Proceedings of the Offshore Technology Conference, Houston, TX, USA, 29 April–2 May 1973.
32. Xuereb, A.; Spiteri Staines, C.; Sant, T.; Mule Stagno, L. *Design of a Linear Electrical Machine for a Wave Generation System in the Maltese Waters*; Sayigh, A., Ed.; Renewable Energy in the Service of Mankind; Springer International Publishing: New York, NY, USA, 2015; Volume I.
33. Evans, D. Maximum wave-power absorption under motion constraints. *Appl. Ocean Res.* **1981**, *3*, 200–203. [CrossRef]
34. Ringwood, J.V.; Bacelli, G.; Fusco, F. Energy-Maximizing Control of Wave-Energy Converters. *IEEE Control Syst. Mag.* **2014**, *34*, 30–55.
35. Garcia-Rosa, P.B.; Kulia, G.; Ringwood, J.V.; Molinas, M. Real-Time Passive Control of Wave Energy Converters Using the Hilbert-Huang Transform. *IFAC-PapersOnLine* **2017**, *50*, 14705–14710. [CrossRef]
36. Nielsen, K.M.; Pedersen, T.S.; Andersen, P.; Ambühl, S. Optimizing Control of Wave Energy Converter with Losses and Fatigue in Power Take off. *IFAC-PapersOnLine* **2017**, *50*, 14680–14685. [CrossRef]
37. Son, D.; Yeung, R.W. Real-time implementation and validation of optimal damping control for a permanent-magnet linear generator in wave energy extraction. *Appl. Energy* **2017**, *208*, 571–579. [CrossRef]
38. Jin, S.; Patton, R.J.; Guo, B. Enhancement of wave energy absorption efficiency via geometry and power take-off damping tuning. *Energy* **2019**, *169*, 819–832. [CrossRef]
39. Rodríguez, C.A.; Rosa-Santos, P.; Taveira-Pinto, F. Assessment of damping coefficients of power take-off systems of wave energy converters: A hybrid approach. *Energy* **2019**, *169*, 1022–1038. [CrossRef]
40. Jiang, X.; Day, S.; Clelland, D. Hydrodynamic responses and power efficiency analyses of an oscillating wave surge converter under different simulated PTO strategies. *Ocean Eng.* **2018**, *170*, 286–297. [CrossRef]
41. Korde, U.A. Preliminary consideration of energy storage requirements for sub-optimal reactive control of axisymmetric wave energy devices. *Annu. Rev. Control* **2015**, *40*, 93–101. [CrossRef]
42. Robyns, B.; François, B.; Delille, G.; Saudemont, C. *Energy Storage in Electric Power Grids*; Wiley-ISTE: Lille, France, 2015.
43. Jin, P.; Zhou, B.; Göteman, M.; Chen, Z.; Zhang, L. Performance optimization of a coaxial-cylinder wave energy converter. *Energy* **2019**, *174*, 450–459. [CrossRef]
44. Anderlini, E.; Forehand, D.; Bannon, E.; Xiao, Q.; Abusara, M. Reactive control of a two-body point absorber using reinforcement learning. *Ocean Eng.* **2018**, *148*, 650–658. [CrossRef]
45. Budal, K.; Falnes, J. Optimum operation of wave power converter. *Mar. Sci. Commun.* **1977**, *3*, 133–150.
46. Babarit, A.; Duclos, G.; Clément, A. Comparison of latching control strategies for a heaving wave energy device in random sea. *Appl. Ocean Res.* **2004**, *26*, 227–238. [CrossRef]
47. Wu, J.; Yao, Y.; Zhou, L.; Göteman, M. Real-time latching control strategies for the solo Duck wave energy converter in irregular waves. *Appl. Energy* **2018**, *222*, 717–728. [CrossRef]
48. Temiz, I.; Leijon, J.; Ekergard, B.; Bostrom, C. Economic aspects of latching control for a wave energy converter with a direct drive linear generator power take-off. *Renew. Energy* **2018**, *128*, 57–67. [CrossRef]
49. Faedo, N.; Olaya, S.; Ringwood, J.V. Optimal control, MPC and MPC-like algorithms for wave energy systems: An overview. *IFAC J. Syst. Control* **2017**, *1*, 37–56. [CrossRef]
50. Hals, J.; Falnes, J.; Moan, T. Constrained Optimal Control of a Heaving Buoy Wave-Energy Converter. *J. Offshore Mech. Arct. Eng.* **2010**, *133*, 011401. [CrossRef]

51. O'Sullivan, A.C.; Lightbody, G. Co-design of a wave energy converter using constrained predictive control. *Renew. Energy* **2017**, *102*, 142–156. [CrossRef]
52. Anderlini, E.; Forehand, D.; Bannon, E.; Abusara, M. Reactive control of a wave energy converter using artificial neural networks. *Int. J. Mar. Energy* **2017**, *19*, 207–220. [CrossRef]
53. Li, L.; Yuan, Z.; Gao, Y. Maximization of energy absorption for a wave energy converter using the deep machine learning. *Energy* **2018**, *165*, 340–349. [CrossRef]
54. Jama, M.; Wahyudie, A.; Noura, H. Robust predictive control for heaving wave energy converters. *Control Eng. Pract.* **2018**, *77*, 138–149. [CrossRef]
55. Zou, S.; Abdelkhalik, O.; Robinett, R.; Korde, U.; Bacelli, G.; Wilson, D.; Coe, R. Model Predictive Control of parametric excited pitch-surge modes in wave energy converters. *Int. J. Mar. Energy* **2017**, *19*, 32–46. [CrossRef]
56. Korde, U.A.; Lyu, J.; Robinett, R.D.; Wilson, D.G.; Bacelli, G.; Abdelkhalik, O.O. Constrained near-optimal control of a wave energy converter in three oscillation modes. *Appl. Ocean Res.* **2017**, *69*, 126–137. [CrossRef]
57. Xiong, Q.; Li, X.; Martin, D.; Guo, S.; Zuo, L. Semi-Active Control for Two-Body Ocean Wave Energy Converter by Using Hybrid Model Predictive Control. In Proceedings of the ASME Dynamic Systems and Control Conference, Atlanta, GA, USA, 30 September–3 October 2018.
58. Burgaç, A.; Yavuz, H. Fuzzy Logic based hybrid type control implementation of a heaving wave energy converter. *Energy* **2019**, *170*, 1202–1214. [CrossRef]
59. Abdelkhalik, O.; Darani, S. Optimization of nonlinear wave energy converters. *Ocean Eng.* **2018**, *162*, 187–195. [CrossRef]
60. Zhan, S.; Wang, B.; Na, J.; Li, G. Adaptive Optimal Control of Wave Energy Converters. *IFAC-PapersOnLine* **2018**, *51*, 38–43. [CrossRef]
61. Senol, K.; Raessi, M. Enhancing power extraction in bottom-hinged flap-type wave energy converters through advanced power take-off techniques. *Ocean Eng.* **2019**, *182*, 248–258. [CrossRef]
62. Song, J.; Abdelkhalik, O.; Robinett, R.; Bacelli, G.; Wilson, D.; Korde, U. Multi-resonant feedback control of heave wave energy converters. *Ocean Eng.* **2016**, *127*, 269–278. [CrossRef]
63. Lekube, J.; Garrido, A.J.; Garrido, I. Rotational Speed Optimization in Oscillating Water Column Wave Power Plants Based on Maximum Power Point Tracking. *IEEE Trans. Autom. Sci. Eng.* **2017**, *14*, 681–691. [CrossRef]
64. Sun, T.; Nielsen, S.R. Stochastic control of wave energy converters for optimal power absorption with constrained control force. *Appl. Ocean Res.* **2019**, *87*, 130–141. [CrossRef]
65. Han, J.; Kitazawa, D.; Kinoshita, T.; Maeda, T.; Itakura, H. Experimental investigation on a cabin-suspended catamaran in terms of motion reduction and wave energy harvesting by means of a semi-active motion control system. *Appl. Ocean Res.* **2019**, *83*, 88–102. [CrossRef]

Article

Impact of Electrical Topology, Capacity Factor and Line Length on Economic Performance of Offshore Wind Investments

Sadik Kucuksari [1,*,†], **Nuh Erdogan** [2,*,†] **and Umit Cali** [3,*,†]

[1] Department of Technology, University of Northern Iowa, Cedar Falls, IA 50614, USA
[2] Marine and Renewable Energy Centre, University College Cork, P43 C573 Cork, Ireland
[3] Department of Engineering Technology and Construction Management, University of North Carolina at Charlotte, Charlotte, NC 28223, USA
* Correspondence: sadik.kucuksari@uni.edu (S.K.); nuh.erdogan@ucc.ie (N.E.); ucali@uncc.edu (U.C.); Tel.: +1-319-273-2753 (S.K.)
† These authors contributed equally to this work.

Received: 10 July 2019; Accepted: 16 August 2019; Published: 20 August 2019

Abstract: In this study, an economic performance assessment of offshore wind investments is investigated through electrical topology, capacity factor and line length. First, annual energy yield production and electrical system losses for AC and DC offshore wind configurations are estimated by using Weibull probability distributions of wind speed. A cost model for calculating core energy economic metrics for offshore wind environment is developed by using a discount cash flow analysis. A case study is then conducted for a projected offshore wind farm (OWF) rated 100 MW and 300 MW sizes situated in the Aegean sea. Finally, a sensitivity analysis is performed for AC and DC OWFs with three different capacity factors (e.g., 45%, 55% and 60%) and various transmission line lengths ranging from 20 km to 120 km. The OWF is found to be economically viable for both AC and DC configurations with the estimated levelized cost of electricity (LCOE) ranging from 88.34 $/MWh to 113.76 $/MWh and from 97.61 $/MWh to 126.60 $/MWh, respectively. LCOEs for both options slightly change even though the wind farm size was increased three-fold. The sensitivity analysis reveals that, for further offshore locations with higher capacity factors, the superiority of AC configuration over the DC option in terms of LCOE reduces while the advantage of DC configuration over the AC option in terms of electrical losses is significant. Losses in the AC and DC configurations range from 3.75% to 5.86% and 3.75% to 5.34%, respectively, while LCOEs vary between 59.90 $/MWh and 113.76 $/MWh for the AC configuration and 66.21 $/MWh and 124.15 $/MWh for the DC configuration. Capacity factor was found to be more sensitive in LCOE estimation compared to transmission line length while line length is more sensitive in losses estimation compared to capacity factor.

Keywords: cost-benefit analysis; DC collection; energy economics; HVDC; HVAC; levelized cost of electricity (LCOE); offshore wind

---

1. Introduction

Technological developments and changes in lifestyles have driven significantly the increase in energy usage worldwide. Global energy consumption increased by 48.3% from 2002 to 2018 [1]. The use of energy is predicted to rise by 28% by 2040 [2]. Accordingly, global electrical energy use is expected to increase by 58% in the next two decades as well. The increase in electrical energy use contributes to CO_2 emissions and creates environmental concerns. Over the past two decades, renewable energy resources have provided alternatives for energy generation, specifically on electrical energy through solar and wind energy utilizations. The share of renewables for electrical energy generation reached

26% in 2018 [2]. Among the renewable energy resources, excluding conventional hydropower, wind energy has the highest share in terms of installed capacity [3]. Global installed wind power capacity has increased about 30 fold from 2000 to 2017, reaching a cumulative capacity of 591 GW at the end of 2018 [4]. Having higher wind energy potential due to less friction on water surfaces makes offshore wind farms more favorable over their onshore counterparts [5]. The global installed offshore wind power capacity has increased from 4177 MW to 23,140 MW from 2011 to 2018 [6]. The installed capacity of 4.5 GW in 2018 has broken the records on increase in a single year. The fast growth in the offshore wind sector deserves special attention with regard to technical, economical and efficient electric power delivery. From an economical point of view, the cost of offshore systems is higher [7] and varies from project to project since the cost of offshore installations is extremely dependent on their site conditions and location as opposed to their on-shore counterparts [8]. To maintain the offshore market growth given the expected increase in turbine sizes and efficiency, cost-effective solutions for the offshore wind energy sector need to be explored.

An offshore wind farm (OWF) configuration typically consists of collection and transmission systems at medium and high voltages, respectively, including an offshore and/or onshore substation. Most of the current offshore wind farms are all alternating current (AC) systems both at collection and transmission systems. Considering the 20% share of transmission system cost over the total cost figure, alternative solutions may reduce the overall cost [9]. One of the growing alternatives for delivering the generated wind power is using high voltage direct current (HVDC) instead of high voltage alternating current (HVAC) for the transmission system since situated further offshore distances started to become more common [10] and HVDC provides benefits over HVAC for longer distances. Even though less charging current occurs over the HVDC cables [11], the total system losses can be higher due to the additional losses associated with the power electronics components of OWFs [12]. Another alternative transmission that appeared is low frequency AC (LFAC) offshore wind systems, which are found to be more costly than their HVAC and HVDC counterparts for longer distances from the coast [13]. Among three alternatives of power transmission, HVDC becomes more attractive over LFAC and HVAC systems due to the higher transmission line losses and costs associated with the AC systems for far further situated offshore locations. While implementing the HVDC system to overcome these challenges, its economical performance assessment needs to be carefully studied [11]. Medium voltage collection system design starts with the micro sitting of wind turbines using optimization methods that consider many parameters such as wind direction and wake effect [14]. As for a high voltage transmission system, DC topology for collection system is being discussed to be an alternative to AC collection system as it may reduce cable losses and costs. The improvement in DC control and protection device technology also will make the DC collection system more attractive over AC collection in which the control and protection devices are well established and used [15]. Currently, there are no OWF with DC collection systems but a few prototypes are being investigated [15,16]. A detailed cost analysis is therefore needed for both complete AC and DC collection and transmission systems since the cost depends on various parameters such as rated power, wind capacity factor, losses, distance from the shore and so forth [5,12].

The cost of the above-mentioned alternative transmission systems is studied to better shape the future direction of OWF configurations. The economic benefits of various offshore network configurations, including HVAC and HVDC, for the coordinated development of interconnection energy flow are presented in Reference [10]. A comparison of incremental operational and investment costs is examined through a cost model using Monte Carlo approach. Results suggest that coordinated multi-terminal HVDC grid with H-grid configuration could offer operational benefits compared to radial connection. However, under certain circumstances the benefits may be reduced. The study is limited with considering only transmission system and focuses more on the interconnection of different configurations. A cost model for HVAC and HVDC cost comparison was redesigned in Reference [12] considering the losses. It is shown that OWFs installed in a larger size and situated further offshore result in reduced costs for HVDC transmission [17]. Xiangyu et. al in Reference [18] presented a techno-economic analysis of voltage source converter (VSC) based HVDC and HVAC transmission systems for OWFs. They showed that the VSC based HVDC is superior in terms of both economic and technical benefits over the HVAC while the latter includes higher electrical losses. A detailed cost-benefit analysis and power loss calculations were performed for HVAC and several HVDC configurations in Reference [11]. It was found that a critical transmission distance of 85 km makes the VSC based HVDC more economical. A cost analysis for three transmission systems (i.e., HVAC. HVDC, LFAC) is presented in Reference [19]. Results indicated that LFAC may become the cost effective option for shorter distances while HVDC is the most cost effective for longer transmission distances. Similar results are presented in Reference [13] in which the LFAC is found to be more competitive at medium distances between 50–200 km from shore compared to its HVDC counterpart. Instead of delivering power from offshore through HVDC, it is even considered to energize offshore oil industry platforms from an onshore power grid [20]. It is proved that receiving electrical energy through a HVDC transmission to offshore platforms can be more economical comparing to local diesel-fired electricity generation. Techno-economic comparison of HVDC and HVAC is presented in Reference [21]. A wind farm rated 300 MW is used for comparison of different wind turbine topologies, power losses, grid requirements and black start capacity of the two topologies. Results show that the break-even distance of the two systems is found to be 80 km and depends upon the wind turbine technology used, the economic superiority of one over another may be different. An economic comparison of HVAC and HVDC topologies for OWFs at varying rated power from 250 MW to 1500 MW in Great Britain is presented in Reference [22]. Unlike the previous literature, the results show that the break-even cost of the two systems can be only achieved for higher wind farm scales with further offshore locations (i.e., a rated power of 500 MW and a transmission line of 160 km). However, the impact of the collection system on the cost calculations is not considered. Technical and economic comparison of HVAC and HVDC, along with the HVDC market size and high-level comparison of HVDC system components, are presented in Reference [23]. It is concluded that HVDC transmission is beginning to dominate the market and multi-terminal DC networks are expected to play a significant role in the future. The economic analysis in this study is not presented in detail and mainly considers the comparison of some existing or case studies.

While the above-mentioned studies focus on the transmission system, few studies have presented the DC collection system for OWF [15,24]. Possible designs, topologies, converter types and platform configurations are given to provide an inside for the possible usage of DC collection systems together with the HVDC transmission [15]. In Reference [24], a new converter topology was proposed to decrease losses within the DC collection system as an alternative to AC topology. The study in Reference [25] presents a cost assessment for the collection system of an OWF. Since the collection system voltage level is not as high as that of transmission system, the AC and DC configurations may have different relations in terms of cost. The length of the collection system considered is very short. The comparison for overall cost and losses shows that the DC collection system has higher costs and losses compared to the AC collection counterpart. However, as the difference is found to be much less, a detailed cost and loss analysis is still needed for different OWF electrical topologies with practical offshore distances. Reference [16] focuses on a collection system with traditional AC and various DC configurations that are both connected to a power system through a HVDC transmission line. Results concluded that DC configuration for larger scale OWFs may not be economically feasible compared to AC systems due to the DC/DC converters' size. Another study on optimizing OWF design and reducing cost is presented in Reference [26], in which different turbine foundations, AC collection and HVDC transmission for OWFs are considered in a sensitivity analysis. Results suggest that the cost factors mainly vary with the turbine size.

The feasibility of offshore wind investments requires a comprehensive analysis of possible electrical topologies in terms of loss and cost perspectives in a decision making process. HVDC transmission systems have been heavily investigated in search of a viable OWF for further offshore installations. However, the use of DC topology for a medium voltage collection system together with

HVDC can bring a new approach to future OWFs for economic viability. This paper, therefore, presents a detailed techno-economic analysis for conventional all AC and emerging all DC OWF configurations. Radial AC and DC OWF topologies are proposed. Considering the Weibull probability distribution function for wind speed, annual energy yield and the losses of OWF electrical system components are estimated in detail and included in cost calculations. Energy economics metrics such as LCOE, net present value (NPV) and discounted pay-back period (DPBP) are calculated using a discounted cash flow analysis. The proposed analysis is implemented for a possible offshore site in the Aegean Sea, which was found to have the highest capacity factor among the 55 possible offshore locations in Turkey in an earlier study [4]. Two OWF sizes of 100 MW and 300 MW are studied to investigate the impact of electrical topologies on OWF economics in terms of installed capacity. A sensitivity analysis is finally performed for three offshore locations with various capacity factors and distances to shore. The rest of this paper is organized as follows. Section 2 describes the electrical topologies considered. Loss estimation is expressed in Section 3 while economic analysis is evaluated for the selected OWF in Section 4. Results and discussion are presented in Section 5. Finally, Section 6 provides the concluding remarks.

2. Description of Electrical Topologies Considered

Several solutions in terms of electrical topology (i.e., AC, DC and hybrid) have been proposed for both collection and transmission systems [11]. Since the best techno-economic solution for a given application depends on total wind power generated and distance from shore, AC and DC topologies will be considered separately for the configuration of OWF collection and transmission systems.

This offshore site is selected as a result of earlier extensive study based on a multi-criteria site selection work that considers many decision criteria in terms of technical (e.g., wind speed and sea depth), social and civil restrictions, that is, territorial waters, military areas, civil aviation, shipping and pipeline routes and environmental concerns [4]. It was also found in another earlier study [8] that the selected site is less favorable in terms of economics compared to other offshore sites in the Aegean sea since it is far from the point of common coupling which makes its electrical system investment cost higher. The examined OWF is considered to be rated 100 MW and 300 MW. It consists of an inner collection system with 25 and 75 turbine system sets, respectively, connected to an onshore substation in the island and high voltage (HV) submarine transmission cables between the onshore substation and the point of common coupling busbar in the mainland.

2.1. AC Offshore Wind Energy System

The AC OWF configuration considered in this study is shown in Figure 1. The collection system consists of several radial branches that are connected to an onshore substation via submarine cables. Each turbine set rated at 4 MW includes AC–DC and DC–AC converters and a step-up transformer rated 0.69 kV/33 kV, 4.5 MVA. The bus voltage at the AC collection system is thus 33 kV. An onshore substation is considered in the island. It consists of 1 and 3 step-up transformers rated 33 kV/154 kV, 100 MVA including reactive power compensation and grid interface control units. The transmission system includes a three-phase HVAC submarine cable that connects the OWF to the power system in the mainland.

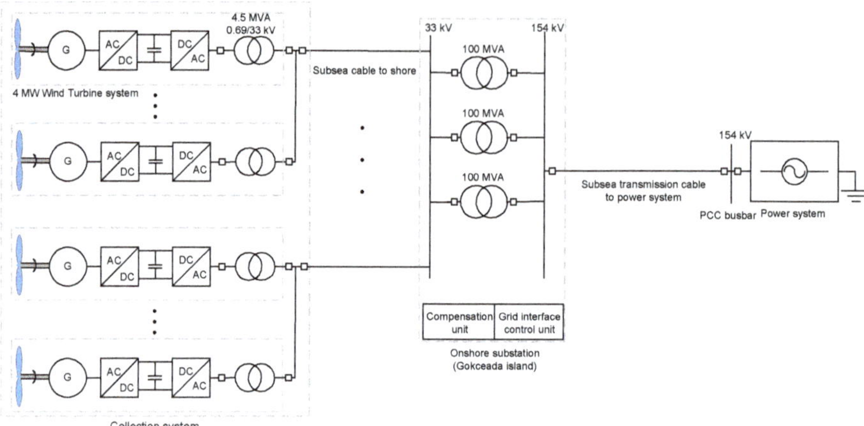

Figure 1. Configuration of proposed AC OWF.

2.2. DC Offshore Wind Energy System

The proposed DC OWF is shown in Figure 2. In this configuration, each turbine is coupled to AC-DC and DC-DC converters that creates a medium voltage (MV) DC bus in the collection system. The collection system is connected to another multi-level step-up DC–DC converter at an onshore platform through monopolar DC 30 kV submarine cables. The onshore platform DC converter is rated at 100 MW and 300 MW. The transmission system includes bipolar HVDC (e.g., 150 kV) submarine cables. Another onshore converter station is placed on the mainland to connect the OWF to the power system. This converter station consists of a multilevel cascaded DC-AC converter including DC capacitors, reactors and filters.

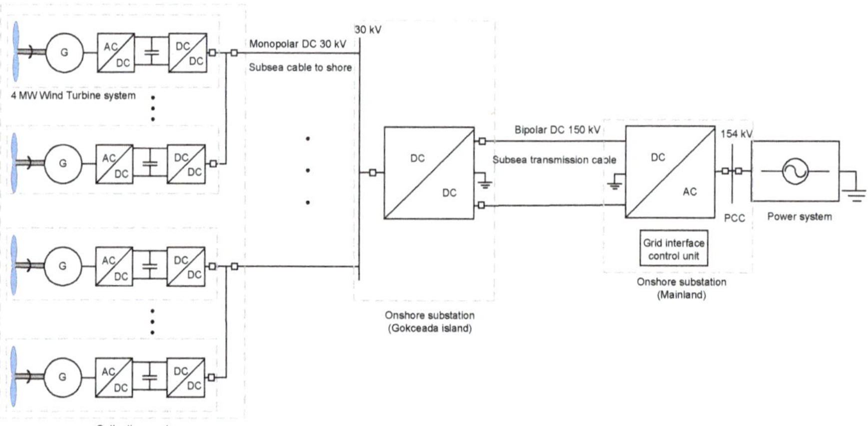

Figure 2. Configuration of proposed DC OWF.

2.3. Electrical Design for Offshore Wind Farm Collection System

The radial design for an OWF collection system was shown to be the most cost-effective option in an earlier study [8]. Herein, based on engineering judgments, radial designs are considered for both the OWFs rated 100 MW and 300 MW as shown in Figure 3. To maximize wind energy usage as well as reduce wake effects, turbines sit perpendicular to the main wind direction and in rows spacing 3.6 rotor diameters (D) within each row and 7 D between rows as recommended in [27].

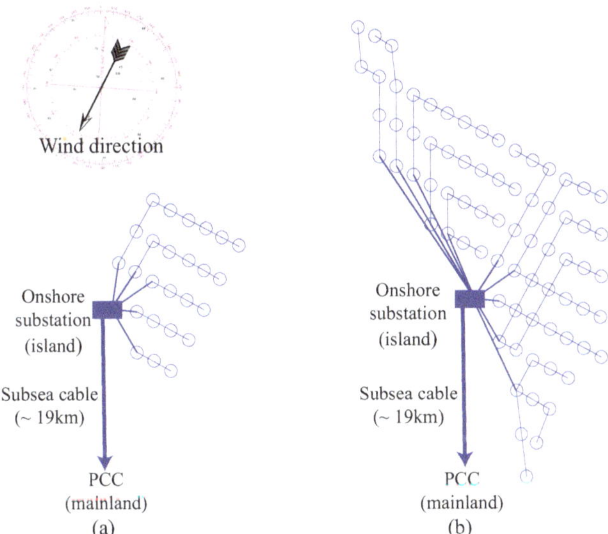

Figure 3. Proposed electrical layouts of the OWF collection system: (**a**) 100 MW. (**b**) 300 MW. Each circle represents the wind turbine system given in Figures 1 and 2.

3. Annual Energy Yield and Electrical Losses Estimation

Estimating wind power output is an integral element of energy business which can be categorized into two clusters—long-term and short-term energy forecasting, respectively. Short-term forecasting tools with hour ahead and day ahead forecast horizons are used for daily operations such as optimal scheduling of the utilities while estimation of annual energy production (AEP) is used for investment decisions for energy generation projects [28]. Within the scope of this article, AEP figures are estimated using an open source renewable energy resource assessment tool named Virtual Wind Farm (VWF) [29]. VMF is based on the weather data which originates from satellite observations and global reanalysis models such as NASA's MERRA (Modern-Era Retrospective Analysis for Research and Applications). The model generates the hourly wind power and wind speed values for the given location for the entire year [30]. The Siemens Wind Turbine SWT-4.0-130 is used in the model [31]. The cut in wind speed (v_{min}) is 5 m/s and the cut out wind speed (v_{max}) is 25 m/s while the rated wind speed is 12 m/s. As wind speed can greatly vary for the cycle of a year, power components (i.e., converters, transformers, etc.) of an OWF are not rated loaded for the most of the time of a year. Using Weibull distribution for power loss of a wind farm achieved more accurate results in Reference [32]. To have more accurate values in estimating losses of each system components, this study therefore considers the Weibull probability distribution function for wind speed. The steady state current values are used in the following calculations.

3.1. Power Electronics Losses

The power electronics (PE) losses for the AC offshore configuration include the losses in the turbine converters within the collection system while the PE losses for the DC counterpart include the losses in the turbine converters, the onshore DC-DC and DC-AC converters. The turbine converter losses are the sum of the switching and conduction losses in the IGBTs(P_{IGBT}) and the freewheeling diodes (P_{FWD}) that are given, respectively, as follows [25]:

$$P_{IGBT-turbine} = N_{sw}(V_{CEO} \cdot I_{c-ave} + R_C \cdot I^2_{c-rms} + (E_{onT} + E_{offT}) \cdot f_{sw}),\qquad(1)$$

$$P_{FWD-turbine} = N_{sw}(V_{DO} \cdot I_{d-ave} + R_D \cdot I^2_{d-rms} + E_{onD} \cdot f_{sw}),\qquad(2)$$

where V_{CEO} and V_{DO} are the on-state voltages (V), I_{c-ave} and I_{d-ave} are the average currents (A), R_C and R_D are the on-state resistances (Ω), I_{c-rms} and I_{d-rms} are the rms currents, for the IGBT and diode, respectively. E_{onT} and E_{offT} are the IGBT's turn-on and off energy losses (W), respectively and E_{onD} is the diode reverse recovery energy loss (W). f_{sw} is the switching frequency which is selected as 1260 Hz. N_{sw} is the number of switches (IGBT or diode). The ABB IGBT modules of type 5SNA 3600E170300 is considered in the turbine converters [33].

To get a high conversion ratio (e.g., 1:5) at MW level, the multilevel step-up DC-DC converter topology proposed in Reference [34] is used for the onshore DC-DC converter. Based on the conventional boost converter configuration, the topology consists of two half-bridges (clamped IGBTs) in the lower position and four chopper (clamped diodes) in the upper position. The ABB IGBT modules of type 5SNA 1200G450300 is used in the onshore DC-DC converter [33]. The switching losses are comprised of the upper and lower stacks IGBT switching losses and upper stack diode reverse recovery losses while the conduction losses comprise the upper and lower stacks IGBT conduction losses and upper and lower stacks diode conduction losses. The total losses are the sum of switching and conduction losses which can be expressed for the IGBT and diode, respectively, as follows [34]:

$$P_{IGBT-HVDC} = 2Nf_{sw}E_{offT} + 4Mf_{sw}E_{onT} + 4Mf_{sw}E_{offT} + 2NV_{CEO}I_N\lambda_{N1} + 2MV_{CEO}I_M\lambda_{M1},\quad(3)$$

$$P_{FWD-HVDC} = 4M \cdot f_{sw} \cdot E_{onD} + 2NV_{DO}I_N\lambda_{N2} + 2MV_{DO}I_M\lambda_{M2},\qquad(4)$$

where $N = 4$ is the number of upper sub modules, $M = 2$ is the number of lower sub modules. I_N and I_M are the average currents in the upper and lower stacks, respectively. λ refers to the ratio of the conduction time to the switching period. The switching frequency of 1 kHz is taken in the calculation.

For the onshore DC-AC converter, the cascaded multilevel converter topology [35] is considered. The topology uses five series connected H-bridges in each phase which creates 11-level line-to-neutral voltage and hence 21-level line-to-line voltages. The ABB IGBT modules of type 5SNA 0750G650300 rated 6500 V is used in the onshore DC-AC converter [33]. The switching frequency of 1 kHz is taken. The associated PE losses in the IGBT modules ($P_{IGBT-HVAC}$) and ($P_{FWD-HVAC}$) can be calculated as given by (1) and (2), respectively. Thus, total PE losses for AC and DC offshore configurations are obtained by (5) and (6), respectively. The parameters of the IGBT modules used are listed in Table 1.

$$P_{PE-AC} = N_{turbine} \cdot (P_{IGBT} + P_{FWD}),\qquad(5)$$

$$P_{PE-DC} = N_{turbine} \cdot (P_{IGBT} + P_{FWD}) + P_{IGBT-HVDC} + P_{FWD-HVDC} + P_{IGBT-HVAC} + P_{FWD-HVAC},\qquad(6)$$

where, $N_{turbine}$ is the total number of turbines within the OWF.

Table 1. Rated values for Parameters of the IGBT modules used in the converters [33].

Parameter	5SNA 3600E170300	5SNA 1200G450300	5SNA 0750G650300
V_{CES} (V)	1700	4500	6500
I_C (A)	3600	1200	750
V_{CEO} (V)	2.5	2.6	2.9
R_C (mΩ)	0.055	0.07	0.07
E_{onT} (mJ)	1100	4350	6400
E_{offT} (mJ)	1600	6000	5300
V_{DO} (V)	1.85	3.2	3.2
R_D (mΩ)	0.094	0.34	0.13
E_{onD} (mJ)	1080	2730	2700

3.2. Transformer Losses

As the transformerless multilevel converter topologies are considered in the DC offshore wind configuration, the transformer losses are therefore a matter of concern for the AC offshore wind configuration. They include the losses of turbine transformers and the onshore substation transformers. The total losses of a transformer at any load level can be obtained by (7) [36],

$$P_{Trfm} = P_0 + I_L^2 \cdot \Re(Z_{pu} \times Z_{base}),$$ (7)

where, P_0, Z_{pu} and Z_{base} are no-load losses, p.u. and base impedances of transformer, respectively, obtained from its nameplate. I_L is transformer primary current.

3.3. Collection and Transmission Lines Losses

The longest collection line is 8 km while the length of the transmission line is 19 km. The short line model can then be used for the modeling of the cables of the collection and transmission lines. It is represented by a series RL circuit. In this case, line losses are calculated by $I^2 \cdot R$. The currents are calculated from operating wind power associated with each wind speed while resistance values are obtained from the underground cable manufacturer's catalog which is selected for rated operation. Herein, the DC resistance values at 20 °C of cables are used for the DC offshore configuration while maximum AC resistance values at 90 °C are used for the AC counterpart. In determining the resistance values, cables are assumed to be directly buried in ground.

3.4. Annual Energy Losses

The total annual energy losses can be found by integrating the losses over the Weibull probability distribution of wind speed for a cycle of a year [25]. Thus, the annual energy losses in the AC and DC offshore configurations become

$$E_{AC-losses} = \int_{v_{min}}^{v_{max}} (P_{PE-AC} + P_{Trfm} + P_{cable-AC}) \cdot f(v_w) \cdot 8760 \, dv_w,$$ (8)

$$E_{DC-losses} = \int_{v_{min}}^{v_{max}} (P_{PE-DC} + P_{cable-DC}) \cdot f(v_w) \cdot 8760 \, dv_w,$$ (9)

where $E_{AC-losses}$ and $E_{DC-losses}$ are the total annual energy losses in the AC and DC offshore configurations, respectively. $P_{cable-AC}$ and $P_{cable-DC}$ are the total collection and transmission lines losses in the AC and DC offshore configurations, respectively. $f(v_w)$ is the Weibull probability distribution function of occurrence of each wind speed for a year obtained using the wind farm model.

4. Economic Analysis

4.1. Major Investment Indicators Considered For Economic Assessment

NPV, LCOE and DPBP are among the major investment energy economic metrics which help the energy sector players to make wise investment decisions [8,37]. Besides, the cash follow diagrams provided contain more information about the entire projected life cycle of an project investment scenarios with yearly resolution. The business of the usual operations of the energy companies shall ideally utilize the above-mentioned techno-economic metrics with higher resolution by considering other dynamic parameters including tax, inflation and risk management components before making their multi-million \$ investments. AEP figures are one of the most important parameters of a cost-benefit analysis where the revenues of the energy economic system are created. NPV indicates the difference between the present value of annual cash inflows (benefits) and annual outflows (expenditures). For the entire project lifespan of an OWF, NPV can be expressed by [8]:

$$NPV = -C_{CAPEX} + \sum_{t=1}^{T} \frac{netCashFlow(t)}{(1+r)^t}, \tag{10}$$

where C_{CAPEX} and T represents the total capital expenditures (CAPEX) and lifespan of OWF, respectively and r is the annual discount rate. The net cash flow for a year is found by subtracting the present value of outflows from the present value of annual cash inflows which includes the annualized operational expenditures (OPEX) and revenues, respectively. In other words, the outflows correspond to the OPEX in the corresponding year while the inflows are related to the AEP of the corresponding year. Revenues of investment calculation are primarily dependent on the AEP figures in energy investments. Positive NPV values represent the economically viable investment option. If there are multiple positive NPV values calculated for various investment options, the options that yield the highest NPV figure shall be selected for the investment. The LCOE is a special per unit cost energy economic metrics which characterizes the NPV of an OWF over its life-cycle and is estimated by [8]:

$$LCOE = \frac{\sum_{t=0}^{T} \frac{C_{CAPEX}(t)+C_{OPEX}(t)}{(1+r)^t}}{\sum_{t=0}^{T} \frac{netAEP}{(1+r)^t}}, \tag{11}$$

where $\forall t \in \{1..T\}$, $C_{CAPEX}(t) = 0$ and initial value of $C_{OPEX}(0) = 0$. $C_{OPEX}(t)$ indicates the OPEX for the year of t. The net AEP is the estimated amount of energy generated by the power plant annually. Pay-back period (PBP) shows the time duration in which the cumulative profit is equal to the cumulative cost. For many investment decision processes, PBP is considered one of the important economic threshold values. However, the PBP does not reflect the time value of money. Thus, using the PBP can be misleading in real life investment decisions where the discount rate is greater than zero percent. In this case, it is essential to utilize the DPBP metrics for more realistic evaluations where the time value of money is also taken in to account. The DPBP is then calculated by [8]

$$DPBP = \frac{ln\left(\frac{1}{1-\frac{r \cdot C_{CAPEX}}{NetCashFlow}}\right)}{ln(1+r)}. \tag{12}$$

The profitability can be measured by using internal rate of return (IRR) metrics that express the discount rate which makes the NPV of an investment equal to zero. The IRR is calculated by dividing the net profit of the investment by the total cost of the investment. The detailed explanation of techno-economic evaluation will be given in the following sections.

4.2. Cost Calculation for AC and DC Offshore Wind Energy Projects

The total investment cost of an OWF project includes CAPEX and OPEX. The parts of each element are expressed below.

4.3. CAPEX

The CAPEX typically have four parts [38]—(i) turbine cost, (ii) support system cost, (iii) electrical system cost, (iv) project development, management and other costs such as insurance. The cost difference in HVAC and HVDC mainly exists due to the different electrical system topologies. The following subsections provide details of each cost component.

4.3.1. Wind Turbine Cost

Wind turbine cost based on the power capacity is provided in Reference [39] for turbines ranging from 2 to 5 MW. Transportation and installation costs of turbines are considered to be 10% of the turbine cost and included into the total cost given by

$$C_{turbines} = 3.245 \cdot 10^3 ln(P) - 412.72 \quad [k\text{€}], \tag{13}$$

where, P is the installed wind power capacity. It is also assumed that this includes the cost of turbine converters and transformer. The given cost function is used both for AC and DC system turbine costs.

4.3.2. Support System Cost

The support system includes foundation and tower. Its cost mainly has three parts—(i) manufacturing, (ii) transportation and (iii) installation. The costs of transportation and installation are included into the manufacturing cost with the assumption of being 50% of the manufacturing cost. The total cost can be formulated as

$$C_{support} = 480 \cdot P(1 + 0.02(d - 8))(1 + 0.8 \cdot 10^{-6}(h(\frac{D}{2})^2 - 10^5)) \quad [k\text{€}/turbine], \tag{14}$$

where d [m] is sea depth, h[m] is hub height and D[m] is rotor diameter. Monopile foundation cost is considered for all the turbines. A sea depth of 45 m is considered; however, the soil properties are not taken into account due to the lack of publicly available data. Both AC and DC systems use the same cost function for support system.

4.3.3. Electrical System Cost

Based on the selected electrical system topology, electrical system cost components vary. Both in AC MV collection and HV transmission systems, the topology is standard AC system topology. In this study, the AC electrical system cost includes inner cable, substation, power factor correction devices and high voltage cables connecting the OWF to the nearest point of common coupling [8]. DC collection and transmission system, on the other hand, may have different configurations both in cables and components. There are two main topologies that exist in a HVDC transmission system, that is, Line-Commuted Converter (LCC) and VSC based topologies. Nowadays, the VCS based converter is the most promising technology that dominates the market [40,41].The overall electrical system cost components for the DC configuration include: (1) medium voltage DC submarine inner cables with installations; (2) onshore substation that includes converters and other necessary components; (3) high voltage DC submarine cables and their installations connecting onshore substation in the island to the mainland; (4) converters in the mainland; and (5) a grid connection unit including other substation components such as reactors.

1. *Collection system cable cost*—the wind turbines are connected to each other as well as to the onshore substation through submarine cables. The core (conductor) of the cables can be either

stranded copper or aluminum. Due to the surrounding sea environment, sufficient electrical insulation is needed around the conductor. The subsea cable insulations are made of different dielectric materials. Among the two common ones are mass impregnated paper and Cross Linked Polyethylene (XLPE) Polymeric cables [42,43]. There are additional layers exist for shielding and mechanical strength purposes. For the DC configuration, XLPE cables are mostly used with VSC topology [44] and were therefore selected for this study. For the AC configuration, 3 core XLPE type cables were selected from available manufacturer datasheets. Reference [45] provides DC cable cost formulation for different voltages. Reference [25] presents a cost function for a 30 kV voltage level cable. The cost model used is given as follows [25]:

$$C_{cable-DC/km} = A_p + B_p \cdot P_n \quad [M\pounds/km], \tag{15}$$

where A_p and B_p are the cost constants of $-0.0256 \cdot 10^6$ and 0.0068 for 30 kV, respectively. P_n rated power of the cable [W]. The cost is converted to €/m and costs for different cross-sections of cables are calculated. The bipole 150 kV cable cost is calculated with the same formula with different constants ($A = -0.1 \cdot 10^6$ and $B = 0.0164$) given in Reference [45]. The calculated cable costs in €/m are given in Table 2.

Table 2. DC cable costs considered for collection and transmission systems.

Voltage (kV)	Topology	Cable Cross-Section (mm^2)	Price (€/m)
30	Monopole	95	50
30	Monopole	120	61
30	Monopole	185	86
30	Monopole	300	124
30	Monopole	400	147
30	Monopole	500	175
30	Monopole	630	207
30	Monopole	1000	279
30	Monopole	1400	339
30	Monopole	1600	368
150	Bipole	150	201
150	Bipole	630	479

Typical cable cross-sections were taken from a manufacturer's [46] publicly available datasheet and depending on the selected voltage and power levels, cable costs are estimated as in Table 2. The cable cross sections were selected such that they carried the maximum power output of the wind turbines. It was considered that the selected radial collector system has a single cable in a row and carries the power of minimum 3 and maximum 13 wind turbine outputs in one feeder. These numbers were determined by considering the physical layout design as well as cable cross sections ampacity levels. An additional 40 m supplementary cable for each turbine was considered as recommended in Reference [38]. The total cable cost for DC collection system was calculated as

$$C_{cable} = C_{cable-DC/km} \times l_{inner-cable} \quad [k\pounds]. \tag{16}$$

The cost function for 3 core XLPE cables given in Reference [39] was used to calculate the cable costs as follows:

$$C_{cable-AC/km} = \alpha + \beta \times e^{\frac{\gamma \cdot I_n}{10^5}} \quad [k\pounds/km], \tag{17}$$

where α, β and γ are constants 52.08, 75.51 and 234.34, respectively. I_n is the current ratings of the cables. Typical cable cross-sections with their current ratings were taken from two manufacturer's available data sheets [47,48].

The cables were considered to be buried under on the seabed. The burying cost in Reference [38] was used as 273 k€/km. The total burying cost for the collection system was calculated by considering all the cable lengths used in the system as

$$C_{burrying} = 273 \times l_{inner-cable} \quad [k€]. \tag{18}$$

2. *Onshore DC/DC converter substation*—one of the factors affecting the DC offshore wind system cost is converter costs. The percentage of the converters over the total cost varies based upon the system topology and transmission distance. Converter costs are found to be around 20% in References [49,50]. The DC configuration becomes more economical over the AC configuration once the converter cost is covered by the cable costs. In DC offshore wind systems, the onshore and offshore substations house the converters and a few other components such as reactors, filters and DC breakers. In addition, the offshore substations cost includes the platform cost as well. In this paper, since there were two onshore substations considered, platform cost was disregarded.

Obtaining the exact cost figures for substation converter stations is very difficult. Reference [51] investigated many studies and proposed € 150/kW. Reference [45,52] presented 1 SEK/VA price for the converters. Reference [53] argues this is because of having different insulation levels at different voltages and proposes three different cost figures for different power ratings. Reference [52] states that this cost includes the cost of valves, filters and other necessary parts. The average of these figures given in the literature was used in this study, which is € 194.23/kVA by assuming that all the substation component costs are included. The onshore substation in the island is a DC/DC converter that steps up the voltage. It is assumed that the converter station includes a series connection of valves for the total power rating. The total price is calculated as

$$C_{substation-DC} - 194.23 \cdot P \quad [k€/km]. \tag{19}$$

3. *Onshore AC substation and power factor correction costs*—since the substation has many components including transformers, switchgeras, backup generators and so forth, the cost is considered as a lump sum that is a function of the installed wind power capacity. Based on the cost model in Reference [38], a cost of 50 k€/MW was used for the calculations. The cost models for power factor correction devices (i.e., SVC, STATCOM, shunt reactors) are given as follows [38]:

$$\begin{cases} C_{shunt-reactor} = 2,556€/MVAr \\ C_{SVC} = 6390€ + 63,900€/MVAr \\ C_{STATCOM} = 128k€/MVAr \end{cases} \tag{20}$$

Thus, the total substation cost is found [8] by

$$C_{substation-AC} = 50[k€]/MW + C_{shunt-reactor} + C_{SVC} + C_{STATCOM}. \tag{21}$$

4. *Transmission line cost*: The total power of 100 MW and 300 MW collected from wind turbines are delivered to onshore substation with the monopole collector system in DC collection system. The DC/DC converter in onshore substation steps up the collector voltage from 30 kV to 150 kV for HV transmission system. A bipole with two conductor system given in Reference [54] is considered for the transmission system from the onshore substation on the island to the other onshore substation on the mainland. The system has 150 kV voltage and two identical cables deliver the power as an underground (1.7 km) and subsea system (17.3 km). No overhead lines are considered for the transmission system. The two cables connecting the two substations share the total power due to being a bipole system and their cross-sections were considered based on the

maximum current. The transmission cable cost that was calculated earlier was used for per km. Additional 100 km supplemental cable was considered and the total cable cost was calculated as:

$$C_{transmission-HVDC} = 2 \cdot C_{cable-DC/km} \times l_{transmission-cable} \quad [k\text{\euro}/km]. \tag{22}$$

Since all the transmission cables are underground and subsea cables, a burying cost of 273 k€/km given in Equation (18) was used. In this study, close laying structure of cables was considered for the bipole system. The two cables were considered to be close to each other and buried together to have a single burying cost. The literature does not present any cost model for a single core XLPE cable used in the HVAC transmission system for high power delivery. It is also difficult to get the cost information from a manufacturer due to it being sensitive information for business operations. Therefore, it was assumed that a single core cable cost is 40% of the 3-core for the same current rating because of better insulation requirements for high voltage. The HVAC transmission system includes three single core cables, hence, the total cost for transmission system cable is tripled.

5. *Onshore DC/AC converter substation*—the onshore substation on the mainland is a DC/AC substation that converts 150 kV DC to 154 kV AC national transmission voltage. Although this is a DC/AC converter, the price of € 194.23 /kVA used for the DC/DC converter in Equation (19) was used for the calculations. Many converters are connected together for the total power as considered earlier. Since the total cost is given as per power, the total cost considered with total power is in Equation (19).

6. *Grid connection cost*—although the cost of the substation in the mainland was considered earlier, it was assumed that there is an additional grid connection cost at the point of common coupling in order to be connected to the 154 kV AC national electric transmission system. The cost for both AC and DC configuration is given as a function of total delivered power in Reference [38] as

$$C_{GC} = 8.047 \times P^{1.66}. \tag{23}$$

4.3.4. Project Development, Management and Other Costs

The project development and management costs including other costs such as insurance and design costs were estimated as $280.38 per MW as given in [55,56].

4.4. OPEX

The OPEX consist of operational, maintenance, administrative, insurance premiums and royalty costs. The sum of all these costs was considered to be 1.9% of the total CAPEX per annum for 20 years lifespan of the project [57].

By considering the aforementioned individual cost models, the total OWF investment cost was calculated as follows:

$$C_{CAPEX} = C_{turbines} + C_{support} + C_{cable} + C_{burrying} + C_{substation} + C_{transmission} + C_{GC}. \tag{24}$$

5. Results and Discussion

5.1. Losses Assessment

The estimated energy losses for each part of the AC and DC configurations for selected wind farm scales are reported in Table 3. It was found that the losses in the AC configuration were slightly higher than those of the DC counterpart for the 100 MW OWF size while they were almost the same for the 300 MW size. In this category, losses are mainly contributed to by the losses of the turbine converters (P_{PE-AC}) with regard to the transformer and onshore converter losses for the AC and DC configurations, respectively. There is a significance decrease in line losses in the case of DC offshore configuration. In this case study, the length of the collection and transmission lines are relatively shorter. The impact of further OWFs from shore on the losses was assessed in the sensitivity analysis.

Table 3. Annual Energy Losses of each components in the AC and DC Offshore Configurations.

AC Configuration	100 MW (kWh)	(%)	300 MW (kWh)	(%)
Collection cables	190,564	0.05	700,393	0.06
Transmission cable	1,422,127	0.36	3,490,676	0.30
Total line losses	1,612,692	0.41	4,191,070	0.36
Power electronics including transformer	13,256,069	3.39	39,768,206	3,39
Total energy losses	14,868,760	3.80	43,959,276	3,75
DC Configuration	**100 MW (kWh)**	**(%)**	**300 MW (kWh)**	**(%)**
Collection cables	129,018	0.03	515,944	0.04
Transmission cable	1,069,279	0.27	2,748,187	0.23
Total line losses	1,198,297	0.31	3,264,131	0.28
Power electronics	13,552,123	3.46	40,697,999	3.47
Total energy losses	14,750,420	3.77	43,962,129	3.75

5.2. Economic Assessment

Revenue of an OWF energy investment mainly depends on the annual energy production and the capacity factor of the power plant. The VWF model estimates the AEP values without considering array (wake effect), electrical and other related losses. Therefore, wind array efficiency parameters were calculated and used to generate realistic net AEP figures. Wind farm array efficiency for this study was assumed to be 96%. The net AEP was calculated by considering wake effects (e.g., 4%) and total electrical losses from the gross AEP estimated by the VWF model. Annual revenues were estimated by multiplying the net AEP and the corresponding Feed-in-Tariff (FIT). The base FIT of 7.3 USD cent/kWh represents the support mechanism which was designed for the wind investments using all exported wind turbines. In addition, the maximum FIT of 11 USD cent/kWh is provided for the investors using domestic wind turbines. The economic calculations in this study were performed in terms of US dollars. The dollar/euro parity of 1.335 is used to convert the CAPEX values to US dollars.

The estimated values for the mean wind speed, OPEX, CAPEX, annual revenues, capacity factors and the net AEP values are summarized in Table 4. This case study is considered as a reference scenario for the sensitivity analysis. The mean wind speed of the selected wind turbine at hub height of 90 m was estimated to be 7.68 m/s. The capacity factor at wind turbines output is constant for all electrical topologies (e.g., 44.7%). Final capacity factor values at the point of common coupling (PCC) that consider wake and electrical losses vary between 41.20% and 41.31% depending on the electrical topology and installed capacity. The net AEP values were estimated between 361.34 GWh and 361.19 GWh and 1,084.59 GWh for the 100 MW and 300 MW sizes, respectively, depending on the final capacity factor. Estimated annual revenue for each configuration ranges from 39.75 million $ to 39.73 million and 119.31 million $ the 100 MW and 300 MW sizes, respectively. CAPEX values were estimated to be in the range of 335.01 million $ and 998.72 million $ for the DC configuration and 300.59 million $ and 998.72 million $ for the AC configuration for the 100 MW and 300 MW sizes, respectively. For the entire project life cycle of 20 year, the OPEX values were estimated to be in the range of 6.37 and 18.98 million $ for the DC configuration and 5.71 and 17.17 million $ for the AC configuration.

Table 4. Cost–benefit analysis components of the OWF for AC and DC configurations.

	AC 100 MW	**DC 100 MW**	**AC 300 MW**	**DC 300 MW**
Capacity factor at PCC (%)	41.24	41.20	41.24	41.24
Net AEP (kWh/year)	361,344,935	361,194,688	1,084,598,234	1,084,598,234
Annual Revenue with base FIT (million$)	26.38	26.37	79,18	79,18
Annual Revenue with max FIT (million$)	39.75	39.73	119.31	119.31
CAPEX (million$)	300.59	335.01	903.88	998.72
OPEX (million$/20 years)	5.71	6.37	17.17	18.98

A comprehensive economic evaluation was performed by using the estimated cost-benefit analysis components. The estimated LCOEs for different discount rates are shown in Figure 4. The results reveal that the LCOE deviates from 88.34 $/MWh (AC 100 MW with 6 % discount rate) to 126.60 $/MWh (DC 100 MW with 10 % discount rate) depending on the electrical topology, interest rates and wind farm size. It can be observed that, although wind farm size is increased by three-fold, LCOEs for the AC and DC options slightly change.

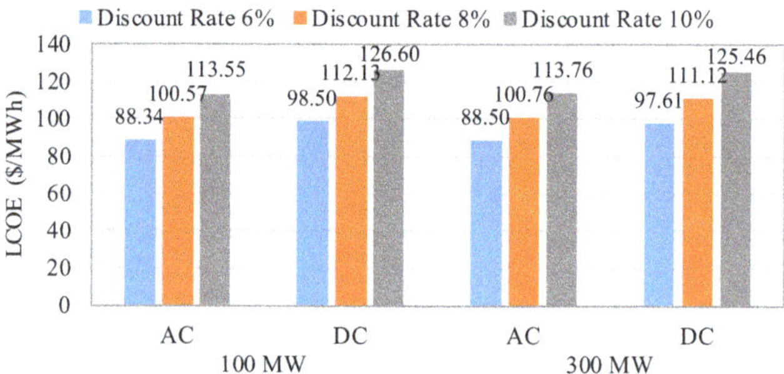

Figure 4. LCOEs of the OWFsfor AC and DC configurations for various discount rates.

Figure 5 shows the NPV distribution over the entire lifespan of the OWF. It is proved that AC and DC options for larger wind farm size (e.g., 300 MW) are delivering a better NPV performance in comparison to the 100 MW wind farm size. DPBP for the AC configuration is found to be 13 years while it is 14 years for the DC options. The one year difference is due to higher CAPEX in DC options which is mainly contributed by the converters' cost. The economic performance of the investigated electrical topologies in terms of NPV for various interest rates with the maximum FIT option is demonstrated in Figure 6. The 300 MW AC configuration under 6 % discount rate yields the best results with a NVP (i.e., 667 million$). Contrarily, the 100 MW DC option under a 10 % discount rate yields the worst performance in terms of NVP (i.e., 37 million). It is also shown that the OWF is economically viable for all interest rates considered for selected configurations and wind farm sizes. Nevertheless, the investigated OWF with both AC and DC electrical topologies for the 300 MW wind farm size is a superior investment option in case the discount rates are in the 6% range with the maximum FIT.

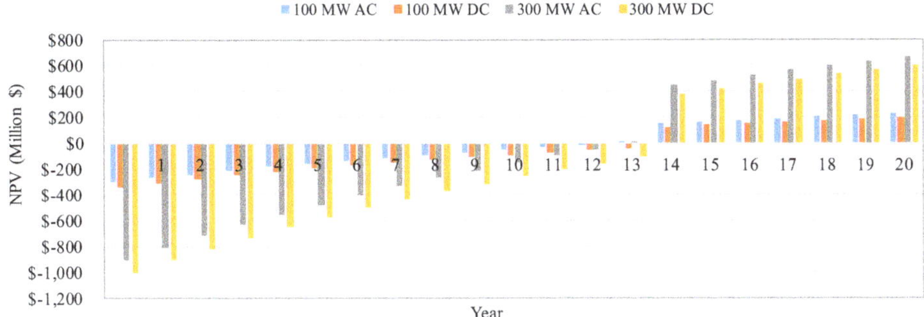

Figure 5. The discounted cash flow of the OWF in terms of years for AC and DC configurations.

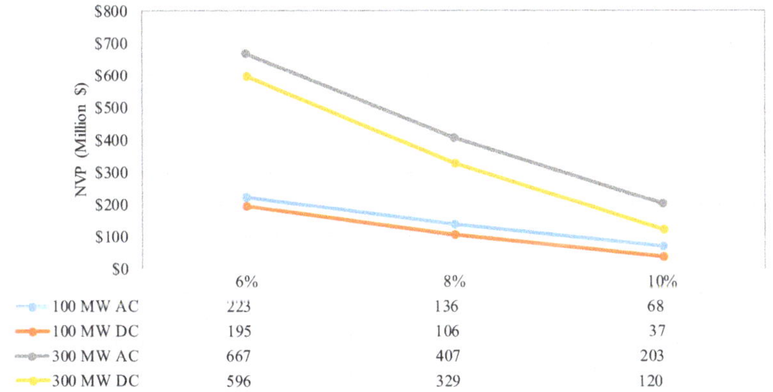

	6%	8%	10%
100 MW AC	223	136	68
100 MW DC	195	106	37
300 MW AC	667	407	203
300 MW DC	596	329	120

Figure 6. The variation of NPV of the OWF in terms of discount rates for AC and DC configurations.

5.3. Sensitivity Analysis

The evaluation was extended with a sensitivity analysis to investigate the influence of capacity factor, transmission line length and the electrical topology with the 100 MW and 300 MW wind farm sizes on the technical and economic viability of OWFs. For this purpose, three different capacity factors with various transmission line lengths ranging from 20 km to 120 km were used. Based on the critical distance of 85 km given in Reference [11] and of 80 km in Reference [21], the transmission line length of 80 km and above distances are considered further offshore locations. Consequently, electrical losses and LCOEs are observed for the estimated values.

The impact of capacity factor and distance to shore on electrical losses is depicted in Figure 7. In terms of electrical losses, the DC configuration is found to be more favorable as the wind farm size, capacity factor and line lengths increase. Losses in the AC configuration range from 3.75% to 5.86% for 45% capacity factor with the transmission line length of the reference scenario (i.e., $l = 19$ km) and 60% capacity factor with a six-fold longer transmission line, respectively. Similarly, losses in the DC configuration vary between 3.75% and 5.34% for 45% capacity factor with the line length of l and 60% with a line length of $6 \times l$, respectively. As the wind farm size, capacity factor and transmission line length increase, the difference between the AC and DC configuration losses becomes significant. It is proved that transmission line length is more sensitive in losses estimation as compared to the capacity factor.

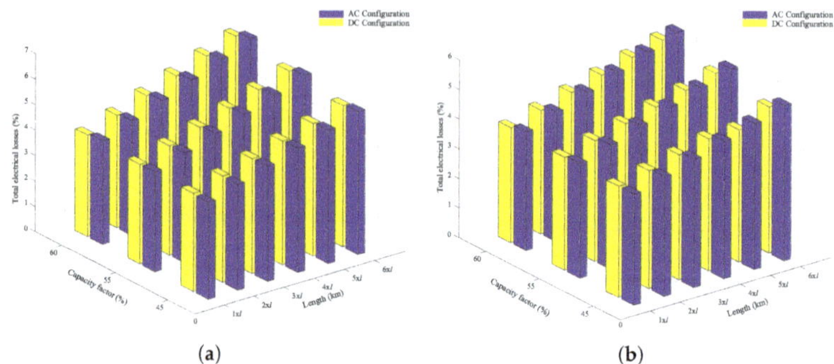

Figure 7. Losses evaluation for AC and DC configurations with respect to capacity factor and transmission line length: (**a**) 100 MW layout. (**b**) 300 MW layout.

Figure 8 illustrates the impact of capacity factor and distance to shore on LCOE based on the 6% discount rate with the maximum FIT. In terms of LCOE, the AC configuration is found to be more favorable for all cases. It is shown that the best LCOE value is estimated to be 59.90 \$/MWh for the AC configuration (i.e., 100 MW wind farm size) with 60% capacity factor and the transmission line of l. The highest LCOE value is, on the other hand, estimated to be 124.15 \$/MWh for the DC configuration (e.g., 100 MW wind farm size) with 45% capacity factor and a transmission line of $6 \times l$. As the wind farm size, capacity factor and distance to shore increase, the difference between the AC and DC configuration LCOEs decreases. Unlike the impact on losses, the capacity factor proved to be more sensitive in LCOE estimation as compared to transmission line length.

It must be noted that this sensitivity analysis considers climatological conditions for the offshore sites with different capacity factors by using the MERRA 2 database for the annual energy yield calculations. However, due to a lack of publicly available data regarding seabed conditions, only one type of foundation structure (e.g., monopole) was considered for the CAPEX component of the foundation cost without any detailed investigation. Considering different types of foundation structures will sway LCOE estimation.

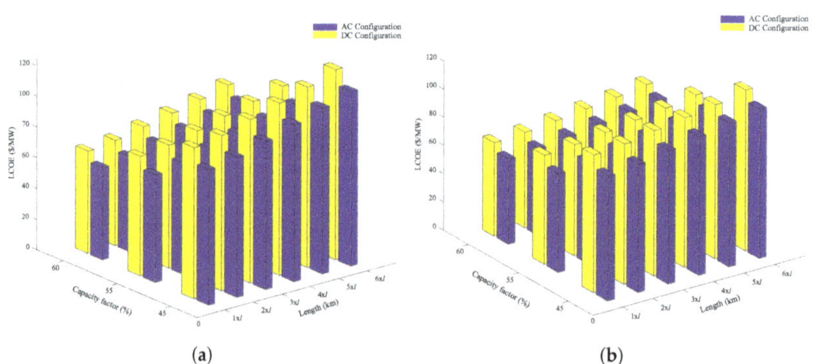

Figure 8. LCOE evaluation for AC and DC configurations with respect to capacity factor and transmission line length: (**a**) 100 MW layout. (**b**) 300 MW layout.

6. Conclusions

In this paper, the economic performance of the AC and DC configurations was investigated for examined OWF with two sizes (e.g., 100 MW and 300 MW). Annual energy yield and electrical losses estimation for AC and DC OWF configurations were detailed. An economic analysis was performed using detailed cost models for the system components. Finally, the impact of capacity factor, transmission line length and electrical topologies on the economic performance of OWF investments was explored through the sensitivity analysis. The analysis has yielded the following conclusions:

- The studied OWF was found to be economically viable for both AC and DC configurations with 13 and 14 years of DPBPs for the AC and DC options, respectively. The estimated LCOEs for the AC and DC OWF configurations range from 88.34 $/MWh to 113.76 $/MWh and from 97.61 $/MWh to 126.60 $/MWh, respectively. LCOEs for both options slightly change even though the wind farm size was increased by three-fold.
- Losses in the AC and DC configurations range from 3.75% to 5.86% and 3.75% to 5.34%, respectively, while LCOEs vary between 59.90 $/MWh and 113.76 $/MWh for the AC configuration and 66.21 $/MWh and 124.15 $/MWh for the DC configuration.
- It was found that the transmission line length parameter is more sensitive in loss estimation while the capacity factor parameter is more sensitive in LCOE estimation.
- It was proved that the superiority of AC configuration over the DC option in terms of LCOE decreases as capacity and transmission line length increase.
- It was also shown that the advantage of DC configuration over the AC option in terms of losses increased as capacity factor and transmission line length increased.

The presented study provides a framework and methodology that can be used to verify the threshold point where LCOE of DC configuration reaches that of AC configuration for a particular wind farm. Although the techno-economic outcomes in this study are unique to the selected offshore sites, the presented methodology can be applied to other specific wind farms with various sizes using the given sensitivity analysis that may result in different break-even LCOE values.

Author Contributions: All authors contributed equally to this work.

Funding: This research received no external funding.

Conflicts of Interest: The authors declare no conflict of interest.

References

1. Enerdata. 2018 Global Energy Trends Report. Available online: http://www.enerdata.net (accessed on 28 May 2019).
2. International Energy Agency (IEA). Available online: https://www.iea.org/geco/electricity/ (accessed on 10 August 2019).
3. Saidur, R.; Islam, M.R.; Rahim, N.A.; Solangi, K.H. A review on global wind energy policy. *Renew. Sustain. Energy Rev.* **2010**, *14*, 1744–1762. [CrossRef]
4. Argin, M.; Yerci, V.; Erdogan, N.; Kucuksari, S.; Cali, U. Exploring the offshore wind energy potential of Turkey based on multi-criteria site selection. *Energy Strategy Rev.* **2019**, *23*, 33–46. [CrossRef]
5. Serrano Gonzalez, J.; Burgos Payan, M.; Riquelme Santos, J. Optimum design of transmissions systems for offshore wind farms including decision making under risk. *Renew. Energy* **2013**, *59*, 115–127. [CrossRef]
6. Global Wind Energy Council (GWEC). Global Wind Report 2018. Available online: http://www.gwec.net (accessed on 30 April 2019).
7. Salo, O.; Syri, S. What economic support is needed for Arctic offshore wind power? *Renew. Sustain. Energy Rev.* **2014**, *31*, 343–352. [CrossRef]
8. Cali, U.; Erdogan, N.; Kucuksari, S.; Argin, M. Techno-economic analysis of high potential offshore wind farm locations in Turkey. *Energy Strategy Rev.* **2018**, *22*, 325–336. [CrossRef]
9. Gardner, P. Offshore wind energy: Resources, technology and grid connection. In *First International Workshop on Feasibility of HVDC Transmission for Offshore Wind Farms*; KTH Royal Institute of Technology: Stockholm, Sweden, 2000.
10. Houghton, T.; Bell, K.R.W.; Doquet, M. Offshore transmission for wind: Comparing the economic benefits of different offshore network configurations. *Renew. Energy* **2016**, *94*, 268–279. [CrossRef]

11. Reed, G.F.; Hassan, H.A.A.; Korytowski, M.J.; Lewis, P.T.; Grainger, B.M. Comparison of HVAC and HVDC solutions for offshore wind farms with a procedure for system economic evaluation. In Proceedings of the IEEE Energytech, Cleveland, OH, USA, 21–23 May 2013; pp. 1–7.

12. Morton, A.B.; Cowdroy, S.; Hill, J.R.A.; Halliday, M.; Nicholson, G.D. AC or DC? Economics of grid connection design for offshore wind farms. In Proceedings of the 8th IEE International Conference on AC and DC Power Transmission (ACDC 2006), London, UK, 28–31 March 2006; pp. 236–240.

13. Ruddy, J.; Meere, R.; O'Donnell, T. Low Frequency AC transmission for offshore wind power: A review. *Renew. Sustain. Energy Rev.* **2016**, *56*, 75–86. [CrossRef]

14. Gonzalez, J.S.; Payan, M.B.; Santos, J.M.R.; Gonzalez-Longatt, F. A review and recent developments in the optimal wind-turbine micro-siting problem. *Renew. Sustain. Energy Rev.* **2014**, *30*, 133–144. [CrossRef]

15. Musasa, K.; Nwulu, N.I.; Gitau, M.N.; Bansal, R.C. Review on DC collection grids for offshore wind farms with high-voltage DC transmission system. *IET Power Electron.* **2017**, *10*, 2104–2115. [CrossRef]

16. Gil, M.D.P.; Domínguez-García, J.L.; Díaz-González, F.; Aragüés-Peñalba, M.; Gomis-Bellmunt, O. Feasibility analysis of offshore wind power plants with DC collection grid. *Renew. Energy* **2015**, *78*, 467–477.

17. Xu, L.; Yao, L. DC voltage control and power dispatch of a multi-terminal HVDC system for integrating large offshore wind farms. *IET Renew. Power Gener.* **2011**, *5*, 223–233. [CrossRef]

18. Kong, X.; Jia, H. Techno-Economic Analysis of SVC-HVDC Transmission System for Offshore Wind. In Proceedings of the 2011 Asia-Pacific Power and Energy Engineering Conference, Wuhan, China, 25–28 March 2011; pp. 1–5.

19. Xiang, X.; Merlin, M.M.C.; Green, T.C. Techno-Cost analysis and comparison of HVAC, LFAC and HVDC for offshore wind power connection. In Proceedings of the 12th IET International Conference on AC and DC Power Transmission (ACDC 2016), Beijing, China, 28–29 May 2016.

20. Hassan, A. Hamdan and Brian Kinsella. Using a VSC Based HVDC Application to Energize Offshore Platforms from Onshore—A Life-cycle Economic Appraisal. *Energy Procedia* **2017**, *105*, 3101–3111.

21. Eeckhout, B.V. The Economic Value of VSC HVDC Compared to HVAC for Offshore Wind Farms. Ph.D. Thesis, Katholieke Universiteit Leuven, Leuven, Belgium, 2007.

22. Elliott, D.; Bell, K.R.W.; Finney, S.J.; Adapa, R.; Brozio, C.; Yu, J.; Hussain, K. A Comparison of AC and HVDC Options for the Connection of Offshore Wind Generation in Great Britain. *IEEE Trans. Power Deliv.* **2016**, *31*, 798–809. [CrossRef]

23. Abdulrahman, A.; Santiago, B.; Omar, E.; Grain, A.; Callum, M. HVDC Transmission: Technology Review, Market Trends and Future Outlook. *Renew. Sustain. Energy Rev.* **2019**, *112*, 530–554.

24. Johnson, M.H.; Aliprantis, D.C.; Chen, H. Offshore wind farm with DC collection system. In Proceedings of the 2013 IEEE Power and Energy Conference at Illinois (PECI), Champaign, IL, USA, 22 February 2013; pp. 53–59.

25. Lakshmanan, P.; Liang, J.; Jenkins, N. Assessment of collection systems for HVDC connected offshore wind farms. *Electr. Power Syst. Res.* **2015**, *129*, 75–82. [CrossRef]

26. Ebenhoch, R. Comparative Levelized Cost of Energy Analysis. *Energy Procedia* **2015**, *80*, 108–122. [CrossRef]

27. Masters, G.M. *Renewable and Efficient Electric Power Systems*; John Wiley & Sons: Hoboken, NJ, USA, 2013

28. Cali, U. *Grid and Market Integration of Large-Scale Wind Farms Using Advanced Wind Power Forecasting: Technical and Energy Economic Aspects*; Kassel University Press GmbH: Kassel, Germany, 2011; pp. 1–156.

29. Renewables.ninja. 2017. Available online: https://www.renewables.ninja/ (accessed on 29 January 2019).

30. Staffell, I.; Pfenninger, S. Using bias-corrected reanalysis to simulate current and future wind power output. *Energy* **2016**, *114*, 1224–1239. [CrossRef]

31. Siemens. Wind Turbine SWT-4.0-130 Technical Specifications. 2015. Available online: https://www.thewindpower.net/scripts/fpdf181/turbine.php?id=957 (accessed on 25 May 2019).

32. Al Ameri, A.; Ounissa, A.; Nichita, C.; Djamal, A. Power Loss Analysis for Wind Power Grid Integration Based on Weibull Distribution. *Energies* **2017**, *10*, 463. [CrossRef]

33. ABB. HiPak IGBT Modules Data Sheet. 2014. Available online: www.abb.com/semiconductors (accessed on 12 January 2019).

34. Zhang, X.; Green, T.C. The modular multilevel converter for high step-up ratio DC–DC conversion. *IEEE Trans. Ind. Electron.* **2015**, *62*, 4925–4936. [CrossRef]

35. Gultekin, B.; Gercek, C.O.; Atalik, T.; Deniz, M.; Bicer, N.; Ermis, M.; Kose, K.N.; Ermis, C.; Ko, E.; Cadirci, I.; et al. Design and Implementation of a 154-kV±50-Mvar Transmission STATCOM Based on 21-Level Cascaded Multilevel Converter. *IEEE Trans. Ind. Appl.* **2012**, *48*, 1030–1045. [CrossRef]

36. Mamane, C. Transformer Loss Evaluation: User-Manufacturer Communications. *IEEE Trans. Ind. Appl.* **1984**, *20*, 11–15. [CrossRef]

37. Lerch, M.; De-Prada-Gil, M.; Molins, C.; Benveniste, G. Sensitivity analysis on the levelized cost of energy for floating offshore wind farms. *Sustain. Energy Technol. Assess.* **2018**, *30*, 77–90. [CrossRef]

38. Gonzalez-Rodrigue, A.G. Review of offshore wind farm cost components. *Energy Sustain. Dev.* **2017**, *37*, 10–19. [CrossRef]

39. Dicorato, M.; Forte, G.; Pisani, M.; Trovato, M. Guidelines for assessment of investment cost for offshore wind generation. *Renew. Energy* **2011**, *36*, 2043–2051. [CrossRef]

40. Wei, Q.; Wu, B.; Xu, D.; Zargari, N.R. IOP Conference Series: Earth and Environmental Science. In Proceedings of the 12th IET International Conference on AC and DC Power Transmission (ACDC 2016), Beijing, China, 28–29 May 2016.

41. Koch, H.; Retzmann, D. Connecting large offshore wind farms to the transmission network. In Proceedings of the IEEE PES T&D 2010, New Orleans, LA, USA, 19–22 April 2010.

42. Europa Cable. An Introduction to High Voltage Direct Current (HVDC) Underground Cables. Technical Brochure. Available online: www.europacable.eu (accessed on 12 January 2019).

43. ENSO. Offshore Transmission Technology. 2011. Available online: https://www.entsoe.eu (accessed on 20 February 2019).

44. Zaccone, E. High voltage underground and subsea cable technology options for future transmission in Europe. In Proceedings of the European Electricity Grid Initiative E-Highway 2050 Workshop, Brussels, Belgium, 15 April 2014.

45. Lundberg, S. *Performance Comparison of Wind Park Configurations*; Technical Report; Chalmers University of Technology: Gothenburg, Sweden, 2003.

46. ABB. HVDC Light. 2011. Available online: www.new.abb.com/docs/default-source/ewea-doc/hvdc-light.pdf (accessed on 20 February 2019).

47. BS6622/BS7835 Three Core Armoured 33kV XLPE Stranded Copper Conductors. Prysmian. 2011. Available online: www.powerandcables.com (accessed on 20 February 2019).

48. BS6622/BS7835 Three Core Armoured 33kV XLPE Stranded Copper Conductors. 6-36kV Medium Voltage Underground Power Cables XLPE Insulated Cables. 2019. Available online: www.nexans.co.uk (accessed on 20 February 2019).

49. Nieradzinska, K.; MacIver, C.; Gill, S.; Agnew, G.A.; Anaya-Lara, O.; Bell, K.R.W. Optioneering analysis for connecting Dogger Bank offshore wind farms to the GB electricity network. *Renew. Energy* **2016**, *91*, 120–129. [CrossRef]

50. Daniel Manrique, O. *Design of the HVDC Connection of a 425 MW Offshore Wind Farm to the German Network*; Technical Report; Universidad Pontificia: Madrid, Spain, 2017.

51. Stamatiou, G. Techno-Economical Analysis of DC Collection Grid for Offshore Wind Parks. Master's Thesis, The University of Nottingham, Nottingham, UK, 2010.

52. Lazaridis, L.P. Economic Comparison of HVAC and HVDC Solutions for Large Offshore Wind Farms underSpecial Consideration of Reliability. Master's Thesis, School of Electrical Engineering, KTH, Stockholm, Sweden, 2005.

53. Stamatiou, G.; Srivastava, K.; Reza, M.; Zanchetta, P. Economics of DC Wind Collection Grid as Affected by Cost of Key Components. In Proceedings of the World Renewable Energy Congress-Sweden, Linköping, Sweden, 8–13 May 2011.

54. Korompili, A.; Wu, Q.; Zhao, H. Review of VSC HVDC connection for offshore wind power integration. *Renew. Sustain. Energy Rev.* **2016**, *59*, 1405–1414. [CrossRef]

55. Heimiller, D.; Ho, J.; Moné, C.; Hand, M. 2015 Cost of Wind Energy Review. In *National Renewable Energy Laboratory/TP-6A20-66861 Technical Report*; 2017; 115p. Available online: https://www.nrel.gov/docs/fy17osti/66861.pdf (accessed on 12 July 2019) .

56.   Kempton, W.; McClellan, S.; Ozkan, D. Massachusetts offshore wind future cost study. *Special Initiative on Offshore Wind*; University of Delaware: Newark, DE, USA, 2016.

57.   Kim, J.-Y.; Oh, K.-Y.; Kang, K.-S.; Lee, J.-S. Site selection of offshore wind farms around the korean peninsula through economic evaluation. *Renew. Energy* **2013**, *54*, 189–195. [CrossRef]

Article

Ocean Renewable Energy Potential, Technology, and Deployments: A Case Study of Brazil

Milad Shadman [1,*], **Corbiniano Silva** [2], **Daiane Faller** [3], **Zhijia Wu** [1,4],
Luiz Paulo de Freitas Assad [2,5], **Luiz Landau** [2], **Carlos Levi** [1] and **Segen F. Estefen** [1]

[1] Ocean Engineering Department, Federal University of Rio de Janeiro, Rio de Janeiro 21941-914, Brazil;
zhijiawu@lts.coppe.ufrj.br (Z.W.); Levi@laboceano.coppe.ufrj.br (C.L.); segen@lts.coppe.ufrj.br (S.F.E.)

[2] Civil Engineering Department, Federal University of Rio de Janeiro, Rio de Janeiro 21941-907, Brazil;
corbiniano@gmail.com (C.S.); luizpauloassad@gmail.com (L.P.d.F.A.); landau@lamce.coppe.ufrj.br (L.L.)

[3] Center for Global Sea Level Change (CSLC), New York University Abu Dhabi (NYUAD),
Abu Dhabi PO Box 129188, UAE; daianecem@gmail.com

[4] China Ship Scientific Research Center (CSSRC), Wuxi, Jiangsu 214082, China

[5] Meteorology Department, Federal University of Rio de Janeiro, Rio de Janeiro 21941-916, Brazil

* Correspondence: milad.shadman@lts.coppe.ufrj.br

Received: 26 July 2019; Accepted: 9 September 2019; Published: 25 September 2019

Abstract: This study, firstly, provides an up-to-date global review of the potential, technologies, prototypes, installed capacities, and projects related to ocean renewable energy including wave, tidal, and thermal, and salinity gradient sources. Secondly, as a case study, we present a preliminary assessment of the wave, ocean current, and thermal gradient sources along the Brazilian coastline. The global status of the technological maturity of the projects, their different stages of development, and the current global installed capacity for different sources indicate the most promising technologies considering the trend of global interest. In Brazil, despite the extensive coastline and the fact that almost 82% of the Brazilian electricity matrix is renewable, ocean renewable energy resources are still unexplored. The results, using oceanographic fields produced by numerical models, show the significant potential of ocean thermal and wave energy sources in the northern and southern regions of the Brazilian coast, which could contribute as complementary supply sources in the national electricity matrix.

Keywords: ocean renewable energy; ocean renewable technologies; ocean source potential; Brazilian ocean energy

---

1. Introduction

Only 14% of the world's primary energy matrix originates from renewable resources (based on the 2016 database), and this value is about 25% for the electrical energy sector [1]. The immediate needs to limit climate change and achieve sustainable growth are two key drivers of global energy transformation. Consequently, it is estimated that the share of renewable energy sources in the electrical energy sector will increase from 25% in 2017 to 85% in 2050 [1], in which ocean renewable energy sources including wave, tidal, thermal, and the salinity gradient will be responsible for the 4% of the total electricity generation. However, new approaches to power system planning, system and market operations, and regulations and public policy will be required to obtain that goal. As the contribution of low-carbon electricity becomes significant and it becomes the preferred energy carrier, the share of electricity consumed in the end-use sectors will need to increase from approximately 20% in 2015 to 40% in 2050 [1]. Electricity generation using coal, oil, gas, hydroelectric, nuclear, and bioenergy is predicted to decline from 2015 to 2050. On the other hand, a rapid evolution associated with the use of renewables like wind, geothermal, solar, ocean renewable energy, and concentrated solar power

(CSP) will likely be observed. International Renewable Energy Agency (IRENA) showed that the sources of renewable electricity in 2050 will be dominated by solar and wind power plants, highlighting significant growth associated with the geothermal, CSP, and ocean renewables.

Although Brazil is currently one of the world's cleanest energy suppliers, there are some concerns associated with the country's energy sustainability. An increasing demand for energy, mainly fossil fuels, expanding oil production, a bioenergy sector struggling with expansion, fast growth of energy-related greenhouse gas emissions, and energy efficiency performance deterioration are the current trends that put the future of the country's sustainable energy performance at risk [2].

Brazil is the world's eighth-largest economy with a population of close to 210 million and a land area expansion the size of about two times the European Union [2,3]. With a domestic energy supply of about 292.1 million tons of oil equivalent (Mtoe) in 2017, it is one of the largest energy producers in the world [4]. The Energy Research Office (EPE) of Brazil estimated a domestic electricity supply of 624.3 TWh in 2017, and this was mainly produced by the hydropower plants.

The Brazilian electrical and energy matrices are predominately based on renewable energy sources, which means that, in addition to having lower operating costs, a much lower greenhouse gas effect is emitted in association with energy production and consumption. For instance, in 2017, the total anthropogenic emissions of the Brazilian energy mix was estimated at approximately 435.8 million tons of equivalent carbon dioxide (Mt CO_2-eq), of which the transport sector emitted the largest part (199.7 Mt CO_2-eq) [4]. Based on the data presented by the International Energy Agency (IEA) [5], each Brazilian issued an average of 2.1 t CO_2-eq, considering the production and consumption of energy in 2017. This is three times less than that of a European or Chinese citizen and about seven times less than an American citizen.

By meeting almost 45% of its primary energy demand from renewable resources, Brazil has the least carbon-intensive energy sector in the world [6]. Figure 1a shows the domestic energy supply breakdown for Brazil for 2017. Petroleum and oil products, with a share of 36.4%, had the largest contribution to energy supply, followed by sugarcane biomass (17%). Natural gas (13%) and hydraulic energy were other players in the energy matrix of Brazil. Black liquor contributed 50.6% of the "other renewables" sector, followed by wind (21.3%), biodiesel (19.7%), other biomasses including rice husk, elephant grass, and vegetable oil (6.5%), charcoal industrial gas (0.4%), biogas (1.1%), and solar energy (0.4%). The Brazilian electrical matrix, as shown in Figure 1b, was dominated by hydropower resources with a contribution of approximately 65.2%. The main Brazilian hydroelectric reservoirs are located in the Paraná River basin, South region, featuring the Itaipu plant, which is the second-largest hydroelectric power plant in the world with a capacity of 14 GW [7]. The hydroelectric power plants in Brazil are mostly concentrated in the Midwest, South, and Southeast regions. Several studies have discussed the benefits and challenges of the hydroelectric plants in Brazil [7–10]. Nevertheless, the remoteness and environmental sensitivity of a large part of the remaining resources are two hurdles that constrain the continued expansion of hydroelectric plants in Brazil [6].

Brazil already has a significant contribution of renewable energy in its energy and electricity matrix; however, there is an inestimable untapped potential for energy supply from the oceans. Although nearly 80% of the Brazilian population lives near the coast, there has been no in-depth survey on the utility of ocean energy and its conversion into electricity. There have only been a handful of studies associated with the ocean renewable energy potential along the Brazilian coastline, and these have mainly focused on the wave and ocean current energy in some specific regions. Some examples of the studies related to the wave and current energy include those in [11–16], which focused on the South and Southeast regions of the Brazilian coast. Moreover, ocean thermal energy conversion (OTEC) resource evaluation of the Southern Brazilian continental shelf is presented in [17]. The EPE, through the National Energy Plan [18,19], established some general roadmaps related to the long-term plan of the Brazilian energy sector. Accordingly, the ocean energy resources, among other alternative energy sources, were suggested as a way to expand the Brazilian energy matrix in the coming decades.

This was also emphasized by the National Agency of Electric Energy through a roadmap project performed by the Center of Management and Strategic Studies of Brazil in 2017 [20].

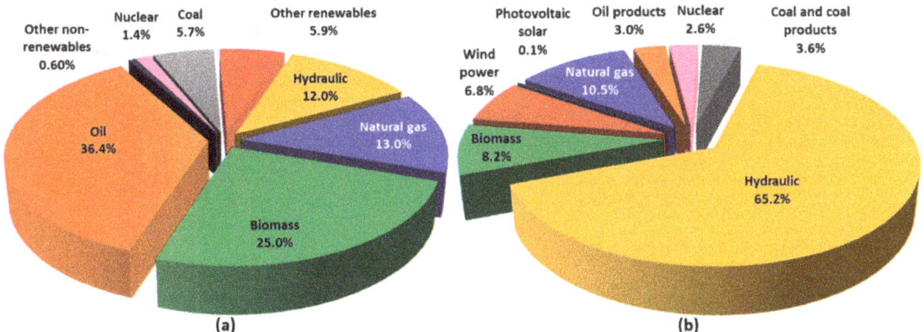

Figure 1. (**a**) Domestic Brazilian energy supply and (**b**) electrical matrix breakdown in 2017.

2. Targets, Materials, and Methods

In Brazil, mapping of the ocean renewable energy resources through a detailed survey of all resources is required to identify potential areas for exploration and, consequently, encourage the development of technologies through the implementation of socio-economically feasible and acceptable projects. Using this perspective, this article firstly presents an overview of the global potential of ocean renewable energy resources and the associated technologies for harnessing such energy. Then, in the second part, the global status of technology maturity is presented through a wide survey of projects, which are at different stages of development. This shows the current global installed capacity for different energy sources, as well as pointing out the more promising technologies through the global interest trend. The third part presents an assessment of the ocean renewable energy resources including ocean currents, waves, and thermal gradients along the Brazilian coastline. This is a preliminary effort aimed at indicating the potential energetic regions. Further detailed works are required to investigate these locations. The methodology applied in this study consists of the use of oceanographic fields produced by hydrodynamic models to estimate the potential of the energy resources. Modeling is performed for a data resolution (one regular horizontal grid) of 1/12° (~9 km). The study reveals the theoretical potential (available energy at sea and not what can be captured) of the resources as well as their seasonal and temporal variability. Finally, the main Brazilian projects are presented, and the challenges are discussed.

2.1. Study Area

The Brazilian coastline is more than 7400 km in length and is situated between 04°52′45″N (Oiapoque River) and 33°45′10″S (Chuí River). The marine areas under Brazilian jurisdiction include the Territorial Sea, with a limit of 12 nautical miles; the Exclusive Economic Zone (EEZ), with 12 to 200 nautical miles; and the Continental Shelf, which comprises the seabed that extends beyond the Territorial Sea, along the natural extension of the land territory off the continental shelf.

The extent of the Brazilian continental shelf varies along the coast, with a few kilometers (~8 km) near Bahia and up to 300 km on the coast of the State of Pará, with a range between 60 and 180 m [21,22]. The Brazilian coastline is characterized by intraseasonal fluctuations in the upper ocean circulation due to several dynamic processes, such as the local forcing dynamics, the remote forcing of winds via waveguide dynamics, the average flow instability, and the resonance as a function of the coastline geometry [23,24]. The ocean circulation is dominated by the Subtropical Turn (Equatorial South Current, SEC) and the Antarctic Circumpolar Current [25]. The SEC is responsible for transporting the water from the Benguela Current to the Brazilian platform (about 10°S and 20°S), where it passes through a fork in the North Brazil Current (NBC) and the Brazil Chain (BC) to the south. Due to this circulation, the western margin of the tropical South Atlantic is a particularly interesting region for the observation of thermohaline circulation [21–24,26].

As illustrated in Figure 2, the study area included the Brazilian coastline inside the EEZ, which is divided into four regions A, B, C and, D, according to both hydrodynamic and atmospheric characteristics. Table 1 shows the regions and the corresponding latitudes.

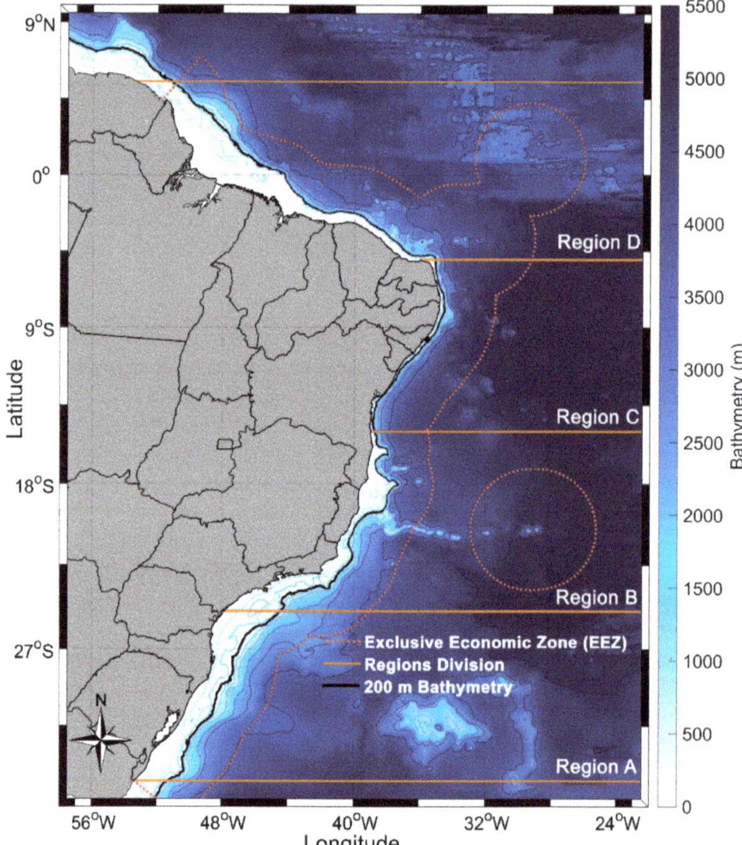

Figure 2. Brazilian coastline and the main marine areas delimited.

Table 1. Regions of the Brazilian coast considered in this study.

Regions	Latitude
A	34°S–25°S
B	25°S–15°S
C	15°S–05°S
D	05°S–05°N

2.2. Model Description

2.2.1. Ocean Current and Thermal Gradient Energy

The datasets for ocean current velocities and temperature were obtained (surface down to 5500 m) from the numerical model product available for the CMEMS (Copernicus Marine Environment Monitoring Service) center. The applied product is a high-resolution global analysis and forecasting system that uses the NEMO (3.1) ocean model [27]. It consists of part of the Operational Mercator global ocean analysis and forecast daily system, which was initiated on December 27, 2006. The dataset has one regular horizontal grid with a 1/12° (~9 km) resolution based on the tripolar ORCA grid [28], 50 vertical levels with 22 layers within the upper 100 m from the surface, bathymetry from ETOPO1 [29], and atmospheric forcings from the ECMWF (European Centre for Medium-Range Weather Forecasts). Additionally, it uses a data assimilation scheme, in which the initial conditions for numerical ocean forecasting are estimated by joint assimilation of the altimeter data, in situ temperature, salinity vertical profiles, and satellite sea surface temperature.

- Ocean current energy

Near-surface (~5 to 50 m) u and v components of velocity from January 1, 2007 to December 31, 2017 were used as a subset of the area corresponding to the Brazilian coastline (25°W–55°W and 6°N–34°S).

The ocean current power can be calculated as the amount of marine-hydrokinetic energy that flows through a unit cross-sectional area oriented perpendicular to the current direction per unit time [30] expressed as follows:

$$P = \frac{1}{2}\rho S^3,$$ (1)

where P is the current power density in (W/m^2), ρ is the density of seawater (defined as 1025 kg/m^3), and S is the flow speed (in m/s). In practice, only a fraction of this energy can be harnessed. The underwater turbine efficiency has a typical range from 35% to 50% [31]. Additionally, a mean peak current of more than 2 m/s is necessary for commercial power generation [32].

- Thermal gradient energy

Gridded daily seawater temperature (°C) model output with 50 vertical layers and ~9 km in horizontal resolution was used to analyze the temperature difference (ΔT (°C)) between the surface warm water and the deeper cold water. It was assumed that the superficial water intake pipe was located at about 20 m and the deepest point in the vertical depth stratification was approximately 1000 m. At specific locations (each grid cell), we calculated the gross power (P_{gross}) following the methodology described by [33,34]. The OTEC gross power can be expressed as the product of the evaporator heat load and the conversion efficiency of the gross OTEC [34]:

$$P_{gross} = \frac{Q_{cw}\rho c_p 3\varepsilon_{tg}\gamma}{16(1+\gamma)T}\Delta T^2,$$ (2)

$$\gamma = \frac{Q_{ww}}{Q_{cw}},$$ (3)

where γ is the flow rate ratio calculated for a 10 MW OTEC plant in which $Q_{ww} = 45$ m^3/s and $Q_{cw} = 30$ m^3/s are the warm surface water and the cold deep water flow rates, respectively [35]. ΔT is the difference in temperature between the surface layers and deeper layers, and T is the absolute temperature at the surface (in Kelvin) (20 m). ρ and ε_{tg} represent the water density, which was equal to 1025 kg/m^3, and the turbo-generator efficiency fixed at 0.75, respectively. c_p is the specific heat of seawater and has a value of 4000 J.kg^{-1}.K^{-1}.

A considerable amount of the P_g is consumed to pump the large seawater flow rates through the OTEC plant. The net power P_{net} should be calculated, which is usually about 30% of the P_g [36,37]. The P_{net} can be expressed by the following equation considering $\Delta T_{design} = 20\ °C$ and the other losses presented in [34]:

$$P_{net} = \frac{Q_{cw}\rho c_p \varepsilon_{tg}}{8T}\left\{\frac{3\gamma}{2(1+\gamma)}\Delta T^2 - 0.18\Delta T^2_{design} - 0.12\left(\frac{\gamma}{2}\right)^{2.75}\Delta T^2_{design}\right\}. \tag{4}$$

2.2.2. Wave Energy

The wave dataset was obtained using the operational global ocean analysis and forecast system of Météo-France that is available for the CMEMS (Copernicus Marine Environment Monitoring Service) center. The model had a horizontal resolution of 1/12° (~9 km) and 3-hourly instantaneous fields of integrated wave parameters. The global wave system of Météo-France is based on the third-generation wave model MFWAM. It uses the computing code ECWAM-IFS-38R2 with a dissipation term [38]. The 2-min gridded global topography data ETOPO2/NOAA were used to generate the model's mean bathymetry. The dataset uses three years of data to estimate the wave climatology along the Brazilian coastline (between 2015 and 2018). The power density P was calculated using the significant wave height H_s and the wave energy period T_e as follows:

$$P = \frac{\rho g^2}{64\pi}H_s^2 T_e \tag{5}$$

where ρ and g represent the seawater density (1025 kg.m^{-3}) and gravity acceleration (9.806 m.s^{-2}), respectively; H_s is the significant height (m); and T_e is the energy wave period (s). This simplified expression uses deep-water approximation [39], which fits well most of the modeled domains; however, more sophisticated techniques as well as in situ measurements are required to precisely determine the shallow water wave climate.

2.3. Metrics

The variability of the available ocean renewable energy in time is an important issue due to its impact on the capacity factor, which, consequently, affects the economy of the ocean energy system. Two different metrics were used to address the seasonal and temporal variability of the Brazilian coastline. The seasonal variability (SV) index [40] can be expressed as follows:

$$SV = \frac{P_{S,max} - P_{s,min}}{P_{year}}, \tag{6}$$

where $P_{s,min}$ and $P_{S,max}$ are the mean wave power of the least and the most energetic seasons, respectively, and the P_{year} is the annual mean power. Greater values of SV imply a larger seasonal variability; however, it should be noted that this is the variability of the energy resources relative to their mean level on a three-month seasonal time scale [40]. The temporal variability of the energy at a site or region can be evaluated by the coefficient of variation (COV) [40], which is expressed as

$$COV(P) = \frac{SD(P(t))}{mean\ (P(t))} = \frac{\left[\overline{\left(P - \overline{P}\right)^2}\right]^{0.5}}{\overline{P}}, \tag{7}$$

where SD is the standard deviation, and the over-bar denotes the time-averaging. A COV equal to zero leads to a fictitious power time series with absolutely no variability, while $COV\ (P) = 1$ and 2 imply that the standard deviation of the time series is equal to and twice the mean value, respectively.

3. Literature on the Issue and State-of-the-Art Technology Related to Ocean Renewable Energy

3.1. Resource Potential

Ocean renewable energy, also referred to as marine renewable energy, is defined as energy captured by technologies that utilize the seawater's motion or potential properties as the driving power or harness its heat or chemical potential. The ocean surface waves, tidal range, tidal current, ocean current, and thermal and salinity gradient are renewable sources of ocean energy that have different origins. Technologies associated with ocean energy convert these renewable energy sources into electricity or other desirable forms of energy.

Other renewable sources of energy can be exploited from the ocean environment that is excluded from the above definition. The production of biofuels from marine biomass, energy harnessing from submarine vents, and offshore wind are some examples that can be considered as forms of bioenergy, geothermal, and wind energy, respectively.

The highest level of potential is theoretical potential, which only takes into account the natural and climatic characteristic limitations. Reducing this potential due to the consideration of the technical limitations, such as the conversion efficiency and storage of electricity, results in technical potential, which varies with the development of technologies.

3.1.1. Wave Power

Temporal variations of the wave condition can be estimated by the use of long-term averages in modeling, applying global databases with reasonably long histories [41]. It can be observed that the most energetic waves exist in the region between latitudes of 30° and 60° of both hemispheres because of the extra-tropical storms [41].

Mørk et al. [42] calculated the theoretical potential of wave energy resources for areas with a wave power larger than 5 kW per meter and a lower latitude than 66.5°. Accordingly, they presented a total theoretical potential of about 3.37 TW (29,500 TWh/yr or 106.2 EJ/yr). An overall technical potential of 500 GW (around 16 EJ/yr) was estimated by Sims et al. [43], assuming an efficiency of 40% for wave energy converters installed in offshore regions with a wave power exceeding 30 kW/m. Krewitt et al. [44] presented a technical potential of 20 EJ/yr. Gunn and Stock-Williams estimated a global theoretical potential of about 2.11 TW, of which 4.6% was predicted to be extractable by deploying a specific wave energy converter (WEC) [45]. They considered the area between 30 nautical miles and the Exclusive Economic Zone (EEZ) for each region. Besides these global studies, some works have assessed the wave energy resource potential at national and regional levels in China [46,47], Italy [48], Spain [49], Ireland [50], and the USA [51].

3.1.2. Tidal Power

Tidal ranges can be forecasted accurately. The world's largest tidal ranges occur in the Bay of Fundy, Canada (17 m), the Severn River Estuary, the United Kingdom (15 m), and the Bay of Monte Saint Michel, France (13.5 m) [52]. In addition, Argentina, Australia, China, India, Russia, and South Korea also have large amounts of tidal power.

The global theoretical potential of tidal energy (tidal ranges and currents) is estimated to be in the range of 500–1000 TWh/yr (1.8–3.6 EJ/yr) [52]. Sims et al. [43] estimated that tidal currents of more than 100 TWh/yr (0.4 EJ/yr) could be converted into electrical energy if major estuaries with large tidal fluctuations could be tapped [43].

The Ocean Energy System (OES) reported the worldwide theoretical potential of tidal energy, including tidal current, to be around 7800 TWh/yr (28.1 EJ/yr) [53]. Some studies of the regional tidal energy resource potential can be found in Scotland [54], Uruguay [55], Ireland [56], Taiwan [57], and Iran [58].

3.1.3. Ocean Current Power

The ocean current is the movement of seawater in the open sea generated by forces acting upon the water, including wind, the Coriolis effect, temperature and salinity differences, and so on. Compared with tidal currents, ocean currents are generally slower, relatively constant, flow in only one direction, and fluctuate seasonally. The currents off South Africa (Agulhas/Mozambique), the Kuroshio Current (off East Asia), the Gulf Stream (off eastern North America), and the East Australian Current are locations with potential ocean currents already identified [59]. Yang et al. [60] estimated a theoretical potential of about 163 TWh/yr for the Gulf Stream system, considering the entire area of the Gulf Stream within 200 miles of the US coastline between Florida and North Carolina as the extraction region. Besides that, they calculated a technical potential of about 49 TWh/yr, assuming a power conversion efficiency of 30%. Chang et al. [61] identified suitable sites for ocean current energy extraction near the coastlines of Japan, Vietnam, Taiwan, and the Philippines. Goundar and Ahmed [62] evaluated marine current resources for Fiji, presenting a peak velocity of about 2.5 m/s.

3.1.4. Ocean Thermal Energy

The resulting temperature difference between the upper layers and the colder layers of seawater—usually at a depth of more than 1000 m—can be converted through different oceanic thermal energy conversion (OTEC) methods [63]. In practice, a minimum temperature difference of 20 °C (or K) is required for the use of the temperature gradient in the generation of electricity. The tropical latitudes (0° to 30°) in both hemispheres, including the western and eastern coasts of the Americas, many islands of the Caribbean and Pacific, and the coasts of Africa and India, are the places with the greatest potential [59]. Although there is a little variation in the temperature gradient from summer to winter, the thermal gradient feature is continuously available. It is estimated that about 44,000 TWh/yr (159 EJ/yr) to 88,000 TWh/yr (318 EJ/yr) of power could be generated through OTEC devices [34,64]. Rajagopalan and Nihous [65] estimated that an annual OTEC net power of about 7 TW could be obtained, considering the small effect on the oceanic temperature field. Thus, ocean thermal energy has the highest potential among ocean renewable energy sources. However, the energy density of OTEC systems is quite low compared with that of other sources, such as waves and tidal currents. This issue may affect the low-cost OTEC operation [63], requiring further investigation.

3.1.5. Salinity Gradient Power

The salinity gradient power (osmotic power) is the potential energy from the difference in the salt concentrations of seawater and freshwater. Energy is released due to the mixing of fresh water with seawater. The entropy of the freshwater–seawater mixture can be exploited as pressure by using the semipermeable membrane. This pressure can be converted into the desired energy form. The freshwater rivers discharging into saltwater are distributed globally, with a volume of about 44,500 km^3/yr. Assuming that only 20% of this discharge can be used for salinity gradient energy generation, the overall potential is approximately 2000 TWh/yr (7.2 EJ/yr) [44]. Skramesto et al. [66] estimated a technical potential of 1650 TWh/yr (5.9 EJ/yr) for the production of salinity gradient energy. Recently, in [67] it was shown that, practically, 625 TWh/yr of salinity gradient energy is globally extractable from river mouths. Some examples of the regional assessment of the salinity gradient potential can be found in Colombia [68], remote regions of Quebec [69], and the hypersaline Urmia Lake of Iran [70]. The potential for salinity gradient energy extraction from some major world rivers is presented in [71].

Figure 3 shows a summary of the potential of ocean renewable energy resources based on the references presented in this paper. The bars illustrate the range of estimated resource potential. Note that the technical potential of ocean current is shown as presented in [60] for the Gulf Stream.

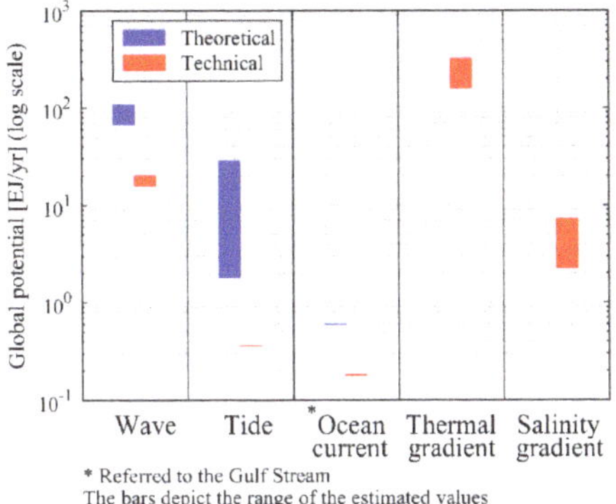

Figure 3. Summary of the global ocean renewable energy resources.

3.2. Conversion Technologies

3.2.1. Wave Energy Conversion

Currently, there are a large number of concepts and patents on the use of wave energy. As illustrated in Figure 4, the process of the wave energy conversion can be divided into three main stages: the primary conversion stage, the secondary conversion stage, and the tertiary conversion stage [72]. In the primary conversion stage, the wave converter captures the kinetic energy of the waves through wave body interactions (e.g., buoy oscillation, air flow, or water flow). The secondary conversion stage converts the body motion energy into electricity through the power take-off (PTO) system, and the tertiary stage adapts the characteristics of the produced power to the grid requirements with power electronic interfaces.

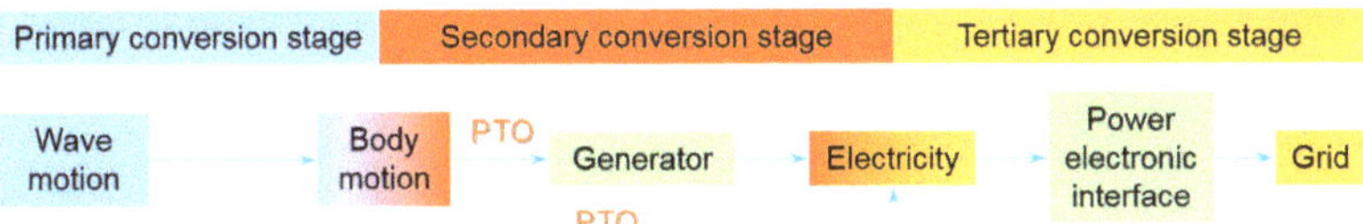

Figure 4. Main stages of wave energy conversion.

Based on the working principles of the WEC, from the primary conversion stage to the secondary conversion stage, the general categories are classified as shown in Figure 5 [72–75]:

- The oscillating water column (OWC), which compresses or decompresses the air in a chamber using the wave elevation to drive a Wells or impulse turbine to convert wave power. Depending on the location of installation, OWC devices can be fixed onshore [76–78], as shown in Figure 6a, or floating, as shown in Figure 6b [79–81].
- Wave activated bodies (WABs), which utilize the wave excitation motions between two bodies to convert wave power into electric power. According to their dimension and orientation, WABs can also be classified as *terminators* [82] (Figure 6c), positioned with large horizontal extensions perpendicular to the wave propagation direction); *attenuators* [83,84] (Figure 6d), which have a large horizontal extension parallel to the wave propagation direction; *point absorbers* [85,86] (Figure 6e), which have small dimensions compared to the predominant wavelength and are usually axisymmetric about their vertical axis; and *submerged pressure differentials* [87] (Figure 6f), which are submerged buoys with large dimensions.

- The overtopping device, which utilizes the overtopping phenomenon to the let the water fall through the turbine to converter the potential energy into electric power [88–90], as shown in Figure 6g.
- Others, describing concepts different from the above categories, e.g., the wave carpet [91] and the *rotating mass* [92,93], which uses the motion of a hull to accelerate and maintain the revolutions of a spinning mass inside.

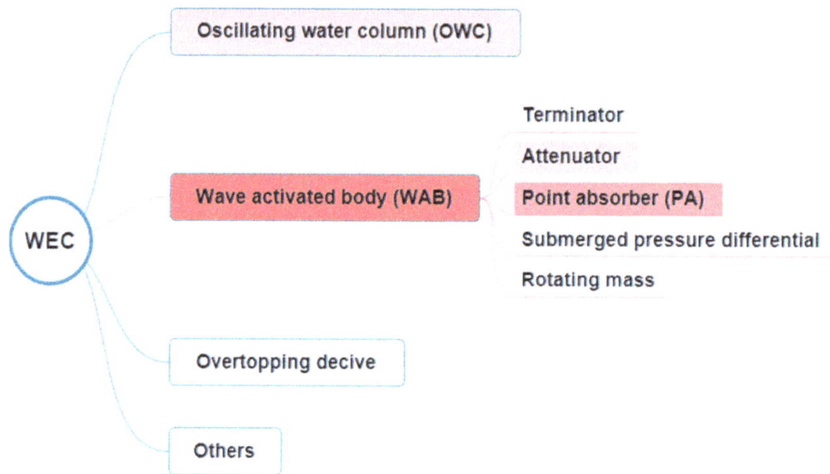

Figure 5. Categories of wave energy converter (WEC) technology.

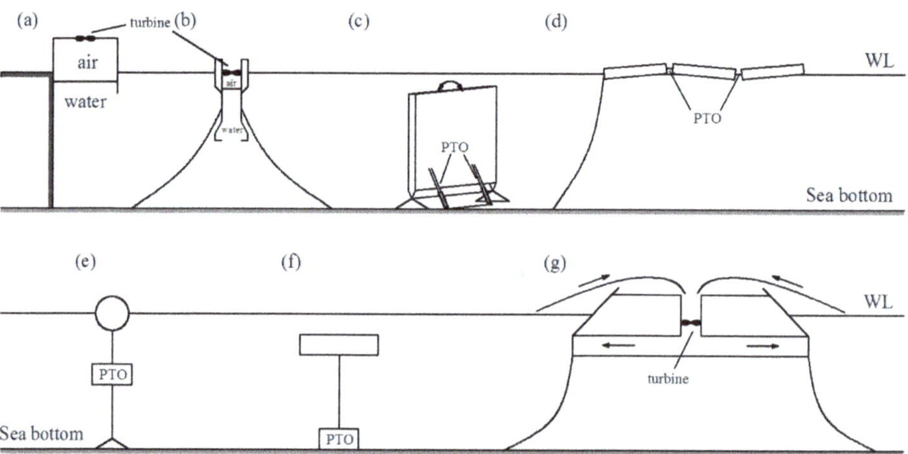

Figure 6. WEC technologies.

There are different types of PTO systems adopted for different WEC devices, e.g., pneumatic [94,95], hydraulic [96–99], direct mechanical drive [100–102], and direct electrical drive [103–105]. An elaborated description of these systems can be found in [106–108].

A time-varying wave climate in real sea may deteriorate the power quality gained from a single device. In practice, some strategies, like short-term energy storage or PTO resistance control, can be utilized to smooth energy production. However, arrays of wave energy converters are more desirable because their cumulative energy generation will be smoother than the energy production of a single device [59]. Efficient power production is strictly dependent on the advanced control systems.

3.2.2. Tidal Range Energy Conversion

Tidal barrage power plants use the difference in the height of the water surface during ebbs and floods (tidal range) to drive common hydro turbines. The greater tidal range results in higher energy extraction by the power plant.

Figure 7 presents the principle of the tidal barrage power plant [109]. The shapes of structures are similar to dams and the structures are built in the estuary or bays to store water at high tide. A difference in water height at the internal and external sides of the dam occurs due to changes in the tidal regime. Three main tidal barrage schemes are

- Flood generation—the power production process starts as the water enters the tidal basin (flood tide);
- Ebb generation—power production starts as the water leaves the tidal basin (ebb tide);
- Two-way generation—the tidal power plant produces power during the flood and ebb tides.

The sluice gates on the barrages are utilized to control water levels and flow rates. In general, ebb generation is generally more efficient than flood generation. This is due to there being more kinetic energy in the upper half of the basin in which ebb generation operates. This is because of the effects of gravity and the second filling of the basin from inland rivers and streams connected to it via the land. The bi-directional tidal turbine generators are generally more expensive and less efficient than unidirectional tidal generators [110]. Some studies related to the tidal barrages' dynamics, power performance, and economy can be found in [111–115].

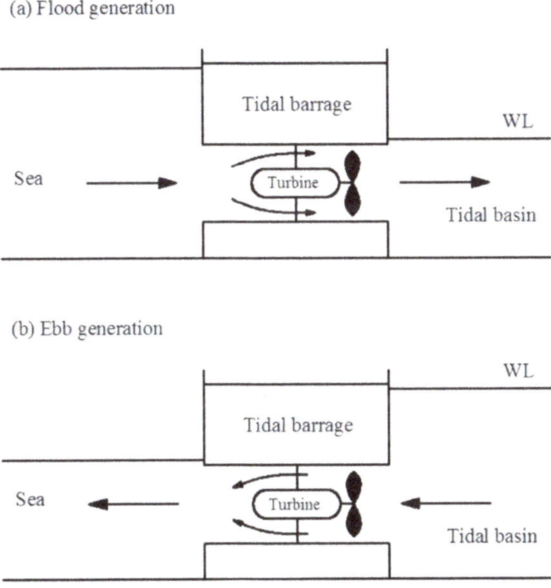

Figure 7. Tidal barrage principles.

3.2.3. Tidal Current and Ocean Current Energy Conversion

The kinetic energy of both tidal and ocean currents can be extracted with hydrokinetic turbines, which have similarities with wind turbines but are still submerged. Due to the higher density of water compared to that of air, the blades of current turbine devices have smaller dimensions and move slower than those of wind turbines while still providing a significant amount of energy. The main difference between tidal and ocean currents is that the tidal currents have bi-directional flows in contrast to the ocean currents which are unidirectional. This has implications for the design of a tidal current turbine, which must act in both directions of the water flow [59].

The four major types of current energy conversion devices are as follows, as shown in Figure 8:

- Horizontal axis turbine [116–118]—blades, driven by current energy, rotate around the horizontal axis;
- Vertical axis turbine [119–124]—blades, driven by current energy, rotate around the vertical axis;
- Oscillating hydrofoil [125–128]—currents passing the hydrofoil result in the lift force, which can drive the motion of the hydraulic system to generate electricity;
- Venturi effect turbines [129]—harness the kinetic energy of the current by amplifying the current velocity by the Venturi effect in the strangulated section of a tube [63].

Additionally, ducted channels are utilized to induce a sub-atmospheric pressure within a constrained area and, consequently, increase the flow velocity around the rotor [130–132]. The tidal kite is a novel technology, which uses wing carrying and pushes a turbine in an "8" shaped trajectory, sweeping a large area with a relative speed of more than the local current speed [133–135]. A review of the tidal current technology is presented in [136].

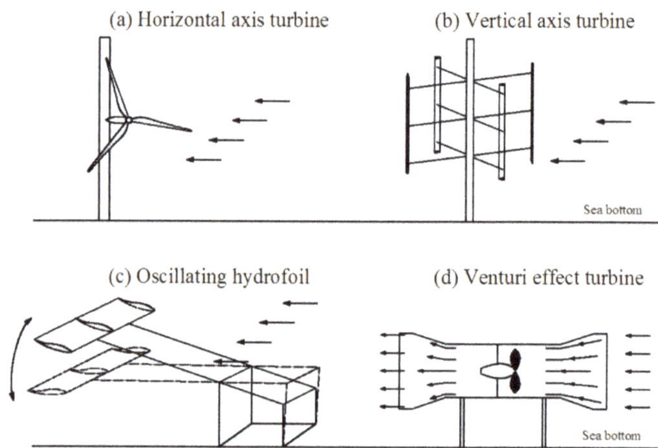

Figure 8. Primary types of tidal and ocean current energy conversion devices.

3.2.4. Ocean Thermal Energy Conversion (OTEC)

OTEC is a technology that converts the difference in temperature between the sea surface and large depths (around 1000 m) for use in heating, cooling, or to generate electricity. Closed-cycle technology requires a minimum temperature difference of 20 °C, which is possible in equatorial marine regions. Warm water from the upper layers is used to vaporize a secondary working fluid (e.g., ammonia), thus driving an electric generator. The resultant steam is then condensed by the cold water, which is brought, using pumps, from the bottom of the ocean and then discarded. Some works that studied the closed-cycle OTEC can be found in [137–139].

Open cycle technology uses the warm surface ocean water as the working fluid, which is drawn into a vacuum vessel, causing the working fluid to vaporize. The main benefit of the open cycle process is that it produces both electricity and desalinated fresh water. An investigation of the performance of a shore-based low temperature thermal desalination using an open-cycle OTEC is presented in [140]. Recently, a novel optimal open-cycle OTEC plant using multiple condensers was designed in [141].

The hybrid cycle combines both closed and open cycle technologies. Similar to the open cycle process, the warm surface water is flash-evaporated into steam. The steam is used to vaporize the ammonia working fluid of a closed-cycle loop and drives a turbine to produce electricity. Afterwards, the steam is condensed through a heat exchanger, providing desalinated water [142]. Such a cycle can realize the generation of both electricity and fresh drinking water simultaneously.

3.2.5. Salinity Gradient Energy Conversion

The salinity gradient takes advantage of the power that can be generated by the mixture of cold and salty water, for instance, at the mouth of a river that flows into the sea. There are two common methods for generating energy from the salinity gradient: pressure-retarded osmosis (PRO) [143–146] and reverse electrodialysis (RED) [147–149].

The PRO method is based on semipermeable membranes that allow only the traveling of water molecules. In this approach, water flows from the diluted solution (freshwater) to the concentrated solution (seawater) to provide a chemical potential equilibrium on both sides of the membrane. This occurs only when the pressure difference between the liquids is less than the difference in osmotic pressure. This flow of water can be used to power turbines that transform mechanical energy into electricity. The RED method is based on the transport of ions (salt) through membranes. Two fluids of different salinities (freshwater and seawater) pass through a series of specific membranes. The difference in chemical potential between membranes results in an electrical voltage. The brackish water is then discarded into the sea.

Additionally, some hybrid processes, such as the production of electricity from thermal energy using a closed-loop RED heat engine, low-energy desalination by integrating RED with desalination facilities, and the use of microbial RED cells with boosted power performance, have been proposed to facilitate energy extraction from salinity gradient resources [149,150].

3.3. *Global Status of Development*

3.3.1. Installed Capacity

Despite decades of development efforts, a large amount of the ocean's renewable energy sources are still untapped [151]. In the last 10 years, the use of ocean energy sources has experienced significant growth globally. Since 2009, many devices have been deployed worldwide to capture the energy from currents, tidal ranges, thermal and salinity gradients, and waves. This progress is noticeable by the gradual increase in installed capacity in some continents, as shown in Figure 9 [152], demonstrating an expansion of marine energy in the world energy matrix. Globally, this growth has more than doubled from 244 MW in 2009 to 532 MW in 2018. However, more than 90% of this operating capacity is represented by two tidal barrages in La Rance, France and Sihwa Lake, South Korea.

The contribution of Asia is led by China and South Korea, where extraordinary progress has been made since 2011, mainly because of the development of tidal barrage facilities. This is due to government support, through the adoption of economic policies, the reduction of tariffs and exemption, which includes financial subsidy policies to encourage scientific research and development, the development of new renewable energy technologies, prototype demonstration, and development of the renewable energy industry [153]. After France developed the La Rance tidal barrage, the United Kingdom led the way in terms of installation capacity followed by Spain, Sweden, the Netherlands, Norway, Portugal, and Italy. In North America, mainly in Canada and the United States, the development of these energy sources is in advanced stages, with the implementation of demonstration-scale commercial projects. Africa, Central America and the Caribbean, South America, and Oceania are in the early stages of deploying ocean resources as energy sources, with incipient projects and installed capacity.

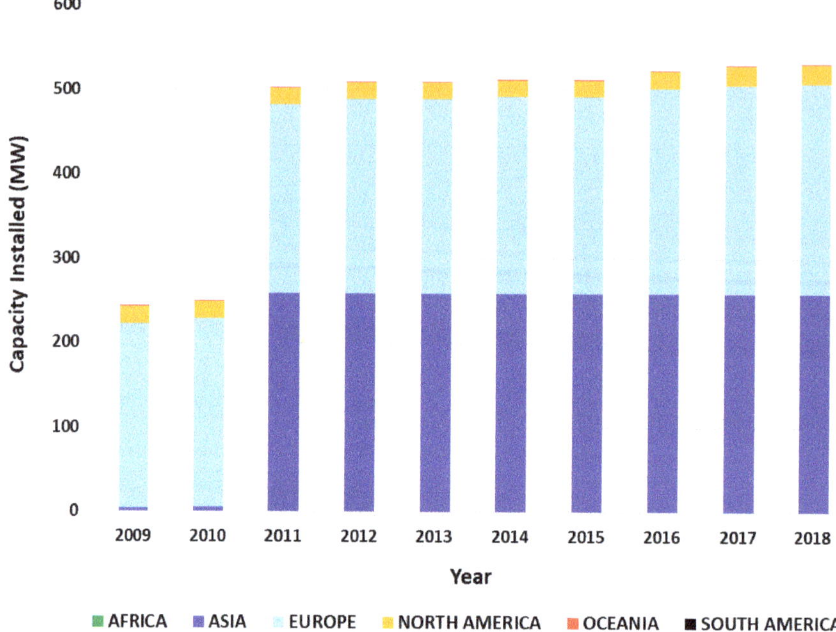

Figure 9. Marine energy: capacity installed by continents, according to the International Renewable Energy Agency (IRENA) (2019) [152].

3.3.2. Technology Status

The technology readiness level (TRL) presented by NASA has been adapted for ocean renewable energy technologies and presented by [154]. The TRL system quantifies the development of ocean energy devices from an initial stage of research and development (R&D) to industrial roll-out, which involves the mass production of off-the-shelf components and devices. Accordingly, the ocean renewable energy technologies are still at the conceptual, R&D, or demonstrative prototype stages. Nevertheless, in the case of waves and tidal currents, global commercial application is expected in the near to medium future. Based on reports, the extraction of energy from tidal ranges is still at the precommercial stage and the use of energy from tidal currents and waves is at the demonstration and prototype stages, respectively.

3.3.3. Deployed Devices

Large-scale (half- or full-scale) testing at sea is necessary for the pre-commercial stage of development. However, only a few devices have been constructed and tested at full scale. Oscillating water column devices have been employed as fixed onshore and floating offshore structures. Some examples of the fixed OWC prototypes have been deployed in Japan [155], Italy [156], Portugal [157], and Norway [158].

The Pelamis Wave power is the most mature wave-activated body device. It was installed for the first time in 2004 at the European Marine Energy Centre (EMEC) wave test site at Billia Croo. Later, its second generation, which comprised five connected sections, giving a total length of 180 m and a rated power of 750 kW, was installed at the same test site in 2010. AquaBuoy [159], Interproject Service (IPS) buoy [160], and Wave Bob [161] are examples of the wave-activated body devices that produce energy from the relative motion of two connected oscillating bodies.

Overtopping devices can be floating or fixed structures. The Wave Dragon [162] and the Seawave Slot-Cone Generator [163] are prototypes that represent overtopping with floating and fixed structures, respectively.

In contrast to the other sources, tidal energy can provide predictable and stable power to the electrical network. However, the tidal-current-generated electricity price is not yet competitive with the current wholesale prices of electricity due to the technical challenges associated with building, installing, operating, and maintaining the plant which affect the cost of the produced energy in energetic sub-sea environments [164].

Some important deployments of the horizontal axis tidal turbines are as follows: The Shetland tidal array was the first deployed tidal array. It includes three turbines of 100 kW each. Meygen, which is the largest operational tidal current array with four turbines of 1.5 MW, was developed by ANDRITZ HYDRO Hammerfest and Atlantis Resource Limited. The Sabella D10 tidal turbine has a capacity of 1 MW. The Cape Sharp Tidal project consists of two turbines of 2 MW and was developed by OpenHydro/Naval Energies.

The two largest tidal barrages are the 240 MW Rance barrage (1966) in France [165] and the 254 MW Sihwa Lake tidal barrage (2011) in South Korea [166,167]. Other countries, such as China, Russia, and the UK [168], are also focusing on tidal barrage technology. In particular, the former two countries have operated tidal barrage power plants in a mean range of 2.4 m with modern low-head turbines [169], which proves that relatively low tides (lower than 5 m, which is considered necessary for tidal barrages [170]) can also be utilized economically.

The extraction of energy from the ocean's thermal gradient is being pursued by some countries including the United States, China, Japan, France, Taiwan, South Korea, India, and the Netherlands. However, two main projects that achieved the prototype phase are the onshore Okinawa 100 kW [171,172] and Hawaii 105 kW OTEC [173] power plants. The former prototype is a hybrid OTEC, developed by Saga University, which uses mixed water/ammonia as "working fluid". It was installed in 2013 in Okinawa, Japan. The latter is a closed-cycle OTEC that was developed by Makai Ocean Engineering. It was installed and connected to the US electrical grid in 2015.

Salinity gradient power is still a concept under development [149,150]. The first PRO (pressure retarded osmosis) power plant was developed by Statkraft in 2006. The main project is a 5 kW RED pilot project that was developed by the REDStack and Fujifilm in 2005. They deployed a 50 kW RED pilot project in "Afsluitdijk" (the sea defense site and major causeway) in 2013.

3.3.4. Status of the Projects

- Project development phases

In general, numerous projects have been implemented in all continents, and some regions present a relatively high TRL compared with others. To address the current status of ocean renewable energy technologies, we integrated the data of 455 projects from five different databases as follows: OES 2019 (Ocean Energy System), EMODnet 2017 (The European Marine Observation and Data Network), UKMED 2019 (UK Marine Energy Database), and OpenEI 2019 (Open Energy Information) provided by the US Department of Energy's Marine and Hydrokinetic Technology Database and PNNL 2019 (Pacific Northwest National Laboratory) [174–178]. Note that, in some cases, mainly related to the wave and current energy, each project may include more than one device unit, forming a farm; however, the numbers that are shown in this section represent the quantity of projects and not the number of the employed technology units. Since each database classifies projects based on its defined categories, unification of the stages of the project development is required. In this work, to unify the classifications of the databases, four different categories were defined based on the "guidelines for project development in the marine energy industry" presented by the EMEC [179]. Accordingly, each marine energy project was divided into seven stages, labelled 0, 1, 2, 3, 4, 5, and 6, associated with the project development strategy, site screening, project feasibility, project design and development, project fabrication and installation, operation and maintenance, and decommissioning, respectively. Table 2 shows a summary of the project stages.

Table 2. Stages of ocean renewable energy project development.

Project Stage	Description
Early concept	The technology is in the early stage of development. The basic principles are observed, and analytical formulations, numerical simulations, and laboratory-scale experimental tests are performed at this stage. (Stages 0, 1, and 2 based on [179]).
In planning	The technology is being used in medium- or large-scale experimental tests in a realistic working environment or in an open sea. The represents preparation for authorized consent. (Stage 3 based on [179]).
Pre-deployment	Consent is authorized by the consent authority and the company or technology developers perform activities such as site preparation, fabrication, and installation. (Stage 4 based on [179]).
Operational	The device is fully operational. In this paper, the operational system can even be connected to a local electrical grid or can provide energy for an isolated center of consumption, such as a marine lighthouse.
Decommissioned	Devices that have been removed from the water after being operational for a certain period.
Dormant	Projects that had site permission or authorized consent or were in the permitting process but were later abandoned.

- Geographical distribution of the projects

Figure 10a,b shows the geographical distribution of the projects over approximately 40 countries. It can be seen that, in terms of quantity, Europe has the largest contribution (about 60.66%), followed by North America, Asia, Oceania, Africa, Central America and the Caribbean, and South America with 17.10%, 13.35%, 5.62%, 1.64%, 0.94%, and 0.7%, respectively. As Figure 10b illustrates, the largest number of projects belong to the wave and current energy sector, and these are mostly located in Europe. Note that in this sector, the current energy projects include technologies that are utilized to harness current energy, independent of its type, including ocean and tidal currents.

It is observed that although the ocean thermal gradient has the largest potential among the energy resources, there is a very low interest in harnessing such energy. This may be due to the technical complexities and high capital cost that decelerate the development process of OTEC technology [180]. Nevertheless, it can be inferred that wave and current energy are considered to be more promising energy resources than others.

Figure 11 illustrates a summary of the global status of ocean renewable energy projects. Approximately half of the projects are in the "planning" and "pre-deployment" stages, and these projects are dominated by the current and wave energy and mostly located in Europe. This means that there will be a significant evolution in ORE deployment in the next 5 years. Europe, Asia, North America, and Africa are the regions with operational projects. Asia may lead the future in OTEC technology having a larger number of "planning" and "pre-deployment" projects compared with other regions. These projects are mostly located on the eastern coast of Asia. Tidal range energy has been harnessed commercially since 1966 (Rance River north-western France); however, to date, only a handful of operational projects have been deployed. On the other hand, wave and current energy projects, which are operational or at an earlier stage of development, represent about 65% of all projects. This implies that tidal range technology has not drawn as much serious developmental interest as wave and current technologies. This may be due to the high cost and ecological impacts of such technology [181].

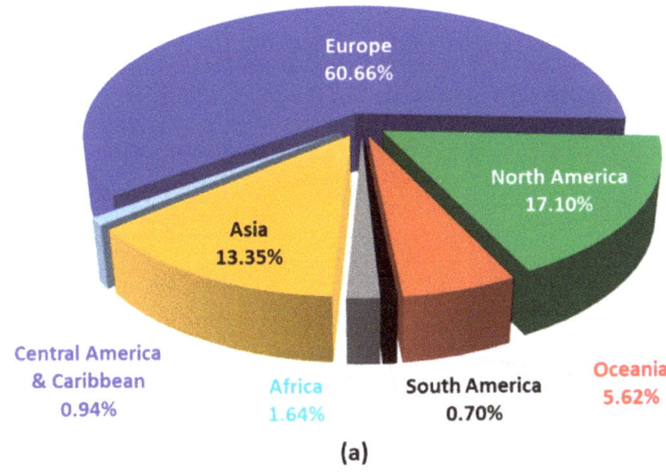

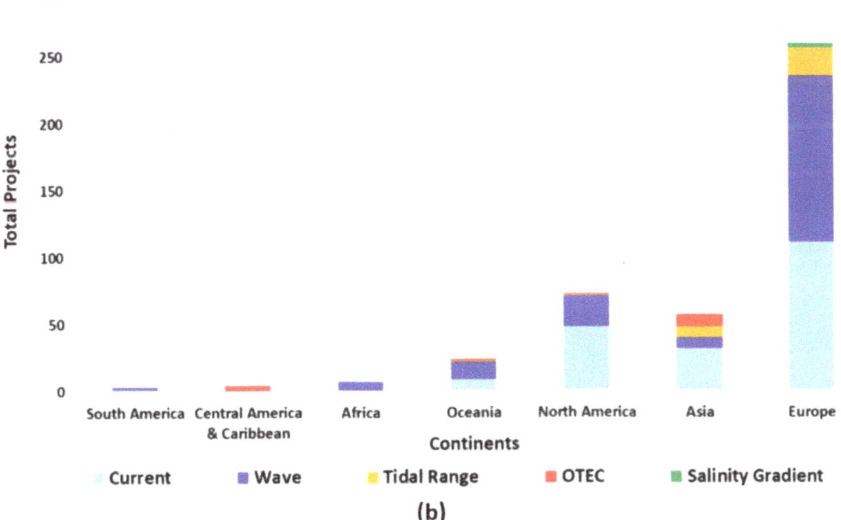

Figure 10. Distribution of the ocean renewable energy projects by (**a**) continent and (**b**) source for each continent.

- Technology distribution

The databases were used to determine the technological distribution of the projects. As Figure 12 illustrates, and as was expected, the most employed technologies are current and wave energy technologies. Current energy technologies are closest to technological maturity showing a significant convergence with the use of horizontal axis turbines. Most projects use horizontal axis turbine technology, followed by vertical axis turbines, tidal kites, and oscillatory hydrofoils. On the other hand, more technology diversity can be observed for wave energy converters, partly due to the diversity of wave resources and the complexity of harnessing wave energy. The wave energy technologies are dominated by point absorber devices followed by OWC, oscillating wave surge converters, attenuators, rotating masses, overtopping devices, and submerged pressure differential devices (see Section 3).

The OTEC technologies are limited to the use of closed and open cycle and hybrid systems with a tendency of deploying the closed cycle method. Tidal range energy is traditionally harnessed using tidal barrages installed in the estuaries, but this method is associated with important environmental issues. The use of tidal lagoons has been proposed and developed in the UK since 2008 as an alternative to reducing such environmental issues [168]. The technology for extracting the salinity gradient energy is still in the conceptual stage of development, and its evolution is highly dependent on membrane enhancement, which will be responsible for 50% to 80% of the total cost [182].

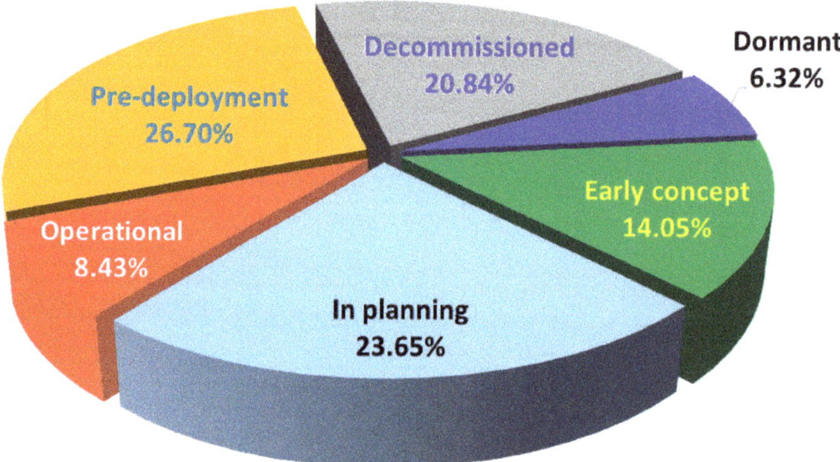

Figure 11. Global status of the ocean renewable energy projects.

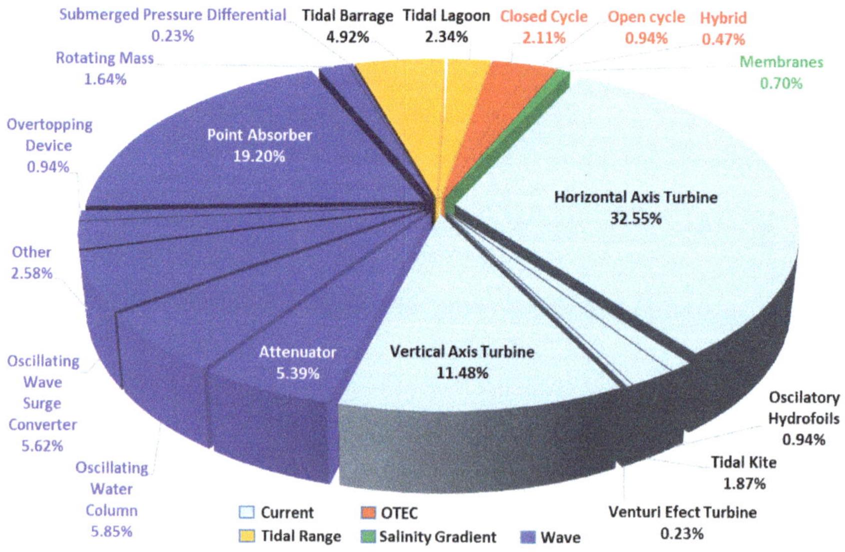

Figure 12. Technology distribution among projects.

4. Case Study of Brazil

4.1. Resource Assessment Results

4.1.1. Ocean Current Energy

Figures 13 and 14 show the spatial variation of the ocean current resources along the Brazilian coast in terms of the annual and seasonal average current speed (m/s) and power density (W/m^2). As shown in Figure 13, the maximum annual average can be observed in the north equatorial margin of Brazil with a velocity of 1.52 m/s. This region is influenced by the NBC. These regions are located at a distance of between 120 and 300 km of the coastline. The same pattern can be observed considering the seasonal distribution. A maximum speed value of 1.67 m/s occurs during autumn. It can be seen that the speed values barely reach 2 m/s, which is recommended for commercial ocean current energy extraction [32]. The current speed values are not significant (less than 0.7 m/s) for the regions A and B. They occur due to the BC flow and are higher during the spring. Figure 14 shows the power density (W/m^2), calculated using Equation (1), across the Brazilian coastline. As expected, following the speed values, region D represents the most energetic area, and power density values higher than 500 W/m^2 can be observed for some areas. Table 3 contains the average values of the seasonal mean power as well as the *SV* and *COV* values. The seasonal average values for regions D and C show that, in contrast to the regions A and B, the current power density is higher during austral autumn and winter compared to austral summer and spring. This is due to the seasonal climatological behavior of the NBC [25]. The fact that the power density of the region D is significantly higher than that of other regions indicates that the values of *SV* and *COV* are less important when the objective is to determine the best region for exploring the ocean current energy. However, these values are important for techno-economic studies of the corresponding energy extracting technologies in the North region.

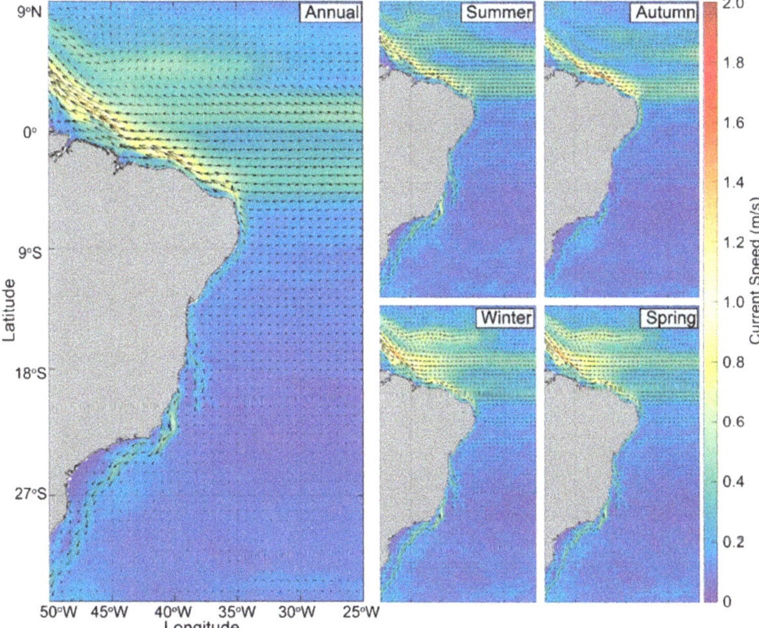

Figure 13. Annual and seasonal (summer, autumn, winter, and spring) mean surface current speed (m/s) at the Brazilian coastline between January 1, 2007 and December 31, 2017.

Table 3. Average values of the ocean current power density (W/m^2), standard deviation ($\pm$), seasonal variability (SV), and coefficient of variation (COV) for each coastline region—A, B, C and D—for each season (summer, autumn, winter, and spring) between January 1, 2007 and December 31, 2017.

Region/Season	Power Density (W/m^2) $\pm$ Standard Deviation				SV	COV
	Summer	Autumn	Winter	Spring		
A	98.9 ($\pm$3.3)	73.67 ($\pm$2.12)	74.35 ($\pm$2.12)	97.85 ($\pm$2.7)	0.269	0.951
B	362.76 ($\pm$9.9)	90.06 ($\pm$3.83)	167.37 ($\pm$5.6)	280.25 ($\pm$7.8)	0.609	1.994
C	193.93 ($\pm$3.84)	399.93 ($\pm$11.1)	379.137 ($\pm$11.06)	216.36 ($\pm$5.06)	0.426	1.460
D	788.21($\pm$39.15)	1416.64 ($\pm$58.28)	1240.21 ($\pm$58.82)	1103.4 ($\pm$52.3)	0.514	1.333

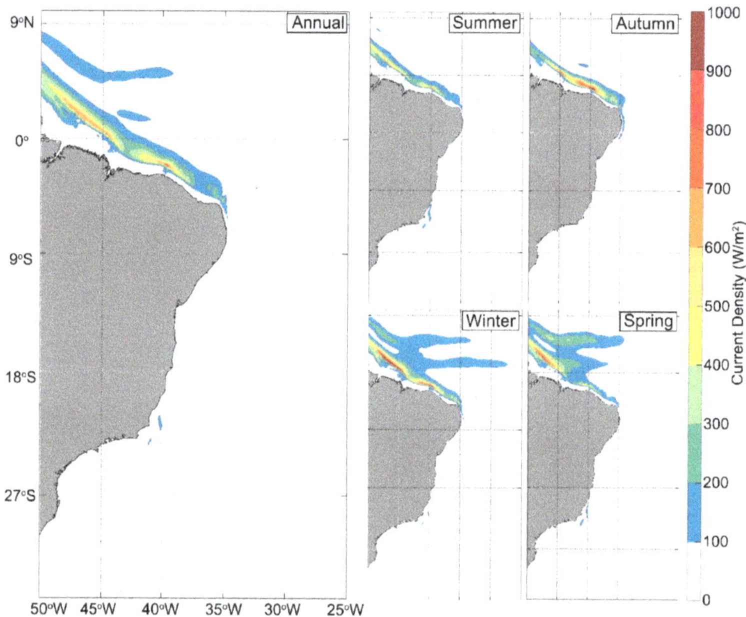

Figure 14. Annual and seasonal (summer, autumn, winter, and spring) mean power density (W/m^2) at the Brazilian coastline between January 1, 2007 and December 31, 2017.

4.1.2. Wave Energy

Figure 15 shows the average annual and seasonal wave power density values. The hindcast shows the variability of the energy resource and provides a holistic view of the wave climate along the Brazilian coast. It can be observed that the most energetic wave areas are located near the regions A and B coasts with a power value between 20 and 25 kW/m. This is intensified during the autumn and winter seasons. This fact is directly related to the increase in the occurrence of extratropical cyclones that generate larger waves that propagate toward these Brazilian regions. The nearshore areas of region A (areas with a water depth of less than 100 m) have values close to 20 kW/m for almost the entire year. This is mainly due to the preponderance of south winds combined with the shoreline orientation that induces strong swells near the coast. The average values of the *SV* and *COV* related to the wave power were calculated and are illustrated in Table 4 for five different bathymetries of 25, 50, 100, 150, and 200 m along the Brazilian coast. It can be observed that, independently of the water depth, the seasonal variability (SV) of the regions A and B is always smaller than that of the regions D and C. On the other hand, the minimum *COV* occurs in the region C, while the region D has a greater *COV* than the other regions. However, the differences between the *COV* values of the region C, when compared with those in the regions A and B, are small and decrease as the water depth increases from 25 to 200 m. The region C has the smallest wave power variability during the year, which may lead to a higher capacity factor, while the regions A and B are the areas with the most energetic waves, allowing the deployment of the devices with higher installed capacity. A trade-off between the WEC nominal power and capacity factor as well as other local characteristics such as water depth should be considered to determine the proper locations for deploying wave farms.

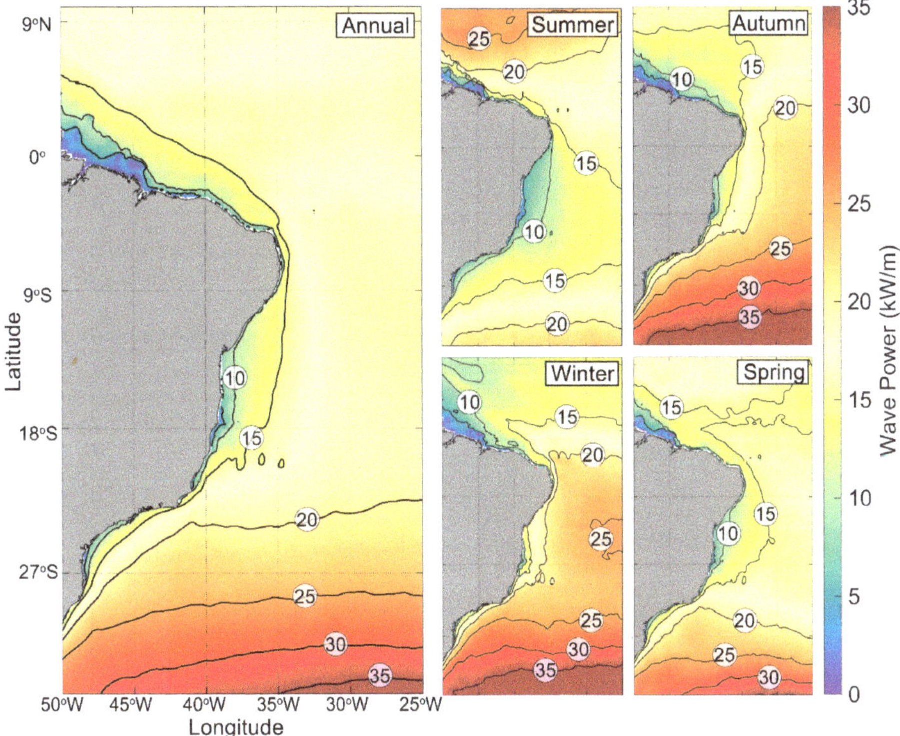

Figure 15. Annual and seasonal (summer, autumn, winter and spring) mean wave power density (k W/m^2) at the Brazilian coastline between January 1, 2015 and December 31, 2017.

Table 4. Average values of wave seasonal variability (SV) and coefficient of variation (COV) considering five different bathymetries (25, 50, 100, 150 and 200 m) for each region—A, B, C and D—between January 1, 2015 and December 31, 2017

Region	Bathymetry 25		50		100		150		200	
	SV	*COV*	*SV*	*COV*	*SV*	*COV*	*SV*	*COV*	*SV*	*COV*
A	0.467	0.533	0.431	0.334	0.482	0.303	0.502	0.285	0.509	0.258
B	0.637	0.515	0.619	0.350	0.595	0.294	0.596	0.283	0.615	0.287
C	0.845	0.375	0.865	0.302	0.856	0.305	0.852	0.274	0.781	0.250
D	0.748	0.685	0.929	0.464	0.929	0.432	0.744	0.301	0.833	0.296

The available wave power for the Brazilian coastline was calculated at an average distance of 128 km from the coast (Table 5). Accordingly, a total available wave power of approximately 91.8 GW was estimated considering a total coastline length of about 7491 km (an approximate value without coastline details). It should be noted that this value is an estimation of the theoretical potential of the Brazilian wave power. In practice, only a fraction of this value can be extracted by the wave energy devices, which depends on different issues such as technical challenges, environmental impacts, economy, deferent use of the sea area, and social impacts. Nevertheless, only one-fifth of this potential is equal to approximately 35% of the Brazilian electricity demand in 2017 [183].

Table 5. Available wave power of the Brazilian coastline.

Parameters	Regions			
	A	**B**	**C**	**D**
Length (km)	~1250	~1952	~1452	~2837
Average power (kW/m)	21.1	12.4	13.8	7.4
Total power (GW)	26.4	24.2	20.1	21.0

4.1.3. Ocean Thermal Energy

Figure 16a shows the annual average ΔT (°C) between the water depth of 20 and 1000 m along the Brazilian coast. The results show that, except for the extreme South below 27 °S, the yearly average ΔT is always about 20 °C or higher along the Brazilian coast. The average gross power of a 10 MW OTEC plant (see Section 2.2.1) was calculated for 12 locations along the coastline. The selected points were located approximately at a distance between 30 and 200 km to the shore and had an annual average ΔT at between 20 and 1000 m of water depth of more than 20 °C. Figure 16b illustrates the annual variation in the gross power for the considered points. A greater average annual gross power, represented by the red solid line, can be observed for the regions D and C comparing to the regions A and B. Moreover, the results show smoother power production for the regions D and C comparing to the regions A and B. Table 6 shows the characteristics of the selected points as well as the P_{gross} and P_{net}.

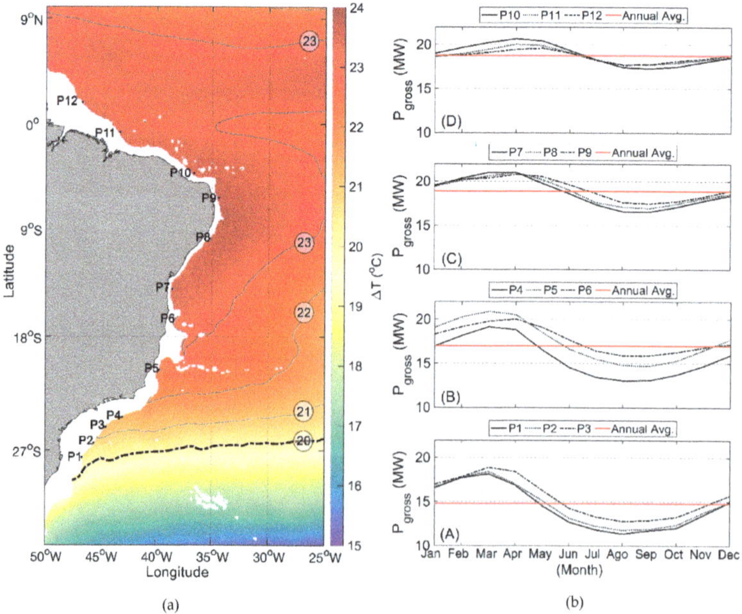

Figure 16. (**a**) Annual mean ΔT (°C) between 20 and 1000 m and (**b**) the annual ocean thermal energy conversion (OTEC) Gross Power Density (P_G in MW) at a depth of 1000 m considering the period between January 1, 2007 and December 31, 2017.

Figure 17 illustrates the seasonal mean ΔT (°C) between the water depths of 20 and 1000 m, for a bathymetry of 1000 m, across the Brazilian coastline. The black line represents a ΔT of 20 °C. It can be observed that, for the regions A and B, the water depth in which the mean ΔT = 20 °C is achieved, varies between 500 m in summer and 700–1000 m in other seasons. On the other hand, in the regions D and C, a mean ΔT of 20 °C can be reached in a water depth of about 500–700 m throughout the year. From a technical point of view, less structural challenges would be expected when bringing the cold water from a depth of 500 m rather than from 1 km.

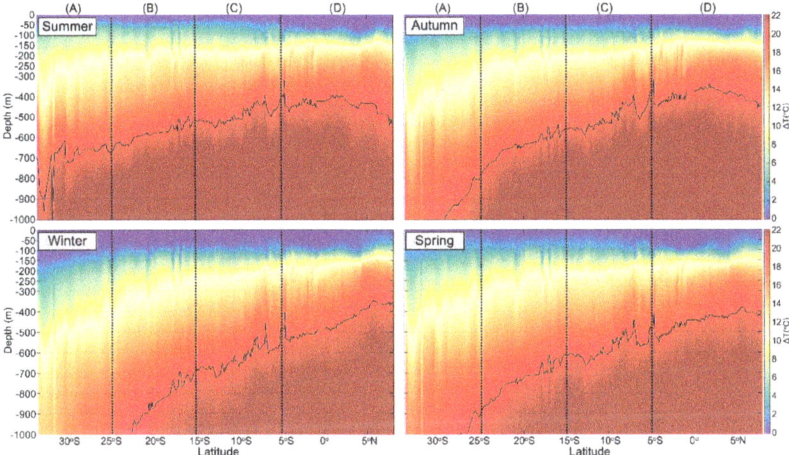

Figure 17. Seasonal (summer, autumn, winter, and spring) mean ΔT (°C) between 20 and 1000 m for the bathymetry of 1000 m (for regions A, B, C and D) across the Brazilian coastline latitude considering the period between January 1, 2007 and December 31, 2017. The black line corresponds to a ΔT of 20 °C.

Table 6. Gross and net power estimation for the selected points along the Brazilian coastline.

Regions	Points	Lat/Lon	Bathymetry (~m)	ΔT (°C)	Annual Average P_{gross} (MW)	Annual Average P_{net} (MW)
A	P1	27.6666°S 46.5833°W	1396	20.17	14.32	10.12
	P2	26.5833°S 45.5833°W	1409	20.37	14.58	10.39
	P3	25.3333°S 44.5000°W	1000	20.98	15.42	11.23
B	P4	24.4167°S 43.0833°W	1396	21.16	15.67	11.49
	P5	20.8333°S 39.7500°W	1422	22.43	17.53	13.36
	P6	16.6667°S 38.4167°W	1005	22.59	17.71	13.55
C	P7	14.1667°S 38.5833°W	1948	23.21	18.88	14.51
	P8	9.9167°S 35.2500°W	2012	23.33	18.84	16.69
	P9	6.5000°S 34.4167°W	2764	23.54	19.15	15.00
D	P10	4.3333°S 36.6667°W	1945	23.37	18.86	14.72
	P11	0.8333°S 43.3333°W	2337	23.29	18.71	14.57
	P12	1.6667°N 46.6667°W	1463	23.24	18.63	14.50

4.2. Deployments in Brazil

The first deployment of an ocean renewable energy converter in Brazil occurred in 1934 when the French engineer Georges Claude used an ocean thermal energy source to produce ice for the residents of Rio de Janeiro. His plant ran into problems and stopped working off the coast of Rio de Janeiro due to fatigue of its long intake tubes [184]. Studies associated with ocean renewable energy in Brazil began in 2001 at the Federal University of Rio de Janeiro (UFRJ), focusing on wave and tidal energy. Some other universities have also started working in this field, such as the Federal University of Maranhão (UFMA), the Federal University of Santa Catarina (UFSC), the Federal University of Pará (UFPA), and the Federal University of Itajubá (UNIFEI).

There are three main ocean renewable energy projects being carried out in Brazil with different technology readiness levels. The first one is the COPPE (The Alberto Luiz Coimbra Institute for Graduate Studies and Research in Engineering) hyperbaric wave converter developed by the UFRJ, which has reached the prototype stage. A full-scale single device of the technology was installed in 2011 in Pecém port of Ceará state located in the northeast of Brazil. The device was decommissioned after 6 months of operation due to the port extension project. The second project is a nearshore wave energy converter, also developed by the UFRJ, which will be installed in relatively shallow water (water depth of 25–30 m) off the Rio de Janeiro coast. The technology is at the R&D stage and is undergoing medium-scale laboratory tests. The last project is the tidal range project of the Bacanga River estuary located in São Luís of Maranhão state in North Brazil. Although the discussion about the tidal energy extraction in this region is relatively old, the project is still at an early stage of development as it is waiting for finance. The following sections describe the characteristics and statuses of the mentioned projects.

4.2.1. COPPE Hyperbaric Wave Converter

As illustrated in Figure 18, this device is composed of a floating body connected to the pumping modules, a hydrodynamic accumulator, a hyperbaric chamber, and a generating unit. The vertical motion of the floating body due to the wave body interactions drives the pump actuator which displaces the water inside the closed circuit to a hydro-pneumatic accumulator. The accumulator is connected to a hyperbaric chamber, which has previously been pressurized. Then, the pressurized water drives a hydraulic turbine coupled to an electrical generator. The hyperbaric chamber works as an energy storage system, which smooth the power fluctuations due to the oscillatory nature of sea waves. The applied pressure is in the range of 250–400 m of water column (m.wc) [185].

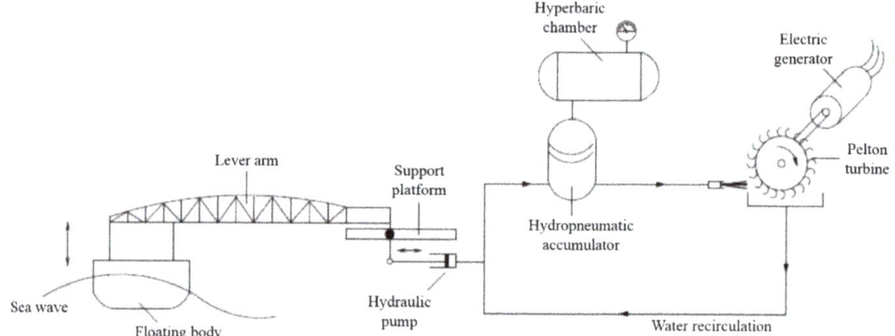

Figure 18. A schematic of the COPPE/UFRJ (Federal University of Rio de Janeiro) hyperbaric wave converter [185].

Additionally, a discrete control scheme was applied to the system to improve power production by adjusting the PTO parameters without wave measurement [186]. The experimental tests were performed at the Ocean Technology Laboratory (LabOceano) of the UFRJ. Figure 19 shows the medium-scale model at a ratio of 1:10 which was tested under regular and irregular wave conditions corresponding to the predominant wave climate at the location of installation [187,188]. As a result of the experimental tests, a capture width ratio of between 19% and 36% was observed for the wave energy converter.

Figure 19. COPPE hyperbaric wave converter: (**a**) medium-scale model with a ratio of 1:10 at LabOceano [187]: (**b**) installed full-scale prototype.

A full-scale prototype with a capacity of 100 kW was deployed at the Pecém port in Northeast Brazil (Figure 19b). The device was installed on two concrete bases, 12 m in length, built on a breakwater. The oscillating part, which consists of a floater, 10 m in diameter, and a mechanical arm, 22 m in length, is connected to two skids mounted on the concrete bases.

4.2.2. COPPE Nearshore WEC

The system is a point absorber WEC type with a capacity of 50 kW that consists of an oscillating body and a bottom-mounted support structure. The oscillating part is a floating conical cylinder which is allowed to move only in the heave direction (Figure 20). The fixed structure consists of four columns with very small diameters relative to the wavelengths (no diffraction). The structure is mounted on the seabed through a concrete base. Eight roller bearings facilitate oscillation of the buoy in the vertical direction (heave). They are placed on the top and bottom of the cylindrical section.

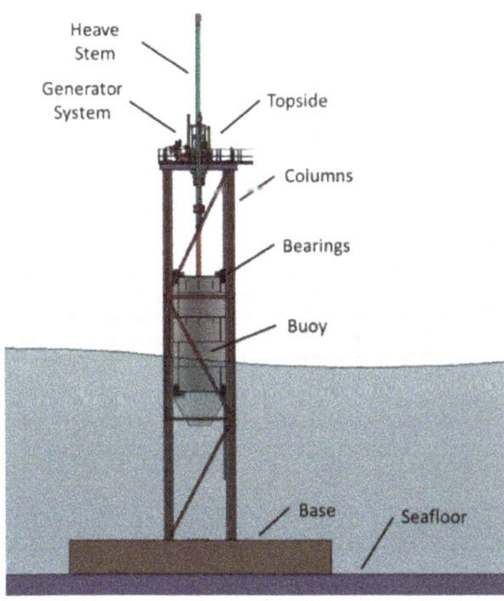

Figure 20. Components of the COPPE nearshore WEC.

The PTO system is located on the topside deck and consists of a gearbox and a rotational generator (Figure 21). The vertical motion of the buoy is transferred through a central rod (heave stem) to the gearbox. Then, the pulley converts the vertical movement into rotation that is adequate for the electrical generator. A backstop system unifies the rotation direction using freewheels. This implies that the buoy can drive the PTO system either upwards or downwards. A solid cylindrical flywheel is used to amplify the rotational inertia as well as smooth the delivered energy to the generator. Additionally, the PTO system includes a gearbox that multiplies the rotational speed so that it is adequate for power generation.

The location that has been considered for installation of the WEC is near to a small island called "Ilha Rasa". The location's water depth is about 20 m, and its distance from shore (Copacabana beach, Rio de Janeiro, Brazil) is about 14 km. The predominant wave climate of the region is a peak period of $T_p = 9.6$ s and a significant height of $H_s = 1.33$ m. Shadman et al. [189] showed that a very large buoy is required to maximize the power absorption in a region like nearshore Rio de Janeiro, where the predominate wave periods are beyond 7 s. This might lead to higher costs, which could make the project economically infeasible. Hence, a specific control called "latching", presented originally by Budal and Falnes [190], was applied on the WEC to overcome this challenge. Latching is a mechanical control method that tunes the natural period of the buoy to the predominate wave period of the sea site by halting and releasing the buoy at its motion extremum. As a result, larger buoy motion amplitude and velocities can be achieved, leading to higher power production. Eventually, the latching control enables a smaller buoy with a smaller natural period to be tuned with such a wave climate [191]. A hydraulic system is designed and tested for latching the oscillating buoy.

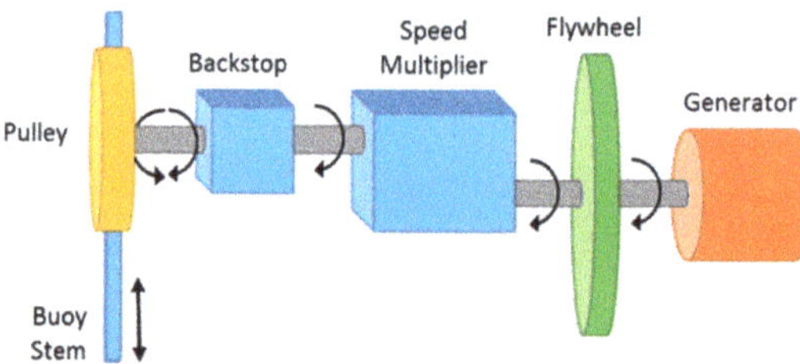

Figure 21. Schematic view of the power take-off (PTO) system.

Experimental tests of small-scale models, shown in Figure 22, were performed in a wave and current channel (LOC) at the COPPE/UFRJ. The hydrodynamic behavior of the buoy was studied by applying different modeling scales including 1:17, 1:20, 1:30, and 1:40. Additionally, a strategy was developed to investigate the effect of latching control on the WEC.

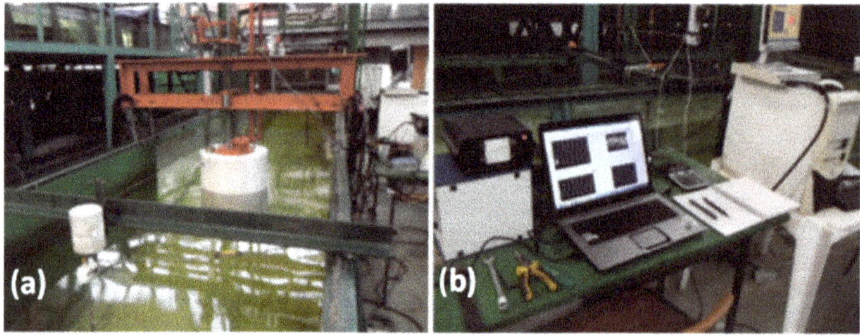

Figure 22. Experimental tests of the COPPE nearshore WEC in a wave channel: (**a**) 1:17 scaled model, (**b**) instruments for data acquisition.

4.2.3. Tidal Power Plant of the Estuary of Bacanga

The largest tidal ranges in Brazil are located on the North coast including the coastal areas of Maranhão, Pará, and Amapá. For instance, a tidal energy potential of 22 TWh/year has been estimated for Maranhão state [192]. Some studies have addressed the exploitation of such energy in Brazil [193,194]. As Figure 23a illustrates, the Bacanga basin is 10,219 ha in size, which includes the estuarine body of water and the Bacanga lake. The reservoir capacity is about 40 million cubic meters at an elevation of +4.5 m, corresponding to the spring tide level [192]. As shown in Figure 23b, the dam includes an 800 m embankment rock which is filled with clay material. Additionally, the dam has two sluice gate systems types of radial and stop-log that were installed in 1974 and 1980, respectively. There are three radial sluice gates with widths of approximately 12.5 m. In the case of a fully open gate, a water height level of 4.5 m is registered for each radial gate. This value is about 3 m for the stop-log gates, which are flat and operate vertically, with widths of 2.85 m.

Figure 23. (a) The Bacanga Estuary, and (b) aerial image of the radial and stop-log sluice gates [192].

Considering some restrictions, including a reservoir water level limit of +2.5 m, Neto et al. [192] proposed a new model for the tidal power plant in which the three radial sluice gates were replaced by the modern and appropriate version for automatic operation, which excluded the necessity of using stop-log type gates to control the reservoir maximum limit. Considering the Kaplan turbine with double regulation provided by ANDRITZ HYDRO [195], they estimated an annual energy production of slightly larger than 14 GWh/yr for the power plant.

5. Discussion and Open Question

The present cost of ocean renewable energy cannot complete with that of grid-connected renewables. The alternative, nowadays, in addition to the development of more optimized projects, is to look for new markets where electricity generation options are either scarce or expensive, for example the oil and gas industry, aquaculture, defense, and the demands from isolated communities. In the particular case of Brazil, there is a concentration of power generation, mostly from hydroelectric plants, located in the South and Southeast regions. It has been demonstrated that a significant amount of ocean renewable energy featuring an ocean thermal gradient is located in the regions D and C (see Figure 2). In these regions, an annual electricity production of 0.8 TWh per year has been calculated, considering only six OTEC plants with 10 MW installed power, as presented in this paper. Accordingly, considering an annual average of 15 MW, 20 OTEC plants would be sufficient to supply approximately 10% of the total residential electricity consumption of the Northeast region of Brazil, which was estimated to be approximately 27.059 TWh in 2017 [183]. This implies that such renewable energy resources could be harnessed as a supplementary alternative for these regions, especially when there is a power generation drop due to seasonal rain shortage. Additionally, the low seasonal and temporal variability of the ocean renewable resources along the Brazilian coast could provide stable power production throughout the year, with substantial capacity. The supply chain associated with ocean renewable technologies is still incipient worldwide. The increasing prototype deployments may promote the association of local supply chains with the global suppliers of specific equipment such as submarine cables, electrical connectors, turbines, and generators. In Brazil, the supply chain would consist of companies already operating in the offshore oil and gas sector. This is a very robust sector, which will be able to meet the demands of the ocean renewable energy sector. The synergy of the long-established offshore oil and gas sector and the new ocean renewable energy sources could represent a crucial factor for the success of the new industry. Updated technologies must be incorporated, especially digital ones associated with artificial intelligence, control, and robotics to provide competitive services for inspection and maintenance, reducing the operational costs. New materials, such as the composites associated with innovative floating structures and installation methods, can also contribute to the competitiveness of the new sector in terms of the electricity cost. In Brazil, the large number of hydropower plants and the complex grid system also present opportunities for the implementation of ocean renewable energy sources. Hydropower plants could be designed as storage components of the whole electrical system, combining a better water supply with clean and efficient power generation throughout the country. The substitution of oil and gas-based power plants for ocean renewables would modify the national energy matrix substantially, reinforcing sustainably oriented electricity generation.

6. Conclusions

This paper, as a preliminary approach, has presented an assessment of ocean renewable energy resources, including wave, ocean current and thermal gradient energy, along the Brazilian coastline. The results show considerable ocean currents, thermal gradients, and wave energy in the regions D, C and A, respectively. A maximum annual average velocity of 1.52 m/s, which represents a power density of approximately 500 W/m^2, was observed for the ocean current energy in the region D near the equatorial margin of Brazil. However, the distance of the resource to the coastline, between 120 and 300 km, is an obstacle to its commercialization. The total theoretical potential of wave energy is estimated to be 91.8 GW along the coastline. The most energetic waves occur in the region A, following by the regions C, B, and D, with average power values of 21.1, 13.8, 12.4, and 7.4 kW/m, respectively. In the region C, the wave resource has the least temporal variability compared with the other regions; nevertheless, the differences are small, and they decrease with an increasing water depth. The results revealed an annual average ocean thermal gradient, between the water depths of 20 and 1000 m, of more than 20 °C for latitudes above 27°S. A mean thermal gradient of 20 °C between the upper layers and water depth between 500 and 700 m can be achieved throughout the year in the regions D and C. This could facilitate the process of bringing cold water from the deep sea, compared with the usual water depth of 1000 m.

The paper also presented an overview of the potential technologies and their statuses of development related to ocean renewable energy sources worldwide. Although available studies indicate different values for the global resource potential, they converge in presenting the ocean thermal gradient as being the most energetic resource followed by waves, salinity gradients, and tides. The TRL and the status of the current projects imply that the global interest tends toward tidal current and wave devices.

Large-scale installations, learning-curves, and innovation are necessary to make the cost of energy competitive with solar and onshore wind energy production. About 27% of the current projects are at the pre-deployment phase and, optimistically, will be deployed in the open sea in the next three years. Apart from tidal range technology, which is already close to the commercialization stage, research, development, and demonstration projects have been led by universities and startups, mostly by taking advantage of public financing. Nevertheless, in the last five years, large industry players and utilities have started carrying out activities and financing in the sector. This is an important step towards speeding up technology commercialization due to the new players' capability to execute utility-scale projects.

Author Contributions: Conceptualization, M.S.; Formal analysis, M.S. and D.F.; Methodology, M.S. and D.F.; Project administration, M.S.; Resources, L.P.d.F.A.; Software, D.F.; Supervision, L.L., C.L. and S.F.E.;

Visualization, C.S. and Z.W.; Writing—original draft, M.S., C.S. and Z.W.; Writing—review & editing, L.P.d.F.A., L.L., C.L. and S.F.E.

Funding: This research received no external funding.

Acknowledgments: The authors acknowledge CNPq, the Ministry of Science, Technology, Innovation and Communication/Brazil, for supporting the research activities of the authors. Additionally, the first author highly appreciates the Instituto Nacional de Ciência e Tecnologia—Energias Oceânicas e Fluviais (INEOF) for supporting his research activities. The third author acknowledges David Holland and the Center for Global Sea Level Change from New York University Abu Dhabi for supporting her research work.

Conflicts of Interest: The authors declare no conflict of interest.

References

1. International Renewable Energy Agency (IRENA). *Global Energy Transformation: A Roadmap to 2050*; International Renewable Energy Agency: Abu Dhabi, UAE, 2018.

2. Luomi, M. *Sustainable Energy in Brazil: Reversing Past Achievements or Realizing Future Potential*; The Oxford Institute for Energy Studies: Oxford, UK, 2014; Volume SP-34, ISBN 978-1-78467-005-4.

3. International Monetary Fund (IMF). IMF Country Information. Available online: https://www.imf.org/en/Countries (accessed on 1 June 2019).

4. Empresa de Pesquisa Energética (EPE). *Brazilian Energy Balance—Summary Report: Year Base 2017*; Empresa de Pesquisa Energética: Rio de Janeiro, Brazil, 2018.

5. International Energy Agency. *World Energy Outlook 2016*; International Energy Agency: Paris, France, 2016.

6. International Energy Agency. *World Energy Outlook 2017*; International Energy Agency: Paris, France, 2017.

7. Dias, V.S.; da Luz, M.P.; Medero, G.M.; Nascimento, D.T.F. An overview of hydropower reservoirs in Brazil: Current situation, future perspectives and impacts of climate change. *Water* **2018**, *10*, 592. [CrossRef]

8. Von Sperling, E. Hydropower in Brazil: Overview of positive and negative environmental aspects. *Energy Procedia* **2012**, *18*, 110–118. [CrossRef]

9. Reggiani, M.C.P. Hydropower in Brazil-Development and Challenges. Master's thesis, Norwegian University of Science and Technology, NTNU, Trondheim, Norway, December 2015. [CrossRef]

10. Rocha Lessa, A.C.; Dos Santos, M.A.; Lewis Maddock, J.E.; Dos Santos Bezerra, C. Emissions of greenhouse gases in terrestrial areas pre-existing to hydroelectric plant reservoirs in the Amazon: The case of Belo Monte hydroelectric plant. *Renew. Sustain. Energy Rev.* **2015**, *51*, 1728–1736. [CrossRef]

11. Beserra, E.R.; Mendes, A.; Estefen, S.F.; Parente, C.E. Wave climate analysis for a wave energy conversion application in Brazil. In Proceedings of the ASME 2007 26th International Conference on Offshore Mechanics and Arctic Engineering, San Diego, CA, USA, 10–15 June 2007; pp. 897–902.

12. Oleinik, P.H.; Marques, W.C.; Kirinus, E.D.P. Estimate of the wave climate on the most energetic locations of the south-southeastern Brazilian shelf. *Defect Diffus. Forum* **2017**, *370*, 130–140. [CrossRef]

13. Kirinus, E.D.P.; Marques, W.C. Viability of the application of marine current power generators in the south Brazilian shelf. *Appl. Energy* **2015**, *155*, 23–34. [CrossRef]

14. Kirinus, E.D.P.; Oleinik, P.H.; Costi, J.; Marques, W.C. Long-term simulations for ocean energy off the Brazilian coast. *Energy* **2018**, *163*, 364–382. [CrossRef]

15. Martins, J.C.; Goulart, M.M.; Gomes, M.N.; Souza, J.A.; Rocha, L.A.O.; Isoldi, L.A.; Santos, E.D. Geometric evaluation of the main operational principle of an overtopping wave energy converter by means of Constructal Design. *Renew. Energy* **2018**, *118*, 727–741. [CrossRef]

16. Barbosa, D.V.E.; Santos, A.L.G.; Santos, E.D.; Souza, J.A. International journal of heat and mass transfer overtopping device numerical study: Openfoam solution verification and evaluation of curved ramps performances. *Int. J. Heat Mass Transf.* **2019**, *131*, 411–423. [CrossRef]

17. Jung, J.Y.; Lee, H.S.; Kim, H.J.; Yoo, Y.; Choi, W.Y.; Kwak, H.Y. Thermoeconomic analysis of an ocean thermal energy conversion plant. *Renew. Energy* **2016**, *86*, 1086–1094. [CrossRef]

18. Empresa de Pesquisa Energética (EPE). *Plano Nacional de Energia—PNE 2030*; Empresa de Pesquisa Energética: Rio de Janeiro, Brazil, 2007.

19. Empresa de Pesquisa Energética (EPE). Plano Nacional de Energia—2050. Available online: http://www.epe.gov.br/pt/publicacoes-dados-abertos/publicacoes/Plano-Nacional-de-Energia-2050 (accessed on 18 April 2019).

20. Agência Nacional de Energia Elétrica (ANEEL). *Volume 3–8 Evolução Tecnológica Nacional no Segmento Geração de Energia Elétrica e Armazenamento de Energia*; Agência Nacional de Energia Elétrica: Brasília, Brazil, 2017.
21. Schmitz, W.J., Jr. On the interbasin-scale thermohaline circulation. *Rev. Geophys.* **1995**, *33*, 151–173. [CrossRef]
22. Stramma, L.; England, M. On the water masses and mean circulation of the South Atlantic Ocean. *J. Geophys. Res. Ocean.* **1999**, *104*, 20863–20883. [CrossRef]
23. Schott, F.A.; Dengler, M.; Zantopp, R.; Stramma, L.; Fischer, J.; Brandt, P. The shallow and deep western boundary circulation of the South Atlantic at 5°–11° S. *J. Phys. Oceanogr.* **2005**, *35*, 2031–2053. [CrossRef]
24. Rodrigues, R.R.; Rothstein, L.M.; Wimbush, M. Seasonal variability of the South Equatorial Current bifurcation in the Atlantic Ocean: A numerical study. *J. Phys. Oceanogr.* **2007**, *37*, 16–30. [CrossRef]
25. Peterson, R.G.; Stramma, L. Upper-level circulation in the South Atlantic Ocean. *Prog. Oceanogr.* **1991**, *26*, 1–73. [CrossRef]
26. Soutelino, R.G.; Da Silveira, I.C.A.; Gangopadhyay, A.; Miranda, J.A. Is the Brazil Current eddy-dominated to the north of 20° S? *Geophys. Res. Lett.* **2011**, *38*, 1–5. [CrossRef]
27. Madec, G. *NEMO Ocean Engine*; Note du Pôle Modélisation, Institut Pierre-Simon Laplace: Paris, France, 2011; pp. 1–332.
28. Madec, G.; Imbard, M. A global ocean mesh to overcome the North Pole singularity. *Clim. Dyn.* **1996**, *12*, 381–388. [CrossRef]
29. Amante, C.; Eakins, B.W. *Arc-minute global relief model: procedures, data sources and analysis. NOAA Technical Memorandum NESDIS NGDC-24*; National Geophysical Data Center NOAA: Boulder, CO, USA, 2018.
30. Lowcher, C.F.; Muglia, M.; Bane, J.M.; He, R.; Gong, Y.; Haines, S.M. Marine hydrokinetic energy in the gulf stream off North Carolina: An assessment using observations and ocean circulation models. In *Marine Renewable Energy: Resource Characterization and Physical Effects*; Yang, Z., Copping, A., Eds.; Springer International Publishing: Cham, Switzerland, 2017; pp. 237–258, ISBN 978-3-319-53534-0.
31. Myers, L.; Bahaj, A.S. Power output performance characteristics of a horizontal axis marine current turbine. *Renew. Energy* **2006**, *31*, 197–208. [CrossRef]
32. Batten, W.M.J.; Bahaj, A.S.; Molland, A.F.; Chaplin, J.R. The prediction of the hydrodynamic performance of marine current turbines. *Renew. Energy* **2008**, *33*, 1085–1096. [CrossRef]
33. Nihous, G.C. An order-of-magnitude estimate of ocean thermal energy conversion resources. *J. Energy Resour. Technol.* **2005**, *127*, 328. [CrossRef]
34. Nihous, G.C. A preliminary assessment of ocean thermal energy conversion resources. *J. Energy Resour. Technol.* **2007**, *129*, 10. [CrossRef]
35. Devis-Morales, A.; Montoya-Sánchez, R.A.; Osorio, A.F.; Otero-Díaz, L.J. Ocean thermal energy resources in Colombia. *Renew. Energy* **2014**, *66*, 759–769. [CrossRef]
36. Nihous, G.C. Conceptual design of a small open-cycle OTEC plant for the production of electricity and fresh water in a Pacific Island. In Proceedings of the International Conference on Ocean Energy Recovery, Honolulu, HI, USA, 28–30 November 1989.
37. Vega, L.A.; Nihous, G.C. Design of a 5 MWe OTEC pre-commercial plant. In Proceedings of the Oceanology International'94 Conference, Brighton, UK, 8–11 March 1994.
38. Ardhuin, F.; Rogers, E.; Babanin, A.; Filipot, J.-F.; Magne, R.; Roland, A.; Van Der Westhuysen, A.; Queffeulou, P.; Lefevre, J.-M.; Aouf, L.; et al. Semi-empirical dissipation source functions for ocean waves: Part I, definition, calibration and validation. *J. Phys. Oceanogr.* **2010**, *40*, 1917–1941. [CrossRef]
39. Nielsen, P. *Coastal and Estuarine Processes, Advanced Series on Ocean Engineering*, 29th ed.; World Scientific: Singapore, 2009; Volume 29, ISBN 978-981-283-711-0.
40. Cornett, A. A global wave energy resource assessment. *Proc. ISOPE* **2008**, *8*, 318–326.
41. Barstow, S.F.; Mørk, G.; Lønseth, L.; Mathisen, J.P. WorldWaves wave energy resource assessments from the deep ocean to the coast. *J. Energy Power Eng.* **2011**, *5*, 730–742.
42. Mork, G.; Barstow, S.; Kabuth, A.; Pontes, M.T. Assessing the global wave energy potential. In Proceedings of the ASME 2010 29th International Conference on Ocean, Offshore and Arctic Engineering, Shanghai, China, 6–11 June 2010; pp. 447–454.
43. Sims, R.E.H.; Schock, R.N.; Adegbululgbe, A.; Fenhann, J.; Konstantinaviciute, I.; Moomaw, W.; Nimir, H.B.; Schlamadinger, B.; Torres-Martínez, J.; Turner, C.; et al. Energy supply. In *Climate Change 2007: Mitigation. Contribution of Working Group III to the Fourth Assessment Report of the Intergovernmental Panel on Climate Change*; Cambridge University Press: Cambridge, UK; New York, NY, USA, 2007; pp. 252–322, ISBN 978-0-511-54601-3.

44. Krewitt, W.; Nienhaus, K.; Kleßmann, C.; Capone, C.; Stricker, E.; Graus, W.; Hoogwijk, M.; Supersberger, N.; von Winterfeld, U.; Samadi, S. *Role and Potential of Renewable Energy and Energy Efficiency for Global Energy Supply*, 18th ed.; German Federal Environment Agency: Dessau-Roßlau, German, 2009; Volume 18.

45. Izadparast, A.H.; Niedzwecki, J.M. Estimating the potential of ocean wave power resources. *Ocean Eng.* **2011**, *38*, 177–185. [CrossRef]

46. Wu, S.; Liu, C.; Chen, X. Offshore wave energy resource assessment in the East China Sea. *Renew. Energy* **2015**, *76*, 628–636. [CrossRef]

47. Mirzaei, A.; Tangang, F.; Juneng, L. Wave energy potential assessment in the central and southern regions of the South China Sea. *Renew. Energy* **2015**, *80*, 454–470. [CrossRef]

48. Mentaschi, L.; Besio, G.; Cassola, F.; Mazzino, A. Performance evaluation of Wavewatch III in the Mediterranean Sea. *Ocean Model.* **2015**, *90*, 82–94. [CrossRef]

49. Silva, D.; Bento, A.R.; Martinho, P.; Soares, C.G. High resolution local wave energy modelling in the Iberian Peninsula. *Energy* **2015**, *91*, 1099–1112. [CrossRef]

50. Penalba, M.; Ulazia, A.; Ibarra-berastegui, G.; Ringwood, J.; Sáenz, J. Wave energy resource variation off the west coast of Ireland and its impact on realistic wave energy converters' power absorption. *Appl. Energy* **2018**, *224*, 205–219. [CrossRef]

51. Ahn, S.; Haas, K.A.; Neary, V.S. Wave energy resource classification system for US coastal waters. *Renew. Sustain. Energy Rev.* **2019**, *104*, 54–68. [CrossRef]

52. Kerr, D. Marine energy. *Philos. Trans. R. Soc. A Math. Phys. Eng. Sci.* **2007**, *365*, 971–992. [CrossRef] [PubMed]

53. Ocean Energy Systems (OES). Tidal & Currents. Available online: https://www.ocean-energy-systems.org/about-oes/what-is-ocean-energy/tidal-currents/ (accessed on 18 April 2019).

54. Neill, S.P.; Vögler, A.; Goward-Brown, A.J.; Baston, S.; Lewis, M.J.; Gillibrand, P.A.; Waldman, S.; Woolf, D.K. The wave and tidal resource of Scotland. *Renew. Energy* **2017**, *114*, 3–17. [CrossRef]

55. Alonso, R.; Jackson, M.; Santoro, P.; Fossati, M.; Solari, S.; Teixeira, L. Wave and tidal energy resource assessment in Uruguayan shelf seas. *Renew. Energy* **2017**, *114*, 18–31. [CrossRef]

56. Lewis, M.; Neill, S.P.; Robins, P.E.; Hashemi, M.R. Resource assessment for future generations of tidal-stream energy arrays. *Energy* **2015**, *83*, 403–415. [CrossRef]

57. Chen, W.B.; Liu, W.C.; Hsu, M.H. Modeling assessment of tidal current energy at Kinmen Island, Taiwan. *Renew. Energy* **2013**, *50*, 1073–1082. [CrossRef]

58. Rashid, A. Status and potentials of tidal in-stream energy resources in the southern coasts of Iran: A case study. *Renew. Sustain. Energy Rev.* **2012**, *16*, 6668–6677. [CrossRef]

59. Lewis, A.; Estefen, S.; Huckerby, J.; Lee, K.S.; Musial, W.; Pontes, T.; Torres-Martinez, J. Ocean energy. In *IPCC Special Report on Renewable Energy Sources and Climate Change Mitigation*; Edenhofer, O., Pichs-Madruga, R., Sokona, Y., Seyboth, K., Matschoss, P., Kadner, S., Zwickel, T., Eickemeie, P., Hansen, G., Schlömer, S., et al., Eds.; Cambridge University Press: Cambridge, UK; New York, NY, USA, 2011; pp. 497–534.

60. Yang, X.; Haas, K.A.; Fritz, H.M. Evaluating the potential for energy extraction from turbines in the gulf stream system. *Renew. Energy* **2014**, *72*, 12–21. [CrossRef]

61. Chang, Y.C.; Chu, P.C.; Tseng, R.S. Site selection of ocean current power generation from drifter measurements. *Renew. Energy* **2015**, *80*, 737–745. [CrossRef]

62. Goundar, J.N.; Ahmed, M.R. Marine current energy resource assessment and design of a marine current turbine for Fiji. *Renew. Energy* **2014**, *65*, 14–22. [CrossRef]

63. Mofor, L.; Goldsmith, J.; Jones, F. *Ocean Energy: Technology Readiness, Patents, Deployment Status and Outlook*; International Renewable Energy Agency (IRENA): Abu Dhabi, UAE, 2014.

64. Pelc, R.; Fujita, R.M. Renewable energy from the ocean. *Mar. Policy* **2002**, *26*, 471–479. [CrossRef]

65. Rajagopalan, K.; Nihous, G.C. An assessment of global ocean thermal energy conversion resources with a high resolution ocean general circulation model. *J. Energy Resour. Technol.* **2013**, *135*, 041202. [CrossRef]

66. Skråmestø, Ø.S.Ø.; Skilhagen, S.S.E.; Nielsen, W.K.W. Power production based on osmotic pressure. *Waterpower XVI* **2009**, 1–9. Available online: https://www.statkraft.com/globalassets/old-contains-the-old-folder-structure/documents/waterpower_xvi_-_power_production_based_on_osmotic_pressure_tcm21-4795.pdf (accessed on 18 April 2019).

67. Alvarez-Silva, O.A.; Osorio, A.F.; Winter, C. Practical global salinity gradient energy potential. *Renew. Sustain. Energy Rev.* **2016**, *60*, 1387–1395. [CrossRef]

68. Alvarez-Silva, O.A.; Osorio, A.F. Salinity gradient energy potential in Colombia considering site specific constraints. *Renew. Energy* **2015**, *74*, 737–748. [CrossRef]

69. Maisonneuve, J.; Pillay, P.; La, C.B. Osmotic power potential in remote regions of Quebec. *Renew. Energy* **2015**, *81*, 62–70. [CrossRef]

70. Emdadi, A.; Gikas, P.; Farazaki, M.; Emami, Y. Salinity gradient energy potential at the hyper saline Urmia Lake—ZarrinehRud River system in Iran. *Renew. Energy* **2016**, *86*, 154–162. [CrossRef]

71. Helfer, F.; Lemckert, C.; Anissimov, Y.G. Osmotic power with Pressure Retarded Osmosis: Theory, performance and trends—A review. *J. Membr. Sci.* **2014**, *453*, 337–358. [CrossRef]

72. Rusu, E.; Venugopal, V. *Offshore Renewable Energy—Ocean Waves, Tides, and Offshore Wind*; Rusu, E., Venugopal, V., Eds.; MDPI: Basel, Switzerland, 2019; ISBN 978-3-03897-593-9.

73. Li, Y.; Yu, Y.H. A synthesis of numerical methods for modeling wave energy converter-point absorbers. *Renew. Sustain. Energy Rev.* **2012**, *16*, 4352–4364. [CrossRef]

74. EMEC European Marine Energy Centre Ltd. Wave Devices. Available online: http://www.emec.org.uk/marine-energy/wave-devices/ (accessed on 18 April 2019).

75. López, I.; Andreu, J.; Ceballos, S.; Martínez De Alegría, I.; Kortabarria, I. Review of wave energy technologies and the necessary power-equipment. *Renew. Sustain. Energy Rev.* **2013**, *27*, 413–434. [CrossRef]

76. Heath, T.V. A review of oscillating water columns. *Philos. Trans. R. Soc. A Math. Phys. Eng. Sci.* **2012**, *370*, 235–245. [CrossRef] [PubMed]

77. Ning, D.Z.; Wang, R.Q.; Zou, Q.P.; Teng, B. An experimental investigation of hydrodynamics of a fixed OWC Wave Energy Converter. *Appl. Energy* **2016**, *168*, 636–648. [CrossRef]

78. Fernandes, M.P.; Vieira, S.M.; Henriques, J.C.; Valério, D.; Gato, L.M. Short-term prediction in an Oscillating Water Column using Artificial Neural Networks. In Proceedings of the 2018 International Joint Conference on Neural Networks (IJCNN), Rio de Janeiro, Brazil, 8–13 July 2018; pp. 1–7.

79. Bull, D.; Jenne, D.S.; Smith, C.S.; Copping, A.E.; Copeland, G. Levelized cost of energy for a Backward Bent Duct Buoy. *Int. J. Mar. Energy* **2016**, *16*, 220–234. [CrossRef]

80. Sheng, W. Motion and performance of BBDB OWC wave energy converters: I, hydrodynamics. *Renew. Energy* **2019**, *138*, 106–120. [CrossRef]

81. Falcão, A.F.O.; Henriques, J.C.C. Model-prototype similarity of oscillating-water-column wave energy converters. *Int. J. Mar. Energy* **2014**, *6*, 18–34. [CrossRef]

82. Dias, F.; Renzi, E.; Gallagher, S.; Sarkar, D.; Wei, Y.; Abadie, T.; Cummins, C.; Rafiee, A. Analytical and computational modelling for wave energy systems: The example of oscillating wave surge converters. *Acta Mech. Sin.* **2017**, *33*, 647–662. [CrossRef] [PubMed]

83. Yemm, R.; Pizer, D.; Retzler, C.; Henderson, R. Pelamis: Experience from concept to connection. *Philos. Trans. R. Soc. A Math. Phys. Eng. Sci.* **2012**, *370*, 365–380. [CrossRef]

84. Zheng, S.; Zhang, Y. Analytical study on hydrodynamic performance of a raft-type wave power device. *J. Mar. Sci. Technol.* **2017**, *22*, 620–632. [CrossRef]

85. Olaya, S.; Bourgeot, J.M.; Benbouzid, M. Optimal control for a self-reacting point absorber: A one-body equivalent model approach. In Proceedings of the 2014 International Power Electronics and Application Conference and Exposition, Shanghai, China, 5–8 November 2014; pp. 1–6.

86. Todalshaug, J.H.; Ásgeirsson, G.S.; Hjálmarsson, E.; Maillet, J.; Möller, P.; Pires, P.; Guérinel, M.; Lopes, M. Tank testing of an inherently phase-controlled wave energy converter. *Int. J. Mar. Energy* **2016**, *15*, 68–84. [CrossRef]

87. Sergiienko, N.Y.; Cazzolato, B.S.; Ding, B.; Arjomandi, M. Three-tether axisymmetric wave energy converter: Estimation of energy delivery. In Proceedings of the 3rd Asian Wave and Tidal Energy Conference (AWTEC 2016), Singapore, 24–28 October 2016; pp. 163–171.

88. Kofoed, J.P.; Frigaard, P.; Friis-Madsen, E.; Sørensen, H.C. Prototype testing of the wave energy converter wave dragon. *Renew. Energy* **2006**, *31*, 181–189. [CrossRef]

89. Musa, M.A.; Maliki, A.Y.; Ahmad, M.F.; Sani, W.N.; Yaakob, O.; Samo, K.B. Numerical simulation of wave flow over the overtopping breakwater for energy conversion (OBREC) device. *Procedia Eng.* **2017**, *194*, 166–173. [CrossRef]

90. Liu, Z.; Shi, H.; Cui, Y.; Kim, K. Experimental study on overtopping performance of a circular ramp wave energy converter. *Renew. Energy* **2017**, *104*, 163–176. [CrossRef]

91. Alam, M.R. Nonlinear analysis of an actuated seafloor-mounted carpet for a high-performance wave energy extraction. *Proc. R. Soc. A Math. Phys. Eng. Sci.* **2012**, *468*, 3153–3171. [CrossRef]

92. Durand, M.; Babarit, A.; Pettinotti, B.; Quillard, O.; Toularastel, J.L.; Clément, A.H. Experimental validation of the performances of the SEAREV wave energy converter with real time latching control. In Proceedings of the 7th European Wave and Tidal Energy Conference, Porto, Portugal, 11–13 September 2007; pp. 1–8.

93. Zhang, Z.; Chen, B.; Nielsen, S.R.K.; Olsen, J. Gyroscopic power take-off wave energy point absorber in irregular sea states. *Ocean Eng.* **2017**, *143*, 113–124. [CrossRef]

94. Camporeale, S.M.; Filianoti, P.; Torresi, M. Performance of a Wells turbine in an OWC device in comparison to laboratory tests. In Proceedings of the 9th European Wave and Tidal Energy Conference (EWTEC), Southampton, UK, 5–9 September 2011; p. 9.

95. Carrelhas, A.A.D.; Gato, L.M.C.; Henriques, J.C.C.; Falcão, A.F.O.; Varandas, J. Test results of a 30 kW self-rectifying biradial air turbine-generator prototype. *Renew. Sustain. Energy Rev.* **2019**, *109*, 187–198. [CrossRef]

96. António, F.D.O. Phase control through load control of oscillating-body wave energy converters with hydraulic PTO system. *Ocean Eng.* **2008**, *35*, 358–366.

97. Zhang, D.; Li, W.; Ying, Y.; Zhao, H.; Lin, Y.; Bao, J. Wave energy converter of inverse pendulum with double action power take off. *Proc. Inst. Mech. Eng. Part C J. Mech. Eng. Sci.* **2013**, *227*, 2416–2427. [CrossRef]

98. Gaspar, J.F.; Calvário, M.; Kamarlouei, M.; Soares, C.G. Design tradeoffs of an oil-hydraulic power take-off for wave energy converters. *Renew. Energy* **2018**, *129*, 245–259. [CrossRef]

99. Zhang, H.; Xu, D.; Ding, R.; Zhao, H.; Lu, Y.; Wu, Y. Embedded Power Take-Off in hinged modularized floating platform for wave energy harvesting and pitch motion suppression. *Renew. Energy* **2019**, *138*, 1176–1188. [CrossRef]

100. Albert, A.; Berselli, G.; Bruzzone, L.; Fanghella, P. Mechanical design and simulation of an onshore four-bar wave energy converter. *Renew. Energy* **2017**, *114*, 766–774. [CrossRef]

101. Liang, C.; Ai, J.; Zuo, L. Design, fabrication, simulation and testing of an ocean wave energy converter with mechanical motion rectifier. *Ocean Eng.* **2017**, *136*, 190–200. [CrossRef]

102. Yin, X.; Li, X.; Boontanom, V.; Zuo, L. Mechanical motion rectifier based efficient power takeoff for ocean wave energy harvesting. In Proceedings of the ASME 2017 Dynamic Systems and Control Conference, Fairfax, VA, USA, 11–13 October 2017; p. 5.

103. Mueller, M.A. Electrical generators for direct drive wave energy converters. *IEE Proc. Gener. Transm. Distrib.* **2002**, *149*, 446. [CrossRef]

104. Polinder, H.; Mueller, M.A.; Scuotto, M.; Goden de Sousa Prado, M. Linear generator systems for wave energy conversion. In Proceedings of the 7th European Wave and Tidal Energy Conference, Porto, Portugal, 11–13 September 2007.

105. Li, W.; Isberg, J.; Engström, J.; Waters, R.; Leijon, M. Parametric study of the power absorption for a linear generator wave energy converter. *J. Ocean Wind Energy* **2015**, *2*, 248–252. [CrossRef]

106. Ozkop, E.; Altas, I.H. Control, power and electrical components in wave energy conversion systems: A review of the technologies. *Renew. Sustain. Energy Rev.* **2017**, *67*, 106–115. [CrossRef]

107. Pecher, A.; Kofoed, J.P. Erratum to: Handbook of ocean wave energy. In *Ocean Engineering & Oceanography*; Pecher, A., Kofoed, J.P., Eds.; Springer: Cham, Switzerland, 2017; p. E1, ISBN 9783319398884.

108. Wang, L.; Isberg, J.; Tedeschi, E. Review of control strategies for wave energy conversion systems and their validation: The wave-to-wire approach. *Renew. Sustain. Energy Rev.* **2018**, *81*, 366–379. [CrossRef]

109. Alternative Energy Tutorials. Tidal Barrage Generation. Available online: http://www.alternative-energy-tutorials.com/tidal-energy/tidal-barrage.html (accessed on 18 April 2019).

110. Waters, S.; Aggidis, G. Tidal range technologies and state of the art in review. *Renew. Sustain. Energy Rev.* **2016**, *59*, 514–529. [CrossRef]

111. Xia, J.; Falconer, R.A.; Lin, B.; Tan, G. Estimation of annual energy output from a tidal barrage using two different methods. *Appl. Energy* **2012**, *93*, 327–336. [CrossRef]

112. Li, Y.; Pan, D.Z. The ebb and flow of tidal barrage development in Zhejiang Province, China. *Renew. Sustain. Energy Rev.* **2017**, *80*, 380–389. [CrossRef]

113. Angeloudis, A.; Falconer, R.A. Sensitivity of tidal lagoon and barrage hydrodynamic impacts and energy outputs to operational characteristics. *Renew. Energy* **2017**, *114*, 337–351. [CrossRef]

114. Angeloudis, A.; Kramer, S.C.; Avdis, A.; Piggott, M.D. Optimising tidal range power plant operation. *Appl. Energy* **2018**, *212*, 680–690. [CrossRef]

115. Harcourt, F.; Angeloudis, A.; Piggott, M.D. Utilising the flexible generation potential of tidal range power plants to optimise economic value. *Appl. Energy* **2019**, *237*, 873–884. [CrossRef]

116. Keysan, O.; McDonald, A.S.; Mueller, M. A direct drive permanent magnet generator design for a tidal current turbine (SeaGen). In Proceedings of the 2011 IEEE International Electric Machines & Drives Conference (IEMDC), Niagara Falls, ON, Canada, 15–18 May 2011; pp. 224–229.

117. Shirasawa, K.; Tokunaga, K.; Iwashita, H.; Shintake, T. Experimental verification of a floating ocean-current turbine with a single rotor for use in Kuroshio currents. *Renew. Energy* **2016**, *91*, 189–195. [CrossRef]

118. Seo, J.; Yi, J.; Park, J.; Lee, K. Review of tidal characteristics of Uldolmok Strait and optimal design of blade shape for horizontal axis tidal current turbines. *Renew. Sustain. Energy Rev.* **2019**, *113*, 109273. [CrossRef]

119. Li, Y.; Calisal, S.M. Three-dimensional effects and arm effects on modeling a vertical axis tidal current turbine. *Renew. Energy* **2010**, *35*, 2325–2334. [CrossRef]

120. Jing, F.; Sheng, Q.; Zhang, L. Experimental research on tidal current vertical axis turbine with variable-pitch blades. *Ocean Eng.* **2014**, *88*, 228–241. [CrossRef]

121. Fernandes, A.C.; Bakhshandeh Rostami, A. Hydrokinetic energy harvesting by an innovative vertical axis current turbine. *Renew. Energy* **2015**, *81*, 694–706. [CrossRef]

122. Wang, S.Q.; Xu, G.; Zhu, R.Q.; Wang, K. Hydrodynamic analysis of vertical-axis tidal current turbine with surging and yawing coupled motions. *Ocean Eng.* **2018**, *155*, 42–54. [CrossRef]

123. Chen, B.; Su, S.; Viola, I.M.; Greated, C.A. Numerical investigation of vertical-axis tidal turbines with sinusoidal pitching blades. *Ocean Eng.* **2018**, *155*, 75–87. [CrossRef]

124. Gorle, J.M.R.; Chatellier, L.; Pons, F.; Ba, M. Modulated circulation control around the blades of a vertical axis hydrokinetic turbine for flow control and improved performance. *Renew. Sustain. Energy Rev.* **2019**, *105*, 363–377. [CrossRef]

125. Kinsey, T.; Dumas, G.; Lalande, G.; Ruel, J.; Méhut, A.; Viarouge, P.; Lemay, J.; Jean, Y. Prototype testing of a hydrokinetic turbine based on oscillating hydrofoils. *Renew. Energy* **2011**, *36*, 1710–1718. [CrossRef]

126. Ma, P.; Yang, Z.; Wang, Y.; Liu, H.; Xie, Y. Energy extraction and hydrodynamic behavior analysis by an oscillating hydrofoil device. *Renew. Energy* **2017**, *113*, 648–659. [CrossRef]

127. Wang, Y.; Huang, D.; Han, W.; YangOu, C.; Zheng, Z. Research on the mechanism of power extraction performance for flapping hydrofoils. *Ocean Eng.* **2017**, *129*, 626–636. [CrossRef]

128. Filippas, E.S.; Gerostathis, T.P.; Belibassakis, K.A. Semi-activated oscillating hydrofoil as a nearshore biomimetic energy system in waves and currents. *Ocean Eng.* **2018**, *154*, 396–415. [CrossRef]

129. Chaudhari, C.D.; Waghmare, S.A.; Kotwal, A. Numerical analysis of venturi ducted horizontal axis wind turbine for efficient power generation. *Int. J. Mech. Eng. Comput. Appl.* **2013**, *1*, 90–93.

130. Khan, M.J.; Bhuyan, G.; Iqbal, M.T.; Quaicoe, J.E. Hydrokinetic energy conversion systems and assessment of horizontal and vertical axis turbines for river and tidal applications: A technology status review. *Appl. Energy* **2009**, *86*, 1823–1835. [CrossRef]

131. Belloni, C.S.K.; Willden, R.H.J.; Houlsby, G.T. An investigation of ducted and open-centre tidal turbines employing CFD-embedded BEM. *Renew. Energy* **2017**, *108*, 622–634. [CrossRef]

132. Tampier, G.; Troncoso, C.; Zilic, F. Numerical analysis of a diffuser-augmented hydrokinetic turbine. *Ocean Eng.* **2017**, *145*, 138–147. [CrossRef]

133. Tsao, C.C.; Han, L.; Jiang, W.T.; Lee, C.C.; Lee, J.S.; Feng, A.H.; Hsieh, C. Marine current power with cross-stream active mooring: Part I. *Renew. Energy* **2017**, *109*, 144–154. [CrossRef]

134. Tsao, C.C.; Han, L.; Jiang, W.T.; Lee, C.C.; Lee, J.S.; Feng, A.H.; Hsieh, C. Marine current power with cross-stream active mooring: Part II. *Renew. Energy* **2018**, *127*, 1036–1051. [CrossRef]

135. Minesto. The Future of Renewable Energy. Available online: https://minesto.com/our-technology (accessed on 18 April 2019).

136. Qian, P.; Feng, B.; Liu, H.; Tian, X.; Si, Y.; Zhang, D. Review on configuration and control methods of tidal current turbines. *Renew. Sustain. Energy Rev.* **2019**, *108*, 125–139. [CrossRef]

137. Faizal, M.; Ahmed, M.R. Experimental studies on a closed cycle demonstration OTEC plant working on small temperature difference. *Renew. Energy* **2013**, *51*, 234–240. [CrossRef]

138. Aydin, H.; Lee, H.S.; Kim, H.J.; Shin, S.K.; Park, K. Off-design performance analysis of a closed-cycle ocean thermal energy conversion system with solar thermal preheating and superheating. *Renew. Energy* **2014**, *72*, 154–163. [CrossRef]

139. Yang, M.H.; Yeh, R.H. Analysis of optimization in an OTEC plant using organic Rankine cycle. *Renew. Energy* **2014**, *68*, 25–34. [CrossRef]

140. Mutair, S.; Ikegami, Y. Design optimization of shore-based low temperature thermal desalination system utilizing the ocean thermal energy. *J. Sol. Energy Eng.* **2014**, *136*, 041005. [CrossRef]

141. Kim, A.S.; Kim, H.J.; Lee, H.S.; Cha, S. Dual-use open cycle ocean thermal energy conversion (OC-OTEC) using multiple condensers for adjustable power generation and seawater desalination. *Renew. Energy* **2016**, *85*, 344–358. [CrossRef]

142. Octaviani, F.; Muslim, M.; Buwono, A.; Faturachman, D. Study of ocean thermal energy conversion (OTEC) generation as project of power plant in West Sumatera-Indonesia. *Recent Adv. Renew. Energy Sources* **2016**, *10*, 64–68.

143. Thorsen, T.; Holt, T. The potential for power production from salinity gradients by pressure retarded osmosis. *J. Membr. Sci.* **2009**, *335*, 103–110. [CrossRef]

144. Han, G.; Zhang, S.; Li, X.; Chung, T.S. Progress in pressure retarded osmosis (PRO) membranes for osmotic power generation. *Prog. Polym. Sci.* **2015**, *51*, 1–27. [CrossRef]

145. Altaee, A.; Zhou, J.; Alanezi, A.A.; Zaragoza, G. Pressure retarded osmosis process for power generation: Feasibility, energy balance and controlling parameters. *Appl. Energy* **2017**, *206*, 303–311. [CrossRef]

146. Altaee, A.; Cipolina, A. Modelling and optimization of modular system for power generation from a salinity gradient. *Renew. Energy* **2019**, *141*, 139–147. [CrossRef]

147. Veerman, J.; Saakes, M.; Metz, S.J.; Harmsen, G.J. Electrical power from sea and river water by reverse electrodialysis: A first step from the laboratory to a real power plant. *Environ. Sci. Technol.* **2010**, *44*, 9207–9212. [CrossRef] [PubMed]

148. Avci, A.H.; Tufa, R.A.; Fontananova, E.; Di, G.; Curcio, E. Reverse Electrodialysis for energy production from natural river water and seawater. *Energy* **2018**, *165*, 512–521. [CrossRef]

149. Tufa, R.A.; Pawlowski, S.; Veerman, J.; Bouzek, K.; Fontananova, E.; Velizarov, S.; Goulão, J.; Nijmeijer, K.; Curcio, E. Progress and prospects in reverse electrodialysis for salinity gradient energy conversion and storage. *Appl. Energy* **2018**, *225*, 290–331. [CrossRef]

150. Mei, Y.; Tang, C.Y. Recent developments and future perspectives of reverse electrodialysis technology: A review. *Desalination* **2018**, *425*, 156–174. [CrossRef]

151. Renewable Energy Policy Network for the 21st Century (REN21). *Renewable 2018: Global Status Report*; REN21: Paris, France, 2018.

152. International Renewable Energy Agency (IRENA). *Renewable Energy Capacity Statistics 2019*; IRENA: Abu Dhabi, UAE, 2019.

153. Wang, S.; Yuan, P.; Li, D.; Jiao, Y. An overview of ocean renewable energy in China. *Renew. Sustain. Energy Rev.* **2011**, *15*, 91–111. [CrossRef]

154. Neill, S.P.; Hashemi, M.R. Introduction. *Fundam. Ocean Renew. Energy* **2018**, *1990*, 1–30.

155. Ohneda, H.; Igarashi, S.; Shinbo, O.; Sekihara, S.; Suzuki, K.; Kubota, H.; Morita, H. Construction procedure of a wave power extracting caisson breakwater. In Proceedings of the 3rd Symposium on Ocean Energy Utilization, Tokyo, Japan, 22–23 January 1991; pp. 171–179.

156. Arena, F.; Romolo, A.; Malara, G.; Ascanelli, A. On design and building of a U-OWC wave energy converter in the Mediterranean Sea: A case study. In Proceedings of the ASME 2013 32nd International Conference on Ocean, Offshore and Arctic Engineering, Nantes, France, 9–14 June 2013; Volume 8.

157. Falcão, A.D.O. The shoreline OWC wave power plant at the Azores. In Proceedings of the 4th European Wave Energy Conference, Aalborg, Denmark, 4–6 December 2000; pp. 42–48.

158. Bønke, K.; Ambli, N. Prototype wave power stations in Norway. In Proceedings of the Utilization of Ocean Waves—Wave to Energy Conversion, San Diego, CA, USA, 16–17 June 1986; pp. 34–35.

159. Weinstein, A.; Fredrikson, G.; Parks, M.J.; Nielsen, K. AquaBuOY-the offshore wave energy converter numerical modeling and optimization. In Proceedings of the Oceans '04 MTS/IEEE Techno-Ocean '04, Kobe, Japan, 9–12 November 2004; pp. 1988–1995.

160. Salter, S.H.; Lin, C.-P. Wide tank efficiency measurements on a model of the sloped IPS buoy. In Proceedings of the 3rd European Wave Energy Conference, Patras, Greece, 30 September–2 October 1998; pp. 200–206.

161. Weber, J.; Mouwen, F.; Parish, A.; Robertson, D. Wavebob—Research & development network and tools in the context of systems engineering. In Proceedings of the 8th European Wave and Tidal Energy Conference, Uppsala, Sweden, 7–10 September 2009; pp. 416–420.

162. Tedd, J.; Peter Kofoed, J. Measurements of overtopping flow time series on the Wave Dragon, wave energy converter. *Renew. Energy* **2009**, *34*, 711–717. [CrossRef]

163. Vicinanza, D.; Frigaard, P. Wave pressure acting on a seawave slot-cone generator. *Coast. Eng.* **2008**, *55*, 553–568. [CrossRef]

164. Magagna, D.; Margheritini, L. *Workshop on Identification of Future Emerging Technologies in the Wind Power Sector*; European Commission, Publications Official of the European Union: Luxembourg, 2018; ISBN 9789279925870. [CrossRef]

165. World Energy Council. *2010 Survey of Energy Resources*; World Energy Council: London, UK, 2010.

166. Bae, Y.H.; Kim, K.O.; Choi, B.H. Lake Sihwa tidal power plant project. *Ocean Eng.* **2010**, *37*, 454–463. [CrossRef]

167. Worldsteel Association. Sihwa Tidal Power Station Worldsteel HR. Available online: https://stories.worldsteel.org/infrastructure/large-scale-tidal-power-reliant-steel/attachment/sihwa-tidal-power-station-worldsteel-hr/ (accessed on 1 March 2019).

168. Hooper, T.; Austen, M. Tidal barrages in the UK: Ecological and social impacts, potential mitigation, and tools to support barrage planning. *Renew. Sustain. Energy Rev.* **2013**, *23*, 289–298. [CrossRef]

169. Hammar, L.; Ehnberg, J.; Mavume, A.; Cuamba, B.C.; Molander, S. Renewable ocean energy in the Western Indian Ocean. *Renew. Sustain. Energy Rev.* **2012**, *16*, 4938–4950. [CrossRef]

170. Hammons, T.J.; Member, S. Tidal power. *Proc. IEEE* **1993**, *81*, 419–433. [CrossRef]

171. Kobayashi, H.; Jitsuhara, S.; Uehara, H. The present status and features of OTEC and recent aspects of thermal energy conversion technologies. In Proceedings of the 24th Meeting of the UJNR Marine Facilities Panel, Honolulu, HI, USA, 4–12 November 2001.

172. Uehara, H. The present status and future of ocean thermal energy conversion. *Int. J. Sol. Energy* **1995**, *16*, 217–231. [CrossRef]

173. Makai Ocean Engineering. Ocean Thermal Energy Conversion. Available online: https://www.makai.com/ocean-thermal-energy-conversion/ (accessed on 9 May 2019).

174. Ocean Energy System (OES). Offshore Installations Worldwide. Available online: https://www.ocean-energy-systems.org/ocean-energy-in-the-world/gis-map/ (accessed on 18 April 2019).

175. EMODnet—The European Marine Observation and Data Network. The European Marine Observation and Data Network. Available online: https://ec.europa.eu/maritimeaffairs/atlas/maritime_atlas/#lang=EN;p=w;bkgd=5;theme=2:0.75;c=1253866.2175874896,7033312.218247008;z=4 (accessed on 1 April 2019).

176. RenewableUK. UK Marine Energy Database (UKMED). Available online: https://www.renewableuk.com/page/UKMED2/UK-Marine-Energy-Database.htm (accessed on 3 April 2019).

177. National Renewable Energy Laboratory NREL. Marine and Hydrokinetic Technology Database. Available online: https://openei.org/wiki/Marine_and_Hydrokinetic_Technology_Database (accessed on 3 March 2019).

178. Pacific Northwest National Laboratory (PNNL). Energy, U.S.D. of Ocean Energy Systems (OES); Wind, I. TETHYS: Marine Energy Content. Available online: https://tethys.pnnl.gov/map-viewer-marine-energy (accessed on 18 April 2019).

179. EMEC (European Marine Energy Centre). *Guidelines for Project Development in the Marine Energy Industry*; EMEC: Stromness, Scotland, 2009.

180. International Renewable Energy Agency (IRENA). *Ocean Thermal Energy Conversion: Technology Brief*; IRENA: Abu Dhabi, UAE, 2014.

181. International Renewable Energy Agency (IRENA). *Tidal Energy: Technology Brief*; IRENA: Abu Dhabi, UAE, 2014.

182. International Renewable Energy Agency (IRENA). *Salinity Gradient Energy: Technology Brief*; IRENA: Abu Dhabi, UAE, 2014.

183. EPE. *Anuário Estatístico de Energia Elétrica 2018 no Ano Base de 2017*; EPE: Rio de Janeiro, Brazil, 2018.

184. Hammar, L.; Gullström, M.; Dahlgren, T.G.; Asplund, M.E.; Goncalves, I.B.; Molander, S. Introducing ocean energy industries to a busy marine environment. *Renew. Sustain. Energy Rev.* **2017**, *74*, 178–185. [CrossRef]

185. Garcia-rosa, P.B.; Paulo, J.; Soares, V.; Lizarralde, F.; Estefen, S.F.; Machado, I.R.; Watanabe, E.H. Wave-to-wire model and energy storage analysis of an ocean wave energy hyperbaric converter. *IEEE J. Ocean. Eng.* **2014**, *39*, 1–12. [CrossRef]

186. Costa, P.R.; Garcia-Rosa, P.B.; Estefen, S.F. Phase control strategy for a wave energy hyperbaric converter. *Ocean Eng.* **2010**, *37*, 1483–1490. [CrossRef]

187. Estefen, S.F.; Esperança, P.T.T.; Ricarte, E.; da Costa, P.R.; Pinheiro, M.M.; Clemente, C.H.P.; Franco, D.; Melo, E.; de Souza, J.A. Experimental and numerical studies of the wave energy hyperbaric device for electricity production. In Proceedings of the ASME 2008 27th International Conference on Offshore Mechanics and Arctic Engineering, Estoril, Portugal, 15–20 June 2008; pp. 811–818.

188. Estefen, S.F.; da Costa, P.R.; Ricarte, E.; Pinheiro, M.M. Wave energy hyperbaric device for electricity production. In Proceedings of the ASME 2007 26th International Conference on Offshore Mechanics and Arctic Engineering, San Diego, CA, USA, 10–15 June 2007; pp. 627–633.

189. Shadman, M.; Estefen, S.F.; Rodriguez, C.A.; Nogueira, I.C.M. A geometrical optimization method applied to a heaving point absorber wave energy converter. *Renew. Energy* **2018**, *115*, 533–546. [CrossRef]

190. Budal, K.; Falnes, J. Interacting point absorbers with controlled motion. In *Power from Sea Waves*; Count, B.M., Ed.; Academic Press: London, UK, 1980; pp. 129–142.

191. Budal, K.; Falnes, J.; Iversen, L.C.; Lillebekker, P.M.; Oltedal, G.; Hals, T.; Onshus, T.; Hoy, A. The Norwegian wave-power buoy project. In Proceedings of the 2nd International Symposium on Wave Energy Utilization, Trondheim, Norway, 22–24 June 1982; pp. 323–344.

192. Neto, P.B.L.; Saavedra, O.R.; de Souza Ribeiro, L.A. Analysis of a tidal power plant in the estuary of Bacanga in Brazil taking into account the current conditions and constraints. *IEEE Trans. Sustain. Energy* **2017**, *8*, 1187–1194. [CrossRef]

193. Ferreira, R.M.; Estefen, S.F. Alternative concept for tidal power plant with reservoir restrictions. *Renew. Energy* **2009**, *34*, 1151–1157. [CrossRef]

194. Leite Neto, P.B.; Saavedra, O.R.; Souza Ribeiro, L.A. Optimization of electricity generation of a tidal power plant with reservoir constraints. *Renew. Energy* **2015**, *81*, 11–20. [CrossRef]

195. Aggidis, G.A.; Feather, O. Tidal range turbines and generation on the Solway Firth. *Renew. Energy* **2012**, *43*, 9–17. [CrossRef]

MDPI

St. Alban-Anlage 66

4052 Basel

Switzerland

Tel. +41 61 683 77 34

Fax +41 61 302 89 18

www.mdpi.com

Energies Editorial Office

E-mail: energies@mdpi.com

www.mdpi.com/journal/energies